RENAISSANCE OF SICKLE CELL DISEASE RESEARCH IN THE GENOME ERA

RENAISSANCE OF SICKLE CELL DISEASE RESEARCH IN THE GENOME ERA

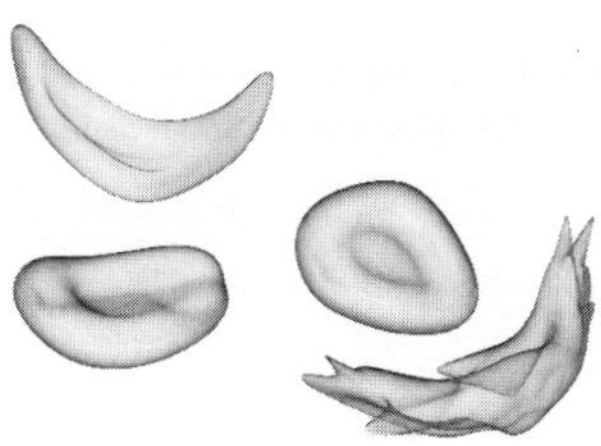

BETTY PACE, MD
UNIVERSITY OF TEXAS AT DALLAS, USA

ICP Imperial College Press

Published by

Imperial College Press
57 Shelton Street
Covent Garden
London WC2H 9HE

Distributed by

World Scientific Publishing Co. Pte. Ltd.
5 Toh Tuck Link, Singapore 596224
USA office: 27 Warren Street, Suite 401-402, Hackensack, NJ 07601
UK office: 57 Shelton Street, Covent Garden, London WC2H 9HE

British Library Cataloguing-in-Publication Data
A catalogue record for this book is available from the British Library.

RENAISSANCE OF SICKLE CELL DISEASE RESEARCH IN THE GENOME ERA

ISBN-13 978-1-86094-645-5
ISBN-10 1-86094-645-3

Typeset by Stallion Press
Email: enquiries@stallionpress.com

Printed by FuIsland Offset Printing (S) Pte Ltd, Singapore

Contents

Contributors

Robert Adams, MD
Regents Professor
Co-Director Cerebrovascular Section
Medical College of Georgia
Department of Pediatrics/Neurology
1429 Harper Street, HF 1154
Augusta, GA 30912, USA

David G. Badman, PhD
Retired, Director of Hematology Programs
National Institute of Diabetes and Digestive and Kidney Diseases
National Institutes of Health
PO Box 302
Poolesville, MD 20837, USA

Lennette J. Benjamin, MD
Professor of Clinical Medicine
Albert Einstein College of Medicine
Clinical Director, Comprehensive Sickle Cell Center
Montefiore Medical Center
111 E. 210th Street
Bronx, NY 10467, USA

Vence L. Bonham, Jr., JD
Associate Investigator
Social and Behavioral Research Branch
Senior Advisor to the Director on Societal Implications of Genomics
National Human Genome Research Institute
National Institutes of Health
Building 31, Room 4B09
31 Center Drive, MSC 3973
Bethesda, MD 20892, USA

Meghan E. Boyer, BS
Research Specialist
University of Wisconsin-Madison
Department of Pharmacology
1300 University Avenue
383 Medical Sciences Center
Madison, WI 53706, USA

Emery Bresnick, MD, PhD
Professor
University of Wisconsin-Madison
Department of Pharmacology
1300 University Avenue
383 Medical Sciences Center
Madison, WI 53706, USA

George R. Buchanan, MD
Professor
Chief, Pediatric Hematology/Oncology Division
Director, Southwestern Comprehensive Sickle Cell Center
The University of Texas Southwestern
Medical Center at Dallas
Children's Medical Center Dallas
Department of Pediatrics, Division of Hematology/Oncology
5323 Harry Hines Blvd
Dallas, TX 75390, USA

Seong-Kyu Choe, PhD
Research Postdoctoral Fellow
Brigham and Women's Hospital
and Harvard Medical School
Department of Medicine, Division of Hematology
Karp Family Research Building, Room 06.004
One Blackfan Circle
Boston, MA 02115, USA

Francis S. Collins, MD, PhD
Director
National Human Genome Research Institute
Genome Technology Branch
31 Center Drive, MSC 2152
Building 31, Room 4B09
Bethesda, MD 20892, USA

Flavia C. Costa, PhD
Postdoctoral Fellow
University of Kansas Medical Center
Department of Biochemistry and Molecular Biology
3901 Rainbow Boulevard
Kansas City, KS 66160, USA

Denise Davis-Maye, PhD, LCSW
Assistant Professor
Auburn University
Department of Sociology Anthropology and Social Work
7018 Haley Center
Auburn, AL 36849, USA

Gregory L. Evans, PhD
Leader
Hemoglobinopathies and Genetics Scientific Research Group
Blood Diseases Program
National Heart, Lung, and Blood Institute
6701 Rockledge Dr, MSC 7950, Room 10152
Bethesda, MD 20892, USA

Frank A. Ferrone, PhD
Professor
Drexel University
College of Arts and Sciences
Department of Physics
3141 Chestnut St., Disque Hall 12-922
Philadelphia, PA 19104, USA

Steven N. Fiering, PhD
Associate Professor
Dartmouth Medical School
Department of Microbiology, Immunology and Genetics
1 Medical Center Drive, Rubin Bldg HB 7936
Lebanon, NH 03756, USA

Alan W. Flake, MD
Professor
Director, Childrens Institute for Surgical Science
Childrens Hospital of Philadelphia
Department of Surgery
Abramson Research Center, Rm. 1116B
3615 Civic Center Blvd
Philadelphia, PA 19104, USA

Vanessa Northington Gamble, MD, PhD
Professor
Director, Tuskegee University National Center for Bioethics
in Research and Health Care
44-107 Bioethics Building
Tuskegee, AL 36088, USA

Steven R. Goodman, PhD
Professor
Director, Institute of Biomedical Sciences and Technology
The University of Texas at Dallas
Department of Molecular and Cell Biology
P O Box 830688, F032
Richardson, TX 75083, USA

Jeffrey A. Grass, MS
Research Specialist
University of Wisconsin-Madison
Department of Pharmacology
1300 University Avenue
383 Medical Sciences Center
Madison, WI 53706, USA

Alan Edward E. Guttmacher, MD
Deputy Director
National Human Genome Research Institute
Genome Technology Branch
31 Center Drive, MSC 2152
Building 31, Room 4B09
Bethesda, MD 20892, USA

Johnson Haynes, Jr., MD
Professor
Director, USA Comprehensive Sickle Cell Center
University of South Alabama
Department of Medicine/Pulmonary and Critical Care
2451 Fillingim Street, Room 1505 MCSB
Mobile, AL 36617, USA

Carlton Haywood, Jr., MA
Doctoral Student
The Phoebe R. Berman Bioethics Institute
Johns Hopkins University
100 North Charles St, Suite 740
Baltimore, MD 21201, USA

Carolyn Hoppe, MD
Associate Hematologist
Children's Hospital Oakland
Department of Hematology/Oncology
747 52nd Street
Oakland, CA 94609, USA

Hogune Im, MS
Graduate Student
University of Wisconsin-Madison
Department of Pharmacology
1300 University Avenue
383 Medical Sciences Center
Madison, WI 53706, USA

Kirby D. Johnson, PhD
Assistant Research Scientist
University of Wisconsin-Madison
Department of Pharmacology
1300 University Avenue
383 Medical Sciences Center
Madison, WI 53706, USA

Clinton Joiner, MD, PhD
Professor
Director, Cincinnati Comprehensive Sickle Cell Center
Children's Hospital Medical Center
University of Cincinnati College of Medicine
Department of Pediatrics
3333 Burnet Avenue
Cincinnati, OH 45229, USA

Philippe Leboulch, MD
Associate Professor
Harvard Medical School
Brigham and Women's Hospital
Department of Medicine/Genetics
NRB 466c, 77 Avenue Louis Pasteur
Boston, MA 02115, USA

Wei Li, MD, PhD
Postdoctoral Fellow
The University of Texas at Dallas
Department of Molecular and Cell Biology
PO Box 830688, FO31
Richardson, TX 75083, USA

Punam Malik, MD
Associate Professor
Cincinnati Children's Hospital Medical Center
Division of Experimental Hematology
3333 Bumet Avenue, ML 7013
TCHRF 6564
Cincinnati, OH 45229, USA

Solomon F. Ofori-Acquah, PhD
Assistant Professor
Department of Cell Biology and Neuroscience
University of South Alabama
MSB 2348, 307 University Blvd
Mobile, AL 36688, USA

Kwaku Ohene-Frempong, MD
Professor
Director, Children's Hospital of Philadelphia
Comprehensive Sickle Cell Center
The Children's Hospital of Philadelphia
Division of Hematology
34th St & Civic Center Blvd.
Philadelphia, PA 19104, USA

Betty S. Pace, MD
Professor
Director, Sickle Cell Disease Research Center
The University of Texas at Dallas
Department of Molecular and Cell Biology
PO Box 830688, FO31
Richardson, TX 75083, USA

Ardie Pack-Mabien, CRNP
Pediatric Nurse Practioner
USA Comprehensive Sickle Cell Center
University of South Alabama
Room 1530 MCSB
Mobile, AL 36617, USA

Barry H. Paw, MD, PhD
Assistant Professor
Brigham and Women's Hospital and Harvard Medical School
Department of Medicine, Division of Hematology
Karp Family Research Building, Room 06.213
One Blackfan Circle
Boston, MA 02115, USA

Richard Payne, MD
Director, Institute on Care at the End of Life
Duke University Divinity School
2 Chapel Drive
Durham, NC 27708, USA

Kenneth Peterson, PhD
Professor and Vice-Chairman
University of Kansas Medical Center
Department of Biochemistry and Molecular Biology
3901 Rainbow Blvd, MS 3030
Kansas City, KS 66160, USA

Rosie Peterson, BS, PAHM
Assistant Director, Sickle Cell Disease Research Center
The University of Texas at Dallas
Department of Molecular and Cell Biology
PO Box 830688, FO31
Richardson, TX 75083, USA

Charles T. Quinn, MD
Assistant Professor
The University of Texas Southwestern Medical Center at Dallas
Southwestern Comprehensive Sickle Cell Center
Children's Medical Center Dallas
Department of Pediatrics, Division of Hematology/Oncology
5323 Harry Hines Blvd
Dallas, TX 75390, USA

Robert I. Raphael, MD
Senior Fellow
Children's Hospital and Research Center, Oakland
Department of Hematology/Oncology
747 52nd Street
Oakland, CA 94609, USA

Clarice Reid, MD
Director, Division of Blood Diseases and Resources (retired)
National Institutes of Health
9715 Fernwood Rd
Bethesda, MD 20817, USA

Griffin Rodgers, MD, MACP
Deputy Director and Chief, Molecular
and Clinical Hematology Branch
National Institute of Diabetes
and Digestive and Kidney Diseases
National Institutes of Health
Bldg 31 Room 9A-52, 10-MSC-1822
Bethesda, MD 20892, USA

Kim Smith-Whitley, MD
Assistant Professor
Associate Director, Sickle Cell Clinical Program
The Children's Hospital of Philadelphia
Division of Hematology
34th St & Civic Center Blvd.
Philadelphia, PA 19104, USA

Martin H. Steinberg, MD
Professor
Director, Comprehensive Sickle Cell Center
Boston Medical Center
88 E. Newton Street, E248
Boston, MA 02118, USA

Shirin V. Sundar, PhD
Postdoctoral Fellow
University of Kansas Medical Center
Department of Biochemistry and Molecular Biology
3901 Rainbow Blvd, MS 3030
Kansas City, KS 66160, USA

Swee Lay Thein, MD
Professor
Department of Hematological Medicine
King's College London School of Medicine
Kings College Hospital
Denmark Hill Campus
London SE5 9PJ, UK

Elliot Vichinsky, MD
Professor
Director, North California Comprehensive Sickle Cell Center
Children's Hospital and Research at Oakland
Department of Hematology/Oncology
747 52nd Street
Oakland, CA 94609, USA

Mark C. Walters, MD
Associate Adjunct Professor
Director, Blood and Marrow Transplant Program
Children's Hospital and Research Center, Oakland
Department of Hematology/Oncology
747 52nd Street
Oakland, California 94609, USA

Foreword

by Francis S. Collins and Alan E. Guttmacher

The "genetic era" might be said to have begun in April 1953, with Watson and Crick's description of the DNA double helix. Sickle cell disease earned an historic distinction in this newly born era with Ingram's description in 1957 of the first molecular basis for a disease, the substitution of valine for glutamic acid at the sixth position of the beta-globin chain, as the cause of sickle cell disease. This knowledge of its genetic nature contributed to many advances in our understanding of the biology of sickle cell disease, such as the pathophysiological importance of hemoglobin polymerization. However, despite the historical prominence of sickle cell disease in genetics, genetics-based knowledge has thus far led to only modest progress in prevention and treatment. Sickle cell disease continues to be a significant cause of morbidity and mortality, both in the United States and globally.

Exactly half a century after Watson and Crick's publication, the "genome era" began with the Human Genome Project's production of the finished sequence of the human genome in April 2003. Will prevention of and therapy for sickle cell disease fare any better in the genome era than in the genetic era? That crucial question is, in many ways, the focus of this volume, which reviews the history of our social and biomedical understanding of sickle cell disease, details our current state of knowledge, and highlights promising areas for future discovery.

There are a number of reasons to be optimistic that the tools and approaches of genomics will, indeed, significantly advance our understanding of sickle cell disease and provide more effective clinical strategies that improve the lives of individuals and families with the condition. The genetic era, with its focus on single genes and how they function, provided important knowledge. The genome era, with its focus on all of the 20,000 or so genes in the human genome, should provide much richer and more clinically relevant understanding.

For instance, those who are members of families with sickle cell disease, or have cared for individuals with the disease, see daily proof that the expression of the same single nucleotide substitution can cause widely varying effects in different individuals. Why is this? Genomics should help tell us.

We know that different variants nearby the *HBB* gene, particularly affecting fetal globin expression, can translate into different levels of severity. However, much more is involved in the varied health of those with sickle cell disease than variants in this gene cluster alone. Clearly, the *HBB* gene interacts both with multiple other genes and with environmental factors, and it is these complex, and hitherto mysterious, interactions that determine health impact. Such tools of the genome era as genome-wide association studies, high throughput sequencing, and sophisticated bioinformatics should enable, for the first time, the identification of important modifier genes and gene-environment interactions, and thus help to unravel this crucial mystery.

Moreover, modern genomic approaches offer optimism not just for improved understanding, but for improved therapies. For instance, the combination of high throughput technologies, easily accessible libraries of hundreds of thousands of small molecular compounds, and the human genome sequence's supply of new targets, make "chemical genomic" approaches practical ways to develop novel therapies for such previously intractable clinical problems as sickle cell disease.

One advantage of the genome era should be that it can profit from the sometimes hard-won lessons of the genetic era, if we are only wise enough to learn from them. Sickle cell disease is distinctive in genetics not only for its preeminence as a molecular model of disease, but also for its sobering reminder that the application of genetic science can inflict discriminatory injuries upon individuals, families, and communities living with genetic disease. These lessons have been heeded, and the genomics era has been truly informed, even shaped, by the social and ethical lessons of sickle cell disease research in an earlier time. As a consequence, genomics may indeed be unique among areas of biomedicine for having, almost since its inception, included a focus on ethical, legal, and social implications (ELSI) issues. This sensitivity includes the conclusion that truly effective biomedical research must involve active participants who help shape the research agenda and delineate its features, rather than passive "subjects" on whom research is done. We believe that this attention to ELSI issues is another important genomic principle, one that should play a vital future role in shaping both basic science and applied biomedical approaches to sickle cell disease.

The Human Genome Project has already proven to be a milestone in the history of biology. However, for the genome era to be an equally important milestone in the history of medicine, it must prove that it can lead to important advances in health. We look forward to the path forward discussed so ably in this volume, turning the great promise of the genome era into a reality of better prevention and treatment of sickle cell disease.

Francis S. Collins
Alan E. Guttmacher

Acknowledgments

To my colleagues who chose to make this project a priority, my heartfelt appreciation to you for making invaluable contributions. Special thanks to Harriet Gross, Solomon Ofori-Acquah and Glorias Dixon for your editing efforts, to the administrative staff in the Sickle Cell Disease Research Center for technical recommendations, and to my family for providing loving support and encouragement which enabled me to serve as Editor. *Renaissance of Sickle Cell Disease in the Genome Era* represents a group effort which conveys the message that significant progress has been made to improve survival and treatment options, however genomics research will produce a cure for sickle cell disease in the not too distant future.

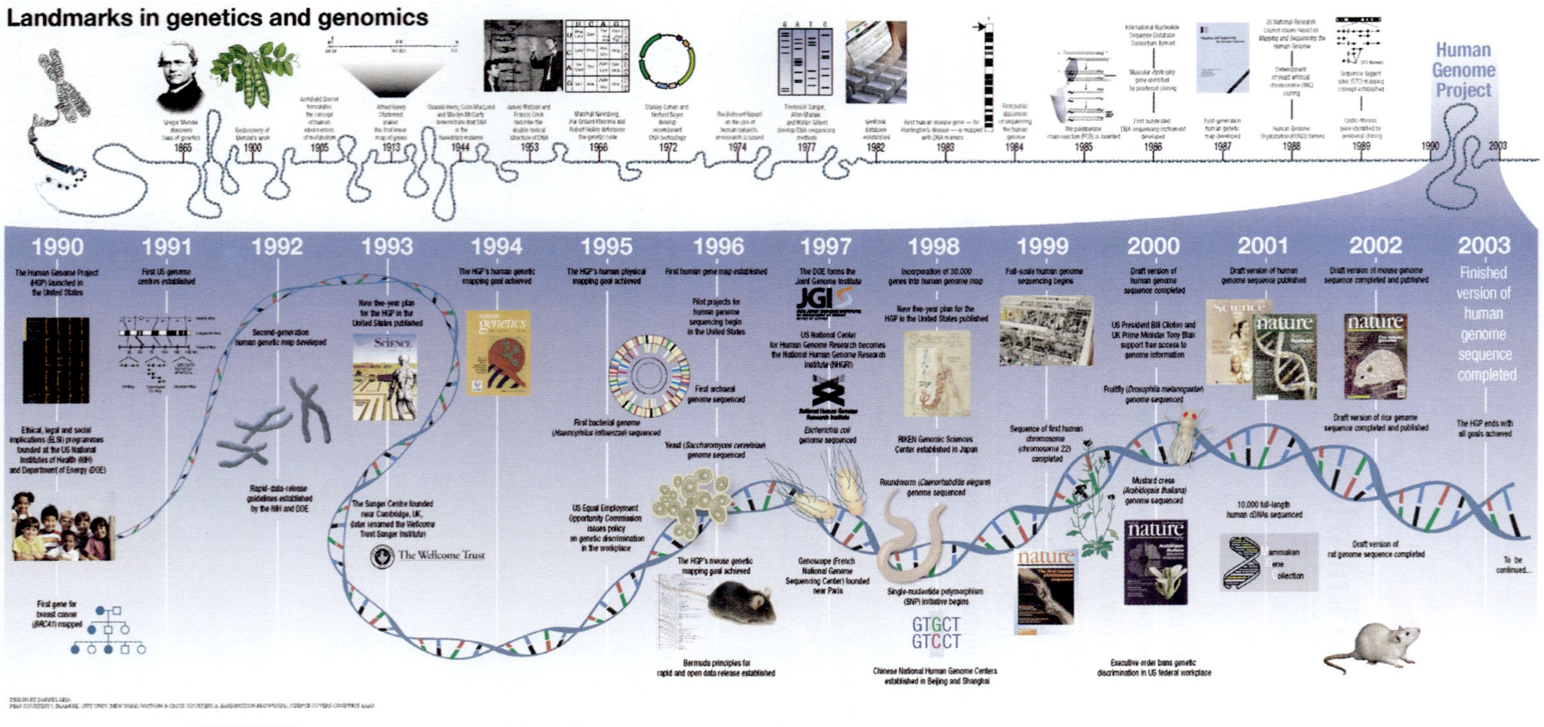

A Vision for the Future of Genomics Research

In contemplating a vision for the future of genomics research, it is appropriate to consider the remarkable path that has brought us to the genome era. The ***rollfold*** shows a timeline of landmark accomplishments in genetics and genomics, beginning with Gregor Mendel's discovery of the laws of heredity material (1866) and their rediscovery in the early days of the twentieth century. Recognition of DNA as the hereditary material (1944), determination of its structure (1953), elucidation of the genetic code (1966), development of recombinant DNA technologies (1972) and establishment of increasingly automatable methods for DNA sequencing in the seventies, made it possible for the Human Genome Project to begin in 1990. The creativity and determination of a legion of talented scientists made this project a success and set the stage for a revolution in biological research. *Use by permission from FS Collins, ED Green, AE Guttmacher and MS Guyer, Nature 422, 835–847, 2003. Larger image available at www.genome.gov/11007524.*

Introduction . . . The Journey Inward

by Betty S. Pace

The true science and study of man, is man himself.

Pierre Charron, 1541–1603

Scientific exploration and discovery are, at their core, a perpetual journey for truth. To know and understand the essence of one's being is the ultimate revelation of that truth. But the boundaries surrounding the quest for this truth have certainly come under great challenge and scrutiny.

Recently, there have been intense discussion and debate around the sanctity of life and the ethical issues raised by efforts to clone humans, or to alter the genetic makeup of children by design. Who would have thought in 1949, when Linus Pauling published the seminal paper declaring "Sickle cell anemia, a molecular disease," followed in 1977 by the demonstration of a point mutation in the β-globin gene, that less than 30 years later the ethics of human cloning would be the focal point of discussion — while we have not yet acquired the knowledge required to cure the *first* genetic disease, sickle cell anemia?

Since that critical discovery of 1977, the scientific community has completed a fantastic voyage to sequence the entire three billion base pairs in the human genome under the leadership of Dr. Francis Collins, Director of the National Human Genome Research Institute, and Dr. Aristides Patronis, Director of the Department of Energy Human Genome Program. The first draft of the entire human genome sequence was published in February 2001, followed by a high-quality, reference sequence in April 2003. Yes, the Human Genome Era was born, bringing new hope for a cure not only for sickle cell anemia, but for other genetic disorders as well. The quest to sequence the human genome was a "journey inward," in the words of Dr. Collins, to the essence of life, so we invite you to join us as we chronicle the journey on the pages of this historic book.

As we embark upon this journey, it is important to stay the course, from beginning to end, for behind every chapter you will find a fresh perspective on research efforts toward improving outcome and producing a cure for sickle cell disease. Four major groups, including family and community, the National Institutes of Health, healthcare providers, and researchers are essential for establishing the inter-disciplinary team that will find the universal cure for any disease. All elements must be acknowledged and balanced to provide a successful healthcare program for individuals affected with genetic disorders. Sickle cell anemia was the *first* genetic disease, and therefore is a perfect paradigm to establish standards for curing monogenic diseases. We will explore sickle cell disease from a contemporary perspective by addressing questions such as the impact of the human genome project on new approaches to an old disease in *the genomic era.*

Here is a sneak preview:

Part I will cover the history of sickle cell disease, the progress and missteps that have been made along the way, the role of the federal government in the success we now enjoy, and future research directions. A picture of progress made with genetic diseases will be painted on the canvas of the Human Genome Project, with a focus on the global implications and impact of this ground-breaking research at home and abroad.

In Part II, we will delve into the heart of the matter: What have we accomplished over the last 50 years to truly improve the quality of life for individuals and families living daily with the burden of sickle cell disease? How can the lessons learned apply to other genetic disorders? Traversing the winding road of genetic research from a clinical perspective, we will discuss everything from the impact of Comprehensive Care Centers to the organized multi-center drug trials that finally led to Hydroxyurea, the *first* specific therapy released by the Federal Drug Administration in 1985 for sickle cell disease. From the very beginning, the control of painful vaso-occlusive episodes was, and remains, a central issue for improving function of those living with sickle cell disease. Indeed, some 50 years later, the passion for progress in meaningful intervention still burns. That passion is inescapably fueled by the voices of those crying out for help. Further, we will take an in-depth look at how state-of-the-art technology has impacted stroke management and transfusion therapy. This will be contrasted with the ongoing efforts to develop more tolerable approaches to control the iron overload secondary to chronic transfusion.

In Part III, we will journey deep into the progress that has been made toward understanding the molecular mechanisms that control globin gene expression during development, to aid efforts to produce a gene therapy cure. It is important to note that a cure for sickle cell disease had been hindered by our inability to accurately isolate bone marrow stem cells prior to the 1990s, or to produce effective gene therapy vectors that transport a normal hemoglobin gene into stem cells and produce protein on a permanent basis. Most fearful are the untoward ill effects associated with random integration of engineered DNA molecules into the human genome, and the potential for activation of deleterious proteins or inactivation of vital genes that may lead to cancer. This risk must be eliminated.

And finally, in Part IV we will explore the progress and great knowledge gained from the perspective of affected individuals, the most important focus of our efforts. These people remain courageous and hopeful despite fear and frustration. We will dare to explore and expose the heart of the African-American community as it relates to sickle cell disease — those fears that may remain based on past experiences with research in America, and the cultural acceptance of genetic therapy. The reader will be enlightened on the recommendations of the Ethical, Legal and Social Implications Committee established by the National Human Genome Research Institute to address concerns over individual rights to privacy, the use of genetic information, forensic profiling, and the impact of genetic testing on healthcare delivery in the United States.

This is a unique document, produced by a diverse group of contributors, with a common goal of improving healthcare and ultimately curing sickle cell disease. We have melded together the successes and failures of the past, the technical and the not-so-technical knowledge, and the human spirit, to produce a complete picture of the impact of genetic diseases in our society. Somehow, the role that sickle cell disease has played in this story has varied from front and center to supporting cast.

However, with the dawn of the Human Genome Era, reawakened interest in the *first* genetic disease will hasten the journey to the final destination, a cure. *Renaissance of Sickle Cell Disease Research in the Genomic Era* will enlighten and empower you to make your personal, inward exploration. You will discover how the Human Genome Project ignited a revolution in genomics and genetics that will change the face of medicine in the 21st century.

1

Sickle Cell Disease: Demystifying the Beginnings

by Clarice Reid and Griffin Rodgers

The Precursors

As a global health problem, sickle cell anemia affects many world populations. Since its initial description in the United States almost a century ago, scientists and medical researchers have continued to better understand the pathophysiology of this inherited disease, while simultaneously attempting to find more effective therapies and ultimately a cure.

The pathology of Sickle Cell Disease (SCD) is complicated, in addition to moderate to severe anemia, it often includes infarction and resorption of the spleen during childhood, which results in diminished immune function. Growth and development are delayed, and the ability to produce offspring is often lessened. Bone and eye disorders may occur. Vaso-occlusive "crises" or painful episodes punctuate the course of the disease in many patients, and they may be characterized by unbearable pain in the legs and arms, back, chest, or abdomen. Also affected are the cardiovascular, pulmonary, and renal systems.

The first case report in United States literature occurred in 1910, when James B. Herrick, a cardiologist at Rush Presbyterian Hospital in Chicago, described the physical attributes of the disease as "peculiar, elongated and sickle-shaped erythrocytes" causing "distortion of red blood cells ... by the aggregation of abnormal hemoglobin molecules." One of his patients, a 20-year-old student at the University of Chicago Dental School, presented years earlier at Rush Presbyterian Hospital with a leg ulcer and, later, pneumonia. On the patient's peripheral blood smear, marked anemia and crescent-shaped erythrocytes were noted. An intern, Ernest Irons, then described these cells and drew pictures of the peculiar sickle-shaped red blood cells; these original drawings provided the classic description of SCD. Herrick postulated that these abnormal cells might play a major role in the patient's illness, which prompted Herrick to report his findings in the Archives of Internal Medicine.[1]

Historically, the disease had been known for generations in Africa, and according to reports of clinical syndromes in Western Africa, the disease had numerous local names. Following Herrick's observations, a number of significant studies were done on these peculiar cells, and several prominent hematologists theorized that the problem might lie within the hemoglobin.

Lemuel Diggs, a pathologist at the University of Tennessee, concluded from autopsy findings that the occlusion of the microvasculature might be a factor in the painful episodes linked to the disease.[2] In 1949, Linus Pauling and his colleagues demonstrated the electrophoretic abnormality in sickle hemoglobin (HbS), which migrated differently than normal hemoglobin in an electric field. Pauling thus recognized that SCD must be a disease of the hemoglobin molecule.[3] In contrast to sickle cell anemia, both sickle hemoglobin and normal hemoglobin were present in the sickle cell trait. This was the first example of understanding a human disease in terms of a specific protein abnormality, and led to the identification of protein abnormalities in several other genetic diseases, especially hemoglobinopathies and enzyme deficiencies.

At about the same time, James V. Neel of the University of Michigan reported that he had examined the parents of 29 patients with sickle cell anemia and found that the red blood cells could be induced to sickle in every parent.[4] James Neel and E. A. Beet independently reported that the mode of inheritance was a recessive gene, which made it clear that hemoglobin was the culprit.[5] In 1957, Vernon Ingram ascribed the molecular abnormality to a substitution of valine, an amino acid, for glutamic acid at the $\beta 6$ position of the globin chain; this discovery was based on the development of a fingerprinting technique used to identify changes in amino acids.[6]

Over the next decade, a number of other abnormal hemoglobins were identified, such as HbC, which in combination with sickle cell hemoglobin is usually a milder form of the disease. Currently, there are over 1,200 abnormal hemoglobins, with most lacking any clinical significance. The field was further advanced by Max Perutz who elucidated the three-dimensional structure of the hemoglobin molecule by X-ray diffraction, for which he received the Nobel Prize for Physiology and Chemistry in 1967.[7]

The 1970s heralded new and important developments in both sickle cell anemia and thalassemia through a revolution in molecular biology brought about by recombinant DNA technology. Also, new techniques of nucleic-acid sequencing were applied, which illustrated that the nucleotide change in the DNA for sickle hemoglobin resulted an $A \rightarrow T$ change in the DNA, a mutation which alters the codon GAG (glu) to GTG (val). Sickle cell anemia thus became the first human disease whose abnormality was recognized at the level of the single nucleotide mutation in the gene. Many of these alterations have also been identified for other genetic diseases, especially thalassemia.

At the national level, sickle cell anemia was the first targeted program for a genetic disease, and offered a point of reference in setting the stage for the incredible progress of the past three decades. These achievements are even more noteworthy given the label of sickle cell anemia as an "orphan disease" in the United States, and the limited support of research by the pharmaceutical industry. Many of these advances will be discussed in detail in later chapters.

The National Sickle Cell Disease Program

The 1970s launched unprecedented attention to SCD with both increased public awareness and mixed enthusiasm. This was sparked in 1971, when President Nixon, in his health message to Congress, targeted sickle cell anemia as a major health problem and encouraged greater support at the national level. An earlier article by Robert Scott, Medical College of Virginia, appearing in the Journal of the American Medical Association, compared federal resources for SCD with less prevalent genetic diseases found predominantly in non-black populations (e.g., cystic fibrosis, muscular dystrophy), further fueling this national debate.[8] An editorial in the same Journal stated that "the level of general ignorance concerning the nature of sickle cell anemia remains depressingly high despite substantial scientific advances." The National Institutes of Health (NIH) support for SCD at that time was approximately one million dollars in three institutes.

In the African-American community, the political climate was one of considerable unrest concerning civil equality. Skepticism and cynicism abounded at the news that a program focused on a virtually unknown genetic disorder in African-Americans had been elevated to national priority level. Further compounding the problem were the prevailing myths and misconceptions about this disorder, the confusion of sickle cell trait with SCD, and the dilemma posed by follow-up counseling.[9] Screening programs proliferated; well-intentioned legislators passed laws mandating premarital and pre-school screening; insurance companies increased premiums for individuals with sickle cell trait; and the Air Force denied trait carriers occupational opportunities as pilots or co-pilots. The community viewed many of these new policies as genocidal, with overtones of political exploitation. These interdictions were soon reversed; the program transcended these barriers and SCD moved to a position of prominence, attracting top scientists throughout the world to work on this genetic disease.[10] This was a major triumph for medicine and science.

The public health response was immediate. Secretary Elliott Richardson of the Department of Health, Education and Welfare (DHEW) appointed the National Sickle Cell Disease Advisory Committee; its professional and lay members recommended a broad-based program of research and service. A national program was established in the Pubic Health Service in 1972, with the NIH designated as lead agency, and program responsibility was assigned to the National Heart, Lung and Blood Institute (NHLBI). Basic research — including studies into globin molecular genetics, red cell metabolism, hematopoiesis and hematopoietic stem cells — was and continues to be supported by the Division of Kidney, Urologic and Hematologic Diseases of the National Institute of Diabetes and Kidney Diseases (NIDDK). NIDDK has also fostered a strong translational research program in iron-chelators and in non-invasive assessment of iron-overload. For the past 33 years, the NHLBI has been the major source of funding support for sickle cell anemia.

Roland Scott, "patron saint" of SCD in the United States, spearheaded many of the initial efforts to obtain federal attention for the disease, and testified at Congressional hearings on this disorder.[11] In 1972, Congress passed the National Sickle Cell Control Act (PL.92-294), mandating federal programs for diagnosis, control, treatment, and research in sickle cell anemia.[12] This was the first federal legislation to focus on a genetic disorder. No appropriations were made under this legislative authorization; funds were from the budget of the NHLBI/NIH. The legislation expired in 1975. John Hercules provided the major leadership for advancing sickle cell disease scientific research programs at the NHLBI.

The Health Services Administration (HSA) carried out the service component, with 19 Screening and Education Clinics funded as demonstration projects to hospitals, universities and community health centers, and to freestanding organizations. All played a key role in lessening hysteria, mobilizing the community to participate in selected screening programs, and providing accurate information and follow-up counseling. Education of the public, including health educators and health care providers, was a high priority for the national program during the early years. In many cases, these programs served as entry points for individuals into the health care delivery system.

In 1978, federal support for the Screening and Education Clinics administered by the Genetics Program at the Health Resources and Services Administration (HRSA) was transferred to the states via State Block Grants. The Genetic Services Branch, Maternal and Child Health Bureau, continued to support services through its appropriation for Special Projects of Regional and National Significance (SPRANS). HRSA played a major role in funding newborn screening and follow-up programs, and in collecting valuable demographic data through the support of the Council of Regional Networks (CORN). In recent years, the Maternal and Child Health Bureau, HRSA, have received a direct appropriation to support a community-based demonstration project for sickle cell newborn screening.

The comprehensive sickle cell centers have been the major component of the national program, established in geographical areas with large at-risk populations. The original center concept focused on the patient, while embracing community involvement and the integration of services in an academic research environment. This was one of the early program models to accommodate the emerging emphasis on prevention, education, and control programs supported by the NIH. In this setting, the previously impersonal, fragmented, and episodic care for sickle cell patients was drastically changed. A cadre of trained personnel — clinicians, nurses, social workers, educators, psychologists, nutritionists, and counselors — worked closely in a team approach to patient care. Emergency room protocols were developed, while parents and caregivers were trained to detect important signs and symptoms of early complications. Sickle cell programs at non-NIH funded medical centers, using similar models, have played a pivotal role in improving overall patient care. A minimum of ten centers was congressionally mandated by the Orphan Drug Act, Public Law 97-414, in 1983. Over the years, the focus of the center program has shifted more toward basic and clinical research, with limited support of non-research and community activities.

The hemoglobinopathy training program at the Centers for Disease Control (CDC) in Atlanta was a prominent part of the early national sickle cell program. It was established to train laboratory personnel in the use of up-to-date technologies to detect abnormal hemoglobin species. Laboratories with federal funding had to participate in the proficiency-testing program of hemoglobin diagnosis; this monitoring of laboratory procedures resulted in better quality assurance and accuracy of laboratory diagnoses. Early focus of the National Program was on emergency room management and the development of treatment protocols to ensure adequate care for all patients. A consensus of experienced clinicians in the field led to the publication of "Management and Therapy," outlining the basics of routine care, and for common medical and surgical problems. This document has been updated over the years and continues to be a handy reference for the management of sickle cell patient.[13]

The Sickle Cell Disease Association of America (SCDAA), founded by Charles Whitten in 1972 as the National Association of Sickle Cell Disease, is the key lay organization directed toward improving the quality of life for sickle cell patients and their families, and ultimately finding a cure. It is the national umbrella organization for 57 community-based associations located across the country that provide public and professional health education, patient services, community outreach, and newborn screening follow-up. Since its inception, through a community-based approach, the SCDAA has been in the forefront of advancing public awareness, correcting misinformation, and providing counseling and patient advocacy for thousands with the disease and their families for the past three decades. These activities had a major impact on changing the earlier climate for sickle cell programming at the local and national levels.

As an ongoing partner with the National Sickle Cell Disease Program, the SCDAA is a strong advocate for obtaining federal funds for sickle cell research and services, and works closely with federal agencies. In its early years, the SCDAA provided research funding to the ten Comprehensive Sickle Centers, and later initiated a Summer Research Apprentice Program that allowed promising high school seniors to gain early exposure to research. The SCDAA has established a Post-Doctoral Research Fellowship Program aimed at developing young investigators beginning their biomedical and psychosocial research careers. In fiscal year 2002, as a result of the legislative efforts of the SCDAA, an unprecedented $4 million was earmarked for the budget of the Maternal and Child Health Bureau, HRSA, to fund sickle cell activities. This support is scheduled to continue to fiscal 2008.

Clinical Advances

Early clinical investigations to improve the outcomes of sickle cell anemia patients targeted the recurrent painful episode that was the clinical hallmark of the disease. In the mid-1960s, Makio Murayama hypothesized that the polymerization of the deoxyhemoglobin S molecule was dependent upon the existence of hydrophobic bonds between molecules.[14] This hypothesis led to *in vitro* studies in 1971 by Robert Nalbandian that demonstrated that the compound urea could interfere with these bonds, and thereby prevent and reverse sickling.[15] A small, uncontrolled study reported efficacy in ameliorating both severity and duration of pain when urea was administered intravenously to sickle cell patients in a solution of invert sugar; however, these findings were not confirmed in a controlled, double-blind, randomized study funded by the NIH. A companion study also revealed that alkali was not beneficial in the treatment of pain, as had been suggested by other reports.[16,17]

Various additional therapies were thus introduced to inhibit polymerization, swell red blood cells, and modify hemoglobin, but with little or no success. Anthony Cerami and James Manning noted that the small amounts of cyanate which formed in urea solutions might have anti-sickling effects after completing *in vitro* and animal studies on cyanate, which showed that anti-sickling properties were due to an increased affinity for oxygen.[18] Clinical trials were initiated in a small number of medical centers; the observed benefits, however, were offset by neurotoxicity and other complications when higher doses were utilized.[19] Thus these studies were terminated. Yet the drug seemed to hold much promise, for it appeared that the drug would be safe if cyanate reacted only to red blood cells, with the extra cyanate being eliminated. Funding from the NIH enabled testing of an instrument similar to a hemodialysis device that delivered cyanate by the extracorporeal route of administration.[20] This process of carbamylation proved cumbersome, and while it demonstrated potential feasibility, problems arose with hyperviscosity. Therefore, the project was discontinued.

These clinical studies on urea and cyanate were quite important in underscoring the necessity for supporting basic clinical research on therapeutic strategies for sickle cell anemia; they also highlighted the need to understand the natural history of the disease in order to assess the therapeutic efficacy for various complications. During this period, a quantum leap in the application of molecular techniques occurred which greatly increased the understanding of sickle cell anemia at molecular, cellular, and clinical levels. Studies on fetal hemoglobin synthesis in erythroid cells catalyzed the field of molecular biology and became a foundation for the development of therapeutic approaches based on increasing the level of fetal hemoglobin in those afflicted with sickle cell anemia.

Another significant development in the 1970s was prenatal diagnosis of the disease by Yuet W. Kan and Andrea Dozy.[21] Earlier techniques for prenatal diagnosis required fetal cells for the biochemical analysis of globin chains to determine the genotype of the fetus. This approach was replaced by gene mapping of DNA fragments using restriction endonuclease enzymes and identification of polymorphisms (restriction fragment length polymorphism or RFLP analysis) adjacent to the β-globin locus. The largest utilization for prenatal diagnosis has been in other populations, especially β-thalassemia. In contrast, the broader application of RFLP has found its way to forensics, agriculture and genomics.

The 1980s ushered in an era of great optimism under the leadership of Clarice Reid, Chief, Sickle Cell Disease Branch, via a series of clinical reports and an abundance of publications containing heretofore unknown clinical data related to sickle cell anemia. The predominant source of these reports was the National Institutes of Health study known as the Cooperative Study of Sickle Cell Disease (CSSCD) organized in 1979. Although there was a great body of knowledge on the molecular basis of SCD, there was a paucity of information on its clinical course and natural history. Much of the information in the medical literature only described the most severe cases, usually isolated in a hospital setting; the reports of clinical outcomes were mostly retrospective or anecdotal. The variability of clinical severity, ranging from mild to debilitating and with the same genotype and hematologic patterns, challenged the clinician and the researcher. For the most part, information on the natural history was only available for the first 10 years within the United States from Darlene Powars and the Jamaican sickle cell anemia pediatric cohort of Graham Sergeant.[22,23]

To improve understanding of the natural history and the variable severity of sickle cell disease, the CSSCD, a large-scale, multi-institutional epidemiological study, became of great importance.[24] Over 4000 patients were recruited, representing the major phenotypes of the disease, from infants to adults in the sixth decade of life. The infants in this study represented an unbiased population that was enrolled based on early diagnosis, and patient recruitment extended beyond hospital-based programs in which patients were the most severely affected; those involved in community-based health departments and under the care of physicians were also included. A number of important papers published by CSSCD investigators resulted from this prospective, longitudinal study and included patient demographics, growth and development patterns in children, pregnancy, transfusion and alloimmunization, infection, pain, mortality, major organ dysfunction, and others. A comprehensive summary of the many CSSCD publications has recently been published by Duane Bonds.[25]

The landmark study from this period was the Prophylactic Penicillin Study led by Marilyn Gaston. Prior to the 1980s, 30% of deaths were due to pneumococcal infections, mostly in children under five years of age. A randomized, controlled clinical trial[26] found that the daily oral administration of penicillin reduced the rates of infection from *S. pneumonia* by 84% in young children with SCD (Fig. 1). There was a strong public health response to these compelling results showing that prophylactic penicillin could save lives. The NIH convened a consensus development conference on Newborn Screening for Sickle Cell Disease and other Hemoglobinopathies to examine the merit of

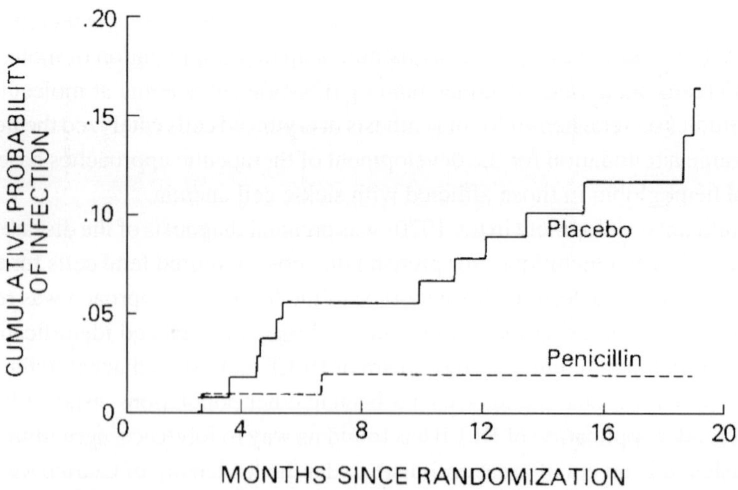

	N									
Pen	105	0.0	102	0.01	91	0.02	76	0.02	64	0.02
Pla	110	0.0	108	0.02	96	0.06	78	0.09	61	0.12

Fig. 1. Cumulative infection rates for all patients in the prophylactic penicillin study. Numbers of patients at risk and infection rates in the penicillin (Pen) and placebo (Pla) groups, at four-month intervals, appear below the curves (one-tailed p value = 0.003). By permission from Gaston MH *et al.* (1991). *N Engl J Med* **314**:1593–1599.

widespread newborn screening and its inherent controversies. The panel included biomedical scientists, clinicians, other health professionals, and representatives of the public. This led to a number of recommendations including that all newborns in the United States be screened for SCD and placed on prophylactic penicillin by the age of three months.[27] This was a major public policy change.

Newborn screening the administration of prophylactic penicillin, and early comprehensive care for infants with SCD are considered as general standards within the medical community, serving as the first steps toward prevention. More than 46 United States states screen at birth for SCD. Although the technology for screening of umbilical cord blood for SCD had been available for many years, the recommendation as policy for newborns was not formally adopted until the findings gained from the penicillin study.[28] Prevention of infection in infants and children now includes routine pneumococcal vaccination. It is safe to discontinue prophylactic penicillin in most children with sickle cell anemia after the age of five years.[29]

For the first time, the CSSCD report on the epidemiology of painful events showed a correlation of pain rates with early death in sickle cell patients over the age of 20 (Fig. 2). The rate of pain was shown to be positively influenced by the levels of fetal hemoglobin (HbF), resulting in decreased pain rate and accompanying decreased morbidity. It was noted that 38% of the patients did not experience a painful event during a given five-year period between 1979 and 1988. In addition, patients with more frequent pain histories (3–10 events per year) represented only 5.2% of the population, but contributed 33% of the painful events to the analysis.[30]

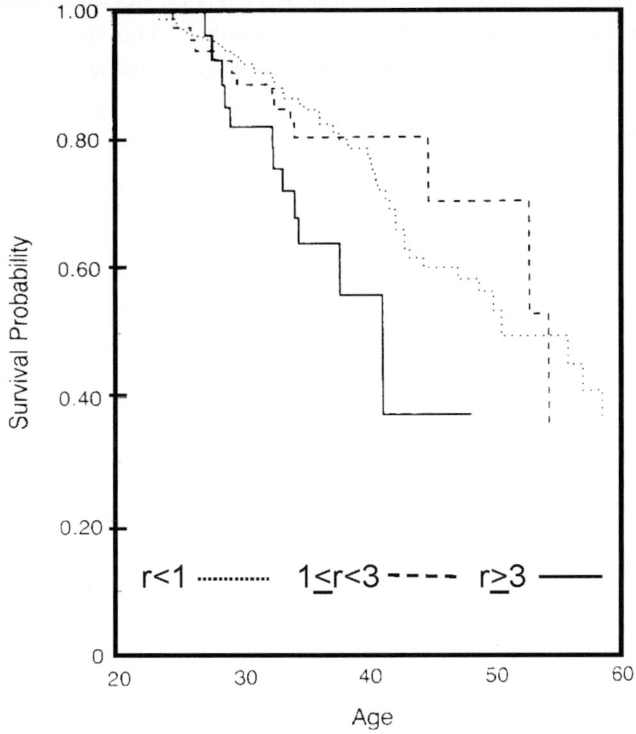

Fig. 2. Survival of patients with sickle cell anemia (≥20 years old at entry) who had different pain rates. The letter "r" denotes the number of episodes of paid per patient-year. By permission from Platt OS *et al.* (1991). *New Engl J Med* **325**:11–16.

Fortunately, the CSSCD data provided encouraging mortality information on both pediatric and adult patients with sickle cell anemia. Early hematology literature declared that patients rarely lived to adulthood. Survival curves, stratified by gender and the age of patients with SS and SC diseases in the CSSCD, indicated that the median life span was about 42 years for SS males, 48 years for SS females, 60 years for SC males, and 68 years for SC females.[31] Although the medians differ by many years from other African-Americans, they reflect a significant improvement in survival for this population. The life expectancy of sickle cell patients has doubled since passage of the National Anemia Sickle Cell Control Act (Fig. 3). More recently published survival data[32] on children with SCD concluded that "childhood mortality is decreasing, the mean age at death is increasing, and a smaller proportion of deaths are from infection." This improvement allows long-range planning for education, careers, professional development, job opportunities, and the prospects of children and families.

With the advent of the 1990s, molecular medicine evolved from the bench to the bedside; basic research on globin gene regulation and how it functions also greatly improved.[33] This classic example of translation research was evidenced in the use of pharmacologic agents to reactivate dormant genes and ameliorate pain in severely affected sickle cell anemia patients. HbF is known to have a sparing effect on sickle hemoglobin, and those with high levels of fetal hemoglobin also were mildly affected. The findings by Joseph DeSimone, Timothy Ley and others that 5-azacytidine increases gamma globin synthesis led to the search for other less toxic cell-cycle specific agents exhibiting the same effects.[34–36] One of the agents known to increase HbF is hydroxyurea.[37] Following small but promising proof-of-concept studies in non-human primates and in patients, Griffin Rodgers, Arthur Neinhuis and colleagues performed a pivotal phase I/II study on sickle cell patients admitted to the clinical center at the NIH for periods of three to four months.[38] This study demonstrated the safety and efficacy of hydroxyurea, and defined the optimal dosages to achieve HbF augmentation. After

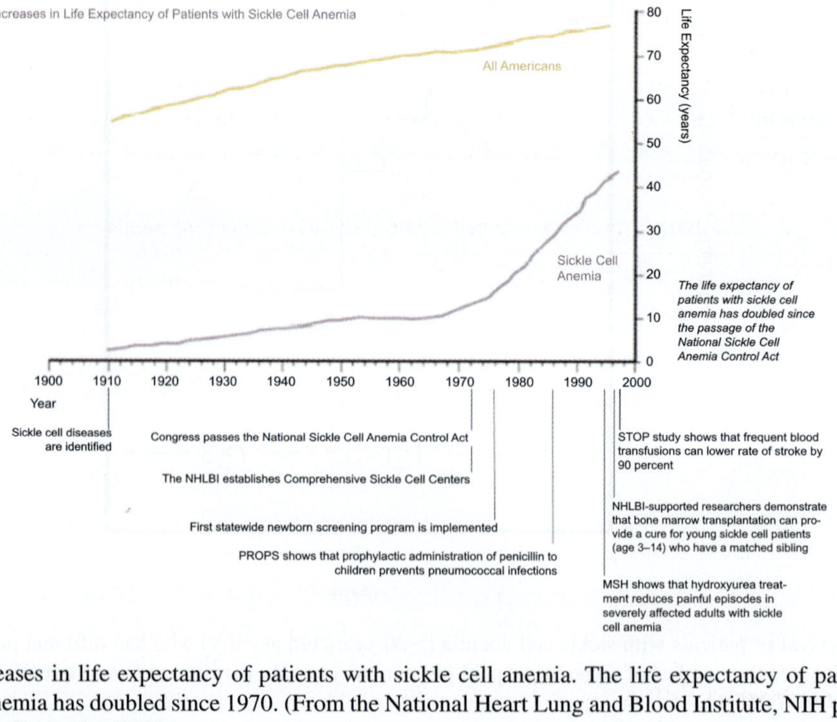

Fig. 3. Increases in life expectancy of patients with sickle cell anemia. The life expectancy of patients with sickle cell anemia has doubled since 1970. (From the National Heart Lung and Blood Institute, NIH publication No. 02-5214, September 2002.)

other phase II reports corroborated the NIH safety and clinical efficacy results, a prospective, multicenter, randomized clinical trial involving 299 patients using hydroxyurea was undertaken through the efforts of clinical investigator Samuel Charache, and Duane Bonds from the NHLBI. It demonstrated a 50% reduction in vaso-occlusive episodes in adults with sickle cell disease who were treated with this agent. There were also significant reductions in transfusion requirements, hospitalization, and acute chest syndrome.[39] Follow-up data available for up to nine years in 233 of the original subjects indicated, among other findings, that taking hydroxyurea was associated with a 40% reduction in mortality.[40]

Hydroxyurea in children has been investigated in a safety and dosing study showing that the drug could be used in children between the ages of 5 and 15 years, accompanied by a significant increase in HbF and hemoglobin concentration.[41] Subsequent reports noted that hydroxyurea was well tolerated in an even younger population of sickle cell infants with a median age of 15 months.[42] The definitive Phase III trial of hydroxyurea in children to determine clinical efficacy in preventing end organ damage has been initiated at the NHLBI.

The investment in basic research and controlled clinical studies has proven to be of enormous benefit to adults with SCD. In addition, the study of this disease and other globin disorders has illuminated broader principles applicable to other genetic diseases, and has facilitated technological developments with applications to many more disorders. A testament to this conclusion is demonstrated by the many "firsts" associated with the study of SCD (Table 1) in which studies illuminated broader principles that resulted in more extensive applications to other fields of science and medicine.

Many of the young children with SCD have crippling central nervous system complications, including stroke, which was the second cause of death in the CSSCD pediatric population. Chronic blood transfusions to maintain the level of sickle hemoglobin below 30% have been successful in preventing recurrent strokes in children.[43] A major goal for preserving cerebro-vascular competence is to prevent the first stroke and its debilitating sequelae. The Stroke Prevention Trial in Sickle Cell Anemia (STOP) study demonstrated that such prevention is possible using non-invasive techniques to detect large cerebral vessel stenosis, identify children at-risk for stroke, and treatment with regular blood transfusions.[44] An additional challenge to the clinician is to prevent the problems of iron-overload that are secondary to treatment with transfusions in these patients.

Table 1. "Firsts" associated with the study of sickle cell anemia.

Authors	Year	Seminal contribution
J. Herrick	1910	First Clinical Report in the US
L. Pauling	1949	Application of protein electrophoresis
J. Neel	1949	Genetic nature of disease
M. Perutz	1951	X-ray crystallography of large protein
A. Allison	1954	Concept of "balance polymorphisms"
V. Ingram	1957	Two dimensional gel "fingerprint"
Y. W. Kan and A. Dozy	1976	Pre-natal diagnosis of a disease
A. Deisseroth	1977	Chromosomal localization of globin genes
Y. W. Kan and A. Dozy	1978	Restriction Fragment Length Polymorphism (RFLP) and DNA Haplotypes
J. DeSimone and P. Heller	1982	Reactivation of fetal (dormant) gene expression
R. Saiki and K. Mullis	1985	Clinical application of PCR

Summary

A number of seminal developments marked the pathway from the early clinical description of sickle cell anemia in 1910 to the unparalleled advances of the past three decades. The first federal program for a genetic disorder, launched in a political climate of civil unrest in the 1970s, continues to impact SCD research and services at the national level. This achievement is noteworthy and reflects the leadership role of the NIH in partnership with other federal agencies and the SCDAA. The commitment to invest in basic and clinical research has led to major changes in public health policy and other interventions to prevent and/or decrease serious complications and save lives.

Tremendous progress has been made at all levels of SCD over the past three decades, and provides the catalyst for even more vigorous research. There is an optimistic outlook for patients, their families, the caring physician, and the vigilant scientist. Yet there remains an imbalance between what is known about SCD and translating this knowledge into effective therapies. There is only one approved drug for treatment, and a cure is not currently available for most patients. A limited number of primary care physicians and hematologists are caring for adult sickle cell patients. Nonetheless, comprehensive care and improved management have extended the quality of life and longevity of these patients. The absence of established genetic modifiers and biomedical markers of clinical severity limit the ability to identify patients who would benefit most from specific therapies, which becomes very important when a therapy has potentially serious or even life-threatening side-effects.

Improvements in preparative regimens for hematopoietic stem cell transplantation from sibling donors and the ability to use unrelated matched HLA donors in the future offer the best prospect to cure a limited number of patients with sickle cell disease. Gene therapy for SCD is the "holy grail" of cure, yet its application has been formidable due to the difficulty in transducing hematopoietic stem cells, and the necessity for erythroid specific, high level, and balanced globin gene expression. Despite these challenges efforts to move forward with gene therapy continue. Current genomic studies should provide insights on more pre-emptive strategies to resolve these therapeutic challenges.

References

1. Herrick JB (1910). Peculiar elongated and sickle-shaped red blood corpuscles in a case of severe anemia. *Arch Int Med* **6**:517–521.
2. Diggs L, Ching RE (1934). Pathology of sickle cell anemia. *South Med J* **27**:839–845.
3. Pauling L, Itano H, Singer SJ, Wells IC (1949). Sickle cell anemia, a molecular disease. *Science* **110**: 543–548.
4. Neel JV (1949). The inheritance of sickle cell anemia. *Science* **110**:164–166.
5. Beet EA (1949). The genetics of the sickle cell trait in a Bantus tribe. *Ann Eugen* **14**:279–284.
6. Ingram VM (1957). Gene mutations in human hemoglobin: the chemical difference between normal and sickle cell hemoglobin. *Nature* **180**:326–328.
7. Perutz MF, Rossman MG, Cullins AF, Muirhead H, Will G, North ACT (1960). Structure of haemoglobin: a three dimensional Fourier synthesis at 5.5, a resolution obtained by X-ray analysis. *Nature* **185**:416–422.
8. Scott R (1970). Health care priority and sickle cell anemia. *JAMA* **214**:731–734.
9. Whitten CF (1973). Sickle cell programming — an imperiled promise. *N Engl J Med* **288**:318–319.
10. Culliton B (1972). Sickle cell anemia: the route from obscurity to prominence. *Science* **178**:140.
11. Scott R (1983). Historical review of legislative and national initiatives for sickle cell disease. *Amer J Pediatric Hematology/Oncology* **5**(4):349.
12. The National Sickle Cell Disease Control Act (1972). Public Law, 92–294.
13. Management and Therapy of Sickle Cell Disease (2002 Revised). US Dept of Health and Human Services. NIH Publication, Rockville, MD. No. 02-2117.

14. Murayama M (1967). Structure of sickle cell hemoglobin and molecular mechanism of the sickling phenomenon. *Clin Chem* **13**:578.
15. Nalbandian RM, Henry RL, Barnhart MI, Nichols BM, Camp FR Jr, Wolf PL (1971). Sickle reversed and blocked by urea in invert sugar. *Am J Pathol* **65**:405–422.
16. McCurdy PR, Mahmood L (1971). Intravenous urea treatment of the painful crisis of sickle cell disease, *N Engl J Med* **285**:992–994.
17. (No authors listed) (1974). Therapy for sickle cell vaso-occlusive crises. Controlled clinical trials and cooperative study of intravenously administered alkali. *JAMA* **228**:1129–1131.
18. Cerami A, Manning JM (1971). Potassium cyanate as an inhibitor of the sickling phenomenon. *Proc Natl Acad Sci* **68**:1180–1183.
19. Harkness D, Roth S (1975). Clinical evaluation of cyanate in sickle cell anemia. *Progr Hematol* **9**:157–184.
20. Deiderich DA, Trueworthy RC, Gill P, Cader AM, Larsen WE (1976). Hematologic and clinical response in patients with sickle cell anemia after chronic extracorporeal red cell carbamylation. *J Clin Invest* **58**: 642–653.
21. Kan YW, Dozy AM (1978). Antenatal diagnosis of sickle-cell anemia by D.N.A. analysis of amniotic fluid cells. *Lancet* **2**:910–912.
22. Serjeant GR, Grandison Y, Lowrie Y, Mason K, Phillips J, Serjeant BE, Vaidya S (1981). The development of haematological changes in homozygous sickle cell disease: a cohort study from birth to six years. *Br J Haematol* **48**:533–543.
23. Powars DR (1975). Natural history of sickle disease: the first ten years. *Semin Hematol* **12**:267.
24. Gaston M, Rosse WF and the Cooperative Group (1982). The cooperative study of sickle cell disease: review of study design and objectives. *Am J Pediatr Hematol Oncol* **4**:197–201.
25. Bonds DR (2005). Three decades of innovation in the management of sickle cell disease: the road to understanding the sickle cell disease clinical phenotype. *Blood Reviews* **19**:99–110.
26. Gaston MH, Verter JI, Woods G, Pegelow C, Kelleher J, Presbury G, Zarkowsky H, Vichinsky E, Iyer R, Lobel JS (1986). Prophylaxis with oral penicillin in children with sickle cell anemia: a randomized trial. *N Engl J Med* **314**:1593–1599.
27. NIH Consensus Conference (1987). Newborn screening for sickle cell disease and other hemoglobinopathies. *JAMA* **258**:1205–1209.
28. Pearson HA, O'Brien RT, McIntosh S, Aspnes GT, Yang MM (1974). Routine screening of umbilical cord blood for sickle cell diseases. *JAMA* **227**:420–421.
29. Falletta JM, Woods GM, Verter JI, Buchanan GR, Pegelow CH, Iyer RV, Miller ST, Holbrook CT, Kinney TR, Vichinsky E (1995). Discontinuing penicillin prophylaxis in children with sickle cell anemia. *J Pediatr* **127**:685–690.
30. Platt OS, Thorington BD, Brambilla DJ, Milner PF, Rosse WF, Vichinsky E, Kinney TR (1991). Pain in sickle cell disease. Rates and risk factors. *N Engl J Med* **325**:11–16.
31. Platt OS, Brambilla DJ, Rosse WF, Milner PF, Castro O, Steinberg MH, Klug PP (1994). Mortality in sickle cell disease: life expectancy and risk factors for early death. *N Engl J Med* **330**:1639–1644.
32. Quinn CT, Rogers Z, Buchanan G (2004). Survival of children with sickle cell disease. *Blood* **103**: 4023–4027.
33. Grosveld F, Dillon N, Higgs D (1993). The regulation of human globin gene expression. *Balliere's Clinical Hematology* **6**:31–55.
34. DeSimone J, Heller P, Hall L, Zwiers D (1982). 5-Azacytidine stimulates fetal hemoglobin sythesis in anemic baboons. *Proc Natl Acad Sci USA* **79**:4428–4431.
35. Ley TJ, DeSimone J, Noguchi CT, Turner PH, Schechter AN, Heller P, Nienhuis AW (1983). 5-Azacytidine increases gamma-globin synthesis and reduces the proportion of dense cells in patients with sickle cell anemia. *Blood* **62**:370–380.
36. Charache S, Dover G, Smith K, Talbot CC Jr, Moyer M, Boyer S (1983). Treatment of sickle cell anemia with 5 aza-cytidine results in increased fetal hemoglobin production and is associated with nonrandom hypomethylation of DNA. *Proc Natl Acad Sci USA* **80**:4842–4846.
37. Platt OS, Orkin SH, Dover G, Beardsley GP, Miller B, Nathan DG (1984). Hydroxyurea enhances fetal hemoglobin production in sickle cell anemia. *J Cl Invest* **74**:652–656.

38. Rodgers GP, Dover GJ, Noguchi CT, Schechter AN, Nienhuis AW (1990). Hematological responses in patients with sickle cell anemia treated with hydroxyurea. *N Engl J Med* **322**:1037–1045.
39. Charache S, Terrin ML, Moore RD, Dover GJ, Barton FB, Eckert SV, McMahon RP, Bonds DR (1995). Effect of hydroxyurea on the frequency of painful crises in sickle cell anemia. *N Engl J Med* **332**:1317–1322.
40. Steinberg MH, Barton F, Castro O, Pegelow CH, Ballas SK, Kutlar A, Orringer E, Bellevue R, Olivieri N, Eckman J, Varma M, Ramirez G, Adler B, Smith W, Carlos T, Ataga K, DeCastro L, Bigelow C, Saunthararajah Y, Telfer M, Vichinsky E, Claster S, Shurin S, Bridges K, Waclawiw M, Bonds D, Terrin M (2003). Effect of hydroxyurea on mortality and morbidity in adult sickle cell anemia: risks and benefits up to 9 years of treatment. *J Am Med Assoc* **289**:1645–1651.
41. Kinney TR, Helms RW, O'Branski EE, Ohene-Frempong K, Wang W, Daeschner C, Vichinsky E, Redding-Lallinger R, Gee B, Platt OS, Ware RE (1999). Safety of hydroxyurea in children with sickle cell anemia: results of the HUG-KIDS Study, a phase 1/11 trial. *Blood* **94**:1550–1554.
42. Wang WC, Wynn LW, Rogers ZR, Scott JP, Lane PA, Ware RE (2001). A two year pilot of hydroxyurea in very young children with sickle cell anemia. *J Pediatr* **139**:790–796.
43. Russell MO, Goldberg HI, Hudson A, Kim HC (1984). Effect of transfusion therapy on arteriographic abnormalities and on recurrence of stroke in sickle cell disease. *Blood* **63**:162–169.
44. Adams RJ, McKie VJ, Hsu L (1998). Prevention of first stroke by transfusion in children with sickle cell anima and abnormal results on transcranial doppler ultrasonography. *N Engl J Med* **339**:5–11.
45. Saiki RK, Scharf S, Faloma F, Mullis KB, Horn GT, Erlich HA, Arnhein N (1985). Enzymatic amplification of beta-globin genomic sequences and restriction site analysis for diagnosis of sickle cell anemia. *Science* **230**:1350–1354.

2

Sponsorship of Sickle Cell Disease Research by the National Institutes of Health: A Brief History and Projections for the Future

by Gregory L. Evans and David G. Badman

Introduction

This chapter will cover the evolution, over the past three decades, of research funding for sickle cell disease (SCD) at the National Institutes of Health (NIH). In particular, it will provide the history and impact of critical legislative events, such as the United States Congressional mandate that gave rise to the Comprehensive Sickle Cell Centers (CSCCs) at the NIH; also discussed will be seminal research programs and clinical trials sponsored by the National Heart, Lung, and Blood Institute (NHLBI) and the National Institute of Diabetes, Digestive, and Kidney Diseases (NIDDK). There will be a snapshot of the current NHLBI SCD research portfolio, including large clinical trials and the current focus on clinical research networks. A discussion of translational research efforts will be centered on the CSCC program. The non-CSCC basic science efforts will also be highlighted, along with current trends in investigator-initiated research and many NIH-initiated special programs targeted to SCD and co-sponsored by the NHLBI and NIDDK. Lastly and most importantly, this chapter will provide some insights into future NIH programs and their role in the SCD research enterprise, including a summary of the outcome of a 2003 genomics-oriented workshop co-led by the National Human Genome Research Institute (NHGRI) and NHLBI.

SCD is a significant contributor to health disparities in the United States African-American population. As we move into the 21st century and continue to search for a universal cure for serious, debilitating congenital disorders like SCD, we hope the NIH can form strong partnerships with community organizations and improve the health-related quality of life of affected individuals. We ultimately hope to eliminate SCD as a source of health disparity in the United States population.

Sickle Cell Disease

SCD is a worldwide health problem, one of the most common inherited disorders of man. The history, pathophysiology, and genetics of SCD are discussed in Chapters 1 and 4, so they will not be described in great detail here. This genetic blood disorder is probably the best understood disease at the molecular level. We know that SCD is caused by a single amino acid substitution of valine for glutamic acid at position 6 of the beta-globin polypeptide chain, resulting from a point mutation in the beta-globin gene. Sickle hemoglobin is less soluble than normal hemoglobin upon deoxygenation. This abnormal hemoglobin aggregates and forms fibers within the red cells, leading to morphological changes that subsequently affect the ability of the cells to traverse the microvasculature. These morphological defects result in occlusion of small vessels, with acute pain, and chronic organ damage. In addition, sickle red cells exhibit early destruction leading to chronic anemia. The cascade of events caused by this abnormal sickle cell morphology affects the structure and function of the red cells, blood flow through tissues and organs throughout the body, and abnormal interaction of these cells with the microvasculature. Despite its distinction as the first described molecular disease, SCD currently has neither a universal cure nor a universally effective treatment.

The National Institutes of Health

Founded in 1887, the NIH today is one of the world's foremost medical research centers. An agency of the United States Department of Health and Human Services, NIH is the federal focal point for health research, the nation's steward of medical and behavioral research. Its mission is science in pursuit of fundamental knowledge about the nature and behavior of living systems, and the application of that knowledge to extend healthy life and reduce the burdens of illness and disability. The United States Congress budgeted $28.1 billion for NIH in FY 2005. As presently constituted, the NIH includes 27 different institutes and centers, the NHLBI, NIDDK, and NHGRI among them. The historical and present contributions of all three institutes towards SCD research efforts will be highlighted in this chapter.

In his "Special Message to the Congress Proposing a National Health Strategy" of February 18, 1971, President Richard Nixon made research on sickle cell anemia a national priority:

"A . . . targeted disease for concentrated research should be sickle-cell anemia — a most serious childhood disease which almost always occurs in the black population. It is estimated that one out of every 500 black babies actually develops sickle-cell disease. It is a sad and shameful fact that the causes of this disease have been largely neglected throughout our history. We cannot rewrite this record of neglect, but we can reverse it. To this end, this administration is increasing its budget for research and treatment of sickle-cell anemia . . ."

Legislation was subsequently introduced in both the House and the Senate to provide for the control of sickle cell anemia, and 14 months later, on May 16, 1972, the National Sickle Cell Anemia Control Act (P.L. 92-294) was signed into law. P.L. 92-294 provided for the establishment of voluntary sickle cell anemia screening and counseling programs; information and education programs for health professionals and the public; and research and research training in the diagnosis, treatment, and control of sickle cell anemia. Shortly after the act was passed, the Secretary of the Department of Health, Education, and Welfare responded by establishing a National Sickle Cell Disease Program and by assigning to the National Heart and Lung Institute (NHLI) the responsibility for developing and supporting a program of research in sickle cell disease, and for coordinating the overall program. On April 22, 1976, the Health Research and Health Services Amendments passed by the United States Congress restructured provisions governing the programs of the NHLI, placing increased emphasis on blood-related research, and changing the institute's name to the National Heart, Lung, and Blood Institute (NHLBI). They also authorized broad-based genetic diseases research under section 301 of the PHS act, and provided for programs of counseling, testing, and information dissemination about genetically transmitted diseases across the NIH (P.L. 94-278).

The National Heart, Lung, and Blood Institute (NHLBI)

Founded in 1948 as the National Heart Institute, the NHLBI provides leadership for a national program in diseases of the heart, blood vessels, lung, and blood; blood resources; and sleep disorders. Since October 1997, the NHLBI has also had administrative responsibility for the NIH Woman's Health Initiative. The Institute plans, conducts, fosters, and supports an integrated and coordinated program of basic research, clinical investigations and trials, observational studies, and demonstration and education projects. The United States Congress budgeted $2.9 billion for the NHLBI in FY 2005.

The National Institute of Diabetes and Digestive and Kidney Diseases (NIDDK)

In 1950, the Omnibus Medical Research Act established the National Institute of Arthritis and Metabolic Diseases (NIAMD), the precursor to the NIDDK. The new Institute incorporated the laboratories of the Experimental Biology and Medicine Institute and was expanded to encompass the broad spectrum of metabolic diseases such as diabetes, inborn errors of metabolism, endocrine disorders, mineral metabolism, digestive diseases, nutrition, urology and renal disease, and hematology. The United States Congress appropriated approximately $1.7 billion for NIDDK in FY 2005. The Hematology Program of the NIDDK emphasizes basic research on blood and blood-forming organs, hematopoiesis, red blood cell metabolism, and disorders such as SCD, Cooley's anemia (thalassemia), and hemochromatosis.

At the time of the passage of the National Sickle Cell Anemia Control Act in 1972, the NIAMD was the principal supporter of basic sickle cell disease research, primarily related to hemoglobin and red blood cell physiology and pathophysiology. However, with the designation of the NHLI as the lead Institute for sickle cell research, subsequent competing renewals and new applications most relevant to SCD were assigned to the NHLI. The remaining portfolio maintained by the NIAMD emphasized basic research, including abnormalities of hemoglobin metabolism, regulation of hemoglobin synthesis, hemoglobin structure, and chemistry of sickle cell anemia variants. In 1976, the NHLI and NIAMD co-sponsored a Request for Applications (RFA) on the genetic regulation of fetal hemoglobin synthesis as a potential avenue for ameliorating the symptoms of SCD. This led George Stamatoyannopoulos and Arthur Nienhuis to propose a Conference on Fetal Hemoglobin Switching, funded by NIAMD and NHLI and held at the Battelle Institute in Seattle in 1978.

The "Switching Meeting" was immediately successful and began a series of biennial conferences. The 1978 conference, however, suffered from lack of a common language between cell and molecular biologists: In fact, the molecular biologists came, gave their talks and left, followed by the arrival of the cell biologists who talked to each other. In subsequent meetings, George Stamatoyannopoulos organized sessions which helped to educate both groups in each others' areas, and the meetings went on to provide great stimulus to this aspect of SCD research.

The growth of molecular genetics research, spurred by the studies of hemoglobin F and the development of recombinant DNA technology (rDNA), began in earnest around this time. However, certain aspects of molecular genetic research were slowed by issues around rDNA. One of the very first applications for the study of hemoglobin genetic regulation using rDNA technology was submitted by Bernard Forget, but, even with its outstanding study section ranking, funding was delayed for nearly a year because of concerns about the technology. In June 1973, a Gordon Conference was convened to discuss safety issues related to rDNA lab workers; NIH and the National Institute of Medicine were asked to appoint a committee to study the matter. At the same time, noted scientists sent letters to the journals *Science* and *Nature* calling for a temporary halt to rDNA experiments, a request unheard of before in the history of science. NIH then set up a Recombinant DNA Advisory Committee (RAC) at the behest of the scientific committee. Then, in February 1975, the Asilomar Conference was held to discuss relevant issues. The conclusion was that most rDNA work should continue, but appropriate safeguards in the form of physical and biological containment procedures should be put in place. The current inventory of NIDDK grants related to hemoglobin molecular biology stems from this unique situation.

Over the succeeding years, the NIDDK and the NHLBI have shared funding for sickle cell disease research, with NIDDK continuing to emphasize basic studies. The two Institutes have released and funded many joint RFAs in the area, principally related to hemoglobin switching and genetic regulation.

In addition, under the leadership of David Badman, the NIDDK has functioned as the lead Institute for the development of new iron chelators for transfusional iron overload. While directed initially toward patients with thalassemia (Cooley's anemia), it was recognized several years ago that SCD patients, particularly those at risk for stroke, would benefit from reducing their body iron burden. Several new drugs developed by the NIDDK's iron chelator program are in clinical trials. An accompanying effort to improve non-invasive methods for assessing iron overload was begun with the development of a biomagnetic susceptometer in the early 1980s. The device, known as a SQUID, is reliable but expensive and is located in a very few sites. In 2003, the NIDDK began a program to adapt magnetic resonance (MRI) to assess body iron stores. MRI quickly began to show remarkable progress, and undoubtedly will become the method of choice in a short time.

NHLBI Funding of Sickle Cell Disease Research

As noted above, the United States Congress three decades ago charged the NHLBI with funding SCD research. The extent of the Institute's commitment to improving the lives of persons with SCD is reflected in the size of its research investment, which totaled $1.11 billion during the years 1972–2004 (Fig. 1). During this period, the life expectancy for sickle cell disease subjects has increased dramatically — from less than 20 years in 1972 to more than 45 years today. Sadly however, only one drug has been approved for SCD by the United States Food and Drug Administration. While this agent — hydroxyurea — does improve the quality and length of life for most patients by reducing the frequency of painful episodes, acute chest syndrome events, and transfusions, it provides no benefit for a significant fraction of patients, and it does not provide a cure. Full bone marrow transplantation from matched donors in children is curative, but comes with a significant mortality risk linked to conditioning regimens. The vast majority of SCD patients do not have matched donors, and this approach has to date been unsuccessful in adults. A universal cure remains elusive.

The NHLBI funds a comprehensive range of SCD research projects, covering the whole spectrum from basic through translational studies to clinical studies and trials. Research projects range from investigator-initiated pilot project grants (the R21 mechanism) for basic research through large, definitive human clinical trials. The accomplishments of NHLBI's historical and current SCD programs are briefly described below.

Clinical Care Guidelines

The NHLBI has published four editions of a SCD treatment manual for physicians. The fourth edition of *The Management of Sickle Cell Disease* — developed by physicians, nurses, psychologists, and

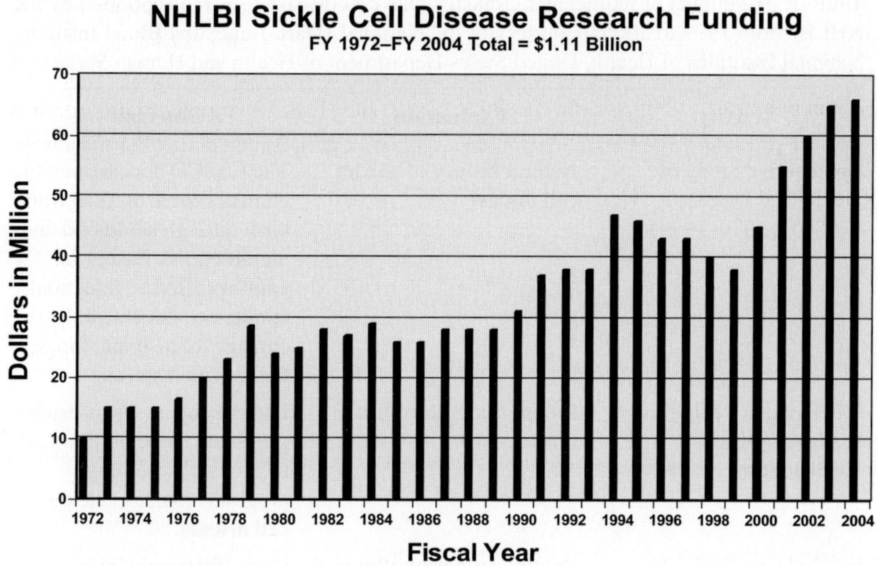

Fig. 1. NHLBI Funding Commitment to Sickle Cell Disease Research. Shown is the level of funding committed to sickle cell disease research between 1972 and 2004. See text for description of research outcomes during this time period. (Courtesy of the National Heart, Lung, and Blood Institute, National Institutes of Health, United States Department of Health and Human Services.)

social workers who specialize in the care of children and adults with SCD — was published in July 2002 and describes the current approach to counseling and management of the major medical complications of SCD. Organized into four parts — Diagnosis and Counseling, Health Maintenance, Treatment of Acute and Chronic Complications, and Special Topics — this book explains the treatment choices available as of its publication date, and serves as an adjunct to recent textbooks that delve more deeply into all aspects of SCD. This is NIH Publication No. 02-2117 (188 pages); it may be viewed free of charge online, or ordered online (http://www.nhlbi.nih.gov/health/prof/blood/sickle/).

Clinical Research — Multicenter Clinical Trials

Because the number of affected SCD patients in the United States (approximately 80,000) is significantly less than the number of individuals affected by common disorders such as cancer, heart disease, and diabetes (tens of millions), and because the larger numbers of patients in other, underdeveloped parts of the world (Africa, Asia) are judged unable to pay for new medicines, the pharmaceutical industry has shown little interest in investing in new drugs for SCD. Thus in the United States, the NIH has funded the vast majority of new drug investment for SCD, and the NHLBI has supported most of the Phases II and III clinical trials for SCD conducted over the last three decades. These are stand-alone trials, in contrast to the clinical research network trials referred to below, and they are listed in Table 1. Of particular, note here are the two Prophylactic Penicillin (PROPS) trials conducted in the 1980s, which showed that pneumococcal sepsis in children (the leading cause of death in children with SCD at the time), could be prevented with penicillin prophylaxis up to age 5. Currently there is one large SCD trial being supported by another institute at NIH — the National Institute of Neurological Disorders and Stroke (NINDS). This is the Silent Infarct Transfusion Study (SITS) trial,

Table 1. Highlights of multicenter clinical sickle cell disease research supported by the NHLBI from 1979 to 2005. (Courtesy of the National Heart, Lung, and Blood Institute, National Institutes of Health, United States Department of Health and Human Services.)

Activity	Evaluation	Outcome
Cooperative Study of Sickle Cell Disease (CSSCD; 1979–1999)	Natural history of sickle cell disease	The CSSCD documented the clinical course of SCD from birth through adulthood and identified risk factors for painful episodes, infection, stroke, early death, and damage to the brain, lungs, kidneys, and spleen.
Prophylactic Penicillin Study (PROPS; 1983–1986)	Prophylactic penicillin in infants	Pneumococcal sepsis can be prevented by penicillin. Led to neonatal screening programs to detect sickle cell disease.
Prophylactic Penicillin Study II (PROPS II; 1987-1994)	Prophylactic penicillin in children	Penicillin prophylaxis can be safely stopped at age 5.
Multicenter Bone Marrow Transplantation Study (1991–1999)	Matched related, ablative bone marrow transplantation in children with symptomatic SCD	Fifty children received allografts with Kaplan-Meier probabilities of survival and event-free survival of 94% and 84% respectively.
Multicenter Study of Hydroxyurea in Sickle Cell Anemia (MSH Trial; 1991–1995)	Hydroxyurea in adults with severe sickle cell anemia	Hydroxyurea lowered the rate of painful episodes, blood transfusions, acute chest syndrome, and hospitalizations by 50%.
MSH follow-up study (1998–present; scheduled to run through 2008)	Hydroxyurea in adults with severe sickle cell anemia	Hydroxyurea decreases mortality. Monitoring of clinical wellness and quality of life is ongoing.
Phase I/II Study of Hydroxyurea in Children (PED HUG; 1994–1999)	Hydroxyurea in children	Hydroxyurea decreases sickle cell complications in children aged 5 to 15 and can be given safely to them.
Pediatric Hydroxyurea Phase III Clinical Trial (BABY HUG; 2000–present; scheduled to end in 2008)	Hydroxyurea in infants	Ongoing. The study is expected to show whether hydroxyurea prevents organ damage when given to infants who have sickle cell anemia.

Table 1. (*Continued*)

Activity	Evaluation	Outcome
Stroke Prevention Trial in Sickle Cell Anemia (STOP Trial; 1994–1997)	Blood transfusions in children	Blood transfusions prevent overt stroke in children who are at high risk of stroke, and trans-cranial Doppler ultrasonography is validated as a screening method for stroke risk.
Stroke Prevention Trial II (STOP II Trial; 2000–2004)	Blood transfusions in children	Transfusions to prevent stroke cannot be safely stopped after a minimum of 30 months of therapy.

*Co-sponsored with the National Cancer Institute.

whose goal is to evaluate transfusion versus observation for prevention of secondary silent stroke (infarction) in children with SCD. Neurocognitive function will be studied as a secondary endpoint, and a linked DNA repository will also be established.

The New NHLBI SCD Clinical Research Network

In addition to supporting large stand-alone SCD clinical trials, the NHLBI has supported a Cooley's Anemia Clinical Research Network (CRN) since 2000, a Phase II SCD CRN since 2003, and will soon support a new Phase III SCD CRN beginning in 2006. The concept behind CRNs is that the network supports infrastructure (multiple clinical centers and a data coordinating center) through which multiple clinical trials can be run, some concurrently and some in succession. This is in contrast to stand-alone trials (like the PROPS trials mentioned above) where the infrastructure is built to complete one trial only. NHLBI launched the first CRN for SCD in April 2003 within the CSCC program; known as the Comprehensive Sickle Cell Centers Clinical Trials Consortium (CSCC CTC). The CSCC program, spearheaded by Gregory Evans, is focused on Phase II trials (see below). In April 2006 the NHLBI will launch a second CRN — to be known as the SCD Clinical Research Network (SCD CRN) — to be focused on Phase III clinical trials. A SCD CRN has been publicly recommended by the NHLBI Sickle Cell Disease Advisory Committee in the past few years, and such a program was also strongly recommended by participants in the recent NHGRI-led workshop held at NIH in 2003 (see below for details). The SCD CRN will, in addition to running Phase III trials, also set up a DNA–genotype–phenotype registry to serve as a public resource for all interested investigators facilitating human genetic studies. Of note, the SCD CRN will also require each funded center to establish and/or maintain strong ties with community organizations — churches, and/or SCD advocacy groups such as local chapters of the Sickle Cell Disease Association of America (SCDAA).

NHLBI-Supported Translational SCD Research — Pulmonary Complications of SCD, and Programs of Excellence in Gene Therapy

In addition to the CSCC program described in detail below, the NHLBI also currently supports by RFA two additional translational programs considered a significant part of the NIH SCD portfolio (Fig. 2).

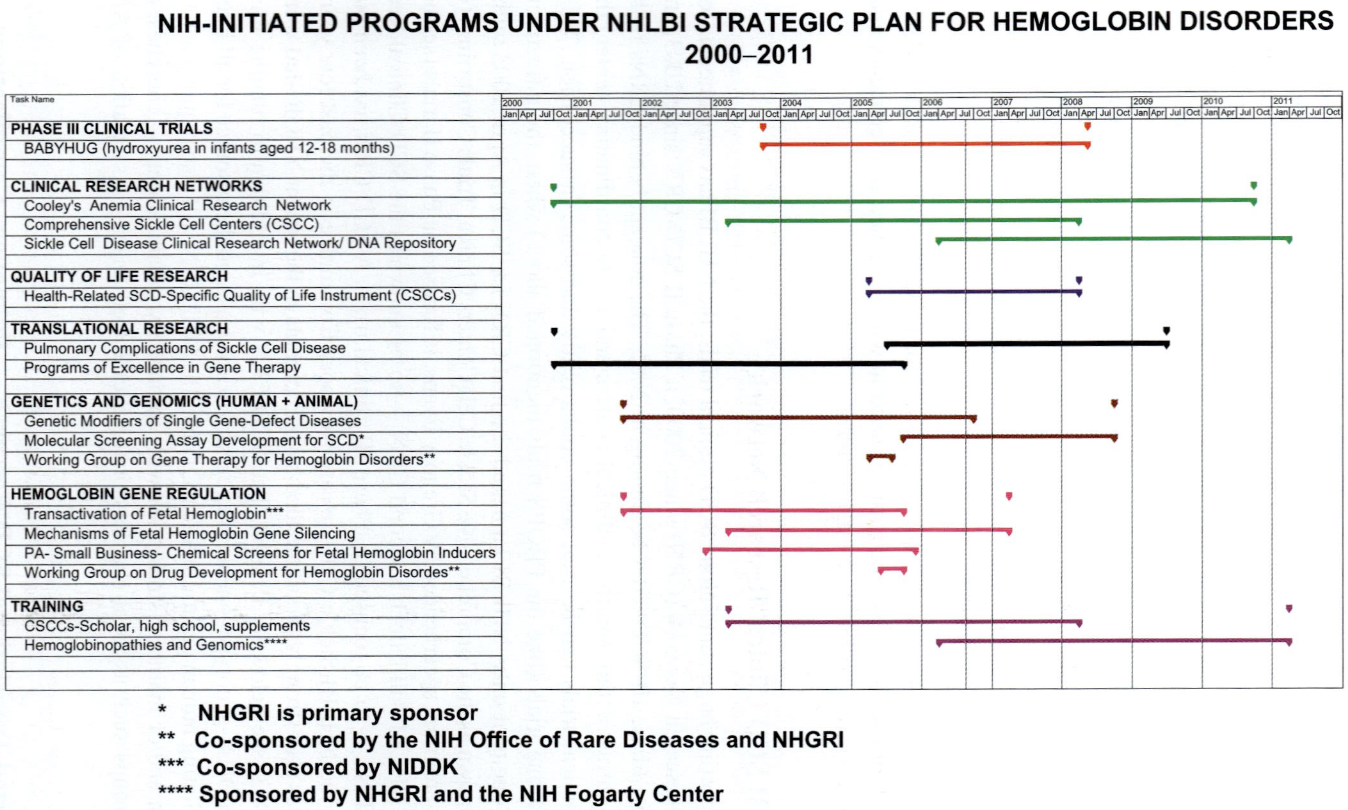

Fig. 2. NHLBI strategic planning for hemoglobin disorders. Summarized are the different programs initiated since 2000 related to innovative research related to SCD. A wide range of topics from clinical to basic research are currently supported. Note new programs on the horizon. (Courtesy of the National Heart, Lung, and Blood Institute, National Institutes of Health, United States Department of Health and Human Services.)

The first is an RFA on Pulmonary Complications of SCD, launched in July 2005. This program supports collaborative research between hematologists, pulmonologists, and basic researchers to study and develop new therapies for the major pulmonary complications of SCD — acute chest syndrome, pulmonary hypertension, and oxyhemoglobin desaturation. The second is focused on gene therapy — the Programs of Excellence in Gene Therapy (PEGT); this program, launched in September 2000, includes four centers in the United States (see PEGT website at *http://pegt.med.cornell.edu/* for geographic locations) and six National Service Cores that provide gene therapy resources (e.g. human CD34+ cells) free of charge to either NIH-funded or NHLBI-funded investigators, depending on the core. We hope that this program, which supports some of the leading investigators in the globin gene transfer field, will play a significant role in supporting the early United States human gene therapy trials for hemoglobin disorders that are yet to come.

Translational Research — NHLBI Comprehensive Sickle Cell Centers

Structure and Main Purpose

The centerpiece of the NHLBI SCD research program is the CSCC program, first established by the NIH in 1972 in response to a Presidential initiative and Congressional mandate. The CSCCs were initially comprehensive research and care centers that also provided hemoglobin diagnosis, genetic counseling, education of patients and care providers, and strong community outreach efforts. After an open competition, ten Centers were funded in 1972 and five additional centers in 1973. Subsequent RFAs were funded in 1977, 1978, 1983, 1988, 1993, 1998, and 2003. Today's CSCC program has the same overall goals as the initial program, while the research component has become stronger. Applications for this program are received every 5 years (23 were received the last time, in 2001), and go through two levels of peer review — an initial merit review by a special panel of experts, and a second by the NHLBI National Advisory Council. The present budget for this program is approximately $22 million per year, equivalent to about one-third of the current annual NHLBI expenditures on SCD research. As presently constituted, NHLBI CSCCs offer interactive, state-of-the-art programs in translational SCD research, carrying out research focused on the development of cures or significantly improved treatments for SCD. This includes basic research efforts, inter-center collaborative clinical research, and local clinical research, all three focused on the most promising therapeutic modalities on the horizon today, and each interactive with the others. While their primary focus is research, CSCCs also support career development of young investigators in sickle cell disease research, summer research experiences for high school students, and patient service activities focused on implementation into clinical practice of the best current models of care and treatment for SCD.

Clinical Research — Change in Focus Toward Multicenter Clinical Trials

For many years prior to 2003, the CSCC program featured 10 Program Project — like grants where there was limited collaborative clinical activity in the form of multicenter trials. Independent advisory panels told the NHLB that: (1) the CSCC program was not meeting its potential in this area; and (2) there was a critical need for such work to move basic research into clinical application. Thus, effective with the 2003–2008 funding cycle, the CSCC program now contains a CRN component known as the CSCC CTC (as mentioned above), where each center receives funding on a per-enrolled-patient basis for each of several active clinical trials. The CSCC CTC has 10 core centers and 23 core clinical sites in total, and it is possible for a limited number of sites outside the funded program (i.e. non-core sites) to participate in these trials. The number of sites participating in any given trial is expected to vary from 5 to 20 (see the public CSCC website at *http://www.rhofed.org/sickle/index.htm* for geographic listing of sites). At present, the CSCC CTC has two active trials: one for oral arginine supplementation in children and adults; and a second for neurocognitive function, neuroimaging, and transfusion as an intervention to evaluate for improvement of neurocognitive function. Additional studies under consideration in the pipeline include a DNA–genotype–phenotype registry, epidemiology of priapism, hydroxyurea and magnesium for hemoglobin SC disease, decitabine for SS disease (sickle cell anemia), dexamethasone for acute chest syndrome, and collaborative development of an SCD-specific health-related quality-of-life instrument.

In addition to multicenter network studies, the CSCC program also supports 1–3 single-center clinical studies at each CSCC. These are generally Phase I interventions or epidemiology studies conducted at 1–2 clinical sites. The complete list of clinical research topics in 15 currently funded projects (available via the NIH CRISP database of all funded NIH grants, http://crisp.cit.nih.gov/) follows:

- Dipyridamole/Magnesium to Improve Sickle Cell Hydration
- *In Vivo* Hydration Changes in HbSS and HbSC Cells
- Combination Treatment with Hydroxyurea and Magnesium (Phase I)
- A Phase II Trial of Hydroxyurea and Pulse Arginine Butyrate
- Biologic Markers in Infants and Young Children with SCD
- Phospholipase A2 Inhibitor Safety Trial for Acute Chest Syndrome
- Nutrition supplementation studies in SCD
- Incidence of Infection in Young Children with SCD in Africa
- Pathogenesis of Pneumococcal Infection in SCD
- Attributes of Sickle Pain in Infants and Young Children
- Keterolac versus Ibuprofen for the Painful Crisis of SCD
- Controlled Pain Interaction for Adolescents with SCD
- Priapism in Boys and Young Men — Incidence and Prevalence
- Non-myeloablative Transplantation Therapy for SCD
- Magnetic Resonance Imaging and Near Infrared Spectroscopy in SCD Patients and Mice

Training Programs

The CSCC program provides a rich training environment for new and young scientists interested in SCD research. Three types of training awards are offered. First, the CSCC Scholar awards support one junior faculty-level investigator at each of the 10 CSCCs. Second, the CSCC SCD Summer for Science program supports three high school students each year at each of the 10 CSCCs. Third, new in 2005 are flexible general training supplements, where each CSCC can support 1–2 trainees at any career level from high school through junior faculty.

Basic Research

The CSCC program supports 1–4 basic science research projects (mini-R01s) at each of the 10 funded CSCCs. The complete list of basic research topics in 20 currently funded projects (available via the NIH CRISP database of all funded NIH grants, http://crisp.cit.nih.gov/) follows:

- Single Molecule Analysis of Sickle Erythrocyte Adhesion

- Activation of cAMP-Mediated Sickle Cell Adhesion
- Sickle Cell Adhesion
- Endothelial Receptors for Phosphatidylserine — Positive Sickle Red Cells
- Regulation of Sickle Cell Phospholipid Organization
- Gamma-Globin Gene Therapy Using *In Vivo* Selection
- Pre-Clinical evaluation of gene therapy for SCD
- Ribozyme — Mediated Repair of Sickle Beta-Globin RNA and DNA
- Genetic Therapy of SCD
- KCl Cotransport Regulation in Red Cells
- KCl Cotransporter Gene Expression
- Deregulation of the Sickle Cell Membrane Skeleton
- Novel Approaches to Increase Gamma — Globin Expression in SCD
- Molecular Mechanisms of Globin Gene Expression
- Oxidant State and Nitric Oxide Metabolism in Acute Chest Syndrome
- Role of Endothelin-1 in Sickle Acute Chest Syndrome
- Identification of Structure-Based Antisickling Peptides that Inhibit HbS Polymerization
- Construction of SCD Mice Carrying Chromosomes from the Four Human SCD Haplotypes
- Mechanistic Basis of Beta-Globin mRNA Stability
- Fetal Tolerance, Chimerism, and SCD

Other NIH RFAs and Program Announcements

Aside from the CSCC program, the NHLBI, NIDDK, and NHGRI have also had in the recent past several other Institute-initiated programs (RFAs and Program Announcements [PAs]) with research opportunities in SCD and related areas. These include two active PAs for R21 pilot projects (basic or clinical in nature); two soon-to-expire RFAs on iron removal and imaging secondary to blood transfusions as well as basic iron biology; two active RFAs on gamma-globin gene regulation (one RFA on transactivation, one on silencing); one RFA on genetic modifiers for single gene disorders; one PA (targeted to small businesses) for chemical screens for new fetal hemoglobin inducers, and two approved but not yet funded RFAs on chemical genomics targeted to SCD, and on training for application of genomics to hemoglobinopathies. Two 2005 Working Group meetings (one for clinical gene therapy, and one for resources for drug development for hemoglobin disorders) are planned. These may lead to future RFAs. A partial listing of current and future programs (with timelines) is shown in Fig. 2, and a complete list is given below, broken out by NIH Institute:

National Heart, Lung, and Blood Institute

- HL-97-013, Clinical Research on Cooley's Anemia (included iron projects very relevant to SCD research), funded 8/98-8/03, co-sponsored by NIDDK.
- HL-01-013, Transactivation of Fetal Hemoglobin Genes for Treatment of Sickle Cell Disease and Cooley's Anemia, funded 9/01-9/05, co-sponsored by NIDDK.
- HL-01-001, Genetic Modifiers of Single Gene-Defect Diseases, funded 9/01-9/06, co-sponsored by NIDDK.
- PA-03-049, Chemical Screens for New Inducers of Fetal Hemoglobin (SBIR/STTR), active 12/02-12/05, co-sponsored by NIDDK.

- HL-02-015, Mechanisms of Fetal Hemoglobin Gene Silencing for Treatment of Sickle Cell Disease and Cooley's Anemia, funded 4/03-4/07.
- PA-03-171, Exploratory and Developmental Research Grants for Investigations in Rare Diseases (R21), active 9/03-9/06.
- HL-04-015, Pulmonary Complications of Sickle Cell Disease, funded 7/05-6/09.

National Institute of Diabetes and Digestive and Kidney Diseases

- PA-97-036, Basic Research on the Metabolism of Iron Chelation, active.
- DK-99-009, Biology of Iron Overload and New Approaches to Therapy, funded 7/00-7/05, co-sponsored by NHLBI.
- PA-03-150, Erythroid Lineage Molecular Toolbox.
- PA-04-144, Pilot and Feasibility Program in Hematologic Diseases (R21), active 8/01-11/07.
- DK-03-007, Noninvasive Measurement of Iron by Magnetic Resonance Imaging, funded 9/03-9/07, co-sponsored by the National Institute of Biomedical imaging and Bioengineering (NIBIB).

National Human Genome Research Institute

- HG-05-001, Molecular Screening Assay Development for Sickle Cell Disease, to be funded 9/05-9/08, co-sponsored by NHLBI.
- HG-05-002, Training in Genomics and Hemoglobinopathies, to be funded 4/06-3/11.
- NOT-HG-05-002, ELSI Small Grant Research Program (R03). This NIH Notice is focused on SCD, and linked to an active NHGRI PA.

Investigator-Initiated Grants

In addition to Institute-initiated programs, the NHLBI and NIDDK support a large number of investigator-initiated training grants (Fs and Ks), research project grants (R01s, multiproject P01s), and small business grants for research related to SCD. In fact, the majority of NIH expenditures for research on hemoglobin disorders are in these categories. The NHLBI currently supports approximately 15 F and K awards, six Program Project grants (P01), 25 R01 grants, and four small business grants (R41–44) focused on SCD. Important topics in these projects include gene therapy, vascular adhesion and pathology, fetal hemoglobin induction, hemoglobin S structure–function, and noninvasive imaging of tissue iron. The NHLBI also currently supports eight F and K awards, six P01s, and 25 R01s in areas related to but not focused on SCD. Important topics in these projects are hematopoietic stem cell gene transfer and gene correction, hematopoietic stem cell biology, the regulation of beta-like globin expression, erythropoiesis, red cell ion channels and membrane proteins, iron transport/storage/metabolism, and nitric oxide biology.

The NIDDK currently supports five K awards related to SCD, including projects on hemoglobin synthesis, gene therapy, and red cell adhesion. Research projects related to transfusional iron overload include three R01s for iron chelator development and four R01s to develop improved technology to assess body iron, as well as 15 projects devoted to understanding the process of iron transport and storage. Twenty-one R01s support basic research on hemoglobin gene regulation and modifying the switch from fetal to adult hemoglobin. Several SBIR grants are devoted to evaluating various fetal hemoglobin modifiers. One R01-supported project has supported the continuing development of protocols for the use of dietary magnesium to alleviate the severity of SCD symptoms. Other aspects of NIDDK research programs include projects not directly related to SCD, but could produce helpful therapies. These include gene therapy research, hematopoietic and human embryonic stem cell studies, red blood cell membrane adhesiveness and stability, and renal complications of SCD.

Joint Efforts between NHGRI, NHLBI, NIDDK, and other NIH Institutes and Centers — 2003 Conference

The conference "New Directions for Sickle Cell Therapy in the Genome Era" was held at NIH in November 2003, organized and supported by NHGRI, NHLBI, NIDDK and the NIH Office of Rare Diseases, the Fogarty International Center, and the Foundation for the National Institutes of Health. Over 120 experts in all aspects of SCD research, primarily from the United States but also from abroad, attended this invitation-only meeting. Conference goals were: first, to reassess what is known about the pathophysiology of SCD and second, to consider how genomics might be applied to develop more effective therapeutic and preventive strategies for the disease.

Over the course of three days, plenary sessions and breakout groups were formed in five areas: (1) genetic modifiers of SCD severity; (2) hemoglobin gene regulation; (3) gene therapy; (4) the social and cultural context of SCD research; and (5) high throughput drug screening for SCD. At the conference's end, the sobering conclusion was that 50 years after we first knew of the molecular basis of SCD, only one drug (hydroxyurea) has been approved by the United States Food and Drug Administration, and it neither cures the disease nor benefits all patients. Much remains to be learned about pathophysiology, and much remains to be done in the therapeutics arena. With this in mind, the following eight recommendations were strongly endorsed by attendees:

(1) Create a clinical research network with a central prospective registry and DNA sample repository with information on several thousand well-phenotyped patients.
(2) Take steps to train the next generation of SCD researchers to integrate genomics, proteomics, and high-throughput drug screening expertise into SCD research.
(3) Use the patient registry and DNA sample repository developed in (1) above to define the genetic basis of clinical SCD severity, using strict, standardized criteria for clinical complications.
(4) Take all necessary steps to move toward a clinical trial of gene therapy for SCD.
(5) Improve resources for drug development by academic investigators, and increase the involvement of contract research organizations and small businesses in this area.
(6) Put in place a pilot grant mechanism (with a special ad hoc peer review group) for high risk, high return research, to make it easier to get novel approaches funded.
(7) Apply new chemical genomic methods of high-throughput screening to identify interactions between families of drugs and proteins that might be exploited for therapeutic use in SCD.
(8) Stimulate research into the ethical, legal and social issues surrounding clinical research in SCD.

After the workshop, these recommendations were included in the NHLBI Strategic Plan for Hemoglobin Disorders, and independently a Trans-NIH staff group — The Trans-NIH Sickle Cell Disease Therapies Working Group — was formed to evaluate their feasibility. We are happy to report (as described above) that NHLBI will implement recommendation 1 (a SCD Clinical Research Network for Phase III trials) in April 2006, and will address recommendations 4 and 5 by holding small working group meetings in 2005. Meanwhile, NHGRI will implement recommendations 2, 7, and 8 in 2005–2006, and has already set up an e-mail listserv for SCD research, in response to the conference attendees' call for a convenient means for sharing news about SCD research.

To join the Sickle Cell Disease Research Listserv (SCDR-L), send e-mail to *listserv@list.nih.gov.* Type as body of the message: subscribe SCDR-L your name. Example: subscribe SCDR-L Jane Doe. You should receive a confirmation e-mail within a few minutes. As of this writing, the Sickle Cell Disease Research Listserv has 110 subscribers. The balance of the recommendations is under study for implementation in the future.

The Future

NHLBI's Strategic Plan for Hemoglobin Disorders (Fig. 2) calls for continued support of definitive clinical trials and clinical research networks (the new Phase III network and the existing Phase II network in the CSCCs). In the face of what is expected to be a nearly flat NIH budget over the next several years, and in the interests of circumventing some of the limitations in working with rare diseases, it may be desirable for the clinical research efforts for SCD and Beta-Thalassemia (Cooley's Anemia) to be combined where feasible, for example in the fetal hemoglobin induction, transfusional iron overload, and gene transfer areas. The NHLBI-funded Cooley's anemia and SCD CRNs may ultimately merge into one CRN for hemoglobinopathies. On the NIH genomics front, it is likely that current efforts to develop genotype–phenotype registries for SCD will reach fruition in the next five years, thus setting the stage for a large-scale, focused effort by the SCD research community and NIH to identify genetic modulators of SCD severity.

3

The Human Genome Project

by Betty S. Pace

Introduction

The Human Genome Project (HGP) represents one of the most successful collaborative research efforts of the last millennium. Completion of the HGP in 2003 made genome research a central underpinning of biomedical science, and the fruits of the HGP have paved the way for a revolution in future health care, built on the foundations of improved understanding of the molecular pathogenesis of human disease. New and emerging technologies in bioinformatics and computational biology will allow high-throughput, sophisticated gene-based diagnostics that will change biology and medical practice forever.

The flagship endeavor of the HGP was to determine the three billion base pairs in the human genome, the physical structure of DNA, and the location of the estimated 30,000 genes. Parallel to this effort, the DNA of model organisms was sequenced to provide the comparative date necessary for understanding the function of the human genome. In this chapter, we will review the inception of the HGP, and the challenges and major accomplishments during the 13-year period. We will also review the expanded focus of the HGP to define haplotype blocks in the human genome, and the potential for population association studies to identify disease causing genes. Finally, the impact of functional genomics and genetic profiling on the practice of personalized medicine will be discussed. Data generated by the HGP is expected to be the sourcebook for biomedical science in the 21st century.

The Pre-Genomic Era

DNA: The Molecule of Heredity

The theories of heredity discovered by Gregor Mendel,[1] based on his work with plants, are well known to any student of biology. The short monograph published in 1866 in which Mendel described the principles of inherited traits, has become one of the most influential publications in the history of science. Mendel derived the basic laws of heredity and his work became the foundation for modern genetics. In 1920, Thomas Hunt Morgan[2] proposed a theory of sex-linked inheritance followed by the purification of DNA by Oswald Avery and associates in 1944.[3] The seminal 1949 paper by Linus Pauling[4] describing sickle cell anemia as the first molecular disease had a major impact on biology and medicine. By 1953, James Watson and Francis Crick defined the three-dimensional structure of DNA.[5] Twenty-four years later, the A→T DNA mutation in the β-globin gene that produced sickle hemoglobin was identified by Marotta *et al.*[6] Thus the first disease-causing single nucleotide polymorphism (SNP) was identified.

Molecular biology as we know it was born when Nirenberg published on the genetic code[7] followed by the construction of the first recombinant DNA molecules.[8] The development of new instrumentation for sequencing DNA[9,10] brought the earliest hope that methods could be developed to sequence the human genome. This became a reality in 1986, when the first automated sequencer was invented by Leroy Hood.[11] As technology developed, mapping the full compliment of human genes became a formidable task that was achieved in 2001, when the draft sequence of the human genome was published[12] followed in 2003, on the 50th anniversary of the discovery of the structure of DNA,[5] by the finished human genome sequence.

The National Institutes of Health (NIH)

The human genome program of the NIH was established after Congress appropriated earmarked funds in 1987 to conduct research on mapping and sequencing approaches. In October 1988, the Office of Human Genome Research was established to plan and coordinate NIH genome activities in cooperation with other federal agencies, industry, academia, and international groups. By October 1989, this office became an independent funding unit within the NIH, with authority to award grants and contracts; it was then renamed the National Center for Human Genome Research (NCHGR) and James D. Watson served as its first director.

The Department of Energy (DOE)

The genome program of the DOE started in 1987 on a small scale, receiving earmarked funds for the first time in 1988. Aristides Patrinos, director of the DOE's Office of Biological and Environmental Research provided leadership for a multidisciplinary program at three national laboratories: Lawrence Berkeley; Los Alamos; and Lawrence Livermore. The DOE's long-standing program of genetic research was directed at improving the assessment of the effects of radiation and energy-related chemicals on human health.

Birth of the Human Genome Project

The Human Genome Initiative began as a worldwide research effort; its goals were to understand DNA structure and to locate the 30,000 genes believed to exist. The NIH and DOE entered into a Memorandum of Understanding,[13] which established 13 years of productive collaboration with the international research community as part of the International Human Genome Project Consortium.

The First Five Years: 1990–1994

NIH and DOE advisors met with experts to develop the five-year plan for the genome project which was published in April of 1990.[14] The National Advisory Council for Human Genome Research was established to oversee activities of the HGP, with a budget of $200 million per year for 15 years. The Human Genome Project officially began on the backs of leaders in the scientific community. The three large centers from the DOE teamed with the 10–20 centers that were established by the NIH during the first five years. Some participants focused on the development of physical maps for human chromosomes, while others concentrated on sequencing the complete genome of model organisms, or new technology development.

The International Human Genome Sequencing Consortium

Twenty centers located in six countries — China, France, Germany, Great Britain, Japan, and the United States — produced the finished human genome sequence (Table 1). Five centers produced 60% of the sequence data, which included the Genome Institute in Walnut Creek, the Sanger Institute, Baylor College of Medicine in Houston, Washington University School of Medicine in St. Louis, and the Whitehead Institute in Cambridge. The group adopted the Bermuda Principle, which calls for automatic, rapid release of new sequence data within 24 hours of sequence assemblies of 1 to 2 kb or greater to the freely accessible public domain.

Table 1. International Human Genome Sequencing Consortium.

1. Whitehead Institute/MIT Center for Genome Research, Cambridge, MA, U.S.
2. The Wellcome Trust Sanger Institute, The Wellcome Trust Genome Campus, Hinxton, Cambridgeshire, U.K.
3. Washington University School of Medicine Genome Sequencing Center, St. Louis, MO, U.S.
4. U. S. Department of Energy Joint Genome Institute, Walnut Creek, CA, U.S.
5. Baylor College of Medicine Human Genome Sequencing Center, Department of Molecular and Human Genetics, Houston, TX, U.S.
6. RIKEN Genomic Sciences Center, Yokohama, Japan
7. Genoscope and CNRS UMR-8030, Evry, France
8. GTC Sequencing Center, Genome Therapeutics Corporation, Waltham, MA, U.S.
9. Department of Genome Analysis, Institute of Molecular Biotechnology, Jena, Germany
10. Beijing Genomics Institute/Human Genome Center, Institute of Genetics, Chinese Academy of Sciences, Beijing, China
11. Multimegabase Sequencing Center, The Institute for Systems Biology, Seattle, WA, U.S.
12. Stanford Genome Technology Center, Stanford, CA, U.S.
13. Stanford Human Genome Center and Department of Genetics, Stanford University School of Medicine, Stanford, CA, U.S.
14. University of Washington Genome Center, Seattle, WA, U.S.
15. Department of Molecular Biology, Keio University School of Medicine, Tokyo, Japan
16. University of Texas Southwestern Medical Center, Dallas, TX, U.S.
17. University of Oklahoma's Advanced Center for Genome Technology, Dept. of Chemistry and Biochemistry, University of Oklahoma, Norman, OK, U.S.
18. Max Planck Institute for Molecular Genetics, Berlin, Germany
19. Cold Spring Harbor Laboratory, Lita Annenberg Hazen Genome Center, Cold Spring Harbor, NY, U.S.
20. German Research Center for Biotechnology, Braunschweig, Germany

Source: National Human Genome Research Institute (*http://www.genome.gov/11006939*).

Goals of First Five Years

The major goal of the HGP was to map and sequence the entire three billion base pairs, and the projected 30,000 to 100,000 genes of the human genome contained in 23 pairs of chromosomes. Each chromosome contains DNA, the chemical of which genes are made; it is a double-stranded helical molecule.[5] Each strand is composed of a linear array of nucleotides or bases; the four bases are adenine (A), thymine (T), guanine (G), and cytosine (C). The bases on one DNA strand are precisely paired with complimentary bases on the other strand by hydrogen bonds which maintain the double helix: A pairs with T and G with C. The order of the four bases on the DNA strand determines the amino acid sequence of proteins.

The second major goal of the HGP was to map and sequence the genomes of model organisms for comparative studies. Briefly, mapping is the process of determining the position and spacing of genes or other genetic markers on chromosomes, relative to one another. There are two types of maps, genetic and physical; the methods used to construct maps and the metric used to measure the distance between genes differ. Sequencing is the process of determining the order of the nucleotides, or base pairs, in a DNA molecule.

The remaining goals of the HGP included data collection and distribution; a program to address the ethical, legal, and social implications (ELSI) of the HGP; multidisciplinary research training programs; and technology development. For a detailed discussion of these goals, see the Executive Summary of the United States Human Genome Project.[14]

Budget Allocations

In addition to the projected $200 million per year for research endeavors, money was requested for new construction at participating institutions. Over a five-year period, $121 million was allocated for the latter purpose. From its inception, 3–5% of the HGP budget was dedicated to the development of ELSI programs in order to keep pace with public response to the sequencing endeavor. The breakdown of funding allocations was: 45–55% for human studies; 20–30% for animal model systems; and 20–30% for infrastructure, including maintenance of public databases; research training, ethics, conferences, and administration.

A New Five-Year Plan: 1993–1998

After James Watson resigned as director of the NCHGR, Francis Collins was appointed in 1993. As more sophisticated technology became available, and as a better understanding of what was required to accomplish the HGP evolved, the NIH and DOE held a series of meetings. The outcome was a new set of extended goals, for the 1993–1998 period[15] to cover the first eight years of the 15-year human genome initiative. Technological advances including (i) genetic markers such as microsatellites; (ii) vector systems for cloning large DNA fragments; (iii) the identification of genes within mapped sequences and the definition of sequence tagged site (STS); and (iv) improvement in automated DNA sequencing capabilities, changed the focus of the HGP.

Goals of the Extended Period

The major modifications to the initial goals are briefly summarized below.[15] To produce two to five cM genetic maps and develop technology for rapid genotyping; complete an STS map of the human genome at a resolution of 100 kb; develop approaches to sequence several megabase regions of DNA; high-throughput technology to reach an aggregate of 80 Mb of DNA sequence by 1998; expanded support for improvement of DNA sequencing methods; finish the STS map of mouse, *E. coli*, and *S. cerevisiae* genomes; and develop databases for easy access to data, effective software for large-scale genome analysis, and tools for interpreting genome information. For the first time, a DNA bank was established under leadership of the HGP as a resource for research in human genetic variation.[16] The activities of ELSI were updated to include defining new issues, policy development, information dissemination, fostering the acceptance of genetic variations, and enhanced professional and public education.

With the major contributions of the HGP, the Department of Health and Human Services Secretary Donna E. Shalala signed documents in 1997 giving the NCHGR a new name, the National Human Genome Research Institute (NHGRI), with status among other institutes at the NIH. The new name more accurately reflected the growth and accomplishments of the HGP under the leadership of Francis Collins.

The Final Phase of the Human Genome Project: 1998–2003

The goals set for the initial eight-year period were successfully completed, and a new plan was formulated and published in 1998 for the 1998–2003 period.[17] The major emphasis was DNA sequencing, adopting a philosophy aptly expressed by the popular motto "Let's just do it!" An ambitious schedule was set to complete the full sequence in five years, two years ahead of schedule. Three new goals were added: (1) to produce a working draft of the human genome by 2001, (2) to develop the discipline of functional genomics, and (3) to study human genome variation. DNA sequence variation is the fundamental raw material for evolution[18] and the basis for the different risks observed among individuals for genetically complex human diseases. The ELSI program was charged with the analysis of the ethical, legal, and social issues raised by the expanded focus related to human DNA variations.

Goals for the Last Phase

To celebrate the 50th anniversary of the discovery of the double helix structure,[8] a decision was made to set the date for completion of the HGP at the end of 2003. A summary of the accomplishments of the HGP is shown in Table 2. By 1998, only 6% of the human genome had been sequenced; therefore, researchers in the International Human Genome Project Consortium had their work cut out for them: to produce a draft sequence by 2001 and a finished product, the entire three billion bases of the human genome, by the end of 2003.

The sequencing strategy used by the public HGP effort was based on mapped clones obtained through shotgun determination of most of the sequence by automated methods, then assembled into a product ("working draft" sequence). In the second phase, the gaps were filled in and discrepancies resolved using assembly software.[19–21] The sequencing effort was expected to fall short of 100%. A working draft covering 90% of the genome was released in 2001,[12] and was made freely accessible to the research community.

The biotechnology firm Celera Genomics made public in 1998 its intention to sequence the human genome by 2003 as well using a whole genome shotgun approach.[22] This "healthy" competition was met with greater determination and commitment of the HGP consortium to complete the sequence by 2003. Despite the inability of public and private projects to join forces, there were joint announcements of the working draft in 2001, and simultaneous publications in *Nature*[12] and *Science*.[23]

Table 2. Human Genome Project goals and completion dates 1998–2003.

Area	HGP goal	Standard achieved	Date achieved
Genetic map	Average 2 to 5 cM resolution map	1 cM resolution map (3000 markers)	September 1994
Physical map	30,000 STSs Complete 80 Mb for all organisms by 1998	52,000 STSs mapped 180 Mb human plus 111 Mb non-human	October 1998
Sequencing technology	Evolutionary improvements and innovative technologies	90 Mb/year capacity at ~0.50 per base	November 2002
Human sequence variation	100,000 mapped human SNPs	3.7 million mapped human SNPs	February 2003
Gene identification	Full-length human cDNAs	15,000 full-length human cDNAs	March 2003
Human DNA sequence completed	Working draft by 2001 Complete sequence by end of 2003	99% of gene-containing part of human sequence finished to 99.99% accuracy	April 2003
Model organisms	Complete genome sequences of *E. coli*, *S. cerevisiae*, *C. elegans*, *D. melanogaster*, mouse	Finished genome sequences of target organisms plus whole-genome drafts of several others	April 2003
Technology development for functional analysis	Develop genomic-scale technologies	High-throughput oligonucleotide synthesis	1994
		DNA microarrays	1996
		Eukaryotic, whole-genome knockouts (yeast)	1999
		Capillary array electrophresis validated	2002
		Microfabrication feasible	2002
		Scale-up of two-hybrid system for protein–protein interaction	2002

Abbreviations: STS, sequence tag site; SNP, single nucleotide polymorphism; Used by permission, Collins *et al.*[26]

The Human Genome Draft Sequence

As planned, a *working draft* of the human genome sequence including a description of its structural organization was published. Humankind immediately reaped the benefits of the HGP. The data generated from the draft was used with microarray technology to develop a gene test for identifying inherited breast cancer,[24] and tumor suppressor genes involved in prostate cancer were identified on chromosome 7.[25] During the same year, the NHGRI established the Centers of Excellence in Genomic Science Award, to support multidisciplinary approaches to genomic research. Several other genetic programs were launched including the Genetic and Rare Disease Information Center, which provides free and accurate information to the general public. The NHGRI website was launched in 2002, and Alan E. Guttmacher was named the Second Deputy Director of the NHGRI and Vence L. Bonham, Jr. was appointed Senior Consultant to the Director on Health Disparities.

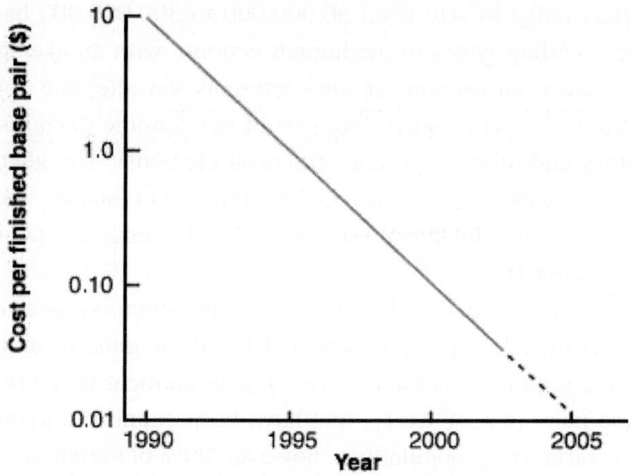

Fig. 1. Decrease in sequencing costs, 1990–2005. Used with permission from Collins *et al.*[26]

Technological Advances

The ability to automate the preparatory steps in the sequencing process through the development of high speed robotic workstations pushed the HGP rapidly forward. Figure 1 illustrates the significant drop is the cost per finished based pair from 1990 to 2003 as the HGP progressed.[26] This trend has continued over the last two years. Major improvements in library production, template preparation, and laboratory information management were essential to the project's success. The advent of capillary electrophoresis sequencing technology[27,28] provided the tools required to achieve a two- to three-fold increase in acquisition of sequencing data from each center.

The finished human genome sequence was produced from map-based and bacterial artificial chromosome (BAC)-based sequencing. BAC libraries were constructed which covered the entire genome. After pre-mapped BAC clones were entered into the data pipeline, the stage was set to produce a finished sequence.[28] By the dawn of the new millennium, the 20 centers participating in the HGP collectively sequenced 1000 base pairs a second, 24 hours a day, a Herculean task. The speed of obtaining accurate DNA sequence was improved using new software programs such as PHRED to assign "base-quality scores"[19] and PHRAP or Consed[20,21] for contig assembly. A dream became reality on 14 April 2003, when the finished human genome sequence was published, under budget ($2.7 billion) and two years ahead of schedule.[28] This unprecedented achievement spawned the genome era.

The Genome Era

The Human Genome

The published sequence consists of a genetic blueprint covering 99% of the approximate three billion base pairs in the human genome with an accuracy of one error every 10,000 base pairs. It is a composite of DNA isolated from many volunteers representing diverse populations, and other sources including established human cell lines. The remaining 1% not sequenced contains ~400 gaps composed of heteochromatic (200 Mb) and euchromatic DNA (25 Mb).[29] These regions are relevant to efforts to identify SNPs in the human genome (discussed below).

Human chromosomes range in size from 50,000,000 to 300,000,000 base pairs. There are an estimated 30,000 protein-coding genes in the human genome with an average coding sequence of 2000 base pairs that produce an average of three proteins via alternate splicing and other post-translational modifications.[30,31] The non-coding part of the genome (98%) is not "junk" DNA, but rather contains regulatory and other important structural elements. The genomic landscape shows marked variations in the distribution of genes, GC content, CpG islands, and recombination rates. Three percent of the cytosines in human DNA are methylated, mostly at CpG islands, a modification consistent with gene silencing.[12]

Fifty percent of the non-coding part of the human genome contains repeated DNA sequences. The most common example is the Alu repeat, present in 10% of the genome and preferentially located in transcriptionally active regions.[32] Another major repeat element is micro satellites or arrays of repeat units of 1–4 base pairs (e.g. $(CAG)_n$) that have been connected to human disease. Tandem repeats increase genetic variation in populations, however, 90% of variation in the human genome is produced by the 1.4 million SNPs identified in the finished sequence, such as those responsible for sickle cell anemia, cystic fibrosis, and approximately 4000 other inherited diseases.

Model Organisms

Experience has shown many times over that information derived from studies of the biology of model organisms is essential to interpretation of data obtained in humans. The HGP supported the completion of the mapping and sequencing of *E. coli*, *S. cerevisiae*, *D. melanogaster*, and *C. elegans* genomes by April 2003.

The National Center for Biotechnology Information (NCBI)

Direct products of the HGP were genome maps and DNA sequences for several organisms. For maximum utility, appropriate computational tools and information systems were developed for the collection, storage, and distribution of the immense amounts of data generated by the HGP. To address these needs, the NCBI, a division of the National Library of Medicine, made freely accessible sequencing data stored in GenBank,[33] a comprehensive public database retrievable through Entrez. GenBank contains DNA, RNA, and protein sequence data for more than 140,000 organisms, obtained primarily through submissions from individual laboratories. As the HGP progressed, there was an exponential rise in the sequence data deposited in GenBank[26] (Fig. 2). Other useful data stored on the NCBI website are expressed sequence tags (ESTs) and STS databases, among others. These databases are comprehensive, up-to-date, and effectively link with one another.

Since the completion of the HGP, there has been continued development of new methods and tools for the analysis and interpretation of genome maps and DNA sequences. A Joint Informatics Task Force was established by the NIH and DOE to help agencies develop detailed informatics programs,[34] to identify uses of the data and establish priorities for both technical and policy areas such as database structures, management, and services; the development of algorithms, software, and hardware for organization; electronic networks for collection and distribution of genome information; and training and education of informatics personnel.

A Vision for the Future

The genome era is now a reality; George Mendel's discovery of the laws of heredity in 1866, the discovery of DNA structure in 1953, and the finished human genome sequence 50 years later are

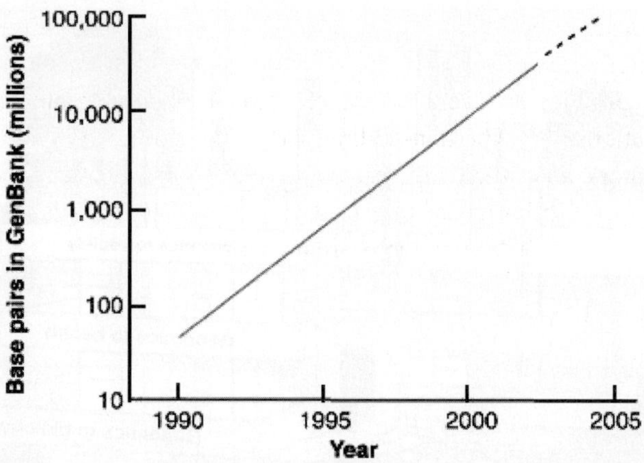

Fig. 2. Increase in DNA sequence stored in GenBank, 1990–2005. Used with permission from Collins *et al.*[26]

three of the greatest accomplishments ever made in science. Genomics has been established as a discipline in biomedical research; an expected benefit of functional genomics is the identification of genes responsible for approximately 4000 diseases with Mendelian patterns of inheritance. The rapid development of genome-wide technology, including microarray chip-based DNA, protein, and SNP analysis, will catapult the field and the benefits predicted for humankind.

Leaders at the NHGRI developed a "vision for the future of genomic research" to provide direction to researchers that will rapidly translate sequence data into improved human health.[35] Three major themes — genomics to biology, genomics to health, and genomics to society — and six cross-cutting elements have been defined to communicate the potential impact of the HGP.

The themes have been captured in a pictorial vision as a three-storey building, with the HGP as the foundation (Fig. 3).[35] Each theme carries with it challenges for the NHGRI and other partners at the NIH, research community and other organizations. To accomplish "genomics to biology," the major challenge is to elucidate the structure and function of genomes by identifying the components encoded in the DNA sequence; to define genetic networks and heritable variations in the genome across species, and to develop public policy for widespread use of genomic data.

The greatest challenge for "genomics to health," is to translate genome-based knowledge into health benefits. The identification of genetic contributions to disease risk, individual drug response, and the use of this information to design new therapeutic approaches is a high priority. Theme three, "genomics to society," pertains to the commitment of the NHGRI to maximizing benefit to humankind and minimizing harm as a result of genomic research. Policy development, one of the main purposes for ELSI, along with understanding the relationship between genomics, race, and ethnicity will be addressed. More important is our understanding of the consequences of uncovering the contributions of genomic variations to human traits, and assessing how to define ethical boundaries for uses of personal genetic information. (See reference 35 for more details.)

The six cross-cutting elements or "Boxes" for the three major themes are summarized briefly as follows. ***Box 1*:** resources development in the form of genome sequences of additional organisms; knock-out and knock-down animal models; tissue banks, large libraries of small molecules; improved protein research; and studies designed to identify genetic contributors to health and disease. ***Box 2*:** technology development to reduce further the cost of sequencing; identification of functional

Fig. 3. The future of genomics rests on the foundation of the Human Genome Project. Used with permission from Collins *et al.*[26]

elements in the genome; improved imaging methods for molecular phenotyping; and linking molecular profiles to biology. ***Box 3*:** computational methods to improve data mining; and improved data management systems, to name a few of the areas identified. ***Box 4*:** training scientists with expertise and computational skills to perform interdisciplinary research. A "different perspective" to increase the number of under-represented minority and disadvantaged individuals that participate in genomic research has been taken by the NHGRI. Several initiatives have already been released to address this issue. ***Box 5*:** continue to address the ethical, legal, and social implications of the HGP through the ELSI program. The goal is to develop models of genomic research and policies to protect human subjects; there will be an ongoing need to evaluate new genetic tests and their ethical implementation. ***Box 6*:** education for the public and health care professionals; information availability on the web; and integration of data from the HGP and ongoing progress into high-school curricula.

The vision has been written and made plain in this seminal paper, to give direction to the research community and general public while generating enthusiasm for the former, and alleviating some of the fears connected to genetic research in the latter group. In the history books, this period will be recognized as a pivotal season in genetic research, catapulted by the accomplishments of the HGP. We will benefit from the fruits of this labor forever.

The Impact of Genome Era Research in Health and Disease

DNA Variations Related to Human Disease

From a genetics perspective, all diseases have a genetic component. There are three general categories of genetic disorders: chromosomal, single-gene (Mendelian), and complex (multi-factorial). Our discussion will center on the latter two causes. Single-gene diseases exhibit patterns consistent with autosomal recessive, autosomal dominant or X-linked inheritance. Approximately 4000 genes causing Mendelian diseases have been identified[36] and are cataloged in the Mendelian Inheritance in Man (OMIM) database (*www.ncbi.nih.gov/omim*).

Genetic variation exists in both health and disease. Any two unrelated individuals share 99.9% of their genome sequence,[12,23,37] but the 0.1% difference translates to over ten million genetic variants. The sequence variations at a locus (gene) are called alleles and there are two alleles at every locus, one for each chromosome. Sites in the DNA sequence where individuals differ at a single DNA base are called single nucleotide polymorphisms (SNPs) which account for 90% of variation in the human genome.[38] The remaining 10% of variant sequences are composed of microsatellites (5%) and insertions and deletions (5%).

Parallel to the efforts of the HGP, a group of scientists from international research centers, pharmaceutical companies, and private foundations teamed up to create The SNP Consortium in 1999.[39] The initial goal of the group was to discover 300,000 SNPs by 2001. This number has been exceeded; in collaboration with the International Human Genome Sequencing Consortium, more than four million SNPs have been cataloged in the public SNP database (dbSNP) maintained on the NCBI website.

The Human Genome Diversity Project,[40] an outgrowth of the HGP, represents a unique scientific opportunity to study natural genetic variations. This effort will help us to identify gene-disease associations, and eventually provide strategies for treating many of the approximately 4000 single-gene diseases that afflict mankind, and polygenic disease with genetic predisposition.

Human Diseases and Haplotypes

The more frequent allele of a SNP is called the major allele, rendering the other one the minor allele. To be useful as a genetic marker to link gene variation with disease, the minor allele must occur at a frequency greater than 10% in the target population and be in linkage disequilibrium with the disease causing gene. This implies that the allelic variation is inherited with the disease gene at a rate higher than expected by normal recombination frequencies. SNPs that occur in gene coding sequence or regulatory elements may produce abnormal protein structure or alter gene transcription respectively. A relevant example is sickle cell anemia caused by an $A \rightarrow T$ mutation in the β-globin gene producing the abnormal hemoglobin *S* protein. SNPs in the γ-globin promoters can produce non-deletional hereditary persistence of fetal hemoglobin. One such example is the -158 $^{G}\gamma$-globin C→T SNP (which creates an *XmnI* restriction site) to produce elevated fetal hemoglobin in African-Americans carrying this mutation.[41] (Reviewed in Chapter 10.)

Sets of nearby SNPs on the same chromosome in "linkage disequilibrium" which are inherited as a block form a haplotype. Blocks may contain a large number of SNPs, but a few termed tag-SNPs can identify the different haplotypes in a block. These chromosomal regions represent stretches up to 35,000 base pairs in length in the human genome.[42] The variation pattern observed might greatly affect an individual's disease risk.

The International HapMap Project

After the release of the draft genome sequence in 2001, a meeting was convened in Washington, D.C., to discuss how haplotype maps could be used to find genes contributing to disease, to establish haplotype structure in populations, and to determine the populations and type of samples that might be considered for a map. Out of this meeting, the Haplotype Map (HapMap) project was initiated.[43] Concerns over the eventual benefits of the HapMap project reaching all ethnic groups have been expressed. Concerns over the potential misuse of DNA sequencing data from individuals in developing countries, without good Institutional Review Board systems, provided the rationale for target samples from the United States, Canada, Europe, and Japan. As a consequence, no phenotypic data, such as medical information, was collected with the 50 samples used to produce the haplotype map of the human genome.

The data obtained by the HapMap project will reduce the number of SNPs required to examine the entire genome for disease-association studies from 10 million to roughly 500,000 tag-SNPs. This will result in genome-wide approaches to make gene-disease associations more comprehensive, eventually allowing individualized SNP typing for genetic risks.

Pharmacogenomics

The efforts put forth by scientists in the SNP and HapMap Consortiums serve as the bedrock of pharmacogenetics, defined as the convergence of pharmacology and genomics to map genetically determined responses to drugs. This new discipline is among the first clinical applications of the HGP. Pharmacogenomics, the broader version of pharmacogenetics, involves the use of genomic-based approaches and inter-individual sequence variation in drug design. Methods for genome-wide genotyping have recently become available in the form of microarray and mass spectrometry technology, which will move the field forward.[44,45]

The genomics "revolution" has already begun; the rapid identification of the new pathogen that causes severe acute respiratory syndrome,[46] the use of gene expression profiling to assess prognosis and guide therapy for breast cancer,[47] genotyping to shed light on response to chemotherapeutic agents,[48] and the use of genomic approaches in the design and implementation of new drug therapies, are a few examples of progress made[49] since the draft sequence was published in 2001. Within a few decades, it will be possible to sequence the entire genome of each person at a cost of approximately $1000, making it feasible to use genetic profiles to establish personalized medicine.

One priority area of pharmacogenetic research is to eliminate serious adverse drug reactions. Differences in drug response among people are common; the current "one size fits all" approach to managing patients will become a thing of the past. Most major drugs on the market in the United States are effective in only 25–60% of patients,[49] and more than two million cases of adverse drug reactions occur annually, including 100,000 deaths.[50] Francis Collins, Director of the NHGRI, stated that "by 2010, it is expected that validated, predictive genetic tests will be available for as many as a dozen common conditions; in the next two decades, all cancer patients will have their malignancies genetically 'fingerprinted' and therapy will be individually designed".[51]

Cytochrome P-450 enzymes (CYPs) are important in the biosynthesis and degradation of endogenous compounds such as steroids, lipids, and vitamins.[52] In humans, 57 cytochrome P-450 genes have been identified, but mainly CYP1, CYP2, and CYP3 family members appear to contribute to 70–80% of drug metabolism.[53] A limited number of these genes have been studied in detail. The CYP2D6 gene was the first example where SNPs were identified that alter individual drug response.[54] Some 78 inherited variant alleles of CYP2D6 have been isolated, which produce three phenotypic sub populations defined by their rate of drug metabolism. The subgroups are comprised of: (1) poor metabolizers with inactive genes; (2) the ultrarapid metabolizer with up to 13 copies of CYP2D6[55] and (3) the intermediate metabolizer group. More than 65 commonly prescribed drugs are metabolized by CYP2D6, for example, the conversion of codeine to morphine, the active metabolite. Codeine is commonly used to control pain in a wide range of clinical settings, including acute vaso-occlusive episodes in sickle cell anemia. Knowledge of a person's CYP2D6 status would be invaluable for choosing the optimal pain medication and dosing regimen. This mechanism might provide a basis for the wide range of responses to morphine observed in sickle cell patients.

CYP3A is abundant in intestinal epithelium and liver where it accounts for 50% of the cytochrome P-450 enzymes. This variant is involved in the metabolism of agents that undergo oxidative degradation. The CYP2C19 family is important in metabolism of proton-pump inhibitors.[56] The Roche AmpliChip© was approved by the Food and Drug Administration in December 2005 for rapid genotyping of 33 common CYP2D6 and CYP2C19 variants.[57] To stimulate increased phamacogenetics research, the same agency released a guidance document encouraging researchers to collect and submit phamacogenomic data (*http://www.fda.gov/cber/gdlns/pharmdtasub.htm*).

Genes that alter responses to cancer chemotherapeutic agents have been studied to determine the impact on drug efficacy which is the most important variable in cancer treatment. Owing to the narrow therapeutic index of many chemotherapeutic agents, research in this area is a high priority. A flagship example of a pharmacogenetic diagnosis is the SNPs found in the TPMT (thiopurine methyltransferase) gene. Three major alleles are currently used to guide dosing for TPMT in the treatment of acute lymphoblastic leukemia.[58] Another example is Genetech's breast cancer treatment "trastuzumab" which produces improved outcomes when the Her-2/neu oncogene is over-expressed.[59] These are two examples of the current use of pharmacogenetics routinely in cancer treatment.

The economics of pharmacogenetics has been studied;[60] the debate is centered on whether the cost favors a genetic test in asymptomatic individuals. Most directly, pharmacogenetics during the development of new drugs may influence decisions about the progression of a compound to clinical trials. Currently, it takes about 20 years and costs millions of dollars to bring a drug to market. After safety assessments (Phase I), efficacy is tested in Phase II trials. If pharmacogenetic studies could identify responders, then subsequent Phase II trials could be smaller and less expensive. Altered drug response often occurs between racial and ethnic groups, based on differences in polymorphisms that influence drug metabolism.[61] Across populations, people can be grouped according to their genetic variants, which move in blocks from generation to generation.[62] This information will be valuable for genetic profiling to establish personalized medicine in the future.

Pharmacogenomics research should be heavily grounded in translational research, and will greatly benefit from scientists who understand both basic research and clinical application. The integration of genetics into healthcare requires different levels of ethical and policy oversight, depending on the way genetics is used; there is currently no federal legislation prohibiting the collection and use of personal genetic information. Bill S1053 was recently passed in the Senate that bars discrimination based on a person's genetic profiles in employment or insurance decisions.

Renaissance of Sickle Cell Disease Research in the Genome Era

Sickle cell anemia was the first genetic disease whose DNA mutation was defined.[63] Many clinicians and basic researchers have given considerable effort to developing new treatments, and ultimately a gene therapy cure (see Chapter 17). Major progress has been made in understanding the biology of this disorder, such as the pathophysiology of hemoglobin polymerization,[64] mechanisms of hemoglobin switching (see Chapter 12), and fetal hemoglobin induction (see Chapter 10).[65,66] Better therapies such as hydroxyurea,[67] butyrate,[68,69] and decitabine[70] have been developed, and a gene therapy cure is in sight (see Chapter 17). However, data from the HGP and the HapMap Consortium will be critical to understanding the effects of β-globin locus haplotypes on disease severity in sickle cell disease (SCD). Despite research advances, significant mortality and morbidity continues to be caused by SCD, in the United States and worldwide.

Improving health outcomes for sickle cell patients has been a focus of attention for the NIH over the last 50 years, as exemplified by a conference convened in November 2003, "New Directions for Sickle Cell Therapy in the Genome Era."[71] The main goal of this meeting was to formulate future directions for SCD research. The conference was organized and supported by the NHGRI; the National Heart, Lung, and Blood Institute; the National Institute of Diabetes, Digestive and Kidney Diseases; the Office of Rare Diseases; the Fogarty International Center; and the Foundation for the NIH. Over 120 individuals from the United States and abroad attended this invitation-only meeting.[71]

The attendees were charged to develop a plan for implementation which took into consideration the social and cultural contexts, past research, and health care delivery in the African-American community. The following is a brief summary of the conference findings. A detailed description is available at *http://www.genome.gov/11509561*.

Conference attendees agreed that genomic era tools and approaches should be used to develop more effective therapies for SCD. In addition, a Sickle Cell Disease Research Network, with a centralized prospective registry of well-phenotyped patients, is greatly needed. Initiatives to identify genetic modifiers in SCD using data from the HapMap project[72] were suggested. Proteomic and microarray-based analysis of hematopoietic cells from a large population will give insight into disease severity. With the slow development of efficacious drugs to treat SCD, small molecule high-throughput screens might be used to push the field forward. Also identified was an immediate need to train physicians to care for adult sickle cell patients.

Core resources of biological materials such as transgenic mice, a DNA construct repository, and so forth were supported. Likewise, the development of new models to study γ-globin reactivation is needed. Finally, improved vectors with greater safety should be developed to ultimately establish a cure for SCD. As with any field, it is important to train the next generation of scientists on the heels of current genome-based research efforts. Renewed excitement about SCD, and innovative NIH-supported multidisciplinary training programs, are needed. The group suggested that new research should be developed in light of the historical, social, economic, and cultural context of past research in the African-American community.

The Genome Era

The completion of the HGP two years ahead of schedule and under budget ushered in the genome era. Yet to come are interpretation of data generated by the HapMap project, which will enhance our understanding of the association of sequence variations with disease risk. The technological ability to affordably screen an individual's entire genome will be available within the next two decades, to make personalized medicine a reality. Education for clinicians is essential, along with smart science policies to move scientific advances of the past decade from the bench to bedside. The importance of an ongoing, balanced debate over issues related to genetic privacy, gene patents, race and genetics, population screening, and access to care cannot be overstated. The ever-vigilant services of the ELSI program is critical, to ensure a beneficial effect of the Human Genome Project for all populations worldwide.

The main roadblock to progress is not a lack of technology, but a lack of samples from sufficient numbers of clinically well-defined individuals.[73] This remains a significant challenge in SCD research, but the Clinical Research Network recently funded by the NIH will help address this hindrance to progress. Immediate attention to the role of genetic modifiers in determining disease severity and increased morbidity and mortality is dire to improvement of lifestyle in severely affected individuals.

We have already begun to witness the fruits of the landmark meeting held in November 2003 at the NIH, in the form of special research initiatives related to SCD, and training for under-represented minorities. A joint effort of NHLBI and NHGRI to establish summer institutes to enhance genomic research on genetic diseases, and to train under-represented scientists, will help ensure success for junior level faculty in academic fields.[74] Equal benefits from genomic era research to improve health outcomes will be expected by the African-American community.

References

1. Mendel G (1866). *Versuche über Pflanzen-Hybrigen. Verhandlungen de Naturforschenden Vereines.* Abhandlungen, Brünn **4**:3–47.
2. Morgan TH (1972). *Heredity and Sex*, 2nd edition. Oxford: Oxford University Press, pp. 70–86.
3. Avery OT, MacLeod CM, McCarty M (1944). Studies on the chemical nature of the substance inducing transformation of pneumococcal types. Inductions of transformation by a deoxyribonucleic acid fraction isolated from pneumococcus type III. *J Exp Med* **149**:297–326.
4. Pauling L, Itano HA *et al.* (1949). Sickle cell anemia a molecular disease. *Science* **110**:543–548.
5. Watson JD, Crick FHC (1953). Molecular structure of nucleic acids: a structure for deoxyribose nucleic acid. *Nature* **171**:737–738.
6. Marotta CA, Forget BG, Cohne-Solal M, Wilson JT, Weissman SM (1977). Human beta-globin messenger RNA. I. Nucleotide sequences derived from complementary RNA. *J Biol Chem* **252**:5019–5031.
7. Nirenberg MW (1963). The genetic code. II. *Sci Am* **208**:80–94.
8. Jackson DA, Symons RH, Berg P (1972). Biochemical method for inserting new genetic information into DNA of Simian Virus 40: circular SV40 DNA molecules containing lambda phage genes and the galactose operon of *Escherichia coli*. *Proc Natl Acad Sci USA* **69**:2904–2909.
9. Sanger F, Coulson AR (1975). A rapid method for determining sequences in DNA by primed synthesis with DNA polymerase. *J Mol Biol* **94**:441–448.
10. Maxam AM, Gilbert W (1977). A new method for sequencing DNA. *Biotechnology* **24**:99–103.
11. Smith LM, Sanders JZ, Kaiser RJ, Hughes P, Dodd C, Connell CR, Heiner C, Kent SB, Hood LE (1986). Fluorescence detection in automated DNA sequence analysis. *Nature* **321**:674–679.
12. The International Human Genome Sequencing Consortium (2001). Initial sequencing and analysis of the human genome. *Nature* **409**:860–921.
13. 1988 Memorandum Sets Foundation for Interagency Cooperation (1990). *Human Genome News*, **2**(1).
14. US Department of Health and Human Services, US DOE, Understanding Our Genetic Inheritance. *The US Human Genome Project: The First Five Years* (1990), No. 90–1590 Bethesda: NIH Publication.
15. Collins F, Galas D (1998). A new five-year plan for the U.S. Human Genome Project. *Science* **262**:43–46.
16. Collins FS, Brooks LD, Chakravarti A (1998). A DNA polymorphism discovery resource for research on human genetic variation. *Genome Res* **8**:1229–1231.
17. Collins FS, Patrinos A, Jordan E, Chakravarti A, Gesteland R, Walters L (1998). New goals for the U.S. Human Genome Project: 1998–2003. *Science* **282**:682–689.
18. Savage SA, Stewart BJ, Eckert A, Kiley M, Liao JS, Chanock SJ (2005). Genetic variation, nucleotide diversity, and linkage disequilibrium in seven telomere stability genes suggest that these genes may be under constraint. *Hum Mutat* **26**:343–350.
19. Ewing B, Green P (1998). Base-calling of automated sequencer traces using phred. II. Error probabilities. *Genome Res* **8**:186–194.
20. Gordon D, Desmarais C, Green P (2001). Automated finishing with autofinish. *Genome Res* **11**:614–625.

21. Gordon D, Abajian C, Green P (1998). Consed: a graphical tool for sequence finishing. *Genome Res* **8**:195–202.
22. Venter JC, Adams MD, Sutton GG, Kerlavage AR, Smith HO, Hunkapiller M (1998). Shotgun sequencing of the human genome. *Science* **280**:1540–1542.
23. Venter JC, Adams MD, Myers EW *et al.* (2001). The sequence of the human genome. *Science* **291**: 1304–1351.
24. Gras E, Cortes J, Diez O, Alonso C, Matias-Guiu X, Baiget M, Prat J (2001). Loss of heterozygosity on chromosome 13q12-q14, BRCA-2 mutations and lack of BRCA-2 promoter hypermethylation in sporadic epithelial ovarian tumors. *Cancer* **92**:787–795.
25. Vaarala MH, Porvari K, Kyllonen A, Lukkarinen O, Vihko P (2001). The TMPRSS2 gene encoding transmembrane serine protease is overexpressed in a majority of prostate cancer patients: detection of mutated TMPRSS2 form in a case of aggressive disease. *Int J Cancer* **94**:705–710.
26. Collins FS, Morgan M, Patrinos A (2003). The Human Genome Project: lessons from large-scale biology. *Science* **300**:286–290.
27. Figeys D, Aebersold R (1998). High sensitivity analysis of proteins and peptides by capillary electrophoresis-tandem mass spectrometry: recent developments in technology and applications. *Electrophoresis* **19**:885–892.
28. Irie T, Oshida T, Hasegawa H, Matsuoka Y, Li T, Oya Y, Tanaka T, Tsujimoto G, Kambara H (2000). Automated DNA fragment collection by capillary array gel electrophoresis in search of differentially expressed genes. *Electrophoresis* **21**:367–374.
29. Eichler EE, Clark RA, She X (2004). An assessment of the sequence gaps: unfinished business in a finished human genome. *Nat Rev Genet* **5**:345–354.
30. Banks RE, Dunn MJ, Hochstrasser DF, Sanchez JC, Blackstock W, Pappin DJ, Selby PJ (2000). Proteomics: new perspectives, new biomedical opportunities. *Lancet* **356**:1749–1756.
31. Brett D, Pospisil H, Valcarcel J, Reich J, Bork P (2002). Alternative splicing and genome complexity. *Nat Genet* **30**:29–30.
32. Deininger PL, Batzer MA (1999). Alu repeats and human disease. *Mol Genet Metab* **67**:183–193.
33. Benson DA, Karsch-Mizrachi I, Lipman DJ, Ostell J, Wheeler DL (2005). GenBank. *Nucleic Acids Res* **3**:D34–D38.
34. Burris J, Cook-Deegan R, Alberts B (1998). The Human Genome Project after a decade: policy issues. *Nat Genet* **20**:333–335.
35. Collins FS, Green ED, Guttmacher AE, Guyer MS (2003). US National Human Genome Research Institute. A vision for the future of genomics research. *Nature* **422**:835–847.
36. Borevitz JO, Liang D, Plouffe D, Chang HS, Zhu T, Weigel D, Berry CC, Winzeler E, Chory J (2003). Large-scale identification of single-feature polymorphisms in complex genomes. *Genome Res* **13**:513–523.
37. Altshuler D, Brooks LD, Chakravarti A, Collins FS, Daly MJ, Donnelly P, International HapMap Consortium (2005). A haplotype map of the human genome. *Nature* **437**:1299–1320.
38. Wang X, Tomso DJ, Liu X, Bell DA (2005). Single nucleotide polymorphism in transcriptional regulatory regions and expression of environmentally responsive genes. *Toxicol Appl Pharmacol* **207**(2 Suppl):84–90.
39. Holden AL (2002). The SNP consortium: summary of a private consortium effort to develop an applied map of the human genome. *Biotechniques* **26**(Suppl):22–24.
40. Cohen JJ (1997). Finishing the bridge to diversity. *Acad Med* **72**:103–109.
41. Peri KG, Gagnon J, Gagnon C, Bard H (1997). Association of −158 (C-->T) (XmnI) DNA polymorphism in G gamma-globin promoter with delayed switchover from fetal to adult hemoglobin synthesis. *Pediatr Res* **41**:214–217.
42. Gabriel SB, Schaffner SF, Nguyen H, Moore JM, Roy J, Blumenstiel B, Higgins J, DeFelice M, Lochner A, Faggart M, Liu-Cordero SN, Rotimi C, Adeyemo A, Cooper R, Ward R, Lander ES, Daly MJ, Altshuler D (2002). The structure of haplotype blocks in the human genome. *Science* **296**:2225–2229.
43. Developing a Haplotype Map of the Human Genome for Finding Genes Related to Health and Disease, Washington, D.C, 18–19 July 2001.
44. Shastry BS (2005). Genetic diversity and new therapeutic concepts. *J Hum Genet* **50**:321–328.
45. Wirtenberger M, Hemminki K, Chen B, Burwinkel B (2005). SNP microarray analysis for genome-wide detection of crossover regions. *Hum Genet* **117**:389–397.

46. Drosten C, Preiser W, Gunther S, Schmitz H, Doerr HW (2003). Severe acute respiratory syndrome: identification of the etiological agent. *Trends Mol Med* **9**:325–327.
47. van't Veer LJ, Dai H, van de Vijver MJ, He YD, Hart AA, Bernards R, Friend SH (2003). Expression profiling predicts outcome in breast cancer. *Breast Cancer Res* **5**:57–58.
48. Yates CR, Krynetski EY, Loennechen T, Fessing MY, Tai HL, Pui CH, Relling MV, Evans WE (1997). Molecular diagnosis of thiopurine S-methyltransferase deficiency: genetic basis for azathioprine and mercaptopurine intolerance. *Ann Intern Med* **126**:608–614.
49. Spear BB (2001). Pharmacogenetics and antiepileptic drugs. *Epilepsia* **42**(Suppl 5): 31–34.
50. Lazarou J, Pomeranz BH, Corey PN (1998). Incidence of adverse drug reactions in hospitalized patients: a meta-analysis of prospective studies. *JAMA* **279**:1200–1205.
51. Eschrich S, Yang I, Bloom G, Kwong KY, Boulware D, Cantor A, Coppola D, Kruhoffer M, Aaltonen L, Orntoft TF, Quackenbush J, Yeatman TJ (2005). Molecular staging for survival prediction of colorectal cancer patients. *J Clin Oncol* **23**:3526–3535.
52. Nebert DW, Russell DW (2002). Clinical importance of the cytochromes P450. *Lancet* **360**:1155–1162.
53. Evans WE, Johnson JA (2001). Pharmacogenomics: the inherited basis for interindividual differences in drug response. *Annu Rev Genomics Hum Genet* **2**:9–39.
54. Kimura S, Umeno M, Skoda RC, Meyer UA, Gonzalez FJ (1989). The human debrisoquine 4-hydroxylase (CYP2D) locus: sequence and identification of the polymorphic CYP2D6 gene, a related gene, and a pseudogene. *Am J Hum Genet* **45**:889–904.
55. Xie HG, Kim RB, Wood AJ, Stein CM (2001). Molecular basis of ethnic differences in drug disposition and response. *Annu Rev Pharmacol Toxicol* **41**:815–850.
56. Furuta T, Shirai N, Sugimoto M, Nakamura A, Hishida A, Ishizaki T (2005). Influence of CYP2C19 pharmacogenetic polymorphism on proton pump inhibitor-based therapies. *Drug Metab Pharmacokinet* **20**:153–167.
57. Food and Drug Administration Registry (2005). **70**(11):865–867.
58. Goldstein DB, Tate SK, Sisodiya SM (2003). Pharmacogenetics goes genomic. *Nat Rev Genet* **4**:937–947.
59. Thor A (2001). HER2 — a discussion of testing approaches in the USA. *Ann Oncol* **12**(Suppl 1):S101–S107.
60. Flowers CR, Veenstra D (2004). The role of cost-effectiveness analysis in the era of pharmacogenomics. *Pharmacoeconomics* **22**:481–493.
61. Tate SK, Goldstein DB (2004). Will tomorrow's medicines work for everyone? *Nat Genet* **36**(11 Suppl):S34–S42.
62. Patil N, Berno AJ, Hinds DA, Barrett WA, Doshi JM, Hacker CR, Kautzer CR, Lee DH, Marjoribanks C, McDonough DP, Nguyen BT, Norris MC, Sheehan JB, Shen N, Stern D, Stokowski RP, Thomas DJ, Trulson MO, Vyas KR, Frazer KA, Fodor SP, Cox DR (2001). Blocks of limited haplotype diversity revealed by high-resolution scanning of human chromosome 21. *Science* **294**:1719–1723.
63. Marotta CA, Wilson JT, Forget BG, Weissman SM (1977). Human beta-globin messenger RNA. III. Nucleotide sequences derived from complementary DNA. *J Biol Chem* **252**:5040–5053.
64. Ferrone FA, Ivanova M, Jasuja R (2002). Heterogeneous nucleation and crowding in sickle hemoglobin: an analytic approach. *Biophys J* **82**(1 Pt 1):399–406.
65. Stamatoyannopoulos G (2005). Control of globin gene expression during development and erythroid differentiation. *Exp Hematol* **33**:259–271.
66. Pace BS, White GL, Dover GJ, Boosalis MS, Faller DV, Perrine SP (2002). Short-chain fatty acid derivatives induce fetal globin expression and erythropoiesis *in vivo*. *Blood* **100**:4640–4648.
67. Charache S, Terrin ML, Moore RD, Dover GJ, Barton FB, Eckert SV, McMahon RP, Bonds DR (1995). Effect of hydroxyurea on the frequency of painful crises in sickle cell anemia. Investigators of the Multicenter Study of Hydroxyurea in Sickle Cell Anemia. *N Engl J Med* **332**:1317–1322.
68. Perrine SP, Ginder GD, Faller DV, Dover GH, Ikuta T, Witkowska HE, Cai SP, Vichinsky EP, Olivieri NF (1993). A short-term trial of butyrate to stimulate fetal-globin-gene expression in the beta-globin disorders. *N Engl J Med* **328**:81–86.
69. Atweh GF, Sutton M, Nassif I, Boosalis V, Dover GJ, Wallenstein S, Wright E, McMahon L, Stamatoyannopoulos G, Faller DV, Perrine SP (1999). Sustained induction of fetal hemoglobin by pulse butyrate therapy in sickle cell disease. *Blood* **93**:1790–1797.

70. Saunthararajah Y, Hillery CA, Lavelle D, Molokie R, Dorn L, Bressler L, Gavazova S, Chen YH, Hoffman R, DeSimone J (2003). Effects of 5-aza-2′-deoxycytidine on fetal hemoglobin levels, red cell adhesion, and hematopoietic differentiation in patients with sickle cell disease. *Blood* **102**:3865–3870.
71. National Institutes of Health (2003). New Directions for Sickle Cell Therapy in the Genome Era. Conference Summary.
72. The International HapMap Consortium (2004). Integrating ethics and science in the International HapMap Project. *Nat Genet* **5**:467–475.
73. Todd JA (1999). Interpretation of results from genetic studies of multifactorial diseases. *Lancet* **354**(Suppl 1):SI15–SI16.
74. National Heart, Lung, and Blood Institute (2005). Summer Institute Program to Increase Diversity in Health-Related Research. RFA-HL-04-035.

4

Sickle Cell Disease: A Phenotypic Patchwork

by Kim Smith-Whitley and Betty S. Pace

Introduction

The first reported case of sickle cell anemia in America was published in 1910 by James Herrick.[1] Forty-seven years later the A → T mutation in the β-globin gene was discovered, to establish the first genetic disease defined at the molecular level. Unfortunately, individuals affected with this dreaded disease lived in isolation and suffered high mortality rates; few efficacious therapeutic regimens were made available such as simple red blood cell transfusions. Furthermore, confusion in the general public about the genetics and clinical symptoms associated with sickle cell trait versus sickle cell disease (SCD), added to the chaos, fear, prejudice and injustice that was perpetrated in the African-American community in the 1960s.

In this setting, researchers designed controlled clinical trials to describe the acute and chronic complications of SCD, identify risk factors for severe disease, and to develop medical interventions to improve survival. To accomplish these goals, the Cooperative Study of Sickle Cell Disease (CSSCD) was initiated in 1979. Data generated from this landmark study which described the clinical phenotypes and complications associated with SCD along with predictors of disease-severity, will be reviewed in this chapter along with the conclusions made about risk factors and expected outcomes; this data — will be linked to our expanded understanding of the pathophysiology of SCD and the impact of genotype on clinical phenotype.

The challenge for the 21st century is to establish reliable prediction models of disease-severity and prognosis, using functional genomics approaches to identify correlative genetic markers. Ultimately, effective therapeutic interventions can be designed and instituted early in life to prevent chronic organ damage and disability in adulthood. The hope is that high-throughput technology developed during completion of the Human Genome Project will make it possible to perform genome-wide mutation analysis, to identify genetic markers that modify the clinical spectrum of SCD. This will facilitate the development of personalized treatment regimens. To bring focus to these issues, we will also review the origin of the sickle mutation, advances made towards diagnosis and identification of genetic markers of disease-severity, and current recommendations for transition to adult care. In subsequent chapters, there will be discussions of newborn screening, infections and preventive measures for children with SCD (Chapter 5), complications and health care maintenance for adults living with SCD (Chapter 6), current approaches to pain management (Chapter 7), and recommendations for transfusion therapy in (Chapter 8). A wealth of information will be presented about clinical research accomplishments over the last 30 years and the impact on quality of life and survival for individuals affected with SCD.

Origin of Sickle Gene

Sickle Mutation

The symptoms related to sickle cell crises were experienced in a Ghanian family in Africa long before they were recognized in the western hemisphere.[2] By 1949, Linus Pauling published the seminal paper declaring SCD a molecular disorder[3] and the amino acid substitution in hemoglobin S (HbS) was demonstrated in 1957 by Ingram and associates.[4] Using restriction endonucleases to identify single nucleotide polymorphisms (SNPs) in the β-globin locus, the inherited chromosome structures (haplotypes) were defined. In Africa the β^S gene was associated with four haplotypes representing regions where independent mutations occurred including Benin, Senegal, Central African Republic (Bantu), and Cameroon.[5,6] A fifth Asian β-locus haplotype was reported in Saudi Arabia and India.[7] The slaves who were exported from various parts of Africa to the United States had β^S-locus haplotypes that were endogenous to their region, but after arrival in America, Jamaica and Brazil, there was considerable admixture of African ethnic groups. Available calculations suggest that the β^S-gene mutation first developed approximately 70,000–150,000 years ago.[8]

Malaria

The regions in Africa where the spontaneous β^S-globin mutation arose (β codon 6 GAG $\rightarrow$ GTG; Glu $\rightarrow$ Val) are endemic for malarial infestation. This observation is consistent with the notion that the high incidence of the β^S mutation is derived from natural selection.[9] This process takes place when a DNA mutation alters the physical characteristics of an organism. If the new trait gives the organism a survival advantage, then it will be represented in the next generation at a higher frequency and the characteristics will be selected. People with one β^S-gene and one normal β^A-gene (sickle cell trait) are more resistant to malaria than people with two normal β^A-genes.

The life cycle of the malarial parasite involves transmission from mosquito to man, then intrahepatic infection is followed by red blood cell (RBC) invasion. Some RBC defenses against malaria produced by natural selection include the Duffy antigen on the membrane used by *Plasmodium vivax* to enter the cell. People lacking this antigen are resistant to this infection.[10] The *P. falciparum* parasite does not survive well in persons with sickle cell trait. The widely accepted theory is that HbS protects against malaria because of sickling at physiological oxygen tension followed by sequestration of parasitized RBCs in the reticuloendothelial system[11]; unfortunately children with SCD have a high fatality rate when malarial infections occur.[12]

Hemoglobin C (β codon 6 GAG $\rightarrow$ AAG; Glu $\rightarrow$ Lys) also protects against malaria. Modiano and associates[13] studied 4000 hemoglobin C trait patients. They had fewer episodes of malaria infections than did HbA controls and even lower rates of infection were observed in HbCC individuals. Two-gene deletion α-thalassemia, protects the host from malaria in part by altering the immune response to parasitized red cells.[14] Finally, glucose-6-phosphate dehydrogenase deficiency produces oxidant stress which is a poor environment for the survival of malaria.[15]

Classification of Sickle Hemoglobin Syndromes

Hemoglobin Genes

Hemoglobin is a globular tetrameric protein consisting of two α-like and two β-like globin chains, and four oxygen-binding heme groups covalently linked to each polypeptide. The β-like gene cluster on chromosome 11 encodes five functional genes including ε, ${}^{G}\gamma$, ${}^{A}\gamma$, δ, and β-globin expressed during development in the order that they appear in the locus.[16] The α-like globins are located on chromosome 16 where the ζ_1, ζ_2, α_1 and α_2 genes are located. Both globin gene clusters contain upstream enhancer elements — the locus control region comprised of five DNAseI hypersensitive sites (HS) and the HS40 enhancer are located in the β-like and α-like clusters respectively.

Hemoglobin Disorders

Over 1300 structural hemoglobin variants exist,[17] the majority of which are clinically benign (Table 1). The hemoglobinopathies are the most common single gene disorders in the world.[18] The globin gene server is a comprehensive public database where mutations are cataloged (*http://globin.cse.psu.edu/*). The generic term sickle cell disease (SCD) is used to refer to the clinically relevant sickling syndromes. In the heterozygote state, the normal HbA symbol is placed first, followed by the variant. Sickle cell trait (AS) is in general, an asymptomatic carrier state. Sixty-five percent of SCD is caused by homozygous HbSS disease (SCD-SS), 25% by the compound heterozygote state HbS and HbC (SCD-SC), nine percent SCD-Sβ thalassemia and the remaining one percent consists of other variants (SCD-SE, SCD-SD^{Punjab}, SCD-SO^{Arab}, and SCD-S^{Lepore}).[19] Mutations also occur in the α-like locus most commonly as gene deletions[17] including the one-gene deletion silent carrier state ($/-\alpha$) and a two-gene deletion ($-\alpha/-\alpha$) producing α-thalassemia trait.

SCD-SS is an autosomal recessive genetic disorder which is inherited according to the rules of Mendelian genetics. If each parent is a carrier (AS), then with each pregnancy, their infant has a 25% chance of inheriting either two defective genes (SS) or two normal genes (AA) and a 50% chance

Table 1. Hemoglobin structural variants.

I. Structural variant by globin gene	Number
A. β-Like Globin Cluster	
${}^{G}\gamma$-globin	54
${}^{A}\gamma$-globin	47
δ-globin	71
β-globin	689
B. α-Like Globin Cluster	
α_1-globin	250
α_2-globin	286
II. Structural variant by types	
Single Nucleotide Polymorphisms	1000
Insertions	50
Deletions	143
Fusions	8
(*http://globin.cse.psu.edu/*)	

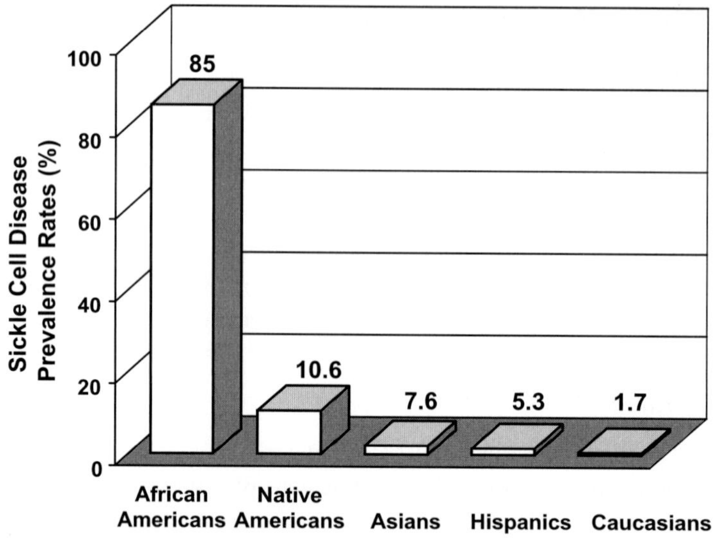

Fig. 1. *Sickle cell disease in the United States*. The most common type of sickle cell disease (SCD) in African-Americans is homozygous SS disease (SCD-SS). Every year, about 1 in 400 African-American infants are born with SCD after inheriting a β^S allele from both parents. Modified by permission from the reported prevalence rates per 100,000 persons, Food and Drug Administration (1999), Publication No. (FDA) 99–1251.

of being a carrier (AS). About 8% of African-Americans are carriers and 72,000 individuals have some form of sickle-hemoglobinopathy. The most current statistics published by the Food and Drug Administration[20] showed that the distribution of SCD among different ethnic groups has become *fluid* due to the high rate of racial admixture in America. The data showed an 85% prevalence rate for all forms of SCD in individuals of African descent (Fig. 1). The second most commonly affected group is Native Americans at a 10.6% prevalence rate; the remaining 4.4% of SCD occurs in Asians, Hispanics and Caucasians.

Diagnosis of Sickle Cell Disease

When a hemoglobinopathy is suspected the workup should include a complete blood count, Hb analysis by electrophoresis, isoelectric focusing or cation-exchange high performance liquid chromatography (HPLC), and the quantification of HbA_2 and fetal hemoglobin (HbF) levels. Family studies are indispensable to making a definitive diagnosis. There may be anemia and characteristic sickle RBCs on the peripheral blood smear depending on the genotype and patient's age (Fig. 2) as well as other RBC abnormalities, such as target cells and crystals (SCD-SC). The peripheral blood smear has 76% sensitivity and 99.7% specificity for making an accurate presumptive diagnosis of SCD.[21] Over the last two decades molecular-based methods have become available to identify less common hemoglobinopathy variants.

Protein-Based Techniques

The gold standard is a protein-based diagnosis of SCD. Hemoglobin electrophoresis depends on the relative charge of various Hb types which allows separation based on mobility at high and low pH conditions. The cellulose acetate and citrate agar Hb electrophoresis methods provide useful

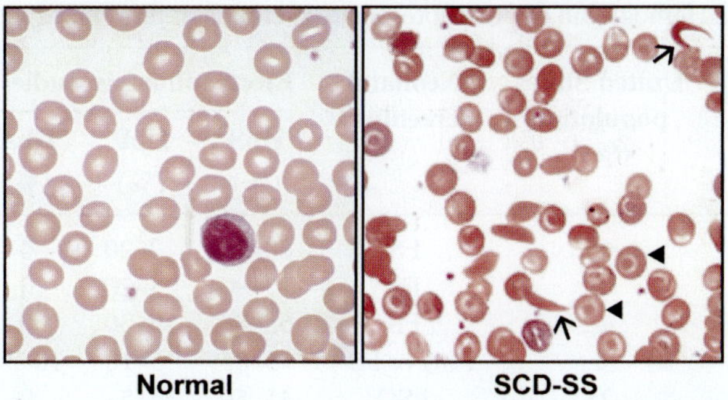

Fig. 2. *Peripheral blood smears.* (Left) Shown is a normal blood smear composed of uniform shaped red blood cells, with a zone of central pallor occupying about 1/3 of the diameter of the cell. Also shown in the field is a mature lymphocyte. (Right) Blood smear from a person with homozygous SS disease (SCD-SS). Shown is the decreased number of red blood cells and the characteristic sickled cells (→) and target cells (▸). By permission from Davidson MW, www.microscopyu.com.

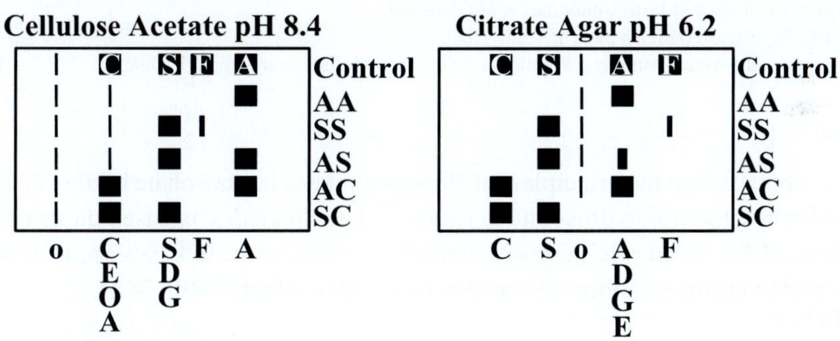

Fig. 3. *Hemoglobin electrophoresis.* When performing hemoglobin electrophoresis, blood samples are lysed, place in the wells at the origin (o) and the different hemoglobin polypeptide chains travel towards the node according to the amino acid composition and charge of the molecule after an electrical current is applied. Samples taken from newborns have high fetal hemoglobin (F) levels therefore they are routinely analyzed in cellulose acetate at alkaline pH to separate F and hemoglobin A (A). Then to separate hemoglobin S (S) from hemoglobin D, G or E (D, G, E) the sample is also analyzed in citrate agar at acid pH. The combined results are used to make a presumptive diagnosis of a hemoglobinopathy or sickle trait at birth.

genotyping data (Fig. 3); both are performed routinely in newborns.[22] The different pH levels allow the separation of HbC from HbA_2 and HbS from HbD definitively. Isoelectric focusing utilizes a pH gradient in which Hb molecules migrate to their isoelectric points. The third method, HPLC is expensive and requires high-level technical expertise; both isoelectric focusing and HPLC have 100% sensitivity and specificity.[23,24]

In healthy adults, the HbA ($\alpha_2\beta_2$) content is about 95%, HbA_2 ($\alpha_2\delta_2$) 2–3% and HbF ($\alpha_2\gamma_2$) 0.8–2% (Table 2). In newborns with FS on Hb electrophoresis (HbF > HbS), the possible genotypes include SCD-SS or SCD-Sβ^0thalassemia however, SCD-Sβ^+thalassemia must also be considered since Hb separation methods may fail to isolate small percentages of HbA at birth. Alternatively, hereditary persistence of HbF may be present therefore, repeat studies between 1 and 2 years of age are recommended to confirm Hb genotype. A diagnosis of SCD is made in a child when HbS predominates after the age of one year. Parental studies help clarify SCD genotypes.

Table 2. Hemoglobin electrophoresis interpretation at birth and 2 years old.

Disorder	United States population (%)[1]	Neonatal screening[2]	Electrophoresis studies at 1–2 years			
			HbS (%)	HbF (%)	HbA (%)	HbA_2/C (%)
SCD-SS	65	FS	80–95	2–20	0	<3.6
SCD-Sβ^0thal	8	FS	75–90	5–20	0	>3.6
SCD-S/HPFH	<1	FS	65–80	15–30	0	<2.5
SCD-Sβ^+thal	1	FSA or FS[3]	5–85	5–10	10–30	>3.6
SCD-SC	25	FSC	45–50	2–5	0	45–50[4]
S Trait	8[5]	FAS	32–45	1–2	52–65	<3.6

Hb = hemoglobin, thal = thalassemia, HbF = fetal hemoglobin.
1. Percent of different types of sickle cell disease for a total of ~72,000 affected individuals in the United States population.
2. Hemoglobin reported in order of quantity (e.g. FSA = F > S > A).
3. Quantity of HbA at birth sometimes is not detected.
4. SDC-SC disease patients have 45–50% HbC.
5. 8% of total African-American Population in the United States carry the β^S allele.

Solubility Test

Solubility tests are based on the principle that HbS precipitates in phosphate buffers at high molarity and neutral pH when sodium hydrosulfite is added.[25] The Sickledex is based on this principle and should not be used for diagnostic purposes since false negative results can occur in newborns. In addition, this method cannot distinguish carriers from those affected with SCD.

Molecular-Based Techniques

The definitive diagnosis of SCD requires DNA analysis. For prenatal diagnosis fetal DNA can be isolated from chorionic villus cells[26] or by amniocentesis in the first and second trimester respectively, however maternal cells can cause contamination. A non-invasive method to obtain DNA from fetal cells[27] or circulating fetal DNA[28] from maternal blood has been developed recently. For newborns, most commonly a protein-based method is used for diagnosis versus prenatal testing where DNA-based methods are required. In the late 1970s restriction endonuclease mapping and blot-hybridization were used to define gene relationships. Kan and associates[29] were the first to describe a restriction fragment length polymorphism (RFLP) created by an *HpaI* restriction site in linkage disequilibrium with the β-globin gene (allele). In the normal β^A allele a 7.6-kb fragment is generated by *HpaI* digestion. Seventy percent of the time the β^S allele was linked to a polymorphism that abolished the *HpaI* site producing a 13.0-kb fragment. RFLP analysis was used for prenatal diagnosis for many years.[29] The next significant breakthrough was direct detection of the β^S mutation by RFLP analysis using *MstII* endonuclease digestion which was produced by southern blot analysis, 1.35-kb and 1.15-kb fragments were deleted in the β^S and β^A alleles respectively.[35]

By 1985, the molecular diagnostic field was revolutionized by the advent of polymerase chain reaction (PCR) technology.[30] Embury *et al.* introduced a PCR-based approach combined with *DdeI* and *HinfI* endonuclease digestion of PCR products and blot-hybridization to diagnose SCD prenatally.[31] Numerous other techniques were also developed, such as allele-specific oligonucleotide (ASO)

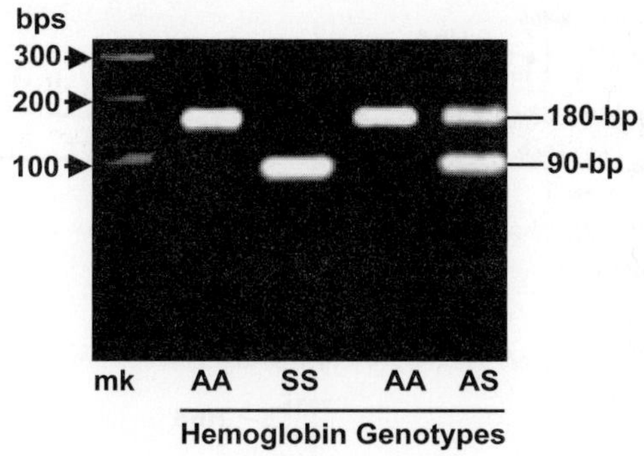

Fig. 4. *Restriction fragment length polymorphism analysis*. To detect the β^S mutation (A → T) in the sixth codon, primers were design to amplify DNA 80-bp in both directions from the mutation to yield a 180-bp PCR fragment. After amplification the DNA was digested with the restriction enzyme *DdeI*, which will only digest in the presence of the β^S mutation to yield a 90-bp fragment. Shown are the molecular weight DNA marker (mk), and the three genotypes identified in a 2% agarose gel after this analysis was completed.

hybridization, allele-specific priming-amplification refractory mutation system (ARMS), amplification created restriction analysis,[32] and gap-PCR. These PCR-based methods were used to identify known SNPs. For mutation discovery, the preferred methods were single-stranded conformation polymorphism (SSCP) and denaturing gradient gel electrophoresis (DGGE), however direct sequencing remains the standard. PCR amplification of the target region was followed by DNA sequencing using the Sanger method.[33] DGGE has been used for large scale population screening for β-thalassemia.[34]

Molecular techniques have provided powerful tools for genetic analysis to identify disease and carrier states. Allele-specific amplification is based on the principle that a perfectly matched primer is more efficient at directing DNA synthesis than a mismatched primer. The ARMS method is based on this principle[35,36] and recently was improved by the development of a single-tube assay.[37] Gap-PCR is mainly used to define structural deletions in DNA.[38] Deletions that cause α-thalassemia trait are detected by gap-PCR.[39] RLFP analysis has been used extensively to identify the β^S mutation that destroys the restriction sites for *MstII*, *MnlI* and *DdeI* and β-cluster haplotypes (see Chapter 11). The β^C-globin mutation does not destroy these restriction sites. For RFLP analysis, the target DNA region is PCR-amplified followed by restriction enzyme digestion and gel electrophoresis. The presence of a SNP in the DNA could create or abolish the restriction site. As shown in Fig. 4, PCR primers were designed to produce a 180-bp product. *DdeI* digest produced a 180-bp fragment in a normal β^A-gene whereas in the β^S-globin allele, an 80-bp fragment is observed because the A → T mutation in β^S creates a *DdeI* site allowing a rapid and accurate diagnosis of SCD-SS or the carrier state, sickle trait (AS).

Genome Era Techniques

The completion of the Human Genome Project ushered in the genome era and new approaches encompassing functional genomics research. Mass spectrometry provides results unsurpassed in quality and speed for mutation detection. The high-throughput MassARRAY system (Sequenom Inc., San Diego, CA) combines primer extension with matrix assisted laser desorption ionization — time of flight (MALDI-TOF) mass spectrometry with data analysis software.[40] The common variants of SCD are accurately detected using this system which validates mass spectrometry as a clinically relevant diagnostic tool (Fig. 5). Recent modifications of this technology allowed efficient detection

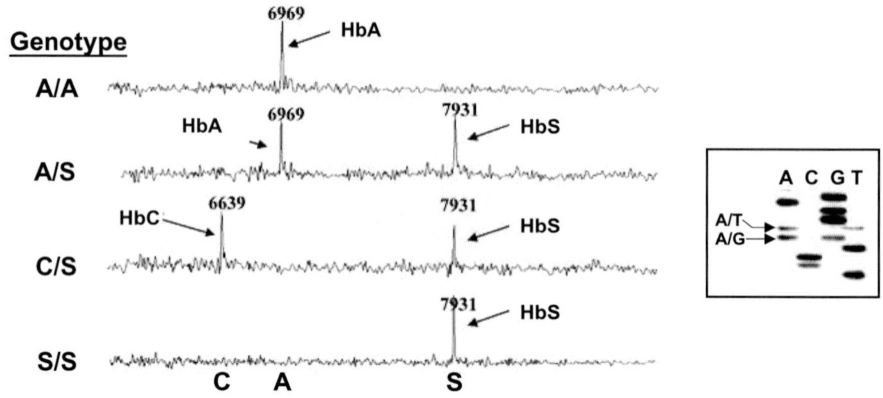

Fig. 5. *β-Globin gene spectra mutation analysis*. Sample spectra from a β-globin assay are shown, depicting the ability of MALDI-TOF mass spectrometry to separate each of the β-globin genes (alleles) and identify accurate genotypes. In contrast to sequencing reactions (gel image to right of spectra) MassARRAY technology is able to distinguish whether two β-globin mutations are on a single chromosome or separate chromosomes. By permission from Braun *et al.* (2003).

of Hb variants using tryptic digest and mass spectrometry–mass spectrometry. Tryptic digestion of globin protein produces a well-characterized series of peptides. In a previous study, the Hb variants S, C, D^{Punjab}, O^{Arab}, and E were correctly identified in 200 blood samples.[41] Heterozygotes and homozygous states can be identified by mass spectrometry in a format suitable for population screening.

In 1999 the SNP Consortium was established to characterize genome-wide SNPs.[42] Chip-based microarray capable of large-scale SNP analysis will allow the identification of genetic markers that impact the phenotypic expression of SCD.[43] The chip-based method, arrayed primer extension (minisequencing) was developed to identify known mutations by Tonisson and associates.[44] The immobilized primers are extended at the 3′-end based on the patient's hybridized DNA which serves as the template to direct insertion of the complementary fluorescent dideoxynucleotide probe by DNA polymerase to terminate chain growth. High-throughput resequencing for SNP detection has been developed by Affymetrix using high-density microarrays.[45] Recently a diagnostic chip to detect mutations in drug metabolizing genes in the cytochrome p-450 system was released.[46] Furthermore, a research SNP-chip was developed to detect common β-gene mutations that cause β-thalassemia. Similar diagnostic tools will be developed for other hemoglobinopathies in the future to facilitate the practice of personalized medicine.

Pathophysiology of Sickle Cell Disease (SCD)

Although the genetic mutation that leads to the production of HbS is known, it is unlikely to be the sole cause of the highly diverse clinical phenotypes. Our understanding of the pathophysiology of SCD, initially thought to be due mainly to vaso-occlusion, has evolved (Fig. 6). Some individuals with SCD never develop strokes, but may have frequent painful episodes and acute chest syndrome (ACS) while others seldom require hospitalization. Therefore epigenetic factors must be involved in the different phenotypes observed in SCD-SS.

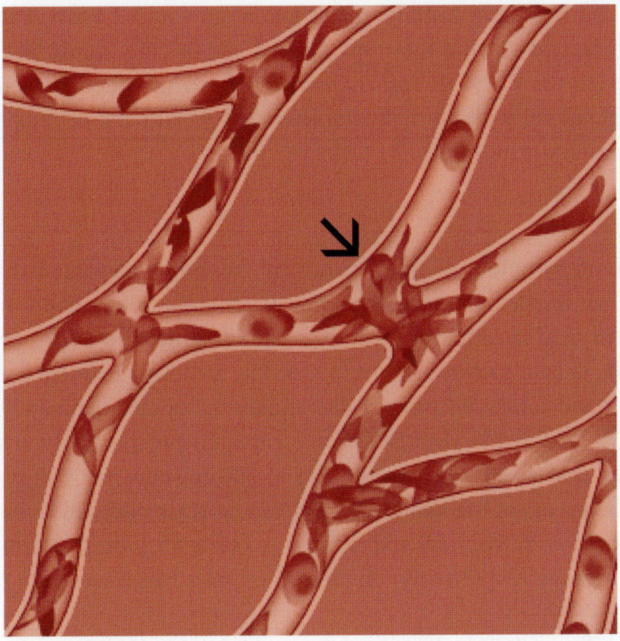

Fig. 6. *Sickle cell vaso-occlusion.* The drawing depicts a capillary containing red blood cells in various stages of deformity, with several sickled cells causing vaso-occlusion (→) which results in sludging in the more proximal vessel. By permission from National Heart, Lung, and Blood Institute, Public Domain.

Intravascular sickling is a primary component of the pathophysiology of SCD. Factors affecting this process have been investigated extensively. However the contributions of non-sickling related mechanisms such as, endothelial dysfunction, activated adhesion molecules, RBC membrane receptors, and vasomotor tone, to clinical disease-severity have not been clearly defined. Organ ischemia may be modulated by reperfusion mechanisms, local tissue resistance, and inflammation. Studies to define organ-specific genetic risk factors for acute and chronic complications will be informative.

Recently, chronic hemolysis was implicated in mechanisms for decreased nitric oxide (NO) availability and impaired vasomotor tone as a precursor to pulmonary hypertension, priapism, and stroke.[47] Since RBC transfusions are common in the treatment of SCD, the benefits of determining genetic factors governing the development of antibodies are vast. Likewise when the genetic basis for the HbF response to hydroxyurea (HU) therapy is defined then personalized medicine can be practiced to develop optimal treatment regimens for individuals with SCD.

Cooperative Study of Sickle Cell Disease (CSSCD)

In 1972, legislation was adopted in the US that provided the economic basis for scientific research to describe the natural history of SCD.[48] Although case reports and single institution studies described clinical phenotypes, less was known about long-term complications and survival. The goals of the CSSCD were to delineate clinical phenotypes, identify predictors of disease-severity, and improve survival through therapeutic interventions.

The CSSCD enrolled 3200 patients with SCD from 23 clinical centers between 1978 and 1981.[49] Enrollment for a newborn cohort, beginning in 1978, included patients identified between 1975 and

1978, and continued throughout the study; mean follow-up was 5.2 years. Patients from urban and rural centers were recruited so that the impact of access to care would not skew morbidity and mortality data. Through the CSSCD, researchers have been provided comprehensive data on the natural history of SCD along with normal parameters for growth and development, blood pressure, and steady-state laboratory tests values.[19]

Painful Episodes

Pain, the most common symptom of SCD, varies by frequency and intensity within and between subgroups. Febrile illnesses, dehydration, stress, and extreme environmental temperatures may trigger painful episodes, but in most cases there is no known precipitating event. The CSSCD defined a painful episode as pain lasting at least two hours, unexplained by any other medical condition, and requiring medical attention;[50] hand-foot syndrome or dactylitis was counted separately.[51] A total of 12,290 painful episodes in 3578 patients with SCD were captured prospectively and occurred mainly in SCD-SS and SCD-Sβ^0thalassemia at a rate of 80 and 100 episodes per 100 pt–years respectively.[51] Children in the first two years of life had the lowest pain rates but dactylitis was highest in this group. Fifty percent of children with SCD-SS and SCD-SC experience a painful episode by 4.9 years and 7.1 years of age on average respectively; children 0–9 years had lower pain rates than those older than 10 years.[50]

The clinical variability of SCD by Hb genotype was highlighted in this seminal study. Interestingly, up to 39% of patients with SCD-SS had no painful episodes requiring medical attention.[50] However, individuals with SCD-SC, a subgroup with lower pain rates, could experience more painful episodes than a person with SCD-SS. Recurrent pain was common across all genotypes. The sub-population with SCD-SS who experienced up to 10 painful episodes per year (5.2%) accounted for 39% of the total pain events in that group.[50]

Globin Gene Effects

The CSSCD was the first study to emphasize the clinical benefit of HbF in SCD-SS. HbF has the ability to inhibit HbS polymerization and prevent complications related to vaso-occlusion. Data from the CSSCD demonstrated that pain, ACS and mortality rates were decreased with incremental increases in HbF. However, increased HbF has not been shown to protect against stroke, or avascular necrosis (AVN). These inconsistencies have made prediction modeling for disease-severity difficult. The haplotypes associated with the β^S-gene were not consistently correlated with HbF or disease-severity.[16] The ability to separate the clinical benefit of HbF from the positive or negative impact of genetic markers associated with β-locus haplotypes has been difficult. Another genetic modifier is α-thalassemia trait ($-\alpha/-\alpha$), thought to be protective for some complications of SCD by decreasing RBC density and increasing Hb levels. In the CSSCD, α-gene status did not alter the frequency of pain episodes or ACS in adults, however in children α-thalassemia trait combined with SCD was associated with higher painful episodes and less ACS; this may be due to sampling error, since less than 50% of children had α-globin gene analysis performed.

Exacerbation of Anemia

Acute exacerbation of anemia occur secondary to acute splenic sequestration, parvovirus B19 infections, sepsis or rarely folate deficiency. The CSSCD defined an acute anemic episode as a reduction in Hb of 30% below steady-state values unrelated to other known SCD-related causes. A splenic sequestration event was defined as a 20% decrease in Hb from steady-state with an increase in spleen size >2 cm.[51] Although acute splenic sequestration decreased substantially in both SCD-SS and SCD-SC cohorts >6 years old, the rate for other causes of acute anemia occurred sporadically in both groups throughout the first 10 years of life. Human parvovirus serology was not available throughout the study period however, the importance of parvovirus B19 as a cause of anemic events has subsequently been established.[53]

Acute Chest Syndrome

The CSSCD defined an episode of ACS as a new infiltrate on chest x-ray or a demonstrated pulmonary perfusion defect on radioisotope scan. Data on 2100 episodes of ACS in 1085 patients followed from 1979 to 1988 were collected prospectively.[54] Fifty-six percent of patients did not have recurrent ACS, but 21% had three or more episodes. The incidence rates were highest in patients with SCD-SS and SCD-Sβ^0thalassemia (12.8 and 9.4 per 100 pt–yrs respectively) compared to 5.2 per 100 pt–yrs in those with SCD-SC and 3.9 for SCD-Sβ^+thalassemia. The incidence rate of ACS per 100 pt–yrs in 2–4 years old children was 25.3 compared to 8.8 in adults, which supports infection as a major cause of ACS in children.[54] Risk factors for the complication of ACS were genotype (SCD-SS and SCD-Sβ^0thalassemia), young age, higher Hb and leukocyte counts, and low HbF.

Because of the high frequency of ACS in young children, a follow-up study was published in 1997.[55] Children who develop ACS often present with signs of infection, whereas adults usually develop ACS after hospitalization for other SCD-related symptoms. A later study reported that the majority of children had no identifiable cause, but infection, pulmonary fat embolism or infarction were most common where a cause could be determined.[56] ACS can also be caused by rib infarcts and injury to underlying lung tissue.[57] Our expanded understanding of the pathophysiology of ACS includes a role for inflammation and vasomotor dysfunction. The association of increased white blood cell (WBC) counts and abnormal nitric oxide (NO) metabolism supports this mechanism. Genes that regulate NO synthesis may regulate exhaled NO levels, which are low in patients with ACS.[58] SNPs in the NO synthesizing enzyme, NO synthetase 3 have been associated with a higher frequency of ACS.[59]

Central Nervous System

The devastating complication of stroke or cerebral vascular accident is associated with increased mortality in children. Cerebral ischemia, primarily caused by progressive vasculopathy, can resolve within the first 24 hours (transient ischemic attack), be associated with motor symptoms that persist beyond 24 hours (ischemic stroke), or may lack overt motor symptoms but neuropsychologic deficits can occur ("silent" CNS infarcts). Hemorrhagic stroke although less common in children, can result from moyamoya or rupture of arteriovenous malformations.

The CSSCD defined stroke as an acute neurologic event due to vaso-occlusion or hemorrhage associated with cerebral ischemia and neurologic signs and symptoms.[75] Stroke age-adjusted prevalence at study entry was highest in patients with SCD-SS and SCD-Sβ^0thalassemia, 4.0% and 2.4% respectively, compared to 0.8% in individuals with SCD-SC and 1.3% in SCD-Sβ^+thalassemia.[75] Eighty-seven events occurred in patients without a history of stroke during the study period and an incidence of 0.61 was observed in SCD-SS patients.

Of patients who survived the initial stroke, 14% had a recurrent event[60] which was highest in individuals less than 20 years old (6.4 events per 100 pt–yrs) versus 1.6 events per 100 pt–yrs for those over 20 years old. The mean times to recurrence ranged from 3 to 22 months. One caveat is the inclusion of patients who started chronic RBC transfusions after their initial stroke and those who were followed off transfusions, therefore recurrence rates may be underestimated.

Risk factors associated with infarctive stroke included a history of transient ischemic attack, ACS within two weeks of the acute stroke or high yearly ACS rates, high systolic blood pressure, low steady-state Hb, and increased WBC counts, whereas only the latter two parameters were associated with hemorrhagic strokes.[60] "Silent" central nervous system infarction is a risk factor for stroke as well.[61] Although α-thalassemia trait is protective for infarctive stroke, the protection for hemorrhagic stroke is marginal.

Progressive vasculopathy is strongly supported as the mechanism of infarctive strokes based on studies demonstrating that children with abnormal trans-cranial Doppler (TCD) studies are at increased risk for stroke.[62] Children between 2 and 16 years old with SCD-SS or SCD-Sβ^0thalassemia and a time-averaged maximum mean velocity (TAMMV) threshold $\geq$200 cm/sec measured in the distal internal carotid artery or the proximal middle cerebral artery have a 10% risk of developing a stroke. RBC transfusion protocols to maintain HbS <30% produced a 92% reduction in stroke risk;[63] α-thalassemia trait was found more frequently in patients with normal TCD studies.[64] It is known that 60% of children with abnormal studies will not develop stroke and 19% with normal studies will develop stroke. This data illustrates the low specificity of TCD studies which supports the continued search for prediction models based on multiple genetic markers to better define individual stroke risk.

Sibling studies demonstrated a genetic predisposition to stroke[65] suggesting that genetic modifiers unrelated to SCD contribute to stroke risk. Studies in the CSSCD newborn cohort showed that the histocompatibility lymphocyte antigen (HLA) markers DPB1*0401 (susceptibility to small vessel disease), DPB1*1701 (protective from small vessel disease), A*0102 (susceptibility to large vessel disease), A*2612 (susceptibility to large vessel disease), and A*3301 (protective from large vessel disease) were associated with stroke subtypes in children with SCD.[67] In a second study over 100 candidate genes involved in diverse metabolic functions were analyzed as stroke risk factors.[68] In the large vessel stroke group, IL4R 503P and HLA-A variants were associated with increased risk, whereas the ADRB2 27E and TNF (−308) A variants were protective. In the small vessel stroke group, VCAM-1 (−1594) C and HLA-DPB1 increased stroke risk whereas LDLR *Ncol I* was protective.[68] More recently, DNA from a CSSCD cohort of 1398 patients with SCD-SS was analyzed and SNPs in the ANXA2.6, BMP6.10, BMP6.12, SELP.14, TGFBR3.10, and ERG.2 genes were associated with increased stroke risk.[69] These studies exemplify how genome-wide genetic markers can be integrated into a prediction model of disease-severity in SCD.

"Silent" Central Nervous System (CNS) Infarcts

"Silent" CNS infarct was defined as ischemic changes on brain magnetic resonance imaging (MRI) scan occurring in patients without a history of stroke however, children with "silent" infarcts have higher rates of abnormal neuropsychological studies.[70] The pediatric CSSCD cohort had routine brain MRI scans starting at six years of age; 17% of children with SCD-SS had evidence of "silent" infarcts.[71] A history of seizure, painful events, WBC counts over 11.8×10^9 per liter, and the Sengal β-cluster haplotype were associated with a higher rate of "silent" infarcts.[72] There were no associations with blood pressure, α-thalassemia trait, or HbF levels however HbF levels were available for only 50% of this cohort.

Priapism

Priapism is a common complication in males with SCD. Priapism was defined as a painful penile erection that lasts longer than one hour, causing the patient to seek medical attention. A cohort of 1737 males with SCD-SS or SCD-SC were followed from 1978–1998.[73] Of these, 273 (15.7%) had one episode of priapism. For the entire cohort and SCD-SS patients alone, high lactate dehydrogenase and reticulocyte and platelet levels were significantly associated with priapism. Specific genetic markers increased the risk for priapism including guanylyl cyclase, a signaling molecule involved in cyclic guanosine monophosphate activation and *KLOTHO*, a gene that encodes a protein involved in the regulation of vascular tone and NO metabolism; penile erection dysfunction may occur when NO levels are low.[53] These findings suggest that pharmacologic agents directed at increasing NO levels might be effective in patients with *KLOTHO* polymorphisms. Responses in males treated with hydroxyurea (HU), an NO donor, are encouraging.[74]

Other Complications

Avascular necrosis (AVN) of the hips is common in patients with SCD, often causing chronic pain syndromes. Although MRI is recognized as the best method for detecting early AVN, a physical examination and plain film radiographs are useful screening tools for advanced disease. The point prevalence of AVN was 9.8% in the CSSCD and involvement of the femoral head was highest in individuals with SCD-SS/[+]α-thalassemia trait, followed by SCD-SS/[−]α-thalassemia trait and those with SCD-SC. Patients with SCD-Sβ^+thalassemia were not protected from AVN[75] but it occurred later in life. Leg ulcers were observed most frequently in patients with SCD-SS and SCD-Sβ^0thalassemia genotypes[75] however α-thalassemia trait appeared to decrease the risk. Dilatation of the heart, particularly of the left side is common in patients with SCD. A subgroup of the CSSCD cohort had echocardiography that showed a correlation between left ventricular dimensions and age in patients with Hb values below 8.0 gm/dl.[76] Therefore, low Hb and increased age are risk factors for cardiac dilatation.

Growth and Development

Somatic growth and pubertal development are delayed in children with SCD, but adolescents ultimately achieve their expected adult height and sexual maturity. This delay is more pronounced in patients with SCD-SS, compared to those with SCD-SC. The onset of menarche is delayed by two to three years with a mean age at menarche of 15–16 years.[77] The differences in height and weight in children with SCD-SS compared to well children, becomes apparent by two years of age but children receiving chronic transfusions may have improved growth.[78] Increased resting energy expenditure and nutritional deficiencies, such as zinc and vitamin A may be responsible for poor growth in children with SCD, but this does not explain the ability of these patients to reach adult parameters.[79]

Genetic Disease Modifiers

As a result of the data generated in the CSSCD, genetic modifiers associated with disease-severity have emerged. They are summarized briefly in this section and in Table 3.

Predictors of Disease-Severity

The data from the CSSCD confirmed that individual with SCD-SS and SCD-Sβ^0thalassemia are more likely to have significant complications while those with SCD-SC and SCD-Sβ^+thalassemia

Table 3. Correlation between clinical phenotype and genotype.

SCD complication	Clinical risk factors	Laboratory risk factors	Genetic risk factors	Investigated no associations
Mortality	ACS Painful episode >20 years old	Pediatric ↓ Hb ↓ HbF	SS	α-thal trait
Painful episode	>10 years old	↓ HbF ↑ Hematocrit	SS $S\beta^0$thal	WBC α-thal trait
Acute chest syndrome (ACS)	Younger age	↑ Hb ↑ WBC ↓ HbF	SS $S\beta^0$thal Polymorphisms: NOS I & 3	
Priapism	Painful episode ACS AVN CVA	↓ Hb ↑ LDH, AST, bilirubin, reticulocytes ↑ WBC ↑ Platelet count	SS[−]α-thal trait Polymorphism: *KLOTHO* (13q12)	β globin haplotype HbF Blood pressure
Cerebral vascular accident*	TIA ACS within 2 weeks ↑ ACS ↑ Systolic BP "Silent" CNS infarcts	↓ Hb** ↑ WBC**	SS[−]α-thal trait Polymorphisms: IL4R 503; HLA-A Polymorphisms: ANXA2.6, BMP6.10, BMP6.12, SELP.14, TGFBR3.10, ERG.2	HbF Platelet count Painful episodes
Avascular necrosis	↑ Pain	↑ Hb ↓ MCV ↓ AST	SS[+]α-thal trait	
"Silent" central nervous system infarcts	↓ Pain episodes Seizures	↑ WBC	Senegalese haplotype Polymorphisms: VCAM-1 (–1594)C HLA-DPB1 HLA homozygosity	α-thal trait Hb HbF Blood pressure
Leg ulcers	Male <10 years old	↓ HbF	SS[−]α-thal trait $S\beta^0$thal	

*Risk factors for infarctive stroke; **Risk factors for hemorrhagic and infarctive stroke.
Abbreviations: AST, aspartate aminotransferase; Hb, hemoglobin; HbF, fetal hemoglobin; MCV, mean corpuscular volume; NOS I, nitric oxide synthase 1; thal, thalassemia; TIA, transient ischemic attack; WBC, white blood cell.

had fewer complications. However, a subpopulation with SCD-SS had a milder phenotype. While total Hb, HbF and reticulocyte count may be informative predictors of disease-severity in SCD-SS, dense cells were more informative in SCD-SC. The lower incidence of SCD-Sβthalassemia genotypes makes clinical predictors difficult to identify in this subpopulation.[75]

Mortality in SCD was determined in pediatric and adult cohorts. The first report included data on 2824 patients enrolled prior to 20 years of age, 640 of whom were enrolled within the first six months of life.[80] The highest death rate occurred between 1 and 3 years of age, 1.66 per 100 pt–years and most had SCD-SS. The probability of surviving to age 20 years was 85% and 95% for children with SCD-SS and SCD-SC respectively.[80] Deaths in the newborn cohort were from infection, acute splenic sequestration, and stroke, in order of frequency[52]; only total Hb and HbF were significantly associated with death rates.[50] The second cohort used to determine survival included 3764 children and adult SCD patients enrolled from 1978 to 1988.[81] The median age at death was 42 years for males and 48 years for females with SCD-SS, and 60 and 68 years for male and females with SCD-SC respectively.[98] Eighteen percent of deaths occurred in patients with chronic organ damage. Risk factors for early death were SCD-SS genotype, low HbF, elevated WBC count, ACS, seizures, and renal insufficiency. There was improved survival in patients with HbF levels above 8.6%.[81]

Having foresight, DNA samples were banked during active recruitment of study cohorts and have been used recently by a few investigators for SNP analysis. The genetic markers indicative of HbF producing capacity or risk for stroke, priapism or AVN have begun to emerge. The CSSCD predictors have been link to the genomic data obtained by these investigators in a first attempt to provide a framework for developing prediction models of disease-severity (Table 3). This powerful scenario allows a retrospective re-evaluation of the original data, so that invaluable information can be gleaned to improve treatment strategies, prognosis and quality of life on the backs of those who labored for many years to complete the CSSCD natural history study.

Newborn Cohort

In an effort to clarify predictors of disease-severity in children, infants less than six months were followed prospectively for SCD-related complications. Death, two or more painful episodes per year, one or more ACS episodes per year, and/or stroke were indicators of severe disease.[82] Included in this study were 380 SCD-SS and 12 SCD-Sβ^0thalassemia patients; 111 patients served as a validation cohort. The risk factors for severe disease were dactylitis in the first year of life and Hb <7.0 gm/dl and elevated WBC counts in the second year of life. HbF, α-thalassemia status, β-globin haplotypes, splenic sequestration, and reticulocyte counts were not correlative (see Chapter 5).

Treatment Strategies

Red cell transfusion (RCT) and HU therapy are interventions used to prevent and treat SCD-related complications while hematopoietic stem cell transplantation (HSCT) remains the only cure. A detailed discussion of these therapeutic interventions for patients with SCD can be found in the NHLBI guidelines for "The Management of Sickle Cell Disease"[83] and Chapter 17 in this textbook.

Hydroxyurea

The Multicenter Study of Hydroxyurea in Sickle Cell Anemia (MSH) demonstrated that HU decreased the rate of painful and ACS episodes and the need for RBC transfusions.[84] Improved survival was observed for patients with SCD receiving HU therapy.[85] Additional studies will be needed to determine whether this drug will be beneficial in other SCD genotypes. Recent clinical trials involving patients with SCD-SC do not show a global benefit.[86] Indications for HU therapy include recurrent painful episodes and ACS in patients with SCD-SS. The role of HU, in stroke prevention for those patients unable to continue chronic RBC transfusion therapy is under investigation. Several groups have reported the effects of HU therapy in children.[87–90] Overall, it has a similar risk to benefit profile in children as was demonstrated for adults showing a decreased rate of painful and ACS episodes. Linear growth and weight are not impaired in children receiving HU therapy.[87] Studies are on-going in the United States to determine whether this agent can prevent chronic organ damage in children; early studies are promising.[90]

Transfusions Therapy

RBC transfusions can be used intermittently or to prevent SCD-related complications, while chronic transfusions are used for prophylaxis. The selection of RBC products, method of transfusion (simple or exchange), antigen matching, and the optimal HbS and total Hb levels desired must be determined. Monitoring for transfusion-related iron overload is recommended however current indications for transfusion therapy are not based on controlled clinical trials.[91] Transfusions are not indicated for uncomplicated painful episodes. Accepted indications for acute transfusions include exacerbation of anemia, hypoxia, and to decrease HbS levels in the management of stroke and ACS. Preoperative simple transfusions are recommended with the goal of increasing total Hb to 10 gm/dl for patients with SCD-SS or SCD-Sβ^0thalassemia having a surgery requiring general anesthesia.[92] Transfusion therapy is discussed further in Chapter 8.

Alloimmunization and Delayed Hemolytic Transfusion Reactions

RBC antibodies and delayed hemolytic transfusion reactions develop in SCD patients. This diagnosis should be suspected when patients develop worsening anemia and pain 7–10 days after transfusion. This diagnosis is often missed since the same signs and symptoms are common in a painful episode. The CSSCD reported RBC alloimmunization at a prevalence rate of 18.6%.[93] The development of alloantibodies was associated with age over 10 years, SCD-SS genotype, a high number of units transfused, and female gender. Most alloantibodies were to the Rh, Kell and Lewis RBC membrane antigens.

Health Maintenance

Medical Home

It is essential that every child with SCD receive care that is coordinated through an appropriate medical home.[94] For many patients, the most appropriate medical home is a multidisciplinary sickle cell clinic that coordinates all aspects of medial care. The comprehensive care team consists of the hematologist, nurse and nurse practitioner, social worker and mental health professional with referral to other subspecialties as needed. In other cases, the medical home may be provided by a knowledgeable pediatrician or family medicine doctor with periodic referrals to the hematologist for comprehensive evaluations and management of severe, life-threatening complications. The location of the medical home and the extent to which the care is provided by the pediatrician versus the hematologist will depend on access to a multidisciplinary team, the frequency and severity of disease manifestations and the expertise of the primary care physician.

Health Maintenance

In addition to rendering routine well-child care[95] the following SCD-related treatments are recommended. All infants with SCD-SS and SCD-Sβ^0-thalassemia should receive penicillin prophylaxis, 125 mg orally, twice a day, by 2 months of age[96] and the dose is increased to 250 mg orally, twice a day at 3 years of age and continued until the fifth birthday.[97] Erythromycin may be used as an alternative for children with a penicillin allergy. The routine use of penicillin prophylaxis for infants and children with SCD-SC and SCD-Sβ^+-thalassemia[98] and folic acid supplementation[99] is controversial. Childhood immunizations are given on the normal schedule.[100] Children with SCD should receive the 7-valent pneumococcal conjugate and 23-valent pneumococcal polysaccharide vaccines[101] and yearly influenza immunizations.[102] The quadrivalent meningococcal polysaccharide vaccine is also recommended.[102]

All patients should have regularly scheduled comprehensive medical evaluations to review disease manifestations, document physical findings and laboratory values, monitor growth and development, and to develop a care plan.[103] School performance should be monitored for evidence of neuro-psychological problems.

Family and Patient Education

Identification of an infant with SCD through neonatal screening provides an opportunity to educate parents about the child's disorder before symptoms develop.[82,103] Initially, the focus should include the genetics and basic pathophysiology of SCD and the importance of regularly scheduled health maintenance visits, penicillin prophylaxis, and immunizations. Education about the need for urgent medical evaluation and treatment of febrile illness, acute splenic sequestration, aplastic crisis, and ACS is critical. Recognition and appropriate management of pain should be reviewed. As the child grows older, other topics such as stroke, enuresis, priapism, cholelithiasis, delayed puberty, retinopathy, AVN, and leg ulcers are introduced. During middle childhood and adolescence, education about the genetic basis of SCD and issues related to education, smoking, contraception, and pregnancy is directed toward the patient.

Transition to Adult Care

Transition from pediatric to adult care is a complex, long-term process involving three important components: assessment, preparation and support, and transfer.[82,104] This process should occur over several years and is best accomplished with the involvement of all members of the comprehensive care teams from pediatrics and adult medicine (see Chapter 6). Education about the transition process should begin around 13 years of age and continue until the patient is ready for transition.[82] Chronologic age should not be the sole indication for transition but other variables such as coping skills, neuro-psychological deficits, education, job readiness and recent severe illness. Most important is the young adult's knowledge of SCD complications which will reinforce the importance of adherence to adult health maintenance programs. Adolescents should be counseled about assuming responsibility for their own medical care. Parents should be encouraged to allow their child to make medical appointments, to increase his/her knowledge of their past medical history and to shift responsibility to the child for compliance with therapeutic regimens.

Adolescents and young adults recently involved in a transition of care can be excellent role models for patients going through this process. Teen groups at pediatric facilities and adult support groups should be utilized when possible. The medical team should recommend hospital alternatives if feasible, that have the facilities and programs required to provide continued medical care. The team must address special needs due to complications of SCD. Adult medical specialists including internal medicine and cardiology among others should be identified for health maintenance and care of complications. Health insurance needs are also considered. When this information is given to the patient and family then an institution that best meets their needs can be selected. Under optimal conditions a visit with the adult team at the pediatric facility before transfer of care is desirable.[82] This also encourages communication between the pediatric and adult teams regarding medical issues that may not be included in the medical records. Patients should be at steady-state prior to transfer to ensure a successful transition to the adult facility. An appointment with the adult team and failure to keep this appointment should be documented at the pediatric facility. An organized, thoughtful transition that ensures patient readiness and the comprehensive transfer of medical information in a supportive environment will increase adherence with adult medical care plans.

Sickle Cell Trait

Sickle cell trait (AS) is routinely identified by newborn screening. Compound heterozygous SCD can be misdiagnosed as sickle cell trait, when unusual globin variants are inherited. Eight percent of African-Americans are carriers of the β^S mutation which translates into about three million people. Genetic counseling is recommended. People with sickle cell trait have a normal peripheral blood smear, Hb, and reticulocyte counts without evidence of hemolysis. Under unusual circumstances serious complications such as gross hematuria, splenic infarction at high altitudes, and sudden death[105] can result from polymerization of deoxy-HbS. The validity of exercise-related death associated with sickle trait aroused heated controversy.[106] This complication was first observed in the Armed Forces during basic training.[107] Twelve cases of exercise-related deaths among healthy young men with sickle trait were reported by 1981 predominantly due to exertional rhabdomyolysis. Recruits with sickle cell trait had a 21-fold higher relative risk of exercise-related death.[108]

Maximal urinary concentrating ability is improved in individuals with sickle cell trait and one or two α-globin gene deletion.[109] Most carriers develop microscopic infarction of the renal medulla because of the extreme hypoxemia, hypertonicity, and acidosis which promote RBC sickling.[110] Studies worldwide established a significant increase in the risk for urinary tract infections[110] and renal medullary carcinoma in African-Americans with sickle trait.[111]

Future Perspectives

SCD comprises a group of Hb disorders in which HbS predominates, causing chronic hemolysis, vascular occlusion, and organ damage. Our understanding of the pathophysiology of SCD, initially thought to be due to vascular occlusion alone, has changed as clinical phenotypes, genetic risk factors for severe disease, and the contribution of non-sickling related mechanisms are defined. Factors such as cellular hydration and adhesion have an important role in modulating the outcome of tissue ischemia secondary to vaso-occlusion. Recently, the relationship between small and large vessel occlusion, and chronic disease-related complications, has emerged. Tissue or organ ischemia may be modulated by reperfusion mechanisms, local tissue resistance to ischemia, and tissue inflammation. Therefore, studies defining clinical or genetic markers that are organ-specific may be the most informative.

The recent emergence of the role of chronic hemolysis in the pathophysiology of SCD in adults provides a rationale for drug development based on decreased bioavailability of NO. This may lead to impaired vasodilatation and the development of pulmonary hypertension, priapism, and stroke. Co-morbid states must be reconsidered as genetic modifiers and taken into account when evaluating responses to therapeutic interventions in SCD. Our ability to link current SCD research with functional genomics will benefit efforts to establish prediction model for disease-severity that integrate the effects of co-morbid states. Finally personalized medicine based on genetic polymorphisms that alter drug metabolism and response to therapeutic agents will impact quality of life and long-term survival. The importance of involving disciplines from multiple subspecialty areas in addition to hematology cannot be overemphasized.

References

1. Herrick JB (1910). Peculiar elongated and sickle-shaped red blood corpuscles in a case of severe anemia. *Arch Int Med* **6**:517–521.
2. Konotey-Ahulu FID. Effect of environment on sickle cell disease in West Africa: epidemiologic and clinical considerations. In: *Sickle Cell Disease, Diagnosis, Management, Education and Research*, Abramson H, Bertles JF, Wethers DL. editors. CV Mosby Co, St. Louis. 1973. 20.
3. Pauling L, Itano H, Singer SJ, Wells IC (1949). Sickle cell anemia: a molecular disease. *Science* **110**:543.
4. Ingram VM (1957). Gene mutations in human haemoglobins: the chemical difference between normal and sickle cell haemoglobin. *Nature* **180**:326.
5. Nagel RL, Fabry ME, Pagnier J, Zohoun I, Wajcman H, Baudin V, Labie D (1985). Hematologically and genetically distinct forms of sickle cell anemia in Africa. *N Engl J Med* **312**:880.
6. Pagnier J, Mears JG, Dunda-Belkhodja O, Schaefer-Rego KE, Beldjord C, Nagel RL, Labie D (1984). Evidence for the multicentric origin of the sickle cell hemoglobin gene in Africa. *Proc Natl Acad Sci USA* **81**:1771–1773.
7. Nagel RL, Fabry ME (1985). The many pathophysiologies of sickle cell anemia. *Am J Hematol* **20**:195.
8. Kurnit DM (1979). Evolution of sickle variant gene. *Lancet* **1**:104.
9. Carlson J, Nash GB, Gabutti V, al-Yaman F, Wahlgren M (1994). Natural protection against severe *Plasmodium falciparum* malaria due to impaired rosette formation. *Blood* **84**:3909.
10. Gelpi AP, King MC (1976). Duffy blood group and malaria. *Science* **191**:1284.
11. Friedman MJ (1994). Erythrocytic mechanism of sickle cell resistance to malaria. *Proc Natl Acad Sci USA* **75**:1994.
12. Fleming HA (1989). Causes of 'natural' sudden death. *BMJ* **298**:120.
13. Modiano D, Luoni G, Sirima BS, Lanfrancotti A, Petrarca V, Cruciani F, Simpore J, Ciminelli BM, Foglietta E, Grisanti P, Bianco I, Modiano G, Coluzzi M (2001). The lower susceptibility to Plasmodium falciparum malaria of Fulani of Burkina Faso (west Africa) is associated with low frequencies of classic malaria-resistance genes. *Trans R Soc Trop Med Hyg* **95**:149–152.
14. Luzzi GA, Merry AH, Newbold CI, Marsh K, Pasvol G (1991). Protection by alpha-thalassaemia against *Plasmodium falciparum* malaria: modified surface antigen expression rather than impaired growth or cytoadherence. *Immunol Let* **30**:233–240.
15. Roth EF Jr, Raventos-Suarez C, Rinaldi A, Nagel RL (1983). Glucose-6-phosphate dehydrogenase deficiency inhibits *in vitro* growth of *Plasmodium falciparum. Proc Natl Acad Sci USA* **80**:298–299.
16. Stamatoyannopoulos G, Grosveld F (2001). Hemoglobin switching. In: Stamatoyannopoulos G, Majerus PW, Perlmutter RM, Varmus H, editors. *The Molecular Basis of Blood Disease*. Vol. 3. Philadelphia: Saunders.
17. Hardison RC, Chui DH, Giardine B, Riemer C, Patrinos GP, Anagnou N, Miller W, Wajcman H (2002). HbVar: a relational database of human hemoglobin variants and thalassemia mutations at the globin gene server. *Hum Mutat* **19**:225–233.
18. Flint J, Harding RM, Boyce AJ, Clegg JB (1998). The population genetics of the haemoglobinopathies. *Baillieres Clin Haematol* **11**:1–51.
19. Bonds DR (2005). Three decades of innovation in the management of sickle cell disease: the road to understanding the sickle cell disease clinical phenotype. *Blood Rev* **19**:99–110.
20. FDA Prevalence SCD.
21. Aluoch JR (1995). The presence of sickle cells in the peripheral blood film. Specificity and sensitivity of diagnosis of homozygous sickle cell disease in Kenya. *Trop Geogr Med* **47**:89–91.
22. Sickle Cell Disease Guideline Panel. *Sickle Cell Disease: Screening, Diagnosis, Management, and Counseling in Newborns and Infants*. Rockville, MD: Agency for Health Care Policy and Research, US Department of Health and Human Services, 1993, AHCPR Publ. No. 93–0562.
23. Sickle Cell Disease Guideline Panel. *Sickle Cell Disease: Screening, Diagnosis, Management and Counseling in Newborns and Infants*. Clinical Practice Guidelines no. 6, Rockville, MD: Agency for Health Care Policy and Research US Public Health Service, 1993, AHCPR Publ. No. 93–0562.
24. Nagel RL, Ranney HM (1990). Genetic epidemiology of structural mutations of the beta-globin gene. *Semin Hematol* **27**:342–359.

25. Nagel RL, Bookchin RM (1973). Simplified automated solubility test for the detection of hemoglobin S applicable to mass screening. *Am J Clin Pathol* **59**:300–305.
26. Old JM, Heath C, Fitches A, Thein SL, Jeffreys AJ, Petrou M, Modell B, Weatherall DJ (1986). Meiotic recombination between two polymorphic restriction sites within the beta globin gene cluster. *J Med Gene* **23**:14–18.
27. Cheung MC, Goldberg JD, Kan YW (1996). Prenatal diagnosis of sickle cell anaemia and thalassaemia by analysis of fetal cells in maternal blood. *Nat Genet* **14**:264–268.
28. Chiu RW, Lau TK, Leung TN, Chow KC, Chui DH, Lo YM (2002). Prenatal exclusion of beta thalassaemia major by examination of maternal plasma. *Lancet* **360**:998–1000.
29. Kan YW, Dozy AM (1978). Polymorphism of DNA sequence adjacent to human beta-globin structural gene: relationship to sickle mutation. *Proc Natl Acad Sci USA* **75**:5631–5635.
30. Saiki RK, Scharf S, Faloona F, Mullis KB, Horn GT, Erlich HA, Arnheim N (1985). Enzymatic amplification of beta-globin genomic sequences and restriction site analysis for diagnosis of sickle cell anemia. *Science* **230**:1350–1354.
31. Embury SH, Scharf SJ, Saiki RK, Gholson MA, Golbus M, Arnheim N, Erlich HA (1987). Rapid prenatal diagnosis of sickle cell anemia by a new method of DNA analysis. *N Engl J Med* **316**:656–561.
32. Jassim N, Al-Arrayed S, Gerard N, Al-Mukharraq H, Al-Ajami A, Ramasawmy R, Krishnamoorthy R (1999). A mismatched-primer polymerase chain reaction-restriction fragment length polymorphism strategy for rapid screening of the polyadenylation signal mutation alpha(T-Saudi) (AATAAA → AATAAG) in the alpha2-globin gene. *Hemoglobin* **23**:213–220.
33. Sanger F, Nicklen S, Coulson AR (1977). DNA sequencing with chain-terminating inhibitors. *Proc Natl Acad Sci USA* **74**:5463–5467.
34. Newton CR, Graham A, Heptinstall LE, Powell SJ, Summers C, Kalsheker N, Smith JC, Markham AF (18989). Analysis of any point mutation in DNA. The amplification refractory mutation system (ARMS). *Nucleic Acids Res* **17**:2503–2516.
35. Old JM, Varawalla NY, Weatherall DJ (1990). Rapid detection and prenatal diagnosis of beta-thalassaemia: studies in Indian and Cypriot populations in the UK. *Lancet* **336**:834–837.
36. Ye S, Dhillon S, Ke X, Collins AR, Day IN (2001). An efficient procedure for genotyping single nucleotide polymorphisms. *Nucleic Acids Res* **29**:E88–8.
37. Faa V, Rosatelli MC, Sardu R, Meloni A, Toffoli C, Cao A (1992). A simple electrophoretic procedure for fetal diagnosis of beta-thalassaemia due to short deletions. *Prenat Diagn* **12**:903–908.
38. Chong SS, Boehm CD, Cutting GR, Higgs DR (2000). Simplified multiplex-PCR diagnosis of common southeast asian deletional determinants of alpha-thalassemia. *Clin Chem* **46**:1692–1695.
39. Losekoot M, Fodde R, Harteveld CL, van Heeren H, Giordano PC, Bernini LF (1990). Denaturing gradient gel electrophoresis and direct sequencing of PCR amplified genomic DNA: a rapid and reliable diagnostic approach to beta thalassaemia. *Br J Haematol* **76**:269–274.
40. Braun A, Roth R, McGinniss MJ (2003). Technology challenges in screening single gene disorders. *Eur J Pediatr* **162** (Suppl 1):S13–6.
41. Daniel YA, Turner C, Haynes RM, Hunt BJ, Dalton RN (2005). Rapid and specific detection of clinically significant haemoglobinopathies using electrospray mass spectrometry–mass spectrometry. *Br J Haematol* **130**:635–643.
42. Masood E (1999). As consortium plans free SNP map of human genome. *Nature* **398**:545–546.
43. Syvanen AC (2005). Toward genome-wide SNP genotyping. *Nat Genet* **37**(Suppl):S5–10.
44. Tonisson N, Kurg A, Kaasik K, Lohmussaar E, Metspalu A (2000). Unravelling genetic data by arrayed primer extension. *Clin Chem Lab Med* **38**:165–170.
45. Warrington JA, Shah NA, Chen X, Janis M, Liu C, Kondapalli S, Reyes V, Savage MP, Zhang Z, Watts R, DeGuzman M, Berno A, Snyder J, Baid J (2002). New developments in high-throughput resequencing and variation detection using high density microarrays. *Hum Mutat* **19**:402–409.
46. Jain KK (2005). Applications of AmpliChiptrade mark CYP450. *Mol Diagn* **9**:119–227.
47. Machado RF, Martyr S, Kato GJ, Barst RJ, Anthi A, Robinson MR, Hunter L, Coles W, Nichols J, Hunter C, Sachdev V, Castro O, Gladwin MT (2005). Sildenafil therapy in patients with sickle cell disease and pulmonary hypertension. *Br J Haematol* **130**:445–453.

48. Gaston M, Rosse WF (1982). The cooperative study of sickle cell disease: review of study design and objectives. *Am J Pediatr Hematol Oncol* **4**:197–201.
49. Gaston M, Smith J, Gallagher D, *et al.* (1987). Recruitment in the Cooperative Study of Sickle Cell Disease (CSSCD). *Control Clin Trials* **8**:131S-140S.
50. Platt OS, Thorington BD, Brambilla DJ, *et al.* (1991). Pain in sickle cell disease. Rates and risk factors. *N Engl J Med* **325**:11–16.
51. Gill FM, Sleeper LA, Weiner SJ, *et al.* (1995). Clinical events in the first decade in a cohort of infants with sickle cell disease. Cooperative Study of Sickle Cell Disease. *Blood* **86**:776–783.
52. Nolan VG, Wyszynski DF, Farrer LA, Steinberg MH (2005). Hemolysis-associated priapism in sickle cell disease. *Blood* **106**:3264–3267.
53. Smith-Whitley K, Zhao H, Hodinka RL, *et al.* (2004). Epidemiology of human parvovirus B19 in children with sickle cell disease. *Blood* **103**:422–427.
54. Castro O, Brambilla DJ, Thorington B, *et al.* (1994). The acute chest syndrome in sickle cell disease: incidence and risk factors. The Cooperative Study of Sickle Cell Disease. *Blood* **84**:643–9.
55. Vichinsky EP, Styles LA, Colangelo LH, Wright EC, Castro O, Nickerson B (1997). Acute chest syndrome in sickle cell disease: clinical presentation and course. Cooperative Study of Sickle Cell Disease. *Blood* **89**:1787–1792.
56. Vichinsky EP, Neumayr LD, Earles AN, *et al.* (2000). Causes and outcomes of the acute chest syndrome in sickle cell disease. National Acute Chest Syndrome Study Group. *N Engl J Med* **342**:1855–1865.
57. Rucknagel DL (2001). The role of rib infarcts in the acute chest syndrome of sickle cell diseases. *Pediatr Pathol Mol Med* **20**:137–154.
58. Sullivan KJ, Goodwin SR, Evangelist J, Moore RD, Mehta P (1999). Nitric oxide successfully used to treat acute chest syndrome of sickle cell disease in a young adolescent. *Crit Care Med* **27**:2563–2568.
59. Sharan K, Surrey S, Ballas S, Borowski M, Devoto M, Wang KF, Sandler E, Keller M (2004). Association of T-786C eNOS gene polymorphism with increased susceptibility to acute chest syndrome in females with sickle cell disease. *Br J Haematol* **124**:240–243.
60. Ohene-Frempong K, Weiner SJ, Sleeper LA, *et al.* (1998). Cerebrovascular accidents in sickle cell disease: rates and risk factors. *Blood* **91**:288–294.
61. Miller ST, Macklin EA, Pegelow CH, *et al.* (2001). Silent infarction as a risk factor for overt stroke in children with sickle cell anemia: a report from the Cooperative Study of Sickle Cell Disease. *J Pediatr* **139**:385–390.
62. Adams R, McKie V, Nichols F, *et al.* (1992). The use of transcranial ultrasonography to predict stroke in sickle cell disease. *N Engl J Med* **326**:605–610.
63. Adams RJ, McKie VC, Hsu L, *et al.* (1998). Prevention of a first stroke by transfusions in children with sickle cell anemia and abnormal results on transcranial Doppler ultrasonography. *N Engl J Med* **339**:5–11.
64. Hsu LL, Miller ST, Wright E, *et al.* (2003). Alpha Thalassemia is associated with decreased risk of abnormal transcranial Doppler ultrasonography in children with sickle cell anemia. *J Pediatr Hematol Oncol* **25**:622–628.
65. Driscoll MC, Hurlet A, Styles L, *et al.* (2003). Stroke risk in siblings with sickle cell anemia. *Blood* **101**:2401–2404.
66. Kwiatkowski JL, Hunter JV, Smith-Whitley K, Katz ML, Shults J, Ohene-Frempong K (2003). Transcranial Doppler ultrasonography in siblings with sickle cell disease. *Br J Haematol* **121**:932–937.
67. Hoppe C, Klitz W, Noble J, Vigil L, Vichinsky E, Styles L (2003). Distinct HLA associations by stroke subtype in children with sickle cell anemia. *Blood* **101**: 2865–2869.
68. Hoppe C, Klitz W, Cheng S, *et al.* (2004). Gene interactions and stroke risk in children with sickle cell anemia. *Blood* **103**:2391–2396.
69. Sebastiani P, Ramoni MF, Nolan V, Baldwin CT, Steinberg MH (2005). Genetic dissection and prognostic modeling of overt stroke in sickle cell anemia. *Nat Genet* **37**:435–440.
70. Wang W, Enos L, Gallagher D, Thompson R, Guarini L, Vichinsky E, Wright E, Zimmerman R, Armstrong FD (2001). In: Cooperative Study of Sickle Cell Disease. Neuropsychologic performance in school-aged children with sickle cell disease: a report from the Cooperative Study of Sickle Cell Disease. *J Pediatr* **139**: 391–397.

71. Moser FG, Miller ST, Bello JA, *et al.* (1996). The spectrum of brain MR abnormalities in sickle-cell disease: a report from the Cooperative Study of Sickle Cell Disease. *AJNR Am J Neuroradiol* **17**:965–972.
72. Kinney TR, Sleeper LA, Wang WC, *et al.* (1999). Silent cerebral infarcts in sickle cell anemia: a risk factor analysis. The Cooperative Study of Sickle Cell Disease. *Pediatrics* **103**:640–645.
73. Nolan VG, Baldwin C, Ma Q, Wyszynski DF, Amirault Y, Farrell JJ, Bisbee A, Embury SH, Farrer LA, Steinberg MH (2005). Association of single nucleotide polymorphisms in klotho with priapism in sickle cell anaemia. *Br J Haematol* **128**(2):266–272.
74. Saad ST, Lajolo C, Gilli S, Marques Junior JF, Lima CS, Costa FF, Arruda VR (2004). Follow-up of sickle cell disease patients with priapism treated by hydroxyurea. *Am J Hematol* **77**:45–49.
75. Koshy M, Entsuah R, Koranda A, *et al.* (1989). Leg ulcers in patients with sickle cell disease. *Blood* **74**:1403–1408.
76. Covitz W, Espeland M, Gallagher D, Hellenbrand W, Leff S, Talner N (1995). The heart in sickle cell anemia. The Cooperative Study of Sickle Cell Disease (CSSCD). *Chest* **108**(5):1214–1219.
77. Platt OS, Rosenstock W, Espeland MA (1984). Influence of sickle hemoglobinopathies on growth and development. *N Engl J Med* **311**:7–12.
78. Wang WC, Morales KH, Scher CD, *et al.* (2005). Effect of long-term transfusion on growth in children with sickle cell anemia: results of the STOP trial. *J Pediatr* **147**:244–247.
79. Leonard MB, Zemel BS, Kawchak DA, Ohene-Frempong K, Stallings VA (1998). Plasma zinc status, growth, and maturation in children with sickle cell disease. *J Pediatr* **132**:467–471.
80. Leikin SL, Gallagher D, Kinney TR, Sloane D, Klug P, Rida W (1989). Mortality in children and adolescents with sickle cell disease. Cooperative Study of Sickle Cell Disease. *Pediatrics* **84**:500–508.
81. Platt OS, Brambilla DJ, Rosse WF, *et al.* (1994). Mortality in sickle cell disease. Life expectancy and risk factors for early death. *N Engl J Med* **330**:1639–1644.
82. Miller ST, Sleeper LA, Pegelow CH, *et al.* (2000). Prediction of adverse outcomes in children with sickle cell disease. *N Engl J Med* **342**:83–89.
83. National Heart Lung and Blood Institute. The Management of Sickle Cell Disease, 4th edition. National Institutes of Health, 2004.
84. Charache S, Barton FB, Moore RD, *et al.* (1996). Hydroxyurea and sickle cell anemia. Clinical utility of a myelosuppressive "switching" agent. The Multicenter Study of Hydroxyurea in Sickle Cell Anemia. *Medicine (Baltimore)* **75**:300–326.
85. Steinberg MH, Barton F, Castro O, *et al.* (2003). Effect of hydroxyurea on mortality and morbidity in adult sickle cell anemia: risks and benefits up to 9 years of treatment. *JAMA* **289**:1645–1651.
86. Miller MK, Zimmerman SA, Schultz WH, Ware RE (2001). Hydroxyurea therapy for pediatric patients with hemoglobin SC disease. *J Pediatr Hematol Oncol* **23**:306–308.
87. Wang WC, Helms RW, Lynn HS, Redding-Lallinger R, Gee BE, Ohene-Frempong K, Smith Whitley K, Waclawiw MA, Vichinsky EP, Styles LA, Ware RE, Kinney TR (2002). Effect of hydroxyurea on growth in children with sickle cell anemia: results of the HUG-KIDS Study. *J Pediatr* **140**:225–229.
88. Zimmerman SA, Schultz WH, Davis JS, Pickens CV, Mortier NA, Howard TA (2004). Sustained long-term hematologic efficacy of hydroxyurea at maximum tolerated dose in children with sickle cell disease. *Blood* **103**:2039–2045.
89. Fung EB, Barden EM, Kawchak DA, Zemel BS, Ohene-Frempong K, Stallings VA (2001). Effect of hydroxyurea therapy on resting energy expenditure in children with sickle cell disease. *J Pediatr Hematol Oncol* **23**:604–608.
90. Hankins JS, Ware RE, Rogers ZR, *et al.* (2005). Long-term hydroxyurea therapy for infants with sickle cell anemia — the Husoft extension study. *Blood* **106**:2269–2275.
91. Ohene-Frempong K (2001). Indications for red cell transfusion in sickle cell disease. *Semin Hematol* **38**:5–13.
92. Vichinsky EP, Haberkern CM, Neumayr L, *et al.* (1995). A comparison of conservative and aggressive transfusion regimens in the perioperative management of sickle cell disease. The Preoperative Transfusion in Sickle Cell Disease Study Group. *N Engl J Med* **333**(4):206–213.
93. Rosse WF, Gallagher D, Kinney TR, *et al.* (1990). Transfusion and alloimmunization in sickle cell disease. The Cooperative Study of Sickle Cell Disease. *Blood* **76**(7):1431–1437.

94. Newborn Screening Task Force (2000). Serving, the family from birth to the medical home. Newborn screening: a blueprint for the future — a call for a national agenda on state newborn screening programs. *Pediatrics* **106**:383–422.
95. American Academy of Pediatrics, Committee on Practice and Ambulatory Medicine. Recommendations for preventive pediatric health care (1995). *Pediatrics* **96**: 373–374.
96. Gaston MH, Verter JI, Woods G, *et al.* (1986). Prophylaxis with oral penicillin in children with sickle cell anemia: a randomized trial. *N Engl J Med* **314**:1593–1599.
97. Sickle Cell Disease Guideline Panel. *Sickle Cell Disease: Screening, Diagnosis, Management, and Counseling in Newborns and Infants*. Rockville, MD: Agency for Health Care Policy and Research, US Department of Health and Human Services; 1993; AHCPR Publ. No. 93–0562.
98. Rogers ZR, Buchanan GR (1995). Bacteremia in children with sickle hemoglobin C disease and sickle beta$^+$-thalassemia: is prophylactic penicillin necessary? *J Pediatr* **127**:348–354.
99. Rodriquez-Cortes HM, Griener JC, Hyland K, *et al.* (1999). Plasma homocysteine levels and folate status in children with sickle cell anemia. *J Pediatr Hematol Oncol* **21**:219–223.
100. American Academy of Pediatrics, Committee on Infectious Diseases (2002). Recommended childhood immunization schedule — United States, 2002. *Pediatrics* **109**:162–164.
101. American Academy of Pediatrics, Committee on Infectious Diseases (2000). Recommendations for the prevention of pneumococcal infections, including the use of pneumococcal conjugate vaccine (Prevnar), pneumococcal polysaccharide vaccine, and antibiotic prophylaxis. *Pediatrics* **106**:362–366.
102. American Academy of Pediatrics (2000). Red Book: Report of the Committee on Infectious Diseases. Pickering LK editor, 25th ed. Elk Grove Village, IL: American Academy of Pediatrics.
103. Pass KA, Lane PA, Fernhoff PM, *et al.* (2000). US newborn screening system guidelines II: follow-up of children, diagnosis, management, and evaluation. *J Pediatr* **37**(suppl):S1–S46.
104. Telfair J, Alexander LR, Loosier PS, Alleman-Velez PL, Simmons J (2004). Providers' perspectives and beliefs regarding transition to adult care for adolescents with sickle cell disease. *J Health Care Poor Underserved* **15**(3):443–461.
105. Sears DA (1978). The morbidity of sickle cell trait: a review of the literature. *Am J Med* **64**:1021–1036.
106. Sears DA. Sickle Cell trait. In: Embury SH, Hebbel RP, Mohandas N, Steinberg MH, editors. *Sickle Cell Disease: Basic Principles and Clinical Practice*. New York, Raven Press, 1994: 381–394.
107. Jones SR, Binder RA, Donowho EM Jr (1970). Sudden death in sickle cell trait. *N Engl J Med* **282**:323–325.
108. Virmani R, Burke AP, Farb A, Kark JA (1997). Causes of sudden death in young and middle aged competitive athletes. *Cardiol Clin* **15**:439–466.
109. Gupta AK, Kirchner KA, Nicholson R, Adams III, JG, Schechter AN, Noguchi CT, Steinberg MH (1991). Effects of α-thalassemia and sickle polymerization tendency on the urine-concentrating defect of individuals with sickle cell trait. *J Clin Invest* **88**:1963–1968.
110. Statius van Eps LW, de Jong PE. Sickle Cell Disease. In: Schrier RW, Gottschalk CW, editors. Disease of the Kidney, 6th edition, Vol. 1, 1997: 2201–2219.
111. Davis CJ Jr, Mostofi FK, Sesterhenn IA (1995). Renal medullary carcinoma. The seventh sickle cell nephropathy. *Am J Surg Pathol* **19**:1–11.

5

Preventive Care and Advances in the Treatment of Sickle Cell Disease

by Charles T. Quinn and George R. Buchanan

Introduction

Sickle cell disease (SCD) was once considered a malady of childhood because so few affected individuals lived to become adults. Overwhelming bacterial sepsis and severe anemia from acute splenic sequestration and aplastic crisis were the prime causes of childhood mortality. As little as 35 years ago, less than half of children with SCD were expected to reach adulthood.[1] Childhood mortality has since markedly decreased because of several specific improvements in medical management, including early diagnosis by newborn screening (NBS), the use of prophylactic penicillin (PCN) to prevent fatal pneumococcal sepsis, parental education about SCD, and a model of comprehensive medical care.[2–5] More than 85% of children with SCD now survive past 18 years of age,[4] and most can expect to live several decades more.[6]

In 1994, the National Institutes of Health (NIH)–sponsored Cooperative Study of SCD (CSSCD) reported a median survival of 42 years for men and 48 years for women with sickle cell anemia (SCD-SS).[6] These data are aging, and more recent survival estimates suggest a median survival into the sixth decade for both sexes.[7] Therapeutic advancements, such as hydroxyurea,[8] stem cell transplantation,[9] and chronic transfusion programs[10] continue to improve survival, although their overall impact is not completely known.[4] This chapter reviews many of these key advances in the treatment of individuals with SCD and summarizes the planned and ongoing clinical trials aimed at decreasing morbidity and mortality and improving the quality of life of those who live with this disease.

Bacterial Infection

During the mid-1960s Robinson and co-workers noted a markedly increased risk of *Streptococcus pneumoniae* infections in infants and children with sickle cell anemia.[11] This report was followed in 1969 by the important observation of Pearson *et al.* that very young infants with sickle cell anemia had impaired splenic function, even when the spleen was markedly enlarged.[12] So called "functional asplenia", which could occur as early as 3–4 months of age, resulted from reticuloendothelial blockage and shunting of blood within the spleen due to sickling in the hypoxic and sluggish flow environment of the red pulp. Functional asplenia was followed by splenic infarction and involution.[13] Age-related defects in humoral immunity against *S. pneumoniae* and other encapsulated pathogens as well as a possible defect in the alternate complement pathway contributed to the 100- to 500-fold increased risk of fatal infection in sickle cell patients.[14]

By the mid-1970s, it was widely recognized that febrile illnesses in sickle cell patients of all ages could rapidly evolve into fatal septicemia and meningitis, especially in young infants whose risk of *S. pneumoniae* bacteremia was particularly high. Therefore, parents of affected infants (ideally diagnosed by NBS) were educated regarding the need for prompt medical attention at the onset of fever so that lifesaving antibiotics could be administered and the child could be closely observed in the hospital. This strategy was shown to be highly effective in reducing deaths due to septicemia.[15,16]

The introduction of the 23-valent polysaccharide pneumococcal vaccine (PPV-23) in the late 1970s was heralded as a major advance and thus recommended for all patients with SCD.[17] A problem, however, was its limited effectiveness in young children in view of its rather poor T-cell dependent antibody responses. Moreover, local reactions and uncertainty regarding the timing of re-vaccination were problematic. Prophylactic antibiotics eventually proved to be a more effective strategy, given the extreme sensitivity (at least until a decade ago) of *S. pneumoniae* to penicillin. Therefore, a few centers began to employ regimens of prophylactic penicillin.[18,19]

The prophylactic penicillin-1 (PROPS-I) study conducted by the NIH in the mid-1980s signaled antibiotic prophylaxis as the standard of care.[20] This placebo-controlled trial that enrolled 225 patients confirmed an 84% reduction in *S. pneumoniae* sepsis and meningitis in penicillin-treated infants. The follow-up study, PROPS-II, which aimed to determine the safety of discontinuing prophylaxis at five years of age, could not provide a definitive answer because it was in retrospect underpowered.[21] However, most centers currently recommend stopping prophylaxis at age 5 years (when a second dose of pneumococcal vaccine is administered) unless the child has previously suffered from pneumococcal sepsis or meningitis.

During the past 10 to 15 years, resistance of as many as 30–40% of *S. pneumoniae* isolates to penicillin may have reduced its effectiveness as prophylaxis. Yet, this emerging problem has been countered, in part, by the approval in the spring of 2000 of the heptavalent protein-conjugate pneumococcal vaccine (PCV-7; Prevnar®, Wyeth Pharmaceuticals). Like its counterpart, the *Haemophilus influenzae* type b conjugate vaccine introduced in the late 1980s, which virtually eradicated invasive infection due to *H. influenzae* type b, it induces a brisk T-cell dependent antibody response.[22] Thus, it provides protection in early infancy and also reduces nasopharyngeal colonization and carriage of the bacteria. Therefore, universal immunization of children with SCD with both conjugate vaccines, continued use of the 23-valent polysaccharide vaccine, and prophylactic penicillin during the first five years of life should minimize the risk of fatal infection. However, penicillin and vaccine failures are still encountered, so prompt medical attention at times of febrile illnesses remain mandatory irrespective of a patient's age. No firm recommendations can be made regarding the use of meningococcal vaccine, given the rarity of invasive infection due to this pathogen in persons with SCD and the suboptimal antibody responses to polysaccharide vaccine preparations.

It is unclear whether all of the practices outlined above should also apply to children and adults with sickle-hemoglobin C disease and sickle-β^+-thalassemia. In these conditions, splenic function is minimally, if at all, impaired during the first five years of life, so twice daily prophylactic penicillin is probably not necessary.[23,24] However, vaccination against encapsulated microorganisms is prudent.

Many older children and adults with hemoglobin SC disease exhibit diminished splenic function; therefore, febrile illnesses must be treated aggressively in these patients, just as they should in persons of all ages with sickle cell anemia and sickle-β^0-thalassemia.

The old strategy of hospitalizing all children with SCD and fever for empiric antibiotic therapy has recently been replaced by outpatient management of selected patients.[25,26] This change in practice was fostered by the introduction of ceftriaxone, an extremely long-acting parenteral antibacterial effective against *S. pneumoniae* and *H. influenzae* type b, and the realization that most febrile children with SCD who indeed had a serious invasive infection were ill on presentation, exhibiting features such as high fever, toxic appearance, leukocyte counts over 30,000/mm^3, and a hemoglobin concentration below the steady-state.[27] Patients who have such high-risk features should be hospitalized, and those who do not can usually be managed safely as outpatients. Outpatient management obviously requires parents who are judged to be adherent to follow-up recommendations by the medical team.

Newborn Screening (NBS)

The ideal way to prevent fatal bacterial infection would be to identify children with SCD at birth to implement preventive measures, such as prophylactic penicillin, before splenic function begins to wane. Prior to the 1970s, NBS for SCD was believed to be neither feasible nor necessary. However, improved laboratory techniques, especially citrate agar gel hemoglobin electrophoresis, made possible the implementation of a pilot screening program at Yale, a statewide program in New York, and a United Kingdom Medical Research Council-sponsored national program in Jamaica.[28,29] Each was fueled by the realization that early death due to *S. pneumoniae* infection and splenic sequestration might be prevented by pre-symptomatic diagnosis, which would allow for both parental education and intervention. By the 1980s, many statewide NBS programs had developed, although structured follow-up of identified infants was often lacking.

In 1987, a landmark Consensus Development Conference held at the NIH soundly ratified the necessity and numerous merits of universal NBS and implementation of comprehensive education and follow-up programs.[30] Within a decade, nearly every state had endorsed universal NBS, and the few states that remain not fully committed have been targeted by the American Academy of Pediatrics for the necessary policy changes.

In the United States and many western European countries, NBS by classic hemoglobin electrophoretic methods has been augmented or replaced by some combination of high pressure liquid chromatography, isoelectric focusing, and DNA-based techniques. Although specific notification mechanisms and follow-up strategies differ by state and country, parents of most babies who are diagnosed with SCD by NBS are now made aware of their child's condition in the first several months of life. This allows for early initiation of prophylactic penicillin, administration of life-saving immunizations, and parental instruction regarding fever and splenic palpation, all of which needs to occur before the physiologic decline in fetal hemoglobin and onset of symptomatic hemolytic anemia or vaso-occlusive complications.

Although no randomized studies have quantified the merits of NBS, a number of well conducted prospective and retrospective analyses have attested to the benefits of screening on mortality during the first several years of life.[3,4] Unfortunately, these advances have had little impact on worldwide mortality, because financial and logistical barriers make screening difficult or impossible in developing countries.

Routine Health Maintenance

NBS programs for SCD cannot be successful without mechanisms for follow-up of patients and delivery of quality pediatric care in a so-called "medical home".[5] This concept refers to a model of primary care that focuses on preventive healthcare, immunizations, and maintenance of good nutrition in a convenient, family-centered office or clinic. An important component of the "medical home" is ready access to physicians who are highly skilled and knowledgeable about SCD, typically pediatric hematologists and a multi-disciplinary team based in a children's hospital or a medical school. One reason for the improved health of children with SCD during the past decade has been the availability of excellent general pediatric care by primary physicians who are far more knowledgeable about SCD than in years past.

Children who have SCD require extra attention beyond the "well child" care needed by all pediatric patients. A recently published statement by the American Academy of Pediatrics (AAP) Sections on Hematology–Oncology and Genetics outlines in specific detail the recommendations for health maintenance for children of various ages.[5] Within the first several months of life, penicillin prophylaxis should begin, along with implementation of the standard immunization schedule, including the conjugate pneumococcal vaccine (PCV-7). In fact, until age 2 years, when the pneumococcal polysaccharide vaccine (PPV-23) is administered as well, recommended immunization strategies in children with SCD are identical to those of hematologically normal infants. Initial physician visits should also emphasize instructions to the parents about spleen palpation, understanding the genetics of the disease, and a review of potential complications occurring in the initial months of life, such as dactylitis and splenic sequestration. Referral to a pediatric hematology center experienced in SCD is highly recommended at this time. The frequency of subsequent follow-up visits to the hematologist depends upon the child's age and clinical severity. Recommendations are provided in the excellent AAP review.[5] An additional reason for patients with SCD of all ages to be followed by an experienced pediatric hematologist is ready access to relevant clinical trials. With the current emphasis on prevention of organ damage and other disease-related complications rather than treating them after they occur, initiation of agents such as hydroxyurea may be possible even within the first 1–2 years of life.[31]

There is disagreement and uncertainty about what monitoring or screening tests are necessary as part of routine health maintenance. In many centers, including ours, hemoglobin concentration, reticulocyte count, and oxygen saturation of hemoglobin measured by pulse oximetry are the only tests performed at regular clinic visits. Others, however, take a much more comprehensive approach to monitoring by routinely ordering fetal hemoglobin concentrations, liver and renal function tests, echocardiograms, pulmonary function testing, and neuroimaging and neuropsychologic testing. In many centers transcranial Doppler measurements are also performed on children with SCD-SS disease between 2 and 16 years of age in order to identify patients at high risk of primary stroke, in accordance with the guidelines recommended by the STOP Study.[32] Further research is necessary to characterize which tests should become standard for all patients versus those that should be restricted to the research setting.

Acute Vaso-Occlusive Complications

The cardinal feature of SCD is the sudden and often alarming array of symptoms related to vascular occlusion and resultant tissue injury. The vaso-occlusive "crisis" or painful episode, the most common of these events, contributes substantial morbidity and predicts early mortality as well. The exact pathophysiology remains uncertain. Previously these events were felt to simply represent a "log jam" of deoxygenated sickled erythrocytes in the microcirculation. During the past 20 years, we have become aware that leukocytes, adhesive proteins, vaso-active peptides, reduced nitric oxide bioavailability, and endothelial dysfunction all contribute to the vaso-occlusive process.[33]

Painful Episodes

Triggers such as stress, dehydration, cold exposure, and swimming can precipitate a painful event, but in most cases the cause is unknown. Thus, preventive strategies are difficult to implement. Little progress has been made during the past several decades in the treatment of a painful episode with hydration and analgesics.[34,35] Adjunctive measures, including supplemental oxygen, corticosteroids, vasodilators, and agents aimed at improving blood rheology have all been explored, but their benefits are either lacking or unclear. Psychosocial support, behavioral modification, and reassurance and encouragement remain important therapeutic measures. Far more research is necessary to better understand and treat this terrible complication, which is the primary reason for the substantially reduced quality of life apparent in many if not most patients with SCD.

Acute Chest Syndrome

Acute chest syndrome, first described by Charache in 1979,[36] is defined as an acute lower respiratory tract illness (with accompanying signs and/or symptoms) associated with a new pulmonary infiltrate. Its causes are diverse, including pneumonia due to viruses, mycoplasma, chlamydia, and bacteria), fat embolism (originating from infarcted bone marrow), *in situ* intrapulmonary sickling, pulmonary edema, and atelectasis.[37] It is now believed to be the most common cause of death in persons with SCD. Rapid progression from a seemingly minor respiratory illness to respiratory failure, requiring intubation and artificial ventilation, may occur within hours. Acute management is supportive, including antibiotics, intravenous fluids, oxygen, and red blood cell transfusions (simple or exchange). A very short course of corticosteroids was beneficial in one randomized clinical trial.[38] This and other novel treatment approaches remain investigational. Recurrent chest syndrome may be associated with asthma and can lead to chronic pulmonary disease.

Acute "Hematologic" Crises

A sudden decline in hemoglobin concentration below the patient's usual steady-state value results from two distinct and well characterized events: acute splenic sequestration crisis (ASSC) and aplastic crisis. The former, the pathophysiology which is poorly understood, is characterized by sudden enlargement of the spleen and pooling or sequestration of much of the patient's blood volume into the splenic red pulp. The sudden decline in the patient's hemoglobin concentration (sometimes to values as low as 2–3 g/dl) may result in hypovolemic shock if it occurs suddenly or congestive heart failure when it evolves over several days. The diagnosis is made easily by demonstrating marked splenomegaly and confirming the presence of reticulocytosis, nucleated red blood cells, and thrombocytopenia on the blood count. Management consists of blood transfusions and careful monitoring for recurrence. Splenectomy is recommended after one life threatening event or multiple episodes requiring transfusion. ASSC is rare in persons with hemoglobin SS after age 5 years but is more common in adolescents and adults with hemoglobin SC disease.[39] The aplastic crisis presents with fever, malaise, syncope, and other manifestations of severe anemia. Here the spleen is generally not larger than usual, and the reticulocyte count is less than 1–2%. Over 20 years ago parvovirus B19 was implicated as its cause. Management consists of contact isolation and slow packed red blood cell transfusion. The patient promptly develops neutralizing antibody which is protective against recurrent events. However, the viral infection is highly contagious, so siblings and other close contacts with SCD are susceptible. Development of a vaccine has been unsuccessful thus far but is currently a high research priority.

Hemoglobin declines below the steady-state value are also commonly encountered in acute chest syndrome and in some patients with seemingly uncomplicated vaso-occlusive pain episodes.

Priapism

One of the most troublesome complications of SCD is priapism, an unwanted painful erection of the penis. Limited data suggest that a majority of boys and men with SCD experience this complication at least once.[40] A comprehensive study of the epidemiology of priapism will soon be underway. Priapism occurs in two forms, so called "stuttering" events lasting just a few hours and "prolonged" episodes which may last for many days. Recurrent stuttering events may precede prolonged episodes, and the latter are associated with impotence. In addition to fluids and analgesics, aspiration and irrigation of the corpus cavernosa is of value for events lasting longer than 2–3 hours but less effective if implemented thereafter,[41] and adrenergic agents such as pseudoephedrine or etilefrine may be effective in preventing or treating stuttering episodes. A multi-center study is planned to explore this latter strategy.

Stroke

Stroke, whether overt or clinically silent (not associated with neurological signs and symptoms), affect 30–40% of children and adolescents with SCD and an unknown number of adults. Much acute and chronic disability results from this complication, which results generally from stenosis or occlusion of large intracranial vessels involving the circle of Willis. Blood transfusions are effective in both primary and secondary stroke prevention.[32,42] Much interest has focused recently on the pathophysiology, risk factors, early identification, treatment, and rehabilitation of affected patients. Further discussion of cerebrovascular complications of SCD is the subject of Chapter 8.

Specific Pharmacotherapy of Sickle Cell Disease

Just two decades ago, no specific therapies were of proven value for children or adults with SCD. Clinicians relied on analgesics, fluid administration, antibiotics, and the periodic use of transfusions to deal with acute complications. Increasingly specific interventions became available in the 1980s, heralded by encouraging small case series and reports and of stem cell transplantation, hydroxyurea, and programs of chronic transfusions. Each of these approaches has evolved rapidly since 1990 through the conduct of multi-center clinical trials and their translation into clinical practice. Investigative groups in the United States and Europe are now exploring in large-scale studies the use of non-myeloablative stem cell transplantation, hydroxyurea in young infants to prevent organ damage, chronic transfusion programs, and the use of erythrocytapheresis for prevention of stroke and other complications.

Although use of hydroxyurea in SCD will soon enter its third decade, its exact mechanisms of action are still incompletely understood or unknown.[43] Induction of fetal hemoglobin, which was initially thought to be the primary reason for the beneficial effects of hydroxyurea, is accompanied by reduced leukocyte cell counts, diminished adhesion of sickle erythrocytes to endothelial cells, and enhancement of nitric oxide production. The degree to which these mechanisms contribute to the therapeutic benefit of hydroxyurea is uncertain. Other agents that increase fetal hemoglobin concentration have also been tested. These include arginine butyrate and, more recently, decitabine.[43,44] The latter, administered by intravenous infusion or subcutaneous injection, seems promising in initial human studies.[44]

Decreased nitric oxide bioavailability may contribute to sickle cell-related complications, and this represents an area of current, intense interest.[45,46] These observations have led to proposed and recently investigated novel treatment strategies to enhance local nitric oxide bioavailability (e.g., using direct nitric oxide inhalation or arginine supplementation). Another class of novel agents reduces the cellular dehydration characteristic of sickle hemoglobin-containing erythrocytes. Pharmacologic blockade of the Gardos channel or the potassium-chloride co-transporter, both of which modulate cellular hydration, is an exciting area of current drug exploration in phase I and phase II studies.[47] Another strategy that has been pursued, although without great success thus far, is the use of agents such as poloxamer 188 and similar compounds aimed at improving blood rheology.[48] In addition to these specific therapeutic interventions, improvements in supportive care are also of intense interest and importance, especially the ongoing development of oral iron chelators, which could render chronic red cell transfusion safer and so increase its use and acceptance.

Clinical Trials

The National Sickle Cell Anemia Control Act passed by Congress in 1972 led to the establishment of ten comprehensive sickle cell centers to conduct research in addition to screening, counseling, and patient care. However, no collaborative interaction between the 10 funded centers was required or even promoted. Instead, NIH-sponsored clinical research involving SCD during the subsequent three decades was conducted under the aegis of the CSSCD or organ-specific or complication-oriented trials conducted by different centers without unity or a common goal. Although important discoveries resulted from these individual efforts (e.g., the National Chest Syndrome Study, Multi-Center Hydroxyurea Study, Prophylactic Penicillin Studies, STOP Study, etc.),[8,20,21,32,37] the concept of a multi-center clinical research network charged with conducting pilot studies and large clinical trials was only recently realized in the United States. Two exciting new NHLBI initiatives now offer much promise to the SCD community. First, the current (2003–2008) Comprehensive Sickle Cell Center grant cycle included a requirement that the 10 designated centers function as a Clinical Trials Consortium to create a national database of affected patients and to design and conduct novel phase 1 and phase 2 studies.[49] Also approved for funding is the separate Clinical Research Network that will include 8 clinical centers charged with conducting large-scale phase 3 trials. It is further anticipated that both of these new multi-center research consortia will interact meaningfully with community-based centers funded by the Talent Bill, the Maternal and Child Health Bureau, or both. These new research groups will provide the opportunity for continuity and collaboration to assure the successful design and conduct of clinical studies to improve the health and well being of individuals with SCD.

Summary

The 20th century delivered an abundance of knowledge about the natural history of SCD and tremendous scientific insight into its complex pathogenesis. However, it was actually the relatively simple and often inexpensive interventions, such as prophylactic penicillin, parental education, and universal newborn screening, that most reduced the morbidity and mortality of the disease. Interestingly, these advances did not stem directly from basic science research, but from the minds of astute and forward-thinking clinicians. Unfortunately, discoveries in the laboratory, while they have impressively out-stripped clinical therapeutic advances, are neither easily nor often translated successfully into clinical practice. Further, the present incarnation of unidirectional bench-to-bedside "translational" research has failed individuals who have SCD. This perilous disconnect between basic and clinical research needs to be solved. The time is ripe for a renaissance in SCD, but this renaissance will only be achieved by devoting the same time, money, and rigor to clinical investigation that is given to basic science research.

References

1. Scott RB (1970). Health care priority and sickle cell anemia. *JAMA* **214**:731–734.
2. Vichinsky EP (1991). Comprehensive care in sickle cell disease: its impact on morbidity and mortality. *Semin Hematol* **28**:220–226.
3. Vichinsky E, Hurst D, Earles A, Kleman K, Lubin B (1988). Newborn screening for sickle cell disease: effect on mortality. *Pediatrics* **81**:749–755.
4. Quinn CT, Rogers ZR, Buchanan GR (2004). Survival of children with sickle cell disease. *Blood* **103**:4023–4027.
5. — (2002). Health supervision for children with sickle cell disease. *Pediatrics* **109**:526–535.
6. Platt OS, Brambilla DJ, Rosse WF, Milner PF, Castro O, Steinberg MH, Klug PP (1994). Mortality in sickle cell disease: life expectancy and risk factors for early death. *N Engl J Med* **330**:1639–1644.
7. Wierenga KJ, Hambleton IR, Lewis NA (2001). Survival estimates for patients with homozygous sickle-cell disease in Jamaica: a clinic-based population study. *Lancet* **357**:680–683.
8. Charache S, Terrin ML, Moore RD, Dover GJ, Barton FB, Eckert SV, McMahon RP, Bonds DR (1995). Effect of hydroxyurea on the frequency of painful crises in sickle cell anemia. Investigators of the Multicenter Study of Hydroxyurea in Sickle Cell Anemia. *N Engl J Med* **332**:1317–1322.
9. Walters MC, Patience M, Leisenring W, Eckman JR, Scott JP, Mentzer WC, Davies SC, Ohene-Frempong K, Bernaudin F, Matthews DC, Storb R, Sullivan KM (1996). Bone marrow transplantation for sickle cell disease. *N Engl J Med* **335**:369–376.
10. Reed W, Vichinsky EP (2001). Transfusion therapy: a coming-of-age treatment for patients with sickle cell disease. *J Pediatr Hematol Oncol* **23**:197–202.
11. Robinson MG, Watson RJ (1966). Pneumococcal meningitis in sickle-cell anemia. *N Engl J Med* **274**:1006–1008.
12. Pearson HA, Spencer RP, Cornelius EA (1969). Functional asplenia in sickle-cell anemia. *N Engl J Med* **281**:923–926.
13. Pearson HA, Gallagher D, Chilcote R, Sullivan E, Wilimas J, Espeland M, Ritchey AK (1985). Developmental pattern of splenic dysfunction in sickle cell disorders. *Pediatrics* **76**:392–397.
14. Barrett-Connor E (1971). Bacterial infection and sickle cell anemia. An analysis of 250 infections in 166 patients and a review of the literature. *Medicine* (Baltimore) **50**:97–112.
15. McIntosh S, Rooks Y, Ritchey AK, Pearson HA (1980). Fever in young children with sickle cell disease. *J Pediatr* **96**:199–204.
16. Powars D, Overturf G, Weiss J, Lee S, Chan L (1981). Pneumococcal septicemia in children with sickle cell anemia. Changing trend of survival. *JAMA* **245**: 1839–1842.
17. Ammann AJ, Addiego J, Wara DW, Lubin B, Smith WB, Mentzer WC (1977). Polyvalent pneumococcal-polysaccharide immunization of patients with sickle-cell anemia and patients with splenectomy. *N Engl J Med* **297**:897–900.
18. Buchanan GR, Smith SJ (1986). Pneumococcal septicemia despite pneumococcal vaccine and prescription of penicillin prophylaxis in children with sickle cell anemia. *Am J Dis Child* **140**:428–432.

19. John AB, Ramlal A, Jackson H, Maude GH, Sharma AW, Serjeant GR (1984). Prevention of pneumococcal infection in children with homozygous sickle cell disease. *Br Med J (Clin Res Ed)* **288**:1567–1570.
20. Gaston MH, Verter JI, Woods G, Pegelow C, Kelleher J, Presbury G, Zarkowsky H, Vichinsky E, Iyer R, Lobel JS (1986). Prophylaxis with oral penicillin in children with sickle cell anemia. A randomized trial. *N Engl J Med* **314**:1593–1599.
21. Falletta JM, Woods GM, Verter JI, Buchanan GR, Pegelow CH, Iyer RV, Miller ST, Holbrook CT, Kinney TR, Vichinsky E (1995). Discontinuing penicillin prophylaxis in children with sickle cell anemia. Prophylactic Penicillin Study II. *J Pediatr* **127**:685–690.
22. Darkes MJ, Plosker GL (2002). Pneumococcal conjugate vaccine (Prevnar; PNCRM7): a review of its use in the prevention of *Streptococcus pneumoniae* infection. *Paediatr Drugs* **4**:609–630.
23. Lane PA, O'Connell JL, Lear JL, Rogers ZR, Woods GM, Hassell KL, Wethers DL, Luckey DW, Buchanan GR (1995). Functional asplenia in hemoglobin SC disease. *Blood* **85**:2238–2244.
24. Rogers ZR, Buchanan GR (1995). Bacteremia in children with sickle hemoglobin C disease and sickle beta(+)-thalassemia: is prophylactic penicillin necessary? *J Pediatr* **127**:348–354.
25. Rogers ZR, Morrison RA, Vedro DA, Buchanan GR (1990). Outpatient management of febrile illness in infants and young children with sickle cell anemia. *J Pediatr* **117**:736–739.
26. Wilimas JA, Flynn PM, Harris S, Day SW, Smith R, Chesney PJ, Rodman JH, Eguiguren JM, Fairclough DL, Wang WC (1993). A randomized study of outpatient treatment with ceftriaxone for selected febrile children with sickle cell disease. *N Engl J Med* **329**:472–476.
27. Platt OS (1997). The febrile child with sickle cell disease: a pediatrician's quandary. *J Pediatr* **130**:693–694.
28. Pearson HA, O'Brien RT, McIntosh S, Aspnes GT, Yang MM (1974). Routine screening of umbilical cord blood for sickle cell diseases. *JAMA* **227**:420–421.
29. Serjeant BE, Forbes M, Williams LL, Serjeant GR (1974). Screening cord bloods for detection of sickle cell disease in Jamaica. *Clin Chem* **20**:666–669.
30. — (1987) Consensus conference. Newborn screening for sickle cell disease and other hemoglobinopathies. *JAMA* **258**:1205–1209.
31. Wang WC, Wynn LW, Rogers ZR, Scott JP, Lane PA, Ware RE (2001). A two-year pilot trial of hydroxyurea in very young children with sickle-cell anemia. *J Pediatr* **139**:790–796.
32. Adams RJ, McKie VC, Hsu L, Files B, Vichinsky E, Pegelow C, Abboud M, Gallagher D, Kutlar A, Nichols FT, Bonds DR, Brambilla D (1998). Prevention of a first stroke by transfusions in children with sickle cell anemia and abnormal results on transcranial Doppler ultrasonography. *N Engl J Med* **339**:5–11.
33. Stuart MJ, Nagel RL (2004). Sickle-cell disease. *Lancet* **364**:1343–1360.
34. Guideline for the Management of Acute and Chronic Pain in Sickle-Cell Disease (1999). Glenview, IL: American Pain Society.
35. Ballas SK (1998). *Sickle Cell Pain*. Seattle: IASP Press.
36. Charache S, Scott JC, Charache P (1979). "Acute chest syndrome" in adults with sickle cell anemia. Microbiology, treatment, and prevention. *Arch Intern Med* **139**:67–69.
37. Vichinsky EP, Neumayr LD, Earles AN, Williams R, Lennette ET, Dean D, Nickerson B, Orringer E, McKie V, Bellevue R, Daeschner C, Manci EA (2000). Causes and outcomes of the acute chest syndrome in sickle cell disease. National Acute Chest Syndrome Study Group. *N Engl J Med* **342**:1855–1865.
38. Bernini JC, Rogers ZR, Sandler ES, Reisch JS, Quinn CT, Buchanan GR (1998). Beneficial effect of intravenous dexamethasone in children with mild to moderately severe acute chest syndrome complicating sickle cell disease. *Blood* **92**:3082–3089.
39. Aquino VM, Norvell JM, Buchanan GR (1997). Acute splenic complications in children with sickle cell-hemoglobin C disease. *J Pediatr* **130**:961–965.
40. Mantadakis E, Cavender JD, Rogers ZR, Ewalt DH, Buchanan GR (1999). Prevalence of priapism in children and adolescents with sickle cell anemia. *J Pediatr Hematol Oncol* **21**:518–522.
41. Mantadakis E, Ewalt DH, Cavender JD, Rogers ZR, Buchanan GR (2000). Outpatient penile aspiration and epinephrine irrigation for young patients with sickle cell anemia and prolonged priapism. *Blood* **95**: 78–82.
42. Scothorn DJ, Price C, Schwartz D, Terrill C, Buchanan GR, Shurney W, Sarniak I, Fallon R, Chu JY, Pegelow CH, Wang W, Casella JF, Resar LS, Berman B, Adamkiewicz T, Hsu LL, Ohene-Frempong K, Smith-Whitley K, Mahoney D, Scott JP, Woods GM, Watanabe M, Debaun MR (2002). Risk of recurrent

stroke in children with sickle cell disease receiving blood transfusion therapy for at least five years after initial stroke. *J Pediatr* **140**:348–354.

43. Steinberg MH (1999). Management of sickle cell disease. *N Engl J Med* **340**:1021–1030.
44. DeSimone J, Koshy M, Dorn L, Lavelle D, Bressler L, Molokie R, Talischy N (2002). Maintenance of elevated fetal hemoglobin levels by decitabine during dose interval treatment of sickle cell anemia. *Blood* **99**:3905–3908.
45. Morris CR, Kato GJ, Poljakovic M, Wang X, Blackwelder WC, Sachdev V, Hazen SL, Vichinsky EP, Morris SM Jr, Gladwin MT (2005). Dysregulated arginine metabolism, hemolysis-associated pulmonary hypertension, and mortality in sickle cell disease. *JAMA* **294**:81–90.
46. Rother RP, Bell L, Hillmen P, Gladwin MT (2005). The clinical sequelae of intravascular hemolysis and extracellular plasma hemoglobin: a novel mechanism of human disease. *JAMA* **293**:1653–1662.
47. Brugnara C, Gee B, Armsby CC, Kurth S, Sakamoto M, Rifai N, Alper SL, Platt OS (1996). Therapy with oral clotrimazole induces inhibition of the Gardos channel and reduction of erythrocyte dehydration in patients with sickle cell disease. *J Clin Invest* **97**:1227–1234.
48. Orringer EP, Casella JF, Ataga KI, Koshy M, Adams-Graves P, Luchtman-Jones L, Wun T, Watanabe M, Shafer F, Kutlar A, Abboud M, Steinberg M, Adler B, Swerdlow P, Terregino C, Saccente S, Files B, Ballas S, Brown R, Wojtowicz-Praga S, Grindel JM (2001). Purified poloxamer 188 for treatment of acute vaso-occlusive crisis of sickle cell disease: a randomized controlled trial. *JAMA* **286**:2099–2106.
49. Peterson CM, Harvath L (2004). Current and future commitment to hematology research. *Hematol Oncol Clin North Am* **18**:1215–1234, vii.

6

Sickle Cell Disease in Adults

by Johnson Haynes, Jr. and Ardie Pack-Mabien

Introduction

The prognosis in sickle cell disease (SCD) has dramatically improved over the past 30 years, with a median survival of 42 years for males and 48 years for females with homozygous SS disease (SCD-SS) and 60 and 68 years, respectively for heterozygous sickle-C disease (SCD-SC) with some patients living into the eighth decade.[1,2] Improved management of infections and stroke complications in childhood, active transition of healthcare maintenance for adults, and improved psychosocial support have contributed to a reduction in morbidity and mortality. The Cooperative Study of Sickle Cell Disease (CSSCD) and other observational studies have defined the prognosis and common complications that occur in adult sickle cell patients.

Improved survival provides unique challenges in healthcare. Risk of early death in adults is associated with acute and chronic complications such as obstructive/restrictive lung disease, pulmonary hypertension and proliferative retinopathy all of which increase in prevalence with age; renal glomerular disease may progress to renal failure and early mortality as well. Chronic pain due to leg ulcers and avascular necrosis (osteonecrosis) of the hips causes disability, requiring social and vocational interventions. Improved survival also provides opportunities to impact the quality of life for patients with SCD. Hydroxyurea has been established as the first pharmacologic intervention that reduces the frequency of painful episodes and acute chest syndrome in adults. Increased frequency and severity of painful episodes associated with menstruation, pregnancy, and menopause as well as geriatric challenges in this population requires investigation. The progress that has been made in defining complications and the standard of care for adults over the last two decades will be discussed.

Medical Complications

Long-term Survival

Neonatal screening, early initiation of penicillin therapy,[3] improved management of infections and childhood stroke, comprehensive care and parental education have all contributed to improved outcome. More than 85% of children with SCD now survive past 18 years of age[4] and most can expect to live for several decades. Health maintenance practices must also address interactions between SCD and other common health problems related to aging in the adult population such as hypertension, diabetes, atherosclerosis, arthritis, cancer, asthma, and other chronic illnesses. Finally geriatric patients with SCD are not well studied therefore there remains much work to do to improve the level of functioning for adults with SCD.

Vaso-Occlusive Episodes

The hallmark of SCD is the vaso-occlusive crisis which presents unique challenges for patients, families, and healthcare professionals. Pain — the most frequent symptom experienced by people with SCD — affects quality of life and function in work, school, play, and social relationships.[5] Many social barriers continue to impede assessment and management of SCD-related pain.[6] Most patients are African-Americans whereas healthcare professionals for the most part are Caucasian. Poor communication between patients and providers often contribute to inadequate treatment.[7] Health disparity and access to healthcare remains a real concern for most sickle cell patients, especially for adults where health insurance and State Medicaid benefits are very limited.[8] These issues required formal intervention to improve healthcare delivery to this population.

Acute Pain

The majority of painful episodes are managed at home.[1] Some patients manage severe pain with strong oral opioids,[9] however the risk for addiction to analgesic medications is overstated in this population[10] and the basic principles of pain assessment are universal.[11] Frequent reassessment is essential to adjust treatment regimens; opioid analgesics are the mainstay of treating acute episodes but it is only one part of an individualized treatment plan. Some experts prefer morphine, or hydromorphone (dilaudid) for frequent or prolonged episodes of pain.[12] Meperidine (Demerol) is not recommended for prolonged treatment of painful episodes because neurological symptoms may occur during use.[13] Meperidine is contraindicated in patients with renal insufficiency, seizures, or a history of having seizures while on meperidine.

Because of the potential side effects of opioids, safer treatments are needed for treating vaso-occlusive episodes in SCD. Newer rapid onset synthetic opioids (fentanyl or hydromorphone) are reasonable alternatives.[14,15] Non-steroidal anti-inflammatory drugs (NSAIDs) may be effective for mild to moderate pain. They include prescription-strength agents such as ketorolac (Toradol; 16) or diflunisal (Dolobid; 17) which are under investigation. Ketorolac may be most helpful for bone pain with a good response in 50% of patients (see Chapter 7). Renal function should be monitored closely when prescribing a NSAID in adults because of frequently unrecognized renal insufficiency. Other agents are being tested as well. Poloxamer 188 (Flocor, RheothRx) is an investigative synthetic surfactant compound. It coats damaged red blood cells (RBCs) and improves blood flow. Clinical trials have been completed to determine the safety and efficacy of poloxamer.[18] Another formulation RheothRx reduced total analgesic requirements, pain intensity and produced a trend toward a shorter duration of painful episodes. When poloxamer was combined with hydroxyurea therapy better pain control was achieved.[18,19]

Clinical trials are needed to test the efficacy of tramadol, a potent oral analgesic with low respiratory depression and addiction potential. Another agent, fructose-1,6-diphosphate (Cordox) reduces pain, inflammation, and protects cells against hypoxia.[20] For postoperative pain, epidural analgesia provides the best control when compared with intravenous patient-controlled analgesia.[21] A variety of compounds are available for treating pain, however individualized treatment plans during acute symptoms are essential to achieve an optimal response.

Chronic Pain

The use of chronic opioids in patients with daily pain is controversial. Many clinicians report improved comfort and function in patients who were formerly debilitated by pain.[22] Others are concerned about perpetuation of the pain syndrome, addiction, symptoms of withdrawal, or exacerbation of concomitant illnesses such as depression. It is recommended that patients who are prescribed daily opioids undergo evaluation to determine if psychosocial factors are contributing to the pain (see Chapter 7).[23]

Acute Chest Syndrome

Acute chest syndrome (ACS), defined as a new infiltrate on chest radiograph involving at least one bronchopulmonary segment, occurs in individuals with SCD-SS, SCD-SC, and SCD-Sβ-thalassemia genotypes. Central to the pathophysiology of ACS is microvascular occlusion due to polymerization of hemoglobin S (HbS). Tissue hypoxia promotes the deoxygenation of HbS and polymer formation which is accelerated by acidosis and hypoxemia.[24] In the CSSCD, 29% of patients with SCD experienced at least one episode of ACS[40]; SCD-SS carries the highest rate with an incidence of 12.83 per 100 patient–years for the entire cohort. The rate of ACS is inversely proportional to age and fetal hemoglobin concentration.[25] This complication may progress to acute respiratory distress syndrome. ACS is second only to painful episodes as the reason for hospitalization of individuals with SCD.[1] Despite recent advances in diagnosis and treatment, ACS remains the most common cause of death in adults.[26]

Etiology of Acute Chest Syndrome

The presenting symptoms in ACS vary significantly with age. Children in the 13–18 age group present with typical symptoms of pulmonary infection, fever and cough, while adults present with chest pain.[27] However, most commonly no symptoms are present at presentation. The primary causes of ACS are pulmonary infection, infarction, and fat embolism. The most definitive study to date on the etiology of ACS was conducted by the National Acute Chest Syndrome Study Group (NACSSG).[26] In this multi-center cohort, 671 episodes of ACS occurred in 538 patients of which, an infectious agent was isolated from about one-third of patients.[26] The most common organisms were *Chlamydia pneumoniae*, *Mycoplasma pneumoniae*, and respiratory syncitial virus. Pneumococcus and *H. influenzae* accounted for a minority of cases in adults. Parvovirus B19 infection has been increasingly recognized as a cause of ACS. This infection can be associated with aplastic crises in sickle cell patients who develop ACS associated with thrombocytopenia and bone marrow necrosis.[28]

The second major cause of ACS is pulmonary infarction due to vaso-occlusion and hypoxic injury to lung tissue. Infarctions are also associated with *in situ* thrombosis and bone marrow emboli.[29] Bone marrow infarction with subsequent fat emboli syndrome (FES) is recognized as a significant cause of ACS.[30–32] Vichinsky *et al.*[31] compared clinical and laboratory characteristics of children with ACS to steady-steady state values in SCD patients and non-SCD controls with other types of lung disease. Lipid laden macrophages obtained through bronchoalveolar lavage was found to support the diagnosis of FES. This finding was significantly increased during episodes of ACS. Another study by Maitre and associates[33] in adult sickle cell patients showed that, a 2 gm/dl drop in hemoglobin, thrombocytopenia and central nervous system symptoms were correlated with FES complicating

ACS; a preceding severe painful episode occurred in 78–100% of patients with FES.[31,33] In the NACSSG, FES caused 8.8% of ACS episodes.[26]

Bronchoscopy[31,33] and/or Swan–Ganz catheterization[34] are better diagnostic tools to demonstrate fat emboli in bronchoalveolar lavage fluid or capillary blood, respectively. These procedures are not utilized routinely because they lack sensitivity and specificity therefore the true incidence of fat emboli in ACS is unknown. A diagnosis of FES should be considered when a patient presents with a severe painful episode, worsening anemia, thrombocytopenia, altered mental status, and later develops ACS.[56,57]

Systemic fat embolism and acute multi-organ failure syndrome can also complicate bone marrow necrosis during a severe vaso-occlusive crisis. The clinical symptoms consist of fever, elevated lactate dehydrogenase, and leukoerythroblastosis.[35] Hiran *et al.*[36] reported 10 episodes of multiorgan failure in sickle cell patients involving the lungs, liver, or kidneys. Respiratory, hepatic, or kidney failure occurred in the study group. Multiorgan failure is a life threatening complication that can be manifested without evidence of chronic organ damage. The treatment is aggressive RBC transfusion therapy.

Pulmonary infarction due to HbS polymerization and vaso-occlusion is a diagnosis of exclusion. Infarction causes ACS in about 16% of cases[26] however some cases are temporally associated with opioids used in pain management or excessive intravenous fluids.[37] The role of hypercoagulability and thromboembolism in pulmonary infarction is not apparent. *In situ* thromboses of small to medium arteries have been described at autopsy in the presence and absence of infarction.[30,38] Pulmonary edema is also commonly observed at autopsy. Hydrostatic pulmonary edema can occur in the presence of underlying hypertension with renal failure and volume overload. In addition, massive hemolysis with severe anemia may result in a "high output" state leading to heart failure. Oxidant damage due to ischemia-reperfusion injury, HbS acting as a Fenton reagent, activated polymorphonuclear (PMN) cell oxidants, FES, and opioids often occurs simultaneously during episodes of ACS.

Treatment

Analgesics, fluids, antibiotics, incentive spirometry, and oxygen for hypoxemia, are the mainstay of treatment of ACS. Early transfusion therapy with sickle cell trait negative, leukocyte-depleted, extended antigen-matched packed RBCs is frequently required to stabilize progressive lung injury. Therapy with hydroxyurea is the treatment of choice for recurrent ACS. A phase I trial was conducted to determine the efficacy of poloxamer 188 in 43 patients with ACS[39] treated by continuous intravenous infusion. No evidence of renal toxicity or other limiting adverse events were observed. Preliminary data suggest that poloxamer 188 may shorten the length of hospitalization for ACS in a dose-dependent manner. The data and safety profile justify further studies with this agent.

Sickle–RBC Pulmonary Vascular Interactions

HbS polymerization does not occur immediately after deoxygenation therefore most sickle RBCs (SRBCs) pass through the capillary bed, making microvascular occlusion uncommon.[40,41] Other factors such as pre-capillary obstruction by dense or irreversible SRBCs and increased adhesion to vascular endothelium[42,43] contribute to vaso-occlusion. Reticulocyte adherence can be correlated with vaso-occlusion severity[44] however the relationship of dense SRBCs to severity of vaso-occlusion episode has not been proven.[45] The vascular endothelium receptors $\alpha_4\beta_1$ integrin, very late activation antigen-4 (VLA-4), and the thrombospondin receptor CD36[46,47] are expressed on stress reticulocytes; VLA-4 binds endothelial cell vascular cell adhesion molecule-1 (VCAM-1). Hypoxia induces VLA-4/VCAM-1 interactions to increase adhesion.[48] Additional studies are required to elucidate the factors involved in the initiation of vaso-occlusion (see Chapter 10).

Leukocytosis has been well documented in SCD related to a chronic inflammatory state.[49] During crises, PMNs isolated from sickle cell patients become more adherent to vascular endothelium.[50,51] Studies support a role for PMN activation in the initiation and propagation of vaso-occlusive in SCD.[52,53] Arachidonic acid (AA) and lyso-platelet activating factor (lyso-PAF) are released from membrane phospholipids of activated PMN by PLA_2.[54,55] Investigators have demonstrated that PMNs from sickle cell patients are activated compared to those from normal healthy controls.[51–53] Styles *et al.* demonstrated that secretory PLA_2 is markedly elevated during acute vaso-occlusion and that PLA_2 was predictive of impending ACS in SCD.[56]

PMNs secrete leukotriene B_4 (LTB_4)[57] and acetylation of the lyso-PAF by acetyltransferase results in the metabolically active product, PAF.[58] Both LTB_4 and PAF are inflammatory mediators and they influence vascular permeability, cell infiltration, and PMN adhesion.[59–62] Recently, Haynes and associates demonstrated in isolated rat lungs, that PMN activation was required to increase the retention of SRBCs in the microcirculation[63] and that PAF and LTB_4 could mimic this effect. PMNs treated with the 5-LO inhibitor, zileuton, prior to activation also attenuated PMN mediated SRBC retention. These findings provide evidence that the secretion of PAF and LTB_4 by activated PMNs increase SRBC retention in the pulmonary circulation of isolated-perfused rat lung. This inflammatory pathway is currently being investigated as a potential target for therapeutic intervention.

In the NACSSG, the mortality rate for ACS was 9% in individuals over 20 years old; 13% of patients require mechanical ventilation, and 11% had neurologic symptoms.[26] ACS is four times more deadly in adults than in children. Acute care consists of supplemental oxygen if hypoxemia is present. With refractory hypoxemia, mechanical ventilation with positive end expiratory pressure may be required. Bronchodilator therapy and antibiotics consisting of a third-generation cephalosporin plus a macrolide or monotherapy with a broad spectrum third- or fourth-generation fluoroquinolone should be initiated. Simple or exchange RBC transfusion should be considered early if progression of ACS symptoms occur.

Sickle Chronic Lung Disease

A prior ACS episode is a risk factor for development of sickle chronic lung disease (SCLD) and early mortality compared to patients without a history of ACS.[64] SCLD is characterized by restrictive and/or obstructive ventilatory impairments that can be progressive and lead to cor pulmonale, and pulmonary artery hypertension (PAH). The time from a diagnosis of stage 1 SCLD (early) to death was reported at 7.1 years versus death in 2.5 years from the time when stage 4 severe disease is diagnosed.[64]

Pulmonary Artery Hypertension

Recurrent vaso-occlusive episodes lead to end-organ damage in adult patients with SCD. The prevalence of PAH and risk factors associated with this complication is life-threatening.[65] Agata and associates tested 60 adult patients, average age of 37 years. The prevalence of PAH was 30% and low levels of fetal hemoglobin and lower systolic blood pressure were associated with this complication.

PAH was the most common echocardiographic abnormality identified in 58% of sickle cell patients in another study.[66] Older age and a prior history of ACS were significantly correlated with an increased prevalence of PAH.

Lactate Dehydrogenase

PAH is associated with a high mortality rate.[67] Intravascular hemolysis leads to impaired bioavailability of nitric oxide (NO) due to scavenging by plasma free hemoglobin. In addition, arginine the precursor of NO, is degraded by arginase released from hemolyzed RBCs. In a cohort of 213 patients evaluated by Gladwin and associates[68] elevated lactate dehydrogenase (LDH) was associated with low hemoglobin and haptoglobin levels and high reticulocytes, bilirubin, plasma hemoglobin, aspartate aminotransferase, arginase and soluble adhesion molecules. LDH elevation identified a subphenotype of SCD consisting of PAH, leg ulcers, priapism, and risk of early death. A syndrome consisting of elevated LDH, hemolysis-associated NO-resistance, endothelial dysfunction and end-organ vasculopathy has been proposed by Gladwin and colleagues.[68]

Endothelial Dysfunction

Endothelial cell adhesion molecules orchestrate the recruitment and binding of inflammatory cells to vascular endothelium. With vascular injury, the levels of endothelial bound and soluble adhesion molecules increase. This process is modulated by NO, a soluble gas that relaxes smooth muscles and dilates blood vessels. In patients with SCD, the levels of adhesion molecules are inversely associated with NO bioavailability.[100] High sVCAM-1 levels were associated with renal and hepatic impairment. Moreover, sVCAM-1, ICAM-1, and E-selectin were independently associated with mortality risk in patients with PAH. Steady state levels of soluble adhesion molecules may be used as markers of PAH.

NO is currently being tested as a treatment option in PAH secondary to SCD. As an inhaled gas it has been used in neonatal pulmonary hypertension and adult respiratory distress syndrome. NO forms a covalent link with free hemoglobin[69] and when inhaled at a concentration of 80 ppm, NO reduced the tendency for HbS to polymerize in sickle cell patients.[70]

Sildenafil

The safety and efficacy of treatment with the NO donor, sildenafil for 6 months was tested in 12 adult patients with SCD and PAH.[71] Sildenafil decreased the pulmonary artery systolic pressure 9 mmHg and no episodes of priapism occurred in men. A Phase II/III placebo-control trial to evaluate the safety and efficacy of sildenafil in SCD is planned in the future.

Arginine

SCD is characterized by a state of NO-resistance and limited bioavailability of l-arginine, an amino acid that is metabolized to NO. Gladwin and associates investigated the role of arginase in dysregulated arginine metabolism and its contribution to endothelial dysfunction.[72] Two hundred and twenty-eight patients with SCD aged 18 to 74 years were enrolled in a multicenter study. Plasma arginase activity was significantly elevated in patients with SCD with highest activity in those with PAH. These data support a mechanism of PAH in which reduced NO bioavailability occurs secondary to the release of erythrocyte arginase. A recent trial suggests that arginine might be an effective treatment for PAH.[73]

Stroke

The age-specific incidence of first stroke in SCD-SS is 2 to 5 years old.[74] A second peak incidence of 1.3% is seen after 50 years. In adults, hemorrhagic strokes occur more frequently than thrombotic strokes.[75] Subarachnoid hemorrhage that involves deep structures in the brain reflects the development of "moya-moya" syndrome years after an earlier thrombotic stroke. Angiography reveals the complex structure of the moya-moya lesion.[76] Although stroke has been studied widely in children, there remains a need for future investigations to define risk factors for intracranial hemorrhage in adult sickle cell patients.

Transfusions are proven to prevent a first stroke and recurrence in children.[77] Transcranial Doppler (TCD) is used to measure blood flow to the brain and is a sensitive method for identifying children at risk for stroke.[74] The role of TCD in assessing stroke risk in adults with SCD is undetermined. Sampaio and associates performed preliminary TCD studies in adults.[78] They observed lower flow velocities in adults than those found in children but higher than normal controls. Therefore age-specific TCD flow rates may also assist in the detection of stroke risk in adults.

Stroke Treatment

At present, there are no prospective-randomized studies that address the efficacy of therapeutic interventions for strokes in adult SCD. Transfusion therapy is recommended for clinical stroke based on studies performed in children. The length of treatment required before transfusions can be stopped safely without stroke reoccurrence is unclear (see Chapter 8). Ware and associated previously reported two young adults successfully treated with HU as prophylaxis for stroke recurrence.[79] Subsequently the same group evaluated the response of 16 children treated with HU after initial transfusions for stroke.[80] Transfusion therapy was discontinued for erythrocyte alloantibodies, recurrent stroke, iron overload, and non-compliance. Three patients (19%) had neurological events indicative of stroke recurrence about 3–4 months after discontinuing transfusions, before maximal HU effect. The findings suggest that some children with SCD and stroke may discontinue chronic transfusions and use HU therapy safely. Other investigators have demonstrated the efficacy of HU in preventing stroke recurrence in children[81] but similar studies are necessary for adults.

Avascular Necrosis

Avascular necrosis (AVN) of cortical bone of the acetabulum, the head of the femur, and the head of the humerus[82] is common in SCD. This complication occurs in patients with all forms of SCD most commonly in the second or third decades, however SCD-SC patients have a higher incidence. AVN is also a common complication of adult SCD in African populations in Cameroon.[83] Out of 84 adult patients a 41.5% experienced at least one episode of AVN with 50% of cases involving the hips and 40% in the lumbar spine; multiple sites of necrosis were observed in six patients.

Deterioration of the joint to a condition of bone-on-bone interface produces significant pain. Non-steroidal anti-inflammatory agents or corticosteroid injections are sometimes combined with decompression with limited success. The decision to perform joint replacement for progressive disease is difficult in such young individuals. Artificial joints are not well-tolerated[84] and up to one-third of sickle cell patients require a second surgery within four years. Increased vulnerability to infections of the orthopedic hardware can destroy the articular interface and produce a flail joint sometimes leaving patients confined to a wheelchair. Early detection of the degenerative process using MRI so that preventive measures can be instituted holds the hope of improved management of this debilitating complication of SCD. Recently Vichinsky and associates completed a multicenter study to establish the validity of the Children's Hospital of Oakland Hip Evaluation Scale (CHOHES) for evaluation of AVN in SCD.[85] Twenty-six patients diagnosed with AVN were given a CHORES score that proved to be a reliable assessment tool for clinical evaluation of AVN in SCD.

Renal Failure

The risk of renal dysfunction is substantial in adult patients with SCD.[86] The high osmolality in the renal medulla increases RBC propensity to sickling producing medullary ischemia and papillary necrosis.[8] Patients usually have low serum creatinine and BUN levels due to the high glomerular filtration rates along with a high rate of creatinine secretion in the distal tubule; BUN values of 7 and creatinine values of 0.5 are typical. A formal evaluation of glomerular filtration rate should be considered if the serum creatinine rises to ≥ 1.0. Microscopic hematuria is a common problem in SCD.[87] No intervention is required unless blood loss is massive. Potentially nephrotoxic drugs such as gentamicin should be avoided and NSAIDs should be used judiciously. Erythropoietin has been used successfully in patients with severe anemia due to renal failure at a typical dose of about 150 U/kg three times per week. Hyperkalemia with a non-anion gap metabolic acidosis is common. Use of a potassium-sparing diet, bicarbonate supplementation and a loop diuretic are often helpful in normalizing the potassium. In rare cases of protein-wasting nephropathy, specific intervention may be required; one report suggested that angiotensin converting enzyme inhibitors may slow the progression of this complication.[88] Limited studies have been performed to determine the efficacy of dialysis[89] or allografts[90] in adult sickle cell patients with chronic renal failure.

People with sickle cell trait sometimes develop massive hematuria[91] from the left kidney. Hydration and alkalization of the urine are commonly used interventions. Anecdotal reports of the use of desmopressin acetate are encouraging.[92] Episilon amino caproic acid (Amicar) has also been used to treat refractory gross hematuria[92]; iron replacement may be necessary. Once considered a benign condition, recently an association between persistence of hematuria and renal medullary carcinoma has been demonstrated in individuals with sickle cell trait[94] therefore an appropriate workup to rule out this diagnosis is warranted.

Retinopathy

Non-proliferative retinopathy is manifested by conjunctival vascular occlusions, iris atrophy, retinal hemorrhages, and retinal pigmentary changes. These findings are diagnosed on ophthalmoscopy and are due to local vaso-occlusive events that rarely have visual consequences. Proliferative retinopathy is a common and insidious problem that does not correlate with vaso-occlusive events. The initiation of proliferative disease is thought to be peripheral retinal arteriolar occlusions, ischemia, and angiogenesis.[95] Goldberg defined five stages of proliferative retinopathy.[96] Briefly: in stage I, peripheral arteriolar occlusion occurs; stage II, involves vascular remodeling and formation of arteriovenous anastomoses; during stage III, neovascularization occurs; in stage IV, vitreous hemorrhage is present; and stage V, the most severe, involves retinal detachment. Retinopathy is found most commonly between 15 and 30 years of age[97] and in people with SCD-SC, but is also common in SCD-SS and SCD-Sβ^0 thalassemia genotypes.[98] Yearly eye examinations are recommended starting in childhood; more frequently if retinopathy is present. Laser treatment is reserved for progressive proliferative retinopathy to prevent loss of vision.

Other Clinical Complications

Infections

Infection remains common in adults with SCD, particularly in the lungs, urinary tract and bones.[99] However, the organisms are different than those found in children. The incidence of pneumococcal infections decreases and those caused by *Chlamydia* and *Mycoplasma pneumoniae* are increased in ACS[26] (see above). In hospitalized patients gram-negative bacteria can cause pneumonias and urinary tract infections.[100] The pneumoccocal vaccine is recommended[101] in adults who have not been immunized against pneumoccocal infections.

Cardiovascular

Chronic hemolysis, anemia, and hypoxia increase the demand on the heart to pump oxygenated blood to body tissues. Eventually, cardiomegaly and an increased risk for myocardial infarctions and cardiac failure occur. Previous studies on cardiac function in patients with SCD demonstrated abnormalities of systolic and diastolic function including elevated left ventricular myocardial performance index.[102] Blood pressure levels in 459 patients with SCD-related crises were compared to African-Americans without SCD.[103] Women had significantly lower blood pressures than men and during uncomplicated vaso-occlusive painful episodes hypertension did not occur. Delclaux *et al.*[104] determined the cardiorespiratory factors associated with dyspnea in 49 patients with SCD-SS prospectively. Evaluations using the Borg 6-min walk test for dyspnea, pulmonary function studies, echocardiography, and biological evaluation were done. Dyspnea and exercise performance were not correlated with any echocardiographic or biological measure including anemia. Lung vascular disease contributed to dyspnea and exercise limitations in adult sickle cell patients.

Priapism

In general about 30–45% of adult men with SCD report having at least one episode of priapism,[105] which is defined as a prolonged and painful penile erection. One group of investigators reported a 90% actuarial probability of at least one episode of priapism will occur by age 21.[106] The condition results from impaired blood egress from the corpus spongiosum of the penis.[107] Priapism lasting more than four hours is a medical emergency.[108] If left untreated from several hours to days, partial or complete impotence will occur in 80% of adult men. Some groups have used exchange transfusions with mixed results.[109] Surgical procedures to shunt or redirect blood flow are sometimes performed but often fail and this procedure can produce impotence.[110] Non-acute cases of repetitive episodes of priapism (stuttering priapism) in severely affected men are sometimes treated with gonadotropin releasing hormone agonist, anti-androgens[111] or vasodilators.[112] Pseudoephedrine at a dose of 60 mg at bedtime may decrease recurrence. A recent study showed that hydroxyurea prevented stuttering priapism in 4 out of 5 patients treated.[113] More aggressive therapy with self-administered intracavernosal injections of metaraminol (Aramine, Merck), a long-acting vasoconstricting amine for recurrent priapism has been successful in recalcitrant cases.[114] Additional research is required to develop better treatment options for priapism in SCD.

Liver and Gallbladder

Enlargement of the liver occurs in over half of sickle-cell patients and acute liver damage is observed in 10% of hospitalized patients. Viral hepatitis secondary to RBC transfusion can occur, however the risk has decreased with improved screening procedures for donated blood.[115] By age 30, 70% of patients with SCD have gallstones. Recurrent or severe pain from gallstones indicates that removal may be required. Abdominal ultrasound is used to confirm a diagnosis of gallstones and minimally invasive procedures such as laparoscopic cholecystectomy reduce surgical complications.

Skin Ulcers

This complication is relatively infrequent in the United States however 10% of sickle-cell patients, usually older than 10 years are affected[116]; in Jamaica skin ulcers occur in 43% of patients with SCD.[117] The most common site of ulcers is over the lateral malleoli or ankles. The ulcerations often have no clear-cut antecedent trauma. With the breakdown in the protection provided by the integument, patients are susceptible to infections and other complications. Treatment of leg ulcers should be conservative.[118] Healing usually takes from several weeks to months and protection against trauma is very important. Skin grafts and transfusions have been helpful in some recalcitrant cases.

Genetic Counseling and Pregnancy

Information to make informed decisions about reproduction should be made available for couples at risk for having a child with SCD. This requires that healthcare providers and trained genetic counselors collect accurate genotyping data from both parents. Counseling should be done in a non-directive manner, providing relevant information for informed decisions so that couples are free to choose the path most comfortable for them. The failure to offer prenatal diagnosis is an abridgment of two fundamental rights: the right to know and the right to decide.[119]

Women with SCD who become pregnant are at higher risk for complications, but serious problems have dropped significantly over the past decades.[120,121] Painful episodes occur in 50% of women and 60% required transfusions. Sun *et al.*[122] reported on 127 deliveries in a cohort of women with SCD-SS and SCD-SC disease. Compared with deliveries among normal women, the SCD group was at increased risk for having infants with intrauterine growth restriction, antepartum hospital admission, and postpartum infection. These women should receive prenatal care at a high-risk obstetrical clinic and take folic acid in addition to multivitamins and iron. The benefits of transfusions to prevent vaso-occlusive crises during pregnancy are unclear[123] but they are given to symptomatic women. In general, the outcome for pregnancy is favorable although pregnancy in a woman with SCD remains high-risk and carries a mortality rate of one percent.[124]

Hemoglobin-SC Disease

Complication rates in individuals with SCD-SC disease are less-well studied. The clinical features of SCD-SC disease in 106 adults (mean age 50 years) followed over 33 years with an average follow-up of 6.8 years was published in 2001.[125] Common clinical features were painful episodes (65%), AVN of the hip (23%), proliferative retinopathy (34%), and splenic sequestration (19%). Obesity (19.8%), essential hypertension (20.7%), and type-2 diabetes mellitus (10.3%) were also common.

Hydroxyurea

The Multicenter Study of Hydroxyurea in Sickle Cell Anemia (MSH) was suspended in 1995 because SCD-SS patients on the hydroxyurea (HU) arm of the study had significantly fewer painful episodes than did controls.[49] This made HU the first and only drug proven to prevent vaso-occlusive episodes. Moreover, 50% fewer episodes of ACS occurred in the HU treated group. Hydroxyurea modifies the characteristics of RBCs by inducing fetal hemoglobin and it has anti-inflammatory effects.[126] Patients 18 years or older with recurrent painful episodes or ACS should be considered for HU therapy. Women are required to use an accepted form of contraception to prevent pregnancy while on treatment. Bimonthly blood counts are performed when patients are started on HU.

Limited clinical evidence exists to support HU as prophylaxis against stroke, priapism, or other complications of SCD. Future trials are needed to address the efficacy of HU in SCD-SC and SCD-Sβ^+thalassemia. The carcinogenic potential with long-term use of HU is unknown, however a study designed to monitor 300 subjects enrolled in the MSH study for long-term side effects is underway. A role for HU in the reduction in mortality in sickle cell patients has been established.

In follow-up studies of patients on HU, those who started treatment at an older age with higher serum BUN and creatinine levels had increased risk for death while on therapy. These patients may represent a subgroup of older patients with more severe disease; the findings argue for early institution of HU. An open-label, non-randomized, study to assess the influence of renal function on the pharmacokinetics of HU in adults with SCD[127] demonstrated that plasma levels increase and urinary recovery decreases as the degree of renal insufficiency worsens. It was recommended that initial dosing of HU be decreased to 7.5 mg/kg/day for sickle cell patients with a creatinine clearance <60 ml/min.

The patterns of HU use by community-based hematologist/oncologists for the treatment of SCD were investigated by Zumberg *et al.*[128] A self-administered survey was received from 70% of 184 eligible physicians. The most common reasons cited for prescribing HU were frequent painful episodes (76%), chronic pain with frequent narcotic use (58%), and ACS (43%). The indications for HU use were consistent with current recommendations.

Other Treatments

Additional therapeutic agents under investigation for the treatment of SCD include omega-3 fatty acid, and the Gardos channel inhibitor ICA-17043.[129] Anti-cellular adhesion therapy has considerable potential however it has yet to be translated into clinical practice. For pulmonary hypertension oral arginine,[73] l-carnitine,[130] and exchange blood transfusion therapy alone or in combination with other agents have limited effectiveness. Oral iron chelators recently have been shown to be effective for iron overload. Deferiprone is licensed in Europe and India,[131] and deferasirox (ICL670)[132] holds promise as well. The oral iron chelator, desferasirox, was recently approved by the Food and Drug Administration for the treatment of iron overload in SCD in the United States.

Preventive Health Maintenance

Adult Healthcare Maintenance

As recently as the 1960s, SCD was considered a disorder for pediatric care since few patients survived into adulthood. As a result, services offered have not kept pace with the needs of this population. In general, comprehensive care programs are not in place for adults with SCD, in contrast to programs available through pediatric departments. As a result, patients frequently continue to receive medical care in pediatric clinics well into adulthood and are reluctant to transition.[133] This issue is being addressed with the advent of adult comprehensive care programs. The teams are very similar to those established in pediatrics so that patients can be confident that medical management and ancillary services will continue at a level they have become accustomed to. The adult comprehensive care programs usually will consist of the following components.

Adult Comprehensive Clinic Components

- Adult multi-disciplinary care team
- Primary care for routine treatment of SCD
- Education focused on health maintenance
- Pain management
- Case management and social services
- Physical and occupational therapy
- Routine dental care

Adult care teams should consist of the physician, nurse/nurse practitioner, social worker, and psychologist. Initially, a number of visits every 1–2 weeks will facilitate the development of rapport and a comprehensive care plan. Routine medical evaluations are scheduled every 2–6 months based on the severity of complications. Blood counts are repeated at each visit, along with pulse oximetry. Routine chemistries and a urinalysis should be repeated annually. A baseline ferritin and hepatitis panel should be established for all patients.[13] Tetanus and hepatitis immunizations are kept up to date and influenza vaccines are administered annually. The recommended schedule for the 23 valent pneumoccocal vaccine in adults is a single revaccination if $\geq$5 years has elapsed since receipt of last dose.[101]

Women should be taught to practice breast self-examination and to have an annual mammogram. Males should be screened for prostatic specific antigen after age 50 and colon cancer based on family history. An ophthalmologic evaluation and echocardiogram are performed annually. Baseline pulmonary function tests and arterial blood gases with O_2 saturation (measured by co-oximetry) are monitored and repeated 8–12 weeks following an ACS episode.

Access to care continues to be a problem for rural populations. The Medical College of Georgia Sickle Cell Center has developed successful telemedicine clinics in rural areas of Georgia[134] to provide an innovative approach for delivering services to this high-risk population.

Unmet Needs

Quality of Life

The Pain in Sickle Cell Epidemiology Study (PiSCES) attempted to develop a biopsychosocial model of SCD pain responses and healthcare utilization in a large, multicenter adult cohort.[135] They concluded that SCD produced a negative impact on health-related quality of life compared to the general population and the scores for people with SCD were most similar to hemodialysis patients. Comparable findings have been reported by another group.[136] The relationship between pain, health-related quality of life (HRQOL), and sleep disturbances was recently studied in adolescents with chronic pain.[137] There was a strong correlation between pain, depression, HRQOL, and sleep. The data suggest that mood is strongly related to sleep and might share a common behavioral origin with chronic pain. Instruments to evaluate self-efficacy are currently available. The psychometric Sickle Cell Self-Efficacy Scale (SCSES) comprised of nine questions relating to the participants' perceptions of their ability to function daily and manage SCD was validated by Edwards and associates.[138] The SCSES showed good internal consistency and predictive validity.

Education and Employment

Individuals with SCD should be encouraged to complete their education and pursue vocations and professional careers. Jobs requiring strenuous physical exertion, exposure to high altitudes or extremes in temperature should be discouraged. Frequent sickle cell crises often interrupt the education process early on[139] and adults are often unemployed because of their illness. Specialized training and intervention can overcome these difficulties and allow each person to fulfill their desire to be productive members of society.

Social and Psychosocial Needs

Psychosocial services are a critical component of health maintenance[140] and are best accomplished if social workers and mental health professionals are integrated into the comprehensive care team. Social workers are invaluable in solving a myriad of social and family problems. Mental health workers can assist in managing psychiatric illnesses and teaching behavioral coping skills for pain management. Some patients function well in life despite frequent and intense painful episodes but those with impaired functioning often display dysfunction by adolescence. Signs of high-risk behavior include increased hospitalizations, school absences or failure, or increased use of analgesics. An adversarial relationship with healthcare professionals can occur as well as depression which is often not appreciated in adults with SCD.[141] Early intervention to improve coping skills has the potential to prevent this outcome.

Workshop on Adults with Sickle Cell Diseases: Meeting Unmet Needs

In 2002, the National Heart, Lung, and Blood Institute (NHLBI) convened a two-day "Workshop on Adults with Sickle Cell Disease: Meeting Unmet Needs" in collaboration with other federal agencies. The goal was to examine the spectrum of issues that affect adult sickle cell patients from the perspectives of consumers, physicians, and researchers.

Sickle Cell Adult Provider Network (SCAPN)

The SCAPN was established in 2002 by providers interested in caring for adult patients. The main goal of the network is to create a forum for communication and a directory of adult care providers. Currently 123 providers from 28 states in the United States, Brazil, Canada, and India participate. The First National Adult Sickle Cell Providers Symposium was held in April 2005 by the University of Cincinnati and SCAPN to address the need of adult sickle cell patients.

A Sickle Cell Disease Clinical Research Network

The Sickle Cell Treatment Act was signed into law by the President in 2004; it provided for inclusion of medical strategies for children and adults with SCD under the Medicaid Program. Subsequently, federal funds were allocated to establish a demonstration program and National Coordinating Center to work on strategies for the prevention and treatment of SCD. A Sickle Cell Disease Clinical Research Network will soon be established as well.

Future Directions

Improved neonatal screening, penicillin prophylaxis, pneumococcal immunization, parental and patient education, red cell transfusion medicine, iron chelation, and hydroxyurea therapy have all contributed to improved survival in patients with SCD. The risk increases with age for complications including acute and chronic pulmonary dysfunction, stroke, avascular necrosis, retinopathy, and chronic pain reinforce the need for standardized treatment protocols. The application of evidence-based medicine to the management of adults with SCD is currently driven by clinical expertise since data from randomized controlled trials (RCT) is limited. The previous trials that impact current medical practices include MSH, pre-operative transfusion, transfusion during pregnancy, and angiotensin-converting enzyme inhibitor therapy for proteinuria. Several issues in adult sickle cell patients will require RCTs in the future since data from children are not predictive of outcomes in adults. The establishment of Comprehensive Care Teams, practice guidelines, and improved access to care will continue to impact long-term survival.

Chronic pain and other medical management problems contribute significantly to the impairment of psychosocial functioning, self esteem, interpersonal relationships, and quality of life in individuals with SCD; early intervention will improve outcome. Research on SCD-related pain is sparse and under-treatment remains a problem to be overcome. The increased development of short-stay or day-treatment facilities will improve management of painful episodes (Chapter 7). In the last decade, pulmonary artery hypertension has emerged as one of the leading causes of morbidity and mortality in adult sickle cell patients. Intravascular hemolysis associated with endothelial dysfunction is caused by reduced nitric oxide bioavailability. Pro-oxidant and pro-inflammatory stress can lead to vasomotor instability and proliferative vasculopathy, the hallmark of pulmonary hypertension in adulthood. Thus therapies aimed at normalizing the vasodilator/vasoconstrictor balance bring new hope for SCD patients and other hemolytic hemoglobinopathies including thalassemia.

References

1. Platt OS, Brambilla DJ, Rosse WF, Milner PF, Castro O, Steinberg MH, Klug PP (1994). Mortality in sickle cell disease: life expectancy and risk factors for early death. *N Engl J Med* **330**:1639–1644.
2. Leikin SL, Gallagher D, Kinney TR, Sloane D, Klug P, Rida W (1989). Mortality in children and adolescents with sickle cell disease. Cooperative study of sickle cell disease. *Pediatrics* **84**:500–508.
3. Gaston MH, Verter JI, Woods G, Pegelow C, Kelleher J, Presbury G, Zarkowsky H, Vichinsky E, Iyer R, Lobel JS, *et al.* (1986). Prophylaxis with oral penicillin in children with sickle cell anemia: a randomized trial. *N Engl J Med* **314**:1593–1599.
4. Quinn CT, Rogers ZR, Buchanan GR (2004). Survival of children with sickle cell disease. *Blood* **103**: 4023–4027.
5. Shapiro BS, Dinges DF, Orne EC, Bauer N, Reilly LB, Whitehouse WG, Ohene-Frempong K, Orne MT (1995). Home management of sickle cell-related pain in children and adolescents: natural history and impact on school attendance. *Pain* **61**:139–144.
6. Shapiro B (1993). Management of painful episodes in sickle cell disease. In: Schechter N *et al.*, editors *Pain in Infants, Children, and Adolescents*. Baltimore: Williams & Wilkins.
7. Levy DR (1985). White doctors and black patients: influence of race on the doctor-patient relationship. *Pediatrics* **75**:639–643.
8. Davidoff A, Kenney G, Dubay L (2005). Effects of the State Children's Health Insurance Program Expansions on children with chronic health conditions. *Pediatr* **116**: e34–42.
9. Ballas SK (2005). Pain management of sickle cell disease. *Hematol Oncol Clin North Am* **19**:785–802.

10. Dampier C, Ely B, Brodecki D, O'Neal P (2002). Characteristics of pain managed at home in children and adolescents with sickle cell disease by using diary self-reports. *J Pain* **3**:461–470.
11. Shapiro B *et al.*, eds. (1993). *Proceedings of the Conference on Sickle Cell Related Pain: Assessment and Management*. Framingham, MA: New England Regional Genetics Group and Maternal and Child Health Bureau.
12. Buchanan ID, Woodward M, Reed GW (2005). Opioid selection during sickle cell pain crisis and its impact on the development of acute chest syndrome. *Pediatr Blood Cancer* **45**:716–724.
13. National Heart Lung and Blood Institute (2004). *The Management of Sickle Cell Disease*, 4th edition. National Institutes of Health.
14. Christensen ML, Wang WC, Harris S, Eades SK, Wilimas JA (1996). Transdermal fentanyl administration in children and adolescents with sickle cell pain crisis. *J Pediatr Hematol Oncol* **8**:372–376.
15. Perlman KM, Myers-Phariss S, Rhodes JC (2004). A shift from demerol (meperidine) to dilaudid (hydromorphone) improves pain control and decreases admissions for patients in sickle cell crisis. *J Emerg Nurs* **30**:439–446.
16. de Franceschi L, Finco G, Vassanelli A, Zaia B, Ischia S, Corrocher R (2004). A pilot study on the efficacy of ketorolac plus tramadol infusion combined with erythrocytapheresis in the management of acute severe vaso-occlusive crises and sickle cell pain. *Haematologica* **89**:1389–1391.
17. Chatterjee R, Iwai Y, Walder RY, Walder JA (1984). Structural features required for the reactivity and intracellular transport of bis(3,5-dibromosalicyl)fumarate and related anti-sickling compounds that modify hemoglobin S at the 2,3-diphosphoglycerate binding site. *J Biol Chem* **259**:14863–14873.
18. Adams-Graves P, Kedar A, Koshy M, Steinberg M, Veith R, Ward D, Crawford R, Edwards S, Bustrack J, Emanuel M (1997). RheothRx (poloxamer 188) injection for the acute painful episode of sickle cell disease: a pilot study. *Blood* **90**:2041–2046.
19. Gibbs WJ, Hagemann TM (2004). Purified poloxamer 188 for sickle cell vaso-occlusive crisis. *Ann Pharmacother* **38**:320–324.
20. Marangos PJ, Fox AW, Riedel BJ, Royston D, Dziewanowska ZE (1998). Potential therapeutic applications of fructose-1,6-diphosphate. *Expert Opin Investig Drugs* **7**:615–623.
21. Wu CL, Cohen SR, Richman JM, Rowlingson AJ, Courpas GE, Cheung K, Lin EE, Liu SS (2005). Efficacy of postoperative patient-controlled and continuous infusion epidural analgesia versus intravenous patient-controlled analgesia with opioids: a meta-analysis. *Anesthesiology* **103**:1079–1088.
22. Brookoff D, Polomano R (1992). Treating sickle cell pain like cancer pain. *Ann Intern Med* **116**: 364–368.
23. Elander J, Lusher J, Bevan D, Telfer P (2003). Pain management and symptoms of substance dependence among patients with sickle cell disease. *Soc Sci Med* **57**:1683–1696.
24. Weil JV, Castro O, Malik AB, Rodgers G, Bonds DR, Jacobs TP (1993). Pathogenesis of lung disease in sickle hemoglobinopathies. *Am Rev Respir Dis* **148**:249–256.
25. Castro O, Brambilla DJ, Thorington B, Reindorf CA, Scott RB, Gillette P, Vera JC, Levy PS (1994). The acute chest syndrome in sickle cell disease: incidence and risk factors. *Blood* **84**:643–649.
26. Vichinsky EP, Neumayr LD, Earles AN, Williams R, Lennette ET, Dean D, Nickerson B, Orringer E, McKie V, Bellevue R, Daeschner C, Manci EA (2000). Causes and outcomes of the acute chest syndrome in sickle cell disease. *N Engl J Med* **342**:1855–1865.
27. Taylor C, Carter F, Poulose J, Rolle S, Babu S, Crichlow S (2004). Clinical presentation of acute chest syndrome in sickle cell disease. *Postgrad Med J* **80**:346–349.
28. Lowenthal EA, Wells A, Emanuel PD, Player R, Prchal JT (1996). Sickle cell acute chest syndrome associated with parvovirus B19 infection: case series and review. *Am J Hematol* **51**:207–213.
29. Oppenheimer EH, Esterly JR (1971). Pulmonary changes in sickle cell disease. *Am Rev Respir Dis* **103**: 858–859.
30. Haupt HM, Moore GW, Bauer TW, Hutchins GM (1982). The lung in sickle cell disease. *Chest* **81**:332–337.
31. Vichinsky EP, Williams R, Das M, Earles AN, Lewis N (1994). Pulmonary fat embolism: a distinct cause of severe acute chest syndrome in sickle cell anemia. *Blood* **83**:3107–3112.
32. Francis RB (1991). Large-vessel occlusion in sickle cell disease: pathogenesis, clinical consequences and therapeutic implications. *Med Hypoth* **35**:88–95.

33. Maitre B, Habibi A, Roudot-Thoraval F, Bachir D, Belghiti DD, Galacteros F, Godeau B (2000). Acute chest syndrome in adults with sickle cell disease: therapeutic approach, outcome, and results of BAL in a monocentric series of 107 episodes. *Chest* **117**:1386–1392.
34. Castella X, Valles J, Cabezuelo MA, Fernandez R, Artigas A (1992). Fat embolism syndrome and pulmonary microvascular pathology. *Chest* **101**:1710–1711.
35. Ataga KI, Orringer EP (2000). Bone marrow necrosis in sickle cell disease: a description of three cases and a review of the literature. *Am J Med Sci* **320**:342–347.
36. Hiran S (2005). Multiorgan dysfunction syndrome in sickle cell disease. *J Assoc Physicians India* **53**: 19–22.
37. Haynes JR, Allison RC (1986). Pulmonary edema complication in the management of sickle cell pain crisis. *Am J Med* **80**:833–840.
38. Oppenheimer EH, Esterly JR (1971). Pulmonary changes in sickle cell disease. *Am Rev Respir Dis* **103**: 858–859.
39. Ballas SK, Files B, Luchtman-Jones L, Benjamin L, Swerdlow P, Hilliard L, Coates T, Abboud M, Wojtowicz-Praga S, Grindel JM (2004). Safety of purified poloxamer 188 in sickle cell disease: phase I study of a non-ionic surfactant in the management of acute chest syndrome. *Hemoglobin* **28**:85–102.
40. Hofrichter J, Ross PD, Eaton WA (1974). Kinetics and mechanism of deoxyhemoglobin S gelation. *Proc Natl Acad Sci USA* **71**:4864–4868.
41. Mozzarelli A, Hofrichter J (1987). Delay time of hemoglobin S polymerization prevents most cells from sickling *in vivo*. *Science* **237**:500–506.
42. Kaul DK, Chen D, Zhan J (1994). Adhesion of sickle cells to vascular endothelium is critically dependent on changes in density and shape of the cells. *Blood* **83**:3006–3017.
43. Kaul DK, Fabry DE, Nagel RL (1986). Vaso-occlusion by sickle cells: evidence for selective trapping of dense red cells. *Blood* **68**:1162–1166.
44. Hebbel RP, Boogaerts MA, Eaton SW, Steinberg MH (1980). Erythrocyte adherence to endothelium in sickle cell anemia: possible determinant of disease severity. *N Engl J Med* **302**:992–995.
45. Ballas SK (1991). Sickle cell anemia with few painful crises is characterized by decreased red cell deformability and increased number of dense cells. *Am J Hematol* **36**:122–130.
46. Joneckis CC, Ackley RL, Orringer EP, Wayner EA, Parise LV (1993). Integrin alpha 4 beta 1 and glycoprotein IV (CD36) are expressed on circulating reticulocytes in sickle cell anemia. *Blood* **82**:3548–3555.
47. Swerlick RA, Eckman JR, Kumar A, Jeitler M, Wick TM (1993). Alpha-4/beta-1 integrin expression on sickle reticulocytes: vascular cell adhesion molecule-1-dependent binding to endothelium. *Blood* **82**: 1891–1899.
48. Setty BNY, Stuart MJ (1996). Vascular cell adhesion molecule-1 is involved in mediating hypoxia induced sickle red blood cell adherence to endothelium: potential role in sickle cell disease. *Blood* **88**:2311–2320.
49. Charache S, Terrin ML, Moore RD, Dover GJ, Barton FB, Eckert SV, McMahon RP, Bonds DR (1995). Effect of hydroxyurea on the frequency of painful crises in sickle cell anemia. Investigators of the Multicenter Study of Hydroxyurea in Sickle Cell Anemia. *N Engl J Med* **332**:1317–1322.
50. Fadlon E, Vordermeier S, Pearson TC, Mire-Sluis AR, Dumonde DC, Phillips J, Fishlock K, Brown KA (1998). Blood polymorphonuclear leukocytes from the majority of sickle cell patients in the crisis phase of the disease show enhanced adhesion to vascular endothelium and increased expression of CD64. *Blood* **91**:266–274.
51. Boghossian SH, Nash G, Dormandy J, Bevan DH (1991). Abnormal neutrophil adhesion in sickle cell anaemia and crisis: relationship to blood rheology. *Br J Haematol* **78**:437–441.
52. Lard LR, Mul FP, de Haas M (1999). Neutrophil activation in sickle cell disease. *J Leukoc Biol* **66**:411–415.
53. Mollapaur E, Porter JB, Kaczmarski R, Linch DC, Roberts PJ (1998). Raised neutrophil phospholipase A2 activity and defective priming of NADPH oxidase and phospholipase A2 in sickle cell disease. *Blood* **91**:3423–3429.
54. Pouliot M, McDonald PP, Krump E, Mancini JA, McColl SR, Weech PK, Borgeat P (1996). Colocalization of cytosolic phospholipase A_2, 5-lipooxygenase and 5-lipooxygenase-activating protein at the nuclear membrane of A-23187-stimulated human neutrophils. *Eur J Biochem* **238**:250–258.
55. Styles L, Schalkwijk CG, Aarsman AJ, Vichinsky EP, Lubin BH, Kuypers FA (1996). Phospholipase A_2 levels in acute chest syndrome of sickle cell disease. *Blood* **87**:2573–2578.

56. Styles L, Aarsman A, Vichinsky E, Kuypers FA (2000). Secretory phospholipase A_2 predicts impending acute chest syndrome in sickle cell disease. *Blood* **96**:3276–3278.
57. Guidot DM, Repine MJ, Westcott JY, Repine JF (1994). Intrinsic 5-lipooxygenase activity is required for neutrophil responsivity. *Proc Natl Acad Sci USA* **91**:8156–8159.
58. Snyder F (1988). Metabolism of platelet activating factor and related ether lipids: enzymatic pathways, subcellular sites, regulation, and membrane processing. *Prog Clin Biol Res* **282**:57–72.
59. Lindström P, Lerner R, Palmblad J, Patarroyo M (1990). Rapid adhesive responses of endothelial cells and of neutrophils induced by leukotriene B_4 are mediated by leucocytic adhesion protein CD18. *Scand J Immunol* **30**:737–744.
60. Minamiya V, Tozawa K, Kitamura M, Saito S, Ogawa J (1998). Platelet-activating factor mediates intracellular adhesion molecule-1-dependent radical production in the nonhypoxic ischemia rat lung. *Am J Respir Cell Mol Biol* **19**:150–157.
61. Kubes P, Suzuki M, Granger DM (1990). Platelet-activating factor-induced microvascular dysfunction: role of adherent leucocytes. *Am J Physiol* **258**:G158–G163.
62. Lewis RE, Granger HJ (1988). Diapedesis and permeability of venous microvessels to protein macromolecules: the impact of leukotriene B_4 (LTB_4). *Microvasc Research* **35**:27–47.
63. Haynes J, Obiako B (2002). Activated polymorphonuclear cells increase sickle red blood cell retention in lung: role of phospholipids. *Am J Physiol Heart Circ* **282**:H122–H130.
64. Powars D, Weidman JA, Odom-Maryon T, Niland JC, Johnson C (1988). Sickle cell chronic lung disease: prior morbidity and the risk of pulmonary failure. *Medicine* **67**:66–75.
65. Ataga KI, Sood N, De Gent G, Kelly E, Henderson AG, Jones S, Strayhorn D, Lail A, Lieff S, Orringer EP (2004). Pulmonary hypertension in sickle cell disease. *Am J Med* **117**:665–669.
66. Ahmed S, Siddiqui AK, Sadiq A, Shahid RK, Patel DV, Russo LA (2004). Echocardiographic abnormalities in sickle cell disease. *Am J Hematol* **76**:195–198.
67. Gladwin MT, Sachdev V, Jison ML, Shizukuda Y, Plehn JF, Minter K, Brown B, Coles WA, Nichols JS, Ernst I, Hunter LA, Blackwelder WC, Schechter AN, Rodgers GP, Castro O, Ognibene FP (2004). Pulmonary hypertension as a risk factor for death in patients with sickle cell disease. *N Engl J Med* **350**:886–895.
68. Kato GJ, McGowan VR, Machado RF, Little JA, Taylor Vi J, Morris CR, Nichols JS, Wang X, Poljakovic M, Morris Jr SM, Gladwin MT (2005). Lactate dehydrogenase as a biomarker of hemolysis-associated nitric oxide resistance, priapism, leg ulceration, pulmonary hypertension and death in patients with sickle cell disease. *Blood* [Epub ahead of print].
69. Gow AJ, Stamler JS (1998). Reactions between nitric oxide and haemoglobin under physiological conditions. *Nature* **391**:169–173.
70. Reiter CD, Gladwin MT (2003). An emerging role for nitric oxide in sickle cell disease vascular homeostasis and therapy. *Curr Opin Hematol* **10**:99–107.
71. Machado RF, Martyr S, Kato GJ, Barst RJ, Anthi A, Robinson MR, Hunter L, Coles W, Nichols J, Hunter C, Sachdev V, Castro O, Gladwin MT (2005). Sildenafil therapy in patients with sickle cell disease and pulmonary hypertension. *Br J Haematol* **130**:445–453.
72. Morris CR, Kato GJ, Poljakovic M, Wang X, Blackwelder WC, Sachdev V, Hazen SL, Vichinsky EP, Morris SM Jr, Gladwin MT (2005). Dysregulated arginine metabolism, hemolysis-associated pulmonary hypertension, and mortality in sickle cell disease. *JAMA* **294**:81–90.
73. Morris CR, Morris SM Jr, Hagar W, Van Warmerdam J, Claster S, Kepka-Lenhart D, Machado L, Kuypers FA, Vichinsky EP (2003). Arginine therapy: a new treatment for pulmonary hypertension in sickle cell disease? *Am J Respir Crit Care Med* **168**:63–69.
74. Adams RJ (2001). Stroke prevention and treatment in sickle cell disease. *Arch Neurol* **58**:565–568.
75. Van Hoff J, Ritchey AK, Shaywitz BA (1985). Intracranial hemorrhage in children with sickle cell disease. *Am J Dis Child* **139**:1120–1123.
76. Peerless SJ (1997). Risk factors of Moyamoya disease in Canada and the USA. *Clinical Neurology and Neurosurgery* **99**:S45–S48.
77. Wang WC, Morales KH, Scher CD, Styles L, Olivieri N, Adams R, Brambilla D (2005). STOP investigators effect of long-term transfusion on growth in children with sickle cell anemia: results of the STOP trial. *J Pediatr* **147**:244–247.

78. Sampaio Silva G, Vicari P, Figueiredo MS, Filho AC, Valadi N, Massaro AR (2005). Transcranial doppler in adult patients with sickle cell disease. *Cerebrovasc Dis* **21**:38–41.
79. Ware RE, Steinberg MH, Kinney TR (1995). Hydroxyurea: an alternative to transfusion therapy for stroke in sickle cell anemia. *Am J Hematol* **50**:140–143.
80. Ware RE, Zimmerman SA, Schultz WH (1999). Hydroxyurea as an alternative to blood transfusions for the prevention of recurrent stroke in children with sickle cell disease. *Blood* **94**:3022–3026.
81. Gulbis B, Haberman D, Dufour D, Christophe C, Vermylen C, Kagambega F, Corazza F, Devalck C, Dresse MF, Hunninck K, Klein A, Le PQ, Loop M, Maes P, Phillippet P, Sariban E, Van Geet C, Ferster A (2005). Hydroxyurea for sickle cell disease in children and for prevention of cerebrovascular events: the Belgian experience. *Blood* **105**:2685–2690.
82. Hernigou P, Bachir D, Galacteros F (1993). Avascular necrosis of the femoral head in sickle-cell disease treatment of collapse by the injection of acrylic cement. *J Bone Joint Surg (BR)* **75-B**:875–880.
83. Bahebeck J, Atangana R, Techa A, Monny-Lobe M, Sosso M, Hoffmeyer P (2004). Relative rates and features of musculoskeletal complications in adult sicklers. *Acta Orthop Belg* **70**:107–111.
84. Moran MC (1995). Osteonecrosis of the hip in sickle cell hemoglobinopathy. *Am J Orthop* **24**:18–24.
85. Aguilar CM, Neumayr LD, Eggleston BE, Earles AN, Robertson SM, Jergesen HE, Stulberg BN, Vichinsky EP (2005). Clinical evaluation of avascular necrosis in patients with sickle cell disease: Children's Hospital Oakland Hip Evaluation Scale — a modification of the Harris Hip Score. *Arch Phys Med Rehabil* **86**: 1369–1375.
86. Flanagan G, Packham DK (1993). Kincaid-Smith P sickle cell disease and the kidney. *Am J Kidney Dis* **21**:325–327.
87. Bruno D, Wigfall DR, Zimmerman SA, Rosoff PM, Wiener JS (2001). Genitourinary complications of sickle cell disease. *J Urol* **166**:803–811.
88. Davis CJ Jr, Mostofi FK, Sesterhenn IA (1995). Renal medullary carcinoma. The seventh sickle cell nephropathy. *Am J Surg Pathol* **19**:1–11.
89. Falk RJ, Mattern WD, Lamanna RW, Gitelman HJ, Parker NC, Cross RE, Rastall JR (1983). Iron removal during continuous ambulatory peritoneal dialysis using deferoxamine. *Kidney Int* **24**:110–112.
90. Gonzalez-Carrillo M, Rudge CJ, Parsons V, Bewick M, White JM (1982). Renal transplantation in sickle cell disease. *Clin Nephrol* **18**:209–210.
91. Sears DA (1978). The morbidity of sickle cell trait: a review of the literature. *Am J Med* **64**:1021–1036.
92. Baldree LA, Ault BH, Chesney CM, Stapleton FB (1990). Intravenous desmopressin acetate in children with sickle trait and persistent macroscopic hematuria. *Pediatrics* **86**:238–243.
93. Moudgil A, Kamil ES (1996). Protracted, gross hematuria in sickle cell trait: response to multiple doses of 1-desamino-8-D-arginine vasopressin. *Pediatr Nephrol* **10**:210–212.
94. Simpson L, He X, Pins M, Huang X, Campbell SC, Yang XJ, Perlman EJ, Bergman RC (2005). Renal medullary carcinoma and ABL gene amplfication. *J Urol* **173**:1883–1888.
95. Aiello LP (1997). Clinical implications of vascular growth factors in proliferative retinopathies. *Curr Opin Ophthalmol* **8**:19–31.
96. Goldberg MF (1971). Classification and pathogenesis of proliferative sickle retinopathy. *Am J Ophthalmol* **71**:649–655.
97. Downes SM, Hambleton IR, Chuang EL, Lois N, Serjeant GR, Bird AC (2005). Incidence and natural history of proliferative sickle cell retinopathy: observations from a cohort study. *Ophthalmology* **112**: 1869–1875.
98. Charache S (1996). Eye disease in sickling disorders. *Hematol/Oncol Clin N Am* **10**:1357–1362.
99. Manci EA, Culberson DE, Yang YM, Gardner TM, Powell R, Haynes J Jr, Shah AK, Mankad VN (2003). Investigators of the Cooperative Study of Sickle Cell Disease. Causes of death in sickle cell disease: an autopsy study. *Br J Haematol* **123**:359–365.
100. Zarkowsky HS, Gallagher D, Gill FM, Wang WC, Falletta JM, Lande WM, Levy PS, Verter JI, Wethers D (1986). Bacteremia in sickle hemoglobinopathies. *J Pediatr* **109**:579–585.
101. *MMWR Morb Mortal Wkly Rep (United States, 2003–2004)*. 2003;**52**:965.
102. Covitz W, Espeland M, Gallagher D, Hellenbrand W, Leff S, Talner N (1995). The heart in sickle cell anemia. The Cooperative Study of Sickle Cell Disease (CSSCD). *Chest* **108**:1214–1219.

103. Ernst AA, Weiss SJ, Johnson WD, Takakuwa KM (2000). Blood pressure in acute vaso-occlusive crises of sickle cell disease. *South Med J* **93**:590–592.
104. Delclaux C, Zerah-Lancner F, Bachir D, Habibi A, Monin JL, Godeau B, Galacteros F (2005). Factors associated with dyspnea in adult patients with sickle cell disease. *Chest* **128**:3336–3344.
105. Bruno D, Wigfall DR, Zimmerman SA, Rosoff PM, Wiener JS (2001). Genitourinary complications of sickle cell disease. *J Urol* **166**:803–811.
106. Mantadakis E, Cavender JD, Rogers ZR, Ewalt DH, Buchanan GR (1999). Prevalence of priapism in children and adolescents with sickle cell anemia. *J Pediatr Hematol Oncol* **21**:518–522.
107. Fowler JE Jr, Koshy M, Strub M, Chinn SK (1991). Priapism associated with the sickle cell hemoglobinopathies: prevalence, natural history and sequelae. *J Urol* **145**:65–68.
108. Mykulak DJ, Glassberg KI (1990). Impotence following childhood priapism. *J Urol* **144**:134–135.
109. Seeler RA (1973). Intensive transfusion therapy for priapism in boys with sickle cell anemia. *J Urol* **110**:360–363.
110. Upadhyay J, Shekarriz B, Dhabuwala CB (1998). Penile implant for intractable priapism associated with sickle cell disease. *Urology* **51**:638–639.
111. Serjeant GR, de Ceulaer K, Maude GH (1985). Stilboestrol and stuttering priapism in homozygous sickle-cell disease. *Lancet* **2**:1274–1276.
112. Baruchel S, Rees J, Bernstein ML, Goodyer P (1993). Relief of sickle cell priapism by hydralazine. Report of a case. *Am J Pediatr Hematol Oncol* **15**:115–116.
113. Saad ST, Lajolo C, Gilli S, Marques Junior JF, Lima CS, Costa FF, Arruda VR (2004). Follow-up of sickle cell disease patients with priapism treated by hydroxyurea. *Am J Hematol* **77**:45–49.
114. McDonald M, Santucci RA (2004). Successful management of stuttering priapism using home self-injections of the alpha-agonist metaraminol. *Int Braz J Urol* **30**:121–122.
115. Ahn H, Li CS, Wang W (2005). Sickle cell hepatopathy: clinical presentation, treatment, and outcome in pediatric and adult patients. *Pediatr Blood Cancer* **45**:184–190.
116. Koshy M, Entsuah R, Koranda A, Kraus AP, Johnson R, Bellvue R, Flournoy-Gill Z, Levy P (1989). Leg ulcers in patients with sickle cell disease. *Blood* **74**:1403–1408.
117. Clare A, FitzHenley M, Harris J, Hambleton I, Serjeant GR (2002). Chronic leg ulceration in homozygous sickle cell disease: the role of venous incompetence. *Br J Haematol* **119**:567–571.
118. Wethers DL, Ramirez GM, Koshy M, Steinberg MH, Phillips G Jr, Siegel RS, Eckman JR, Prchal JT (1994). Accelerated healing of chronic sickle-cell leg ulcers treated with RGD peptide matrix. RGD Study Group. *Blood* **84**:1775–1779.
119. Whitten-Shurney W (2001). Sickle cell. *Clin Perinatol* **28**:435–448.
120. Koshy M, Burd L, Dorn L, Huff G (1987). Frequency of pain crisis during pregnancy. *Prog Clin Biol Res* **240**:305–311.
121. Seoud MA, Cantwell C, Nobles G, Levy DL (1994). Outcome of pregnancies complicated by sickle cell and sickle-C hemoglobinopathies. *Am J Perinatol* **11**:187–191.
122. Sun PM, Wilburn W, Raynor BD, Jamieson D (2001). Sickle cell disease in pregnancy: twenty years of experience at Grady Memorial Hospital, Atlanta, Georgia. *Am J Obstet Gynecol* **184**:1127–1130.
123. Tuck SM, James CE, Brewster EM, Pearson TC, Studd JW (1987). Prophylactic blood transfusion in maternal sickle cell syndromes. *Br J Obstet Gynaecol* **94**:121–125.
124. Rust OA, Perry KG Jr (1995). Pregnancy complicated by sickle hemoglobinopathy. *Clin Obstet Gynecol* **38**:472–484.
125. Koduri PR, Agbemadzo B, Nathan S (2001). Hemoglobin S-C disease revisited: clinical study of 106 adults. *Am J Hematol* **68**:298–300.
126. Bridges KR, Barabino GD, Brugnara C, Cho MR, Christoph GW, Dover G, Ewenstein BM, Golan DE, Guttmann CR, Hofrichter J, Mulkern RV, Zhang B, Eaton WA (1996). A multiparameter analysis of sickle erythrocytes in patients undergoing hydroxyurea therapy. *Blood* **88**:4701–4710.
127. Yan JH, Ataga K, Kaul S, Olson JS, Grasela DM, Gothelf S, Kutlar A, Orringer E (2005). The influence of renal function on hydroxyurea pharmacokinetics in adults with sickle cell disease. *J Clin Pharmacol* **45**:434–445.

128. Zumberg MS, Reddy S, Boyette RL, Schwartz RJ, Konrad TR, Lottenberg R (2005). Hydroxyurea therapy for sickle cell disease in community-based practices: a survey of Florida and North Carolina hematologists/oncologists. *Am J Hematol* **79**:107–113.
129. Okpala IE (2005). New therapies for sickle cell disease. *Hematol Oncol Clin North Am* **19**:975–987, ix.
130. Serjeant BE, Harris J, Thomas P, Serjeant GR (1997). Propionyl-L-carnitine in chronic leg ulcers of homozygous sickle cell disease: a pilot study. *J Am Acad Dermatol* **37**:491–493.
131. Kontoghiorghes GJ, Neocleous K, Kolnagou A (2003). Benefits and risks of deferiprone in iron overload in Thalassaemia and other conditions: comparison of epidemiological and therapeutic aspects with deferoxamine. *Drug Saf* **26**:553–584.
132. Cappellini MD (2005). Iron-chelating therapy with the new oral agent ICL670 (Exjade). *Best Pract Res Clin Haematol* **18**:289–298.
133. Anie KA, Telfair J (2005). Sickle Cell Disease Transition Study Working Group. Multi-site study of transition in adolescents with sickle cell disease in the United Kingdom and the United States. *Int J Adolesc Med Health* **17**:169–178.
134. Woods KF, Johnson JA, Kutlar A, Daitch L, Stachura ME (2000). Sickle cell disease telemedicine network for rural outreach. *J Telemed Telecare* **6**:285–290.
135. Smith WR, Bovbjerg VE, Penberthy LT, McClish DK, Levenson JL, Roberts JD, Gil K, Roseff SD, Aisiku IP (2005). Understanding pain and improving management of sickle cell disease: the PiSCES study. *J Natl Med Assoc* **97**:183–189.
136. Grant MM, Gil KM, Floyd MY, Abrams M (2000). Depression and functioning in relation to health care use in sickle cell disease. *Ann Behav Med* **22**:149–157.
137. Palermo TM, Kiska R (2005). Subjective sleep disturbances in adolescents with chronic pain: relationship to daily functioning and quality of life. *J Pain* **6**:201–207.
138. Edwards CL, Scales MT, Loughlin C, Bennett GG, Harris-Peterson S, De Castro LM, Whitworth E, Abrams M, Feliu M, Johnson S, Wood M, Harrison O, Killough A (2005). A brief review of the pathophysiology, associated pain, and psychosocial issues in sickle cell disease. *Int J Behav Med* **12**:171–179.
139. Farber MD, Koshy M, Kinney TR (1985). Cooperative Study of Sickle Cell Disease: demographic and socioeconomic characteristics of patients and families with sickle cell disease. *J Chronic Dis* **38**:495–505.
140. Edwards CL, Scales MT, Loughlin C, Bennett GG, Harris-Peterson S, De Castro LM, Whitworth E, Abrams M, Feliu M, Johnson S, Wood M, Harrison O, Killough A (2005). A brief review of the pathophysiology, associated pain, and psychosocial issues in sickle cell disease. *Int J Behav Med* **12**:171–179.
141. Barbarin OA, Whitten CF, Bonds SM (1994). Estimating rates of psychosocial problems in urban and poor children with sickle cell anemia. *Health Soc Work* **9**:112–119.

7

Pain in Sickle Cell Disease: A Multidimensional Construct

by Lennette J. Benjamin and Richard Payne

Introduction

Pain is the hallmark phenotypic attribute of sickle cell disease (SCD). The type of pain most commonly encountered by patients of all ages is referred to as acute painful episodes or more commonly as painful crises.[1–3] They are episodic, unpredictable, unpleasant sensory and emotional experiences of abrupt onset involving one or multiple sites of the body. Acute painful episodes last hours to days and are followed by abatement and return to steady state. The protean features of sickle cell disease distinguish it from most common pain states such as dental, postoperative, cancer, and obstetrical pain. Unlike these acquired clinically-related pain states, it occurs in the context of a genetic disease and manifests as dispersions of devastating pain and resulting opioid therapy as early as infancy and throughout a lifetime.

Pain accounts for greater than 90% of emergency outpatient visits and hospital admissions in SCD. In addition, there is considerable morbidity due to pain that occurs in the process of daily living and is managed at home without seeking professional medical care.[4,5] The frequency of vaso-occlusive events and pain-related disability differs markedly among patients within the same sickle hemoglobinopathy group and within the same family. There is however direct correlation between the number of pain episodes requiring visits to the hospital and mortality from this disease.[6] This chapter is a review of current understanding of pain mechanisms with emphasis on the contribution from the disease, environmental factors and pain treatments.

The Nature of Pain in Sickle Cell Disease (SCD)

The pain experience in SCD can be modeled as a multidimensional construct involving genetic and environmental factors that converge and conspire to challenge classical definitions of pain and confound the search for treatment and cure.[7–9] Sickle pain can be acute, chronic or mixed, and can be related to disease severity and/or complications of treatment.[10] It manifests early in life typically as the hand-foot syndrome in infants about six months of age. This initial pain marks the beginning of a lifetime of recurrent, abrupt-onset acute painful episodes (pain crises) and insidiously inflicted and debilitating chronic pain. Pain in the back, legs, hips, pelvis, and rib cage frequently result from bone involvement in the sickling process (see below), however all organ systems are potential targets of pain.

Pathophysiology of Sickle-Related Pain

Biological, social, and psychological perturbations that include emotional stress, mood, and physical stress (such as hypoxia, dehydration, acidosis, infection, and fatigue) precipitate the pathophysiologic processes including hemoglobin S polymerization, sickle cell adhesion and altered deformability that lead to painful episodes.[1] Adhesion of sickle cells to vascular endothelium and log-jamming of poorly deformed cells result in microvascular occlusion, tissue hypoxia, and injury.[7,11] Microvascular occlusion arises in localized areas of the bone marrow, and evokes an inflammatory response and tissue acidosis. Inflammatory mediators activate nociceptive afferent fibers which transmit sensations via the spinothalamic pathways to the thalamus and cortex, where pain is perceived; the pain response and pain modulation are then evoked. Molecular and cytochemical factors influence the wide variability of the pain responses observed clinically.

Appraisal of Risk Factors

Currently, we cannot strictly classify most patients in terms of predicting pain severity. We can, however, identify risk factors of pain, as demonstrated in two large multicenter studies: the Cooperative Study of Sickle Cell Disease (CSSCD), and the Multicenter Study of Hydroxyurea.[12–15]

The hand-foot syndrome in infants is an early prognostic indicator for increased risk of complications in children.[15] Frequency of sickle cell pain episodes, after controlling for other comorbid conditions, is an independent predictor of reduced survival. Adults older than 20 years of age with high frequency of pain ($\geq$3 events per year) have a greater risk of early death than their counterparts with low pain rates ($<$3 events per year). In the CSSCD study, 33% of deaths that occurred were associated with an acute painful episode.[6] Increased pain rates are also associated with decreased fetal hemoglobin, increased hematocrit, and leucocytosis.[12–15]

In general, disease-severity as defined by frequency of vaso-occlusive pain events is associated with the sickle genotype. As a group, patients with homozygous sickle cell disease (SCD-SS) have the highest rate of painful episodes, followed by those with sickle β^0 thalassemia (SCD-Sβ^0 thal), Hb SC disease (SCD-SC), and mild sickle–β^+ thalassemia (SCD-Sβ^+ thal). To date, no correlation has been found between β-globin cluster haplotypes or α-globin gene status and the rate or severity of pain episodes.[16–18] However, more rigorous haplotype analysis using Genomic era approaches may bring clarity to the role of haplotypes in disease-severity. Psychological states such as daily stress and negative mood have been implicated as predictors of pain.[19]

Genetic background affects nociceptic pain sensitivity in animals, and may influence susceptibility to the development of persistent pain in humans.[20] N-Methyl D-aspartate (NMDA) receptors and nitric oxide levels play a prominent role in the risk of inadequately controlled acute pain, progressing to chronic pain. Genetic profiles, gender or expression patterns of opioid receptors may also alter individual responses to analgesic drugs. Metabolism and bioavailability of numerous compounds used for pain control such as codeine is influenced by polymorphisms in the hepatic cytochrome P-450 monoxygenase gene.[21] Characterization of pain and its mechanisms through multidimensional assessment should facilitate identification of patients with multiple risk factors that may resolve some of the complexities in pain management.

Pain Mechanisms and Biopsychosocial Integration

Pain treatment should incorporate neurobiological and psychobiological mechanisms responsible for the pain. There are two major categories for pain mechanisms: nociceptive pain produced by tissue injury or inflammation; and non-nociceptive pain of the neuropathic variety produced by nerve injury, nervous system impairment or idiopathic etiology.[22]

Nociceptive Pain

This type of pain can be somatic and visceral. The most common type encountered in SCD is somatic bone pain. The microanatomy of bone underscores why it is such a common site of pain. Bone receives extrinsic innervations from sensory and sympathetic neurons,[23] which are associated with blood vessels in bone marrow (Fig. 1). Nociceptors in bone and other tissues are multimodal, responding to a wide range of noxious or potentially tissue-damaging stimuli. In particular, the complex extracellular milieu generated by ischemia and inflammation, provides many potent agonists to stimulate nociceptors. Specific chemical factors released that are known-to-excite nociceptors include protons, endothelins, prostaglandins, bradykinin, and nerve growth factor.[24] This microenvironment therefore provides a pervasive molecular and cytochemical background for multiple inter-related pathologies of SCD including sickling, vaso-occlusion, and associated ischemia to generate pain.[11,25]

The earliest manifestation of vaso-occlusive pain, the hand-foot syndrome (Fig. 2A), occurs at about six months of age, presenting as dorsal swelling of the hands and feet. The metacarpal and/or metatarsal bones are infarcted in the setting of a developing blood supply, and associated with a pronounced inflammatory response. Magnetic resonance imaging reveals infarcts in the bone and soft tissue in the lower extremities,[25] and increase levels of acute phase reactants, cytokines, coagulation, and fibrinolytic systems, which support the contribution of ischemia and inflammation in the acute painful episode.[9,11,26]

Long bones are primary sites for pain in children (Fig. 2B), however the axial skeleton is also affected in adult patients (Fig. 3A and B).[27] Fractures occur at sites of bone destruction, which greatly amplifies pain and causes debilitation (Fig. 3C). Head pain can be localized to the lower jaw due to mandibular infarction, or pain in the teeth due to necrosis of the pulp.[28] Moreover, the gingival inflammation around the teeth is thought to also serve as a trigger for painful episodes.

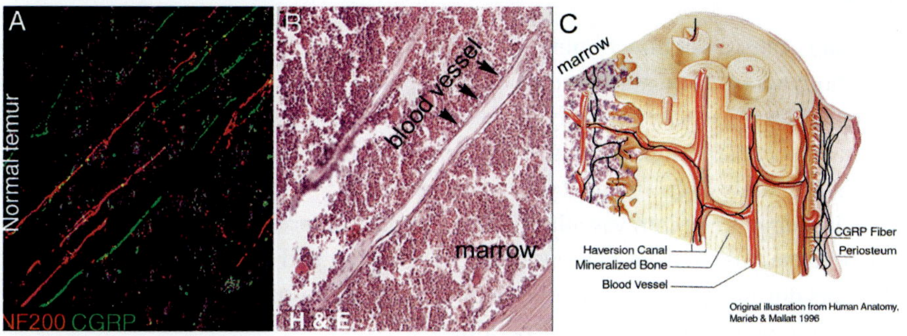

Fig. 1. *Innervation of bone.* Histophotomicrographs of (A) confocal and (B) histologic serial images of normal bone. Note the extensive myelinated (red, NF200) and unmyelinated (green, CGRP) nerve fibers within bone marrow which appear to course along blood vessels (arrowheads, B). (C) Schematic diagram demonstrating the innervation within periosteum, mineralized bone and bone marrow. Used with permission from Sabino and Mantyh.[23]

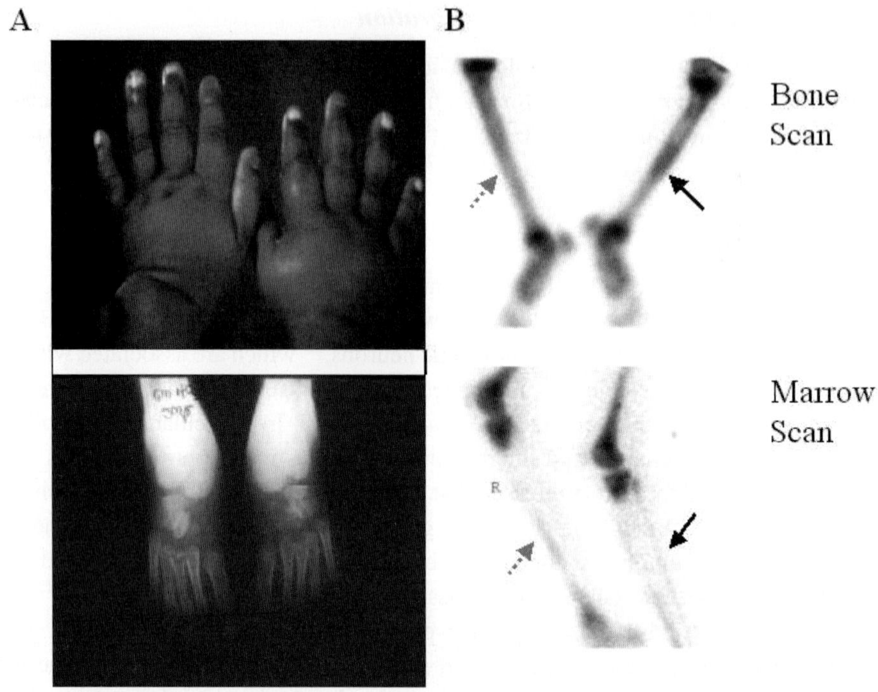

Fig. 2. *Hand foot syndrome.* (A) Photograph of dorsal swelling of hands in an infant and X-rays showing metacarpal infarction in feet. (B) Bone scan and bone marrow scan of infarction (dotted arrow) and infection (solid arrow) in an 8-year-old female with bilateral tibial pain.

Pain at a single site such as the head and/or chest may represent infarction and/or inflammation of the bone. It may also be an early sign of more severe complications of SCD such as stroke or acute chest syndrome and should be correlated with clinical findings.[1,9,29] Bone pain can also occur as a more insidious consequence of vaso-occlusion, resulting in either chronic somatic pain such as avascular necrosis or neuropathic pain.

Neuropathic Pain

Neuropathic pain arises from direct injury or dysfunction in peripheral or central nervous system. It manifests as allodynia (pain on non-noxious touch), spontaneous pain and paresthesias, hyperalgesia (exaggerated pain response to a normally noxious stimulus), sensory deficits, or sympathetic abnormalities such as vasoconstriction. This pain type has not been adequately appreciated in SCD, however it occurs at a sufficiently high frequency to merit research.[10,29] Proposed underlying mechanisms include nerve ischemia from intravascular sickling, nerve compression as a result of bone infarction or vertebral collapse, injury from noxious substances elaborated at the sites of pain, ischemia, and ill-defined events during acute painful crisis and/or as a result of therapy.[10] Allodynia can also be an early sign of opioid neurotoxicity (e.g. meperidine).

Biopsychosocial Integration

In addition to the established nociceptive and neuropathic antecedents to pain, a number of studies examining pain in patients with SCD and other chronic diseases suggest that non-biological factors,

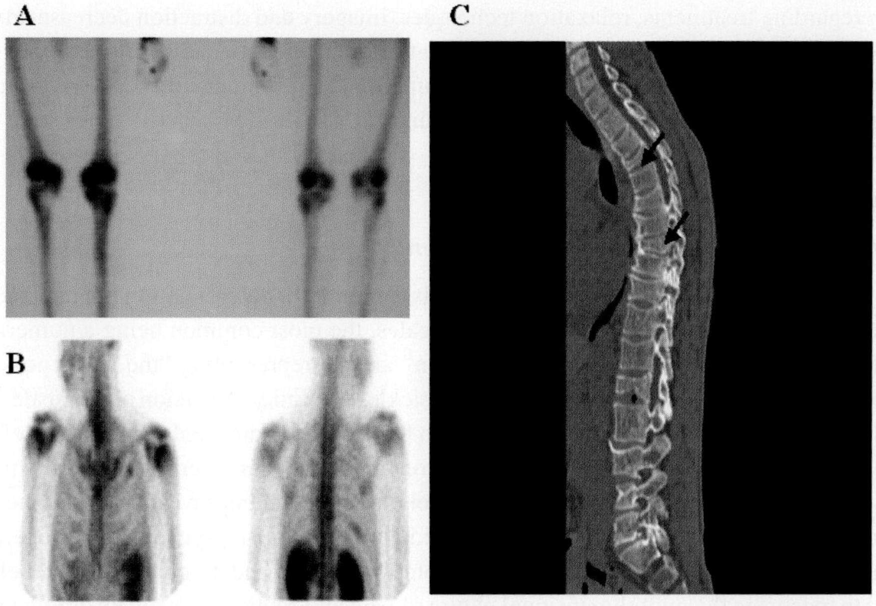

Fig. 3. *Bone and bone marrow pathology in adults*. (A) Bone marrow scan of a 21-year-old man showing infarction of painful right knee. (B) Bone marrow scan showing rib infarct and (C) CT Scan of thoracic and lumbar spine without contrast in a 40-year-old male who experienced multiple fractures (T4, T6, T7, T8, T10, 11, L3, L4, and L5) occurring spontaneously and without a history of trauma. Shown here are the most severe fractures involving T4 and T8.

such as coping skills and social support influence individual pain perception.[8,30,31] Daily stress and negative mood have been shown to predict pain and influence psychosocial and functional outcomes.[19] There is widespread belief that hypoxia, dehydration, infection, acidosis, physical exertion, and exposure to cold, precipitate pain episodes however these assertions have not been substantiated by research findings. Nonetheless, fluid intake, physical exertion, and alterations in temperature have been found to have short-term associations with pain, whereas stress and mood changes have carryover effects to subsequent days. A positive mood has been shown to offset some negative consequences of pain and other illness symptoms.[19]

It is well recognized that differences in human resilience and response to stress are key determinants of vulnerability to psychiatric and complex diseases. Negative emotions such as depression and anxiety correlate with worse perceptions of pain in SCD. In general, stress-related polymorphisms such as the Val^{158}Met polymorphism of the catechol-O-methyl transferase gene, a key modulator of dopaminergic and adrenergic/noradrenergic neurotransmission, influence the experience of pain and may contribute to individual variability in pain adaptation and responses.[31] For instance, the responses to anxiety can be intense to a degree that the levels of cortisol and catecholamine generated supersede those directly provoked by nociception. Importantly, anxiety-related stress response can have additional pathophysiological consequences such as increased blood viscosity, clotting time, fibrinolysis, and platelet aggregation.[32]

Stress induces endothelial dysfunction in experimental animals, and moreover this phenomenon has been demonstrated in humans with cardiovascular disease; hypnosis has been shown to modulate this stress response.[33] Several strategies including preemptive cognitive and behavioral interventions, education regarding treatments, relaxation techniques, imagery and distraction decrease anxiety and accelerate recovery. Recent studies have shown that religious beliefs and spirituality may serve as predictors of psychological well-being and social support, suggesting that these may warrant further study as moderating factors in the experience of pain.[8]

Treatment-Related Issues

Assessment of Pain and Symptom Management

Accurate assessment of pain and related symptoms is essential for determination of pain mechanisms and diagnosis and therapy. Simple self-reporting scales, the most common being a numerical scale ranging from 0 to 10, with 0 representing "no pain" and 10 representing "the worst possible pain imaginable" have proved to be invaluable in several sickle cell clinics. A version of this scale depicting various degrees of facial pain expression in children has been validated and serves as a useful tool in pediatric clinics.[2] Recent data suggest that a reduction of 30% in pain severity is clinically important, and equivalent to categorical ratings of "moderate relief" or "much improved".[34] While the potential limitation of self-report is evident, numerous validation studies of psychometric properties have documented that patients do reliably report symptom intensity and frequency. Nonetheless, it is important to recognize the multidimensional nature of pain and cautiously evaluate dramatic changes in intensity scores, particularly in a single pain episode.

There is little doubt in this field that pain evaluation is a very complex and difficult subject. An artist's graphic rendition of unrelieved pain entitled "Ten redefined" illustrates a deep emotion of pain and suffering that essentially challenges the notion that any given nomenclature can reflect the multiple facets of acute or chronic pain experienced by individuals with SCD (Fig. 4). Indeed, patients and health care practitioners consistently have varying pain scores or ranking for the same pain episode. In addition, the meaning of the level of pain ranking fluctuates from one pain episode to another in the same patient. It is therefore important that both patients and health care practitioners recognize the inherent fluidity in pain assessment given the lack of rigorous standardization. While a pain "episode" is the sensory experience recognized by many clinicians, pain "crisis" has become a perceptual metaphor of suffering and/or grief for the patient. Assessment of the psychological impact of pain, including the meaning that the patient ascribes to the pain during the disease trajectory, is important. A brief assessment of mood is a validated screen for psychological distress;[35] when distress is detected, a subsequent extensive evaluation is essential after the pain is adequately controlled. This evaluation includes an assessment of home and family environment, and the family support system.[36] In addition, family members often have crucial information concerning the extent of functional impairment or disability experienced by the patient, which can be measured by less subjective indices.

Multiple symptoms such as fatigue, anxiety, weakness, and fever often accompany pain and require attention. These symptoms may be associated with a common mechanism related to the effects of cytokines released during the acute phase response, and the "pain" of being sick.[29] These biochemical markers are intriguing targets for development of analgesics that may be more selective than opioids. They offer the possibility of advancing our understanding of the molecular mechanisms of pain and developing rationale pharmacological therapies based on specific pain paradigms.

A thorough and careful physical and neurological examination is essential in medical and pain evaluation. For example, tenderness and swelling at pain sites are usually indicative of nociceptic pain, whilst sensory abnormalities such as allodynia or hyperpathia, with or without subjective

Fig. 4. *"Ten Redefined"*. First in a three-part Sickle Cell Disease Series by Hertz Nazaire. By permission of Hertz Nazaire.

numbness, are highly suggestive of neurologic pain.[37] The finding of nociceptive pain suggests opioid and non-steroidal anti-inflammatory agents (NSAIDS), whereas the finding of neuropathic pain suggests specific pharmacotherapy options involving the use of adjuvant analgesic drugs such as antidepressants or anticonvulsants alone or in combination with opioid analgesics, with or without NMDA receptor antagonists.[1,2] An essential part of a comprehensive assessment is a review of prior radiologic and laboratory studies to gauge disease-severity and to correlate signs of pathology with the acute vaso-occlusive event and other co-morbid symptoms.

Access to Quality Care

Access to medical care and the quality of care given at health centers and at home are inextricably linked to successful pain management. Care given at home, at local pharmacies, the doctor's office, and in hospital outpatient and inpatient facilities, is vital to therapeutic success; therefore the basic principles of pain management must be an integral part of standard practice at each location.[1,2,38–40]

These strategies for managing SCD have been extensively reviewed, including several guidelines that deal with empirically-derived treatments.[2,29,41,42]

Home Care

Several studies in children show the majority of pain is treated at home.[4,5,36] Ethnographics have been employed to recognize the effectiveness of pain management in this setting and the important role that educating family caregivers in pain management principles and practices play in this effort.[36] "Noetic" interventions, including religious thinking and prayer, have been proposed to improve coping, and the management of pain and psychological distress.[8] In a recent study,[40] it was determined that the events at home, and those requiring visits to the hospital constituted a continuum: events of varying pain severity with similar vaso-occlusive pathophysiology. Effective management at home primarily involved mild to moderate pain, while moderate to severe pain was most effectively managed in the hospital.

Outpatient Day Hospital and Day Care Centers

Day Hospitals and Day Care Centers are alternatives to emergency departments for outpatient treatment of moderate to severe acute painful episodes in some facilities in Europe, North America, and the Caribbean.[38,39,41] These dedicated units are access-friendly and apply assessment-determined packages of care that generally include opioids, NSAIDs, anti-histamines, and non-pharmacologic measures. Opioids are the mainstay of therapy. While the evaluation of opioids in SCD has received limited research attention, there are now several studies, from these units and others, reporting the safe and effective use of parenterally-administered opioid analgesics for the management of acute vaso-occlusive pain episodes.[38–43] Pain requirements and responses may be highly variable, therefore titration and individualization of doses that maximize analgesic effect and minimize toxic effects are integral.[1,2,38]

When home and hospital-based outpatient and inpatient care occurs along a continuum with the same principles of care applied relevant to the setting, pain management becomes a manageable aspect of overall care.

Disease-Specific Therapy

It is always desirable to treat the underlying causes of pain, utilizing disease-modifying treatments and agents targeting pain mechanisms. Hydroxyurea therapy instituted by private physicians and in outpatient clinics as prophylactic treatment will decrease the frequency and severity of acute painful episodes and acute chest syndrome in a subset of patients.[14] Since there are no approved disease-modifying therapies during an acute event, only supportive care therapies such as oxygen and hydration are currently available, to manage associated hypoxia and dehydration.[1,2,10] Analgesics are almost always required, not only for acute severe vaso-occlusive pain, but also for chronic intermittent or persistent pain. Adjuvant treatments are used to improve analgesia or for treatment of side effects.[43,44]

Disparities in Access to Care

Health disparity in pain management is a serious issue in SCD. Despite legislative efforts to ensure equality in government-sponsored programs, many persons with SCD do not have access to adequate pain management. Many patients endure unnecessary and long hospital stays due to poor pain control; they are often discharged with persistent moderate to severe pain, and have difficulty getting opioid prescriptions filled at neighborhood pharmacies. Other factors that impact the disparity in access to treatment are the attitudes of nurses, physicians, and other heath care providers concerning race and ethnicity.[45] Studies showing under-medication of persons of Hispanic and African-American ancestry have been reported.[46] These institutional biases are compounded by the findings that pharmacists in inner city neighborhoods, which are typically populated by communities with high gene frequencies of the sickle cell mutation, are less likely to carry opioids.[47] There are also within-group biases towards opioid therapy regarding frequency of pain. In one study of children with SCD, less medication was prescribed when patients expressed a history of frequent pain, as compared to larger dosing if the history reflected infrequent episodes.[48]

Inadequate Pain Treatment: A Major Public Health Problem

Physiological and Pharmacological Consequences

The consequences of inadequate treatment of pain have translated into a major public health problem.[49] Evidence is accruing that suggests outcomes of increased morbidity and mortality when pain is not relieved or, conversely, decreased morbidity and mortality when pain is relieved.[49] Unrelieved pain can lead to more injury and a heightened stress response, affecting cardiovascular, pulmonary, gastrointestinal, neuroendocrine, musculoskeletal, and cerebrovascular systems. With injury, persistent nociceptive input into the central nervous system induces sensitization of the dorsal horn and other neuronal populations.[49,50] Sensitization produces pain hypersensitivity by shifting the response curve to the left, relative to the normal, non-sensitized situation (Fig. 5).[51] Thus, pain prior to treatment has an element of hyperalgesia related to injury and inflammation-induced pain sensitization that must be addressed in treatment plans.[52] For example, non-steroidal anti-inflammatory

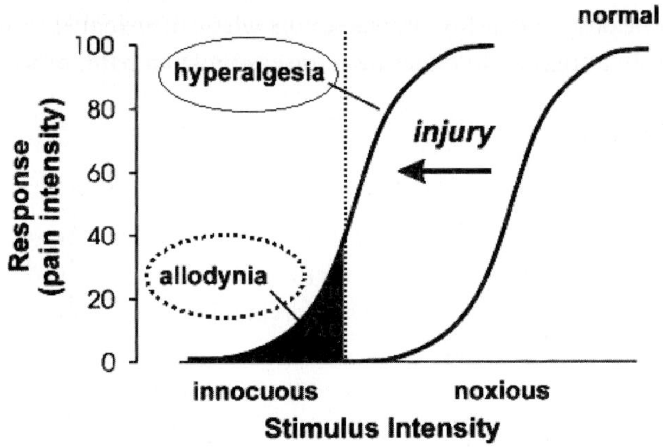

Fig. 5. *Pain hypersensitivity induced by injury.* The normal relationship between stimulus and the magnitude of pain sensation is represented by the curve at the right-hand side of the figure. Pain sensation is only evoked by stimulus intensities in the noxious range (the vertical dotted line indicates the pain threshold). Injury provokes a leftward shift in the curve relating stimulus intensity to pain sensation. Under these conditions, innocuous stimuli evoke pain (allodynia) and the pain sensitivity to noxious stimuli is increased (hyperalgesia). Used with permission from Cervero and Laird.[51]

drugs (NSAIDs) inhibit cyclooxygenase in the spinal cord and periphery with a resultant decrease in post injury inflammation-induced hyperalgesia.

Pain Under Treatment and Pseudoaddiction

The consequences of the medical under treatment of pain were described by Marks and Sachar in a psychiatric population 30 years ago.[53] Under treatment, and some cases of aberrant behaviors, were attributed to the findings that physicians tended to under prescribe analgesic agents; some nurses administered fewer analgesics than prescribed; some patients requested less analgesic medication than they needed; and the "as needed" regimen ensured that the pain, even if controlled, would recur.[54] Approximately 15 years later, this phenomenon of aberrant behaviors (drug-seeking, doctor-shopping, etc.) was evaluated in cancer patients, and found again to be secondary to inadequate treatment of pain. It was given the designation of "pseudoaddiction", an iatrogenic syndrome;[55] its pathogenesis was identified as unrelieved pain. In a recent study in the United Kingdom, pseudoaddiction was found to adversely influence hospital pain management of SCD patients.[56] The aberrant behaviors related to under treatment of pain make patients vulnerable to misperceptions of substance abuse and denial, or decreased access to opioids. These missteps are fueled by fears expressed by patient or family members, who believe that they will become addicted. Thus, contributing factors to pseudoaddiction have been enumerated as: (1) inadequate education about the diagnosis and treatment of pain by health professionals; (2) underutilization of existing pain management principles and techniques; and (3) excessive fear of tolerance and dependence, and their confusion with addiction by health professionals, patients, and the public.

Opioid-Induced Hyperalgesia: An Emerging Neurotoxicity Syndrome

Another form of aberrant behaviors occurs as a consequence of failure to recognize hyperalgesia syndromes related to opioid use.[57] Paradoxically, the chronic administration of opioid analgesics to treat pain may lead to similar confusion, by also contributing to or causing pain.[57–60] Increased sensitivity to pain may be observed in any clinical setting where recurrent acute or chronic pain occurs. This is often erroneously attributed to either more disease-related pain, or to aberrant drug-seeking or addictive behaviors.

NMDA and Opioid Receptor Interactions

It appears that similar pharmacological and molecular mechanisms, relating to interactions between the NMDA and opioid receptors, provide a basis for the paradoxical effects of opioids. These receptor systems are linked through intracellular pathways involving G-proteins, adenyl cyclase, protein kinase C, and cyclic guanosine monophosphate.[61]

As illustrated in Fig. 6, experimental studies with rodent models of persistent nociception have observed that persistent opioid receptor occupancy can lead to down regulation of receptor activity. The NMDA receptor system is crucial for central sensitization through linkages with nitric oxide and cyclic guanosine monophosphate intermediaries. This receptor system provides a potential mechanism to relate the phenomena of physical dependence, opioid hyperalgesia, and pharmacological tolerance. Non-NMDA mediated mechanisms, such as increased binding to proteins or increased clearance of opioids in an SCD transgenic mouse model, could lead to larger dose requirements for a sickle cell patient who has developed tolerance.[62]

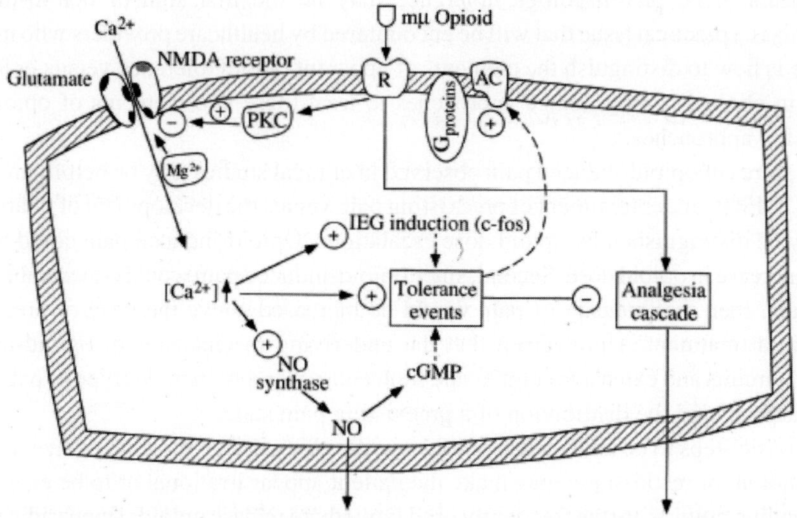

Fig. 6. *Inter-relationship between opioid and NMDA receptor systems.* Opioid receptor activation can influence NMDA receptor activity through the action of protein kinase C (PKC). NMDA receptor activation and calcium influx may also influence nitric oxide synthesis, and, acting through cyclic GMP mediated events, can feed back and down regulate opioid receptor activities. These inter-related pathways have implications for chronic pain states, opioid tolerance and hyperalgesia associated with prolonged exposure to morphine and other opioids. Used with permission from Elliott *et al.*[61]

Physical Dependence and Opioid Withdrawal-Related Pain

Physical dependence is a pharmacologic property of opioids, defined by the opioid withdrawal syndrome seen most commonly when opioid administration is abruptly terminated or reversed by the administration of an opioid antagonist. Precipitation of the opioid abstinence syndrome (so-called opioid "withdrawal") can ensue when persons who have been treated for an acute painful episode are not tapered slowly, or when persons on chronic opioid therapy miss a dose of medication.[2] The peaks of opioid analgesic actions are followed by periods of relative opioid withdrawal three to four times a day; methadone will stabilize those effects.[10,63,64]

Opioid withdrawal-associated hyperalgesia, an exacerbation of pain with heightened pain in the sites of their clinical involvement,[2,10,65] can mimic the pain caused by vaso-occlusion. Furthermore, withdrawal is emotionally and physiologically stressful, associated with activation of the hypothalamic-pituitary-adrenal axis,[63] thereby becoming a factor in the precipitation of a vaso-occulsive event. This can become a vicious cycle, and the consequences of the combination of frequently recurring unrelieved pain and opioid-induced hyperalgesia may become a disease which can inflict further physical and mental damage.[10,57,65–67]

Pharmacologic Tolerance and Tolerance-Associated Hyperalgesia

Diminishing efficacy of opioid analgesic during a course of therapy is often considered a sign of pharmacological tolerance. As such, an opioid dose escalation (increasing the dose or increasing the intervals) has been the conventional approach to restoring opioid analgesic effects. But increasing the opioid dose is not always the answer to ineffective opioid therapy.[68]

What appears to be pharmacologic tolerance may be the first sign of opioid-induced pain sensitivity. This is a practical issue that will be encountered by healthcare providers who manage pain. The challenge is how to distinguish the elements of apparent opioid tolerance versus opioid-induced hyperalgesia in clinical settings, since approaches to treat these two elements of opioid tolerance require opposing approaches.

Several features of opioid-induced pain observed in clinical studies may be helpful in recognizing the syndrome.[68] First, under treatment of preexisting pain versus the development of pharmacological tolerance may be distinguished by opioid dose escalations. Opioid-induced pain could be worsened following an increase in opioid dose. Second, since opioid-induced pain would conceivably exacerbate preexisting pain, then the intensity of pain would be increased above the level of preexisting pain following opioid treatment. Third, given that the underlying mechanisms of opioid-induced pain involve neural circuits and extensive cellular and molecular changes, then this type of pain is expected to be diffuse and beyond the distribution of a preexisting pain state.

There are basic steps to be taken when opioid neurotoxicity exists. First, recognize the syndrome. Delirium, agitation, or restlessness may make the patient appear irrational or to be exaggerating the pain. The offending opioids are the frequently used immediate release opioids (meperidine, morphine, hydromorphone or oxycodone) or very high doses of sustained release formulations of morphine, hydromorphone, and oxycodone.[44,57,66] An early sign may be clonus, which can be seen while the patient is asleep before it becomes clinically overt. Allodynia and hyperalgesia can cause the pain to occur all over and does not necessarily follow a reasonable distribution. Rapidly increasing the opioid makes the pain worse. Second, discontinue the offending opioid and rotate to another drug. Third, add additional non-opioid adjunctive medications. Fourth begin hydration to clear opioid metabolites. Fifth, consider benzodiazepines to decrease neuromuscular irritability but avoid sedation.

NMDA Receptor Antagonists

From a mechanistic standpoint, pharmacological antagonists of the NMDA receptor may reverse hyperalgesia and tolerance.[58] Although the optimal NMDA receptor antagonist has not yet been identified, this has enormous implications for pain management. Understanding the mechanisms involved in these complications will hopefully prevent withdrawal and tolerance-associated hyperalgesia. Treatment with an effective NMDA receptor antagonist will be helpful. Drugs such as ketamine, dextorphan, dextromethorphan, and amantadine have NMDA antagonist properties. Most of these compounds have a very narrow therapeutic index due to unacceptable effects on mental functioning. Studies are underway to evaluate ketamine in sub-anesthesia doses and the antitussive agent dextromethorphan as NMDA receptor antagonists.

Opioid Rotation

Fentanyl or Sufentanil

There are advantages of rotating from morphine or hydromorphone to fentanyl or sufentanil in treating opioid-induced neurotoxicity in a hospital setting. They are in a different class of opioids than morphine and hydromorphone; there are no known active metabolites, and neurotoxicity is uncommon.[68] Also, sufentanil is not used on an outpatient basis therefore tolerance will not be an issue during treatment. Because of its rapid onset, titration is facilitated in difficult circumstances. If recognized before life-threatening neurotoxicity occurs, then methadone is the drug of choice based on the mechanisms that produce symptoms in this syndrome (see below).

Combine Opioid Rotation and NMDA Receptor Antagonist Therapy

Methadone

Co-administration of a new opioid and adjuvant therapy is a reasonable approach to treat opioid-induced neurotoxicity. Switching to the opioid agonist methadone accomplishes both opioid rotation and the use of an NMDA receptor antagonist. Methadone can be used with increasing frequency in patients with difficult pain syndromes such as those patients who become refractory to immediate release and controlled release opioids when used chronically; and for treatment of pharmacological tolerance and opioid-induced pain, allodynia, and hyperalgesia.[69,70] The long half-life of methadone prevents or attenuates withdrawal, but may also be detrimental if dosing and increasing blood levels are not carefully monitored. Many physicians have reported the safe and effective use of methadone, especially in the setting of cancer-related pain.[71]

Methadone has unique pharmacology which makes it a rational choice compared to other μ opioids. It is active at the μ and δ opioid receptors; exerts antagonistic activity at the NMDA receptor, and counteracts the development of opioid tolerance. This may be the basis for lower opioid dose escalation requirements in patients treated with methadone compared to morphine.[72] The antagonistic activity of methadone at the NMDA receptor (a subtype of glutaminergic receptors that play a key role in central sensitization) results in efficacy against allodynia and hyperalgesia, and may explain its greater efficacy against neuropathic pain and other pain states.[70] Finally, methadone has serotonin and norepinephrine reuptake inhibitor properties, the mechanism by which drugs used to treat neuropathic pain work.[70,71] The multiple actions of methadone may also explain the exacerbation of pain and withdrawal sometimes observed when patients are switched or rotated from methadone to other μ receptor antagonist opioids.[70]

Recurring acute pain and recidivism in SCD pose very difficult management problems suggestive of tolerance, hyperalgesia, and withdrawal. Methadone has been evaluated as part of a treatment strategy for such patients.[10] In a single center open label study employing a biopsychosocial approach to subjects with persistent pain, switching or rotation from short-acting opioids to long-acting methadone, and/or tapering resulted in decreased pain, reversal of uncontrolled frequent pain state, improved functioning, and decreased hospital admissions. Methadone was central to reversing the uncontrolled state, with important contributions from adjuvant therapies employing pharmacological, psychological, and physical techniques.

These findings also indicate that in converting from opioids such as morphine, hydromorphone, and oxycodone to methadone, there should be at least five-fold lower dosing than the conversion charts suggest, because of the need to make adjustments for incomplete cross tolerance, prevention of withdrawal and in some cases of hyperalgesia. Similar dosing conversion ratios were determined utilizing cancer-related pain treatment.[10,70] Using these techniques, the average daily dose ranged from 20 to 90 mg, in divided doses ranging from 10 mg every twelve hours to 30 mg every eight hours. No serious adverse effects were noted in the study (see below).

Caution in the Use of Methadone

There are reports indicating the possibility of serious side effects for intravenous or oral dosing that exceeds 200 mg of methadone per day, which prompts a note of caution. Although many experienced clinicians have reported the safe and effective use of methadone,[71] there are recent reports of significant toxicity, including deaths, associated with the use of this analgesic.[73] It is well known that methadone has a long and variable plasma half-life of elimination; the consequence is that if it is administered too frequently, it may accumulate, producing sedation and respiratory depression, especially in opioid-naïve patients. This may be avoided by starting methadone as and when needed rather than a fixed schedule, to minimize drug accumulation associated with frequent dosing.[74] An around-the-clock regimen can be established once the 24-hour dose requirement is known. Recent data on opioid conversion ratios suggests that the higher the morphine dosing, the lower will be the methadone equivalent dosing.[70] Failure to make these conversion adjustments can result in methadone dosing 5–20 times higher than indicated, thereby increasing the likelihood of serious adverse effects.

Fatalities have been reported when inappropriately large oral or intravenous doses have been given. Deaths may be caused by respiratory depression however other mechanisms of death have been reported. Methadone is potentially cardiotoxic — prolongation of the QTc interval on electrocardiograms leading to a potentially fatal ventricular arrhythmia, torsades de pointes, has been reported with large oral doses (>200 mg/day), or with continuous intravenous infusion of methadone formulated with chlorbutanol as a solvent.[75] A recent non-fatal case of torsades de pointes was reported in a sickle cell patient receiving greater than 200 mg of methadone per day;[76] this patient also had a negative screen performed as a result of a history of substance abuse, known to cause prolongation of the QTc interval. A current study reviewed electrocardiograms in 111 patients treated with methadone with doses up to 1200 mg per day; prolongation of the QTc interval was also noted.[77] None had QTc intervals that were 500 msec or more, the effect most often associated with torsades de pointe.

Finally, measures must be taken to limit toxicities so that the benefits of this multimechanistic drug can be exploited to benefit persons with life quality markedly impaired because of ineffective pain management. Current strategies derived from the conversion studies should result in dosing levels that do not approach those reported to cause cardiac toxicities. A baseline electrocardiogram should be obtained prior to treatment, irrespective of the dose of methadone. Intravenous usage should be avoided, and the use of other agents that are known to prolong the QTc interval, such as clopromazine, clarithromycin, and haloperidol should be monitored carefully.[70,77] The QTc interval is also prolonged by illicit drugs such as cocaine. Methadone is clearly a unique opioid analgesic that has been shown to be beneficial in difficult pain syndromes. With a resurgence of methadone usage in pain management, approaches to safely rotate to or from this agent in sickle cell patients require more detailed studies.

Practical Matters: Post Script

In the challenge to conquer pain, it is disconcerting as practitioners to know that the toxic effects of pain treatment regimens can produce more pain. Educating and encouraging healthcare providers to be aggressive in pain management, with an awareness of how to recognize pain management syndromes will decrease treatment limitations due to opioid-induced toxicities. Conquering pain involves meeting the challenge to clearly understand the nature of pain, to apply principles and best practices, and to prevent, recognize or reverse the untoward consequences of under treatment and over treatment of pain.

Summary

The Congress of the United States declared the current decade (2001–2010), as the Decade of Pain Research and Control,[78] and yet too little research attention has been directed at sickle pain; success at controlling this phenotypic hallmark of SCD remains extremely variable. A sophisticated effort directed at discovering targets for new analgesics will be required in the genomic era to bridge the gap between neurobiology of pain and clinical pain therapy. The implications of these observations, in composite for sickle cell pain, are that the indiscriminate use of opioid analgesics and the under treatment of pain promote prolonged pain and increase opioid requirements; begetting more pain and complications from treatment in an ever-increasing spiral. Opioids need to be titrated aggressively to control pain and then tapered to no drug with subsidence of acute pain or to a level that maintains control of persistent pain. Strategies to limit opioid exposure, such as opioid rotation, or combining non-opioid analgesics should be employed to prevent hyperalgesia. NMDA receptor antagonists such as ketamine and dexthomethorphan are especially attractive however they require further investigation in sickle cell-related pain.

Clinical approach to pain management should be reevaluated. There is a great need to change from empirical management strategies and to adopt mechanistically driven approaches to develop novel pain management therapies. The concept of personalized medicine is most apt in this regard, and heralds a major paradigm shift in pain management that is also tailored to individual capabilities or peculiarities in drug metabolism.

The future availability of technology developed during the mapping and sequencing of the human genome should hasten our success at effective pain treatment regimens. Microarray and high-throughput genomic technologies may permit clear definition of risk factors and molecular classification of these factors. Gene profiling can be used as a fingerprint for diagnostic, mechanistic, and therapeutic assessment of markers related to sickle-related pain. The investigation of psychosocial determinants of pain tolerance and pain behaviors must also continue unabated since variability in pain pathology is determined to a large extent by environmental factors. The success of the Human Genome Project, and advances in pain research provide hope for individuals with SCD. Knowledge of the molecular-genetic basis of pain-related traits and environmental factors may facilitate the development of novel analgesic strategies targeting specific mechanisms, and improve pain management using conventional therapies.

References

1. US Department of Health and Human Services, National Institute of Health, National Heart, Lung, and Blood Institute, Division of Blood Diseases and Resources (2002). *The Management of Sickle Cell Disease*, 4th edition. NIH Publication, No. 02–2117.
2. Benjamin LJ, Dampier CD, Jacox AK *et al.* (1999). *Guideline for the Management of Acute and Chronic Pain in Sickle-Cell Disease*, APS Clinical Practice Guidelines Series No 1. Glenview, IL.
3. Payne, R (1997). Pain management in sickle cell anemia. *Anesth Clin N Am* **15**:306–318.
4. Gil KM, Porter L, Ready J *et al.* (2000). Pain in children and adolescents with sickle cell disease: an analysis of daily pain diary. *Children's Health Care* **29**:225–241.
5. Dampier C, Ely B, Brodericki D, O'Neal P (2002). Characteristics of pain managed at home in children and adolescents with sickle cell disease by using diary self-reports. *J Pain* **3**:461–470.
6. Platt OS, Brambilla DJ, Rosse WF *et al.* (1994). Mortality in sickle cell disease. Life expectancy and risk factors for early death. *N Engl J Med* **330**:1639–1644.
7. Stuart MJ, Nagel RL (2004). Sickle Cell Disease. *Lancet* **364**(9442):1343–1360.
8. Harrison OM, Edwards CL, Koenig HG *et al.* (2005). Religiosity/Spirituality and pain in patients with sickle cell disease. *J Nerv Mental Dis* **193**(4):250–257.
9. Ballas SK (1998). Sickle cell pain. In: *Progress in Pain Research and Management*, Vol. 11. Seattle, WA: IASP Press.

10. Benjamin LJ (2001). The nature and management of acute painful episode in sickle cell disease. In: Forget BG, Higgs D, Nagel RL, Steinberg MH, editors. *Disorders of Hemoglobin*. Cambridge, UK: Cambridge University Press, pp. 671–710.
11. Hebbel RP, Osarogiagbon R, Kaul D (2004). The endothelial biology of sickle cell disease: inflammation and a chronic vasculopathy. *Microcirculation* **11**(2):129–151.
12. Bonds DR (2005). Three decades of innovation in the management of sickle cell disease: the road to understanding the sickle cell disease clinical phenotype. *Blood Rev* **19**:99–110.
13. Platt OS, Thorington BD, Brambilla DJ *et al.* (1991). Pain in sickle-cell disease: rates and risk factors. *New Engl J Med* **325**:11–16.
14. Charache S, Terrin ML, Moore Rd *et al.* (1995). Effect of hydroxyurea on the frequency of painful crises in sickle cell anemia. Investigators of the Multicenter Study of Hydroxyurea in Sickle Cell Anemia. *N Engl J Med* **332**:1317–1322.
15. Miller ST, Sleeper LA, Pegelow CH *et al.* (2000). Prediction of adverse outcomes in children with sickle cell disease. *N Engl J Med* **342**(2):83–89.
16. Steinberg MH, Hsu H, Nagel RL *et al.* (1995). Gender and haplotype: effects upon hematological manifestations of adult sickle cell anemia: effects of haplotype in sickle cell anemia. *Am J Hematol* **48**:175–181.
17. Powars DR *et al.* (1994). Beta-S gene cluster haplotypes modulate hematologic and hemorrheologic expression in sickle cell anemia. Use in predicting clinical severity. *Am J Pediatr Hematol Oncol* **16**:55–61.
18. Alexander N, Higgs D, Dover G, Sergeant GR (2004). Are there clinical phenotypes of homozygous sickle cell disease? *Br J Haematol* **126**(4):606–611.
19. Gil KM, Carson JW, Porter LS *et al.* (2004). Daily mood and stress predict pain, health care use and work activity in African American adults with sickle cell disease. *Health Psychol* **23**(3):267–274.
20. Niv D, Dvor M (1998). Transition from acute to chronic pain. In: Arnoff GM, editor. *Evaluation and Treatment of Chronic Pain*, 3rd edition. Baltimore: Williams and Wilkins.
21. Eap CB, Budin T, Bauman P (2002). Interindividual variability of the clinical pharmacokinetics of methadone: implications for the treatment of opioid dependence. *Clin Pharmacokinet* **41**:1153–1193.
22. Scholz J, Woolf CJ (2002). Can we conquer pain? *Nat Neurosci* **5**(Suppl): 1062–1067.
23. Sabino MA, Mantyh PW (2005). Pathophysiology of bone cancer pain. *J Support Oncol* **3**(1):15–24.
24. Julius D, Basbaum AI (2001). Molecular mechanisms of nociception. *Nature* **413**(6852):203–210.
25. Watkins LR, Maier SF (2000). The pain of being sick: implications of immune-to-brain communications for understanding pain. *Ann Rev Psychol* **51**:29–57.
26. Mankad VN, Williams JP, Harpen MD *et al.* (1990). Magnetic resonance imaging of bone marrow in sickle cell disease: clinical, hematologic, and pathologic correlations. *Blood* **75**(1):274–220.
27. Kim SK, Miller JH (2002). Natural history and distribution of bone and bone marrow infarction in sickle hemoglobinopathies. *J Nucl Med* **43**(7):896–900.
28. Kavadia-Tsatala S, Kolokytha O, Kaklamanos EG, Antoniades K, Chasapopoulou E (2004). Mandibular lesions of vasocclusive origin in sickle cell hemoglobinopathy. *Odontology* **92**(1):68–72.
29. Yaster M, Byerly S, Maxwell L (2000). The management of pain in sickle cell disease. *Pedia Clin North Am* **47**:699–710.
30. Fuggle P, Shand PA, Gill LJ, Davies SC (1996). Pain, quality of life, and coping in sickle cell disease. *Archives of Diseases in Childhood* **75**:199–203.
31. Zubieta J-K, Heitzeg MM, Smith, YR *et al.* (2003). COMT val158met genotype affects u-opioid neurotransmitter responses to a pain stressor. *Science* **299**:1240–1244.
32. Coda BA, Bonica JJ (2001). General considerations of acute pain. In: Loeser JD, Butler SH, Chapman CR, Turk DC, editors. *Bonica's Management of Pain.* Philadelphia: Lippincott Williams & Wilkins, Ch. 9, pp. 222–240.
33. Jambrik Z, Santarcangelo EL, Rudisch T, Varga A *et al.* (2005). Modulation of pain-induced endothelial dysfunction by hypnotisability. *Pain* **116**:181–186.
34. Farrar JT, Portenoy RK, Berlin JA, Kinman JL, Strom BL (2000). Defining the clinically important difference in pain outcome measures. *Pain* **88**:287–294.
35. Fishman B, Pasternak S, Wallenstein SL *et al.* (1987). The Memorial Pain Assessment Card: a valid instrument for the evaluation of cancer pain. *Cancer* **60**:1151–1158.

36. Beyer JE, Simmons LE (2004). Home treatment of pain for children and adolescents with sickle cell disease. *Pain Manage Nurs* **5**(3):126–135.
37. Backonia MM, Stacey B (2004). Neuropathic pain symptoms relative to overall pain rating. *J Pain* **5**: 491–497.
38. Benjamin W, Swinson GI, Nagel R (2000). Sickle cell anemia day hospital: an approach for the management of uncomplicated painful crisis. *Blood* **95**:1130–1137.
39. Ware M, Hambleton I, Ochaya I, Sergeant GR (1999). Day care management of sickle cell painful crisis in Jamaica: a model applicable elsewhere? *Br J Haematol* **104**:93–96.
40. Dampier C, Ely B, Broderick D *et al.* (2004). Vaso-Occlusion in children with sickle cell disease: clinical characteristics and biologic correlates. *J Pediatr Hematol/Oncol* **26**(12):785–790.
41. Wright J, Bareford D, Wright C *et al.* (2004). Day case management of sickle pain: 3 years experience in a UK sickle cell unit. Br J Haematol **126**(6):878–880.
42. Rees DC, Olujohungbe AD, Parker NE *et al.* (2003). Guidelines for the management of the acute painful crisis in sickle cell disease. *Br J Haematol* **120**(5):744–752.
43. Marlone KF, Chicella MF (2002). Treatment of sickle cell pain. *Pharmacotherapy* **22**(4):484–491.
44. Bruera E, Pereira J (1997). Neuropsychiatric toxicity of opioids. In: Jensen TS, Turner JA, Wissenfeld-Hallin Z, editors. *Proceedings of the Eighth World Congress on Pain*, Vol. 8.
45. Rouse CM (2004). Paradigms and politics: shaping health care access for sickle cell patients through the discursive regimes of biomedicine. *Cult Med Psychiatry* **28**(3):369–399.
46. Bonham VL (2001). Race, ethnicity, and pain treatment: striving to understand the causes and solutions to the disparities in pain treatment. *J Law Med Ethics* **29**:52–68.
47. Morrison RS, Wallenstein S, Natale D, Senzel RS, Huang LL (2000). "We don't carry that" — failure of pharmacies in predominantly nonwhite neighborhoods to stock opioid analgesics. *N Engl J Med* **342**: 1023–1026.
48. Armstrong ED, Pegelow CH, Gonzalez JC, Martinez A (1995). Impact of children's sickle cell history on nurse and physician ratings of pain and medication decisions. *J Pediatric Psychol* **71**:651–664.
49. Melzack R (1990). The tragedy of needless pain. *Sci Am* **262**(2):27–33.
50. Pasero C, Paice JA, McCaffery M (1999). Basic mechanism underlying the causes and effects of pain. In: McCaffery M, Pasero C, editors. *Clinical Manual.* Mosby, Inc., pp. 15–34.
51. Cervero F, Laird JMA (1996). Mechanisms of touch-evoked pain (allodynia): a new model. *Pain* **68**:13–23.
52. Wolff CJ, Salter MW (2000). Neuronal plasticity: increasing the gain in pain. *Science* **288**:1765–1769f.
53. Marks RM, Sacher EJ (1973). Undertreatment of medical inpatients with narcotic analgesics. *Ann Intern Med* **78**:173–181.
54. Bennett P, St. Marie B (2002). *The Epidemiology of Pain*, Core Curriculum for Pain Management Nursing. Philadelphia: W.B. Saunders Company, pp. 45–53.
55. Weissman DE, Haddox JD (1989). Opioid pseudo-addiction: an iatrogenic syndrome. *Pain* **36**:363–366.
56. Elander J, Lusher J, Bevan D, Telfer P, Burton (2004). Understanding the causes of problematic pain management in sickle cell disease: evidence that pseudoaddiction plays a more important role than genuine analgesic dependence. *J Pain Symptom Manage* **27**(2):156–169.
57. Mercante S, Ferraera P, Villari P, Arcuri E (2003). Hyperalgesia: an emerging Iatrogenic syndrome. *J Pain Symptom Manage* **26**:769–775.
58. Mao J, Sung B, Ji RR, Lim G (2002). Chronic morphine indices down regulation of spinal glutamate transporters: implication in morphine tolerance and abnormal pain sensitivity. *J Neurosci* **22**(18): 8312–8323.
59. Ballantyne JC, Mao J (2003). Opioid therapy for chronic pain. *N Engl J Med* **349**(20):1943–1953.
60. Mayer DJ, Mao J, Holt J, Price DD (1999). Cellular mechanisms of neuropathic pain, morphine tolerance, and their interactions. *Proc Natl Acad Sci USA* **96**:7731–7736.
61. Elliott K, Kest B, Inturrisi CE (1995). NMDA receptors, mu and kappa opioid tolerance and perspectives on new drug development. *Neuropsychopharmacology* **13**:347–356.
62. Nagar S, Remmel RP, Hebbel RP, Zimmerman CL (2004). Metabolism of opioids is altered in liver microsome of sickle cell transgenic mice. *Drug Metab Dispos* **32**(1): 98–104.
63. Kreek MJ (1992). Rationale for maintenance pharmacotherapy of opiate dependence. In: O'Brien CP, Jaffe JH, editors. *Addictive States*. New York: Raven Press, pp. 205–230.

64. Kreek MJ (1996). Opiates, opioids and addiction. *Mol Psych* **1**:232–254.
65. Compton P, Athanasos P, Elashoff D (2003). Withdrawal hyperalgesia after acute opioid physical dependence in nonaddicted humans: a preliminary study. *J Pain* **4**:11–519
66. Linklater DR, Pemberton L, Taylor S, Zeger W (2005). Painful dilemmas: an evidence-based look at challenging clinical scenarios. *Emerg Med Clin North Am* **23**(2):367–392.
67. Ballas SK, Lusardi M (2005). Hospital readmission for adult acute sickle cell painful episodes: frequency, etiology, and prognostic significance. *Am J Hematol* **79**(1):17–25.
68. Mao J (2002). Opioid-induced abnormal pain sensitivity: implications in clinical opioid therapy. *Pain* **100**:213–217.
69. Rapamonti C, Dickerson ED (2001). Strategies for the treatment of cancer pain in the new millennium. *Drugs* **61**(7):955–977.
70. Manfredi PL, Houde RW (2003). Prescribing methadone: a unique analgesic. *J Support Oncol* **1**:216–220.
71. Rosen SH, Zeldin G, Davidson EM (2001). Methadone is safe for treatment of hospitalized patients with severe pain. *Can J Anest* **48**(11):1109.
72. Davis AM, Inturrisi CE (1999). d-Methadone blocks morphine tolerance and N-methyl-D-aspartate (NMDA)-induced hyperalgesia. *J Pharmacol Exp Ther* **289**:1048–1053.
73. US Dept HHS (2004). *Methadone-Associated Mortality: Report of a National Assessment.*
74. Foley KM, Houde RW (1998). Methadone and cancer pain management: individualize dose and titrate to effect. *J Clin Oncol* **16**:3213–3215.
75. Kornick CA, Kilborn MJ, Santiago-Palma J *et al.* (2003). QTc interval prolongation associated with intravenous methadone. *Pain* **105**:499–506.
76. Porter BO, Coyne PJ, Smith WR (2005). Methadone-related Torsades de Pointes in a sickle cell patient treated for chronic pain. *Am J Hematol* **78**:316–319.
77. Cruciani RA, Sekine R, Homel P, Lussier D *et al.* (2005). Measurement of QTc in patients receiving chronic methadone therapy. *J Pain Symptom Manage* **29**:385–391.
78. Lippe PM (2000). The decade of pain control and research. *Pain Med* **1**(4):286–289.

8

Transfusion Therapy in Sickle Cell Disease

by Carolyn Hoppe, Robert Adams and Elliot Vichinsky

Introduction

Several major technological advances have changed best practices for routine clinical care in sickle cell disease (SCD). The impact of these advances on individuals with SCD who are at increased risk for stroke will be discussed through a review of the study conception, design and results of the Stroke Prevention Trial in Sickle Cell Anemia (STOP). The effect of the STOP study on the natural history of other SCD-related complications will be presented, and advances in transfusion medicine and the technology for the spectrum of transfusion options now available will be reviewed.

Stroke in Sickle Cell Disease

In 1923, Sydenstricker reported the case of a five-year-old sickle cell patient with hemiparesis and recurrent seizures. Twelve years later, Arena described another sickle cell child with headache, stupor, and hemiplegia, and additional case reports of SCD and stroke were soon published.[1] In 1972, Stockman and colleagues documented large cerebral vessel occlusions; since then, large vessel cerebrovascular disease has received considerable attention.[2] What causes these predisposing lesions within cerebral arteries and promotes the worsening of these lesions to the point of critical stenosis or occlusion, and what finally triggers the onset of symptomatic ischemia in SCD, are not known. It is also not clear why the intracranial arteries are particularly targeted in a disease that primarily involves the microvasculature.[3]

Epidemiology of Stroke

The Baltimore–Washington Cooperative Young Stroke Study retrospectively identified all cases of ischemic and hemorrhagic stroke among children and young adults within the Baltimore–Washington area between 1988 and 1991, and reported an overall incidence of childhood stroke of 1.29 per 100,000 per year.[4] A 200-fold greater incidence was found in the subgroup of children with SCD, with an annual rate of 285 per 100,000. This estimate agrees with the Cooperative Study of Sickle Cell

Disease (CSSCD), which reported an annual incidence during childhood of 460 per 100,000 for all SCD genotypes (HbSS, HbSC, and HbS/-thalassemia), and a rate of 610 per 100,000 in those with sickle cell anemia (HbSS).[5] The incidence of first stroke was found to be age-dependent, peaking in children between 2 and 5 years, followed by children between 6 and 9 years, with annual incidence rates of 1020 per 100,000 and 680 per 100,000, respectively. A second peak in ischemic stroke occurred in adults greater than 30 years old; in contrast, hemorrhagic stroke predominated in young adults between the ages of 20 and 29 years. The risk of recurrence, typically within the first three years after the first stroke, is as high as 67% in untransfused patients.[6]

Diagnosis of Stroke

Stroke is a clinical syndrome characterized by focal symptoms, such as hemiparesis or hemisensory deficits. Symptoms depend on the location, size, and extent of vascular territory involved (large artery or microvascular).[2,7,8] Neuroimaging with cranial-computed tomography (CT) is typically the initial study performed to rule out alternative diagnoses, and to distinguish cerebral hemorrhage from infarction. Magnetic Resonance Imaging (MRI), with the recent addition of diffusion-weighted imaging (DWI), allows early detection of brain ischemia within an hour after stroke onset.[9] Magnetic resonance angiography (MRA) adds information on the status of the cerebral vasculature, and has replaced intra-arterial catheter angiography as an accurate and non-invasive technique to detect cerebral artery lesions.[10]

Newer imaging techniques that assess brain physiology are now being used to identify ischemic damage prior to the development of lesions on conventional imaging studies. Abnormalities in glucose metabolism and microvascular blood flow, particularly in the frontal lobes, have been demonstrated in SCD patients using positron emission tomography (PET).[11–13] Perfusion magnetic resonance (dynamic susceptibility contrast MRI) and blood oxygen level-dependent (BOLD) MRI are additional imaging modalities to assess cerebral blood flow and perfusion. Using PMR, Kirkham *et al.*[14] reported perfusion abnormalities related to various neurological symptoms, including headache, seizures, TIA, and stroke, in children with SCD. Although the adjunctive use of these functional imaging techniques appears promising, further clinical investigations are needed.

Neurocognitive Functioning and Silent Infarction

Almost 25% of children with SCA (HbSS) exhibit "silent" cerebral infarcts, appearing on MRI as punctate hyperintensities in the deep white matter of the brain. These ischemic lesions are a result of small vessel occlusion, and usually occur in arterial border zones. Children with silent infarcts identified by MRI may appear asymptomatic, but perform significantly lower on neuropsychological tests than their counterparts with a normal MRI.[7] Data from the CSSCD revealed an association between silent infarcts and future stroke;[15,16] the prevalence of silent infarcts in asymptomatic adults with SCD has been reported to be as high as 52%,[17] but data are limited. Autopsy studies have confirmed subclinical ischemic brain injury in many cases.[18–21]

"Silent" infarcts are often associated with irreversible and progressive neurocognitive impairments that are reflected in standardized tests used to predict employability and higher-level skills of daily life,[7,16,22,23] and may contribute to lower quality of life in SCD.[24–26] Red cell transfusion likely prevents the progression of silent infarcts, but the definitive clinical trial Silent Infarction Treatment Trial (SITT) is still in progress.[27]

Intracranial hemorrhage represents a smaller, but clinically significant, stroke phenotype in SCD, often manifesting with dramatic and non-focal symptoms including severe headache or coma. Intracranial hemorrhage may be caused by hemorrhagic conversion of a large brain infarction, rupture of friable moya-moya vessels, or cerebral aneurysms.[5,28] Presumably, the same pathophysiologic pathways responsible for the development of large vessel occlusion continue to cause progressive weakening of large intracerebral vessels, and ultimately hemorrhage due to ruptured intracerebral vessels.[29] Alternatively, subarachnoid hemorrhage (SAH) may result from rupture of berry aneurysms, and is associated with high morbidity and mortality. To avoid missing potentially correctable anomalies such as aneurysms or arteriovenous malformations, catheter arteriography should be performed on all SCD patients with intracranial hemorrhage, with the exception of those who have a CT and clinical picture that suggest hemorrhagic conversion of a brain infarction.[30]

Pathophysiology of Stroke

Cerebral infarction usually occurs in the distribution of the large vessels comprising the anterior circle of Willis, most often as a result of stenosis or occlusion in the area of the bifurcation of the carotid artery.[31] Arteriography or MRA often demonstrates progressive narrowing of these vessels, with collateral vessel development in a pattern very similar to Moyamoya disease.[32] Pathologic studies of these segments suggest that these lesions contain areas of intimal proliferation, as well as fragmentation and reduplication of the internal elastic lamina without significant lipid or cellular content.[21,33] Stroke is caused by perfusion failure due to critical limitation of blood flow in areas of stenosis, or to arterial embolization, but the pathophysiology underlying the development of the complex vascular lesions leading to brain injury is not fully understood. Factors that have been implicated include: (1) The sickle red blood cell itself, and its potential to adhere to the endothelium; (2) leukocyte-mediated injury to the endothelium; (3) thrombophilia, as part of a generalized hypercoagulable state; (4) inflammation mediated by cytokines; and (5) quantitative and qualitative platelet abnormalities.[34–38]

Lack of an appropriate animal model and inaccessibility to the pathologic lesions in SCD patients with stroke make it difficult to establish which of these mechanisms is most crucial to the development of cerebrovascular disease. One fact that emerges, however, is that regular red cell transfusions are effective in the prevention of both primary and recurrent ischemic stroke. The specific mechanisms by which regular transfusion therapy ameliorates cerebrovascular disease in SCD is not entirely clear, but it may do so in part by improving oxygen saturation and alteration of oxygen affinity.[39]

Risk Factors for Ischemic Stroke

The CSSCD provided the best evidence for risk factors leading to brain infarction and hemorrhage through a population study performed in the 1980s. Prior transient ischemic attack, a low steady-state hemoglobin level, hypertension, and a history of acute chest syndrome (ACS) were the only variables found to be significantly associated with cerebral infarction on multivariate analyses.[5] Interestingly, a high fetal hemoglobin level, which has been shown to have an ameliorative effect on almost all other complications of SCD, did not significantly correlate with stroke. Multivariate analysis identified a low baseline hemoglobin level and an elevated white blood cell count as the only significant predictors of hemorrhagic stroke. Although informative, none of the risk factors found in the CSSCD could sufficiently predict which patients were at highest risk for stroke to test for potential treatment interventions.

A genetic predisposition to stroke is supported by studies documenting specific risk-conferring β-globin gene haplotypes[40] and protection from coexistent alpha thalassemia.[41] Associations with thrombophilic genetic traits outside of the β-globin locus have also been reported.[42] Some small studies investigating stroke associations with polymorphisms in the factor V, methylene tetrahydrofolate reductase (MTHFR) glycoprotein IIIA, prothrombin, low-affinity Fc-gamma receptor and tumor necrosis factor alpha (TNFA) genes have been negative;[43–48] others have shown an increased stroke risk with specific angiotensinogen gene GT repeat alleles[49] and decreased protein C and S activity levels.[50] The influence of elevated homocysteine on stroke risk in SCA remains unproven.[44,51,52] Particular HLA associations with stroke have been documented in children previously enrolled in the CSSCD.[53] Several other risk-associated genes, including VCAM1, were subsequently identified in the same cohort.[54] Taylor *et al.* also reported stroke associations with VCAM1, suggesting that several polymorphisms within this locus may jointly influence stroke risk.[54,55]

While the findings from these studies suggest a number of plausible genetic risk associations with stroke, confirmatory studies are needed. To this end, powerful bioinformatics tools have recently been applied to the study of stroke in SCD. Using a machine-learning method based on Bayesian networks, Sebastiani and coworkers developed a multigenic model to predict individual stroke risk in SCD.[56] As additional candidate genes and clinical markers are identified, such prediction algorithms will undoubtedly become an invaluable tool in genetic association studies of SCD.

Primary Prevention of Ischemic Stroke in Children

Transcranial Doppler ultrasound (TCD), a non-invasive method of measuring flow velocities in the intracranial arteries, is the single most important predictor of stroke in children with SCD.[57] In the early 1990s, Adams *et al.* first demonstrated the effectiveness of TCD screening for cerebrovascular abnormalities and its potential use for risk stratification in SCD.[57,58] An elevated time-averaged mean of the maximum (TAMM) flow velocity, measured in the distal internal carotid artery (dICA) or the proximal middle cerebral artery (MCA), indicates either increased cerebral blood flow (appearing as widespread increases in velocity) or arterial stenosis (focal increase in velocity). Due to anemia, children with SCD generally have higher baseline TCD flow velocities (130–140 cm/sec) than children without SCD (90 cm/sec).[59]

The TCD stroke risk model demonstrated that risk increased as a function of flow velocity without any apparent threshold effect. In two independent studies, a TAMM flow velocity threshold greater than 200 cm/sec identified children with a 10% risk of stroke for the ensuing one to three years, if not treated with prophylactic transfusions.[57] This led to a multicenter randomized clinical study, the Stroke Prevention in Sickle Cell Anemia Trial (STOP),[60] in which TCD was used to screen over 2000 asymptomatic children with SCD (HbSS or HbS/β^0 thalassemia) aged 2–16 years. Of the 200 children with a high risk TCD, 130 were randomized to receive either regular transfusions (to keep Hb%S $< 30\%$) ($n = 63$) or standard care which could include episodic transfusions ($n = 67$). The trial was stopped early, as transfusion was shown to decrease the risk of stroke by 90%.[61] Based on the findings of the STOP study, a Clinical Alert recommending TCD screening and chronic blood transfusions for children with sickle cell disease at high stroke risk was issued by the National Heart, Lung, and Blood Institute in 1997.

Despite the success of the STOP trial, the study also documented that nearly 40% of screened children had subclinical infarcts on MRI.[62] Although baseline MRA data were incomplete, a surprisingly low prevalence of severe stenosis was found, and generally limited to those children with very elevated TCD flow velocities (TAMM > 250 cm/sec). Unlike the children with lower TCD TAMM velocities (200–250 cm/sec) without MRA lesions, these children did not revert to low risk TCD with transfusion and were at highest risk.

It was not possible to sort out whether MRA or the very high TCD best predicted stroke, since the two variables were tightly linked.[63] Figure 1 shows the evolution of abnormalities in a STOP study subject who was randomized to receive transfusion, with subsequent normalization of her TCD studies. At the close of the trial, this patient decided to discontinue regular transfusion therapy, at which point her TCD study again became abnormal. Some months later, she had a stroke. She restarted transfusion therapy, but the TCD failed to normalize, and she opted for hydroxyurea therapy as an alternative to transfusion therapy.

The STOP study represents the first randomized clinical trial investigating chronic transfusion therapy for any indication in SCD and, as such, several other salutary effects of transfusion could be proven for the first time. In the group randomized to transfusion, growth parameters improved and there were fewer cases of ACS and vaso-occlusive pain episodes therapy.[64] There were no cases of blood-borne infection, and the rate of alloimmunization was the lowest reported in this setting.[65]

Although the STOP study was not designed to determine how long transfusion therapy should be continued, growing concerns surrounding the problem of secondary iron overload led to a second randomized study. With the aim of "optimizing" stroke prevention, the STOP II study attempted to discontinue regular transfusions in children whose TCD result had reverted to normal. Children who had two consecutive normal TCD results after 30 months of regular transfusions, and no MRA evidence of moderate or severe vessel lesions, were randomized to discontinue chronic transfusion therapy (experimental arm) or to continue transfusion therapy (standard care arm). The primary endpoint was a composite of clinical stroke or TCD reversion to high risk. The hypothesis was that the absolute difference between the percent of children with endpoints in the two treatment arms would not exceed 50%. The age at randomization was 12 years. The study was halted early, after 79 of a planned 100 children were randomized.[66]

Sixteen endpoints, 14 TCD reversions, and two strokes were observed on study, exceeding the interim monitoring limits. All endpoints were in children removed from transfusion, and occurred soon after halting transfusion (mean 4.5 months). There was one death related to acute chest syndrome in the standard care arm (continued transfusion therapy). Only the average of the last two TCDs prior to starting transfusion approached significance as a predictor of return to high risk after halting transfusion. Discontinuation of transfusion resulted in an increased rate of reversion to high risk, and stroke was not completely prevented with intensive TCD surveillance.

Prevention of Recurrent Stroke

In patients who develop symptomatic stroke, the risk of recurrent stroke approaches 70%.[6] The role of transfusion therapy in preventing stroke recurrence has not been studied systematically, but given the historical data showing a high recurrence rate without transfusion, there likely will never be a controlled trial of this nature.[67] However, data from observational studies exist. Sixty subjects with a history of overt stroke who were placed on chronic transfusion therapy were prospectively followed for a total of 191.7 patient–years, during which eight (13.3%) had recurrent stroke.[68] In an effort to identify children for whom transfusion can be safely stopped after an initial stroke, Scothorn *et al.* assessed the frequency of recurrent stroke after a minimum five years of transfusion therapy in a retrospective cohort of 137 children with SCD.[69] None of the children whose initial stroke was temporally

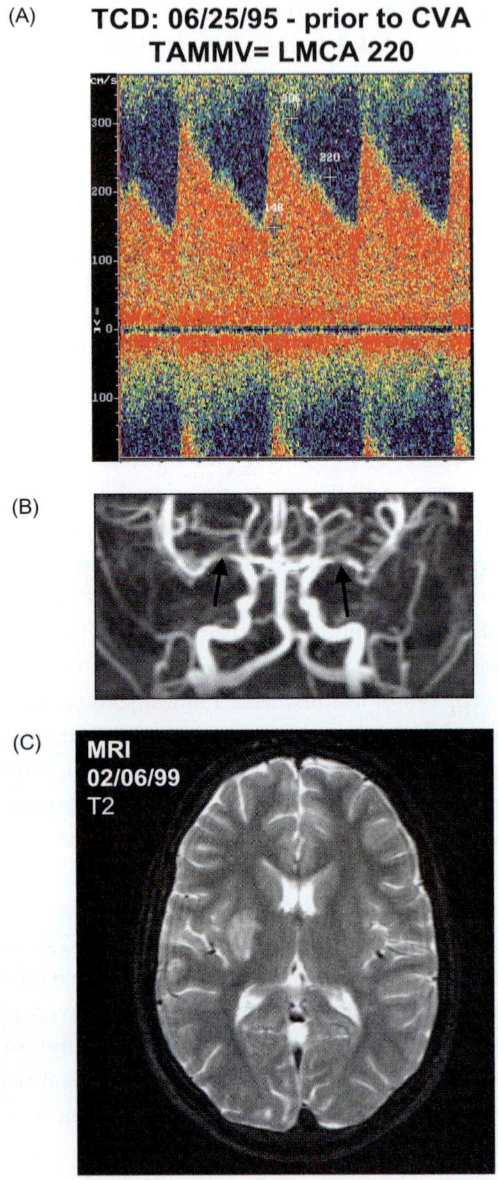

Fig. 1. Transcranial Doppler Images. Panel (A) Transcranial Doppler from a 10-year-old with SCD. On the basis of this and a confirmatory she enrolled in STOP and randomized to transfusion. At the end of STOP (9/97) she elected to discontinue transfusions and in 1999 she presented with left sided weakness. Panel (B) Shown is the magnetic resonance arterial image with bilateral middle cerebral artery stenosis (arrows). Panel (C) A T2 weighted MRI showing cortical infarction on the right is shown. The patient restarted transfusion after a stroke. (Images shown with permission from Adams RJ.[30])

associated with a clinical event (e.g. ACS, infection) had a recurrent stroke beyond the first two years of starting transfusion therapy. In contrast, those children whose initial stroke occurred as an isolated medical event remained at an increased risk of recurrent stroke beyond a five-year period of transfusion therapy. In children with a history of silent infarction and abnormal TCD flow velocities, transfusion therapy lowers the risk of new silent infarcts.[27] Case series showing a low rate of recurrent stroke with reductions in transfusion intensity, or treatment with hydroxyurea (HU) alone, have suggested that transfusion therapy may be discontinued or modified, but this needs systematic study.[70,71]

Alternative Treatments for Stroke

Given the lack of data to support indefinite transfusions, as well as the substantial morbidity associated with a chronic transfusion program, many patients choose to discontinue long-term transfusion therapy.[62] Antiplatelet agents and anticoagulants have been used, but there are no controlled studies to prove the efficacy of this approach in SCD.[72] Patients with ruptured aneurysms have been managed acutely with surgical clipping or coil embolism;[73–75] surgical revascularization procedures may be an alternative for selected patients who cannot tolerate continued transfusions.[76] Children with SCD, stroke and a moyamoya pattern of collateralization may be candidates for encephaloduroarteriosynangiosis, but controlled data on efficacy are not available.[29]

Hydroxyurea (HU) therapy is another potential alternative to transfusion therapy, but has not been compared to transfusion in a randomized fashion. Based on results from the double blind, placebo-controlled Multicenter Study of Hydroxyurea (MSH), HU was approved in adults with SCD to decrease the frequency of vaso-occlusive episodes and blood transfusion requirements.[77] Follow-up data from the MSH have confirmed a reduced mortality after nine years of HU therapy;[78] however, HU did not protect patients from cerebral vascular accidents in this study.

In a pilot study investigating HU for primary stroke prevention, Wang and colleagues[79] documented one stroke and one transient ischemic attack among 28 children followed for two years on therapy;[79] no recurrent events were reported in a small series of five patients with a history of stroke or transient attack, followed for 42 to 112 months on HU therapy.[80] Bernaudin *et al.* stratified stroke risk in children with SCD based on TCD, augmented by MR and catheter angiography to select for treatment with bone marrow transplantation, transfusion, hydroxyurea, or observation.[81] The results of this study are encouraging, but a randomized trial has not been reported.

Ware and colleagues[82] treated 35 children with a history of stroke, and placed them on hydroxyurea for a period of 42 ± 30 months. The overall stroke risk was 5.7 per 100 patient–years, but only 3.6 per 100 patient–years in those who had overlapping transfusion and drug therapy until reaching the maximum tolerated dose of HU. Based on these encouraging data, a randomized controlled clinical trial of HU as a maintenance therapy for secondary stroke prevention, following an initial period of regular transfusions for at least three years, has been proposed. Patients receiving HU require monthly monitoring of blood counts, and although the risk of leukemic transformation due to HU therapy is likely very low, other long-term risks are not known.

Bone marrow transplantation (BMT) is an option for secondary stroke prevention in selected individuals. Of 59 children with SCD followed for a median of 35 months after myeloablative HLA-matched sibling BMT, 30 had a history of prior overt stroke, and one had documented MRI/MRA lesions and neurocognitive deficits.[83] Engrafted patients with a history of stroke had no recurrences after BMT, with a stroke-free survival of 93% and stable or improved cerebral MRI/MRA evaluations. A single patient who experienced graft rejection after BMT had a second stroke when the HbS fraction rose to 60%. Lack of an eligible HLA-compatible sibling donor and potential transplant-related complications remain substantial barriers to BMT in SCD. Novel conditioning regimens that minimize transplant-associated toxicity and alternative stem cell sources show promise for the wider application of BMT in SCD.

Advances in Transfusion Medicine

Most individuals with SCD will require a transfusion during their lifetimes, and a growing fraction of the aging SCD population is presently receiving chronic transfusion therapy.[84] With few effective treatment options available for the majority of patients with SCD, transfusion has become a relatively safe and acceptable alternative. The clinical indications for transfusion therapy in SCD are summarized in Table 1. Despite the many benefits of red cell transfusions in SCD, blood-borne infection, alloimmunization, and iron overload remain significant causes of morbidity and mortality in multiply transfused patients.[85,86] However, major advances have been made in the prevention and treatment of these transfusion-induced complications, and are now being applied in the clinical setting.

Transfusion-Transmitted Infection

The use of leukocyte-depleted red cell transfusions has had a dramatic effect on reducing the transmission of intracellular viruses such as cytomegalovirus (CMV), human lymphotrophic virus (HTLV),

Table 1. Indications for transfusion therapy.

Acute transfusion	Intermediate (~6 mos)	Chronic transfusion	Unknown benefit from transfusion
Ischemic stroke	Recurrent VOC[1]	Abnormal TCD study[2]	Silent infarcts
ACS[3]	Severe ACS[3]	Ischemic stroke	Hemorrhagic stroke
Splenic sequestration[4]	Splenic sequestration[5]	Recurrent or severe ACS[6]	PHT
Aplastic crisis		Recurrent VOC[6]	Priapism[7–10]
Hepatic sequestration		Heart disease	Renal failure
Retinal artery occlusion		Complicated pregnancy[11–12]	AVN
Pre-operative prevention of ACS, VOC[13]			Leg ulcers

Abbreviations: ACS, acute chest syndrome; AVN, avascular necrosis; PHT, pulmonary hypertension; TCD, transcranial Doppler ultrasound; VOC, vaso-occlusive crises.

1. Miller ST, *J Pediatr, 2001*
2. Adams R, *NEJM, 1998*
3. Vichinsky E, *NEJM, 2000*
4. Topley JM, Rogers DW, Stevens MCG, Serjeant GR: *Acute splenic sequestration and hypersplenism in the first five years in homozygous sickle cell disease. Arch Dis Child 56:765, 1981*
5. Kinney TR, Ware RE, Schultz WH, Filston HC: *Long-term management of splenic sequestration in children with sickle cell disease. J Pediatr 117:194, 1990*
6. Miller ST, *J Pediatr, 2001*
7. McCarthy LJ, *Ther Apher, 2000*
8. Rackoff WR, *J Pediatr, 1992*
9. Siegel JF, *J Urol, 1993*
10. Miller ST, *J Urol, 1995*
11. Cunningham FG, *Obstet Gynaecol, 1983*
12. Koshy M, *NEJM, 1988*
13. Vichinsky E, *NEJM, 1995.*

Epstein-Barr virus (EBV) and human herpes virus.[87] Improved donor selection criteria and screening of banked units have also reduced the transmission of human immunodeficiency virus (HIV), hepatitis B (HBV) and C (HCV), and human T-cell leukemia/lymphoma virus-1 (HTLV).[88] Serologic testing of donor units is limited by the inability to detect very early infection when antibodies are not yet present. These donations presently account for almost all of the infectious risk from known pathogens.

Though rates of viral transmission are extremely low, several strategies have been employed to further reduce this risk. In 1999, United States blood centers began testing all donated blood units using molecular methods, or nucleic acid testing (NAT), to detect viral antigenemia during the seronegative window period.[89] Nucleic acid testing is particularly useful for the detection of HCV, as this pathogen has a prolonged seronegative window period. For the detection of HIV, NAT may even eliminate window-period transmission altogether.[90] Strategies to reduce the risk of transfusion-transmitted infection may have the most significant impact in developing countries, where hemoglobin disorders are most prevalent and NAT is not affordable or logistically feasible. A simple and rapid assay to detect and quantify HCV core antigen (trak-C, Ortho Clinical Diagnostics) has recently been developed, providing a viable and inexpensive alternative to NAT.[91]

Alloimmunization

Alloimmunization was not recognized as a significant problem in SCD until the mid-1970s. In 1961, Elizabeth Giblet from the Puget Sound Blood Bank championed the universal use of blood products, regardless of racial disparities between donor and recipient, arguing that the risk of alloimmunization was negligible. Sixteen years later, Dr. Giblet published her findings showing an increased rate of alloimmunization to Kell, Duffy and Kidd antigens in transfused individuals. Recent studies have documented alloimmunization rates as high as 47% in adult and 27% in pediatric transfused SCD patients, respectively.[92–95] In one study, multiple antibodies were found in over 50% of alloimmunized patients, and 44% had delayed transfusion reactions.[94]

Comparison of RBC phenotypes in SCD patients and blood donors revealed significant differences that were attributable to minor blood group incompatibilities (Rh, Kell, Duffy, and Kidd) in racially mismatched blood. The use of phenotypically-matched blood in the multicenter STOP trial resulted in lower rates of alloimmunization and hemolytic transfusion reactions than previously reported.[65] Other strategies to minimize alloimmunization include polyethylene glycol (PEG)-coating of red cells to mask red cell antigens from antibodies,[96] and artificial blood substitutes such as perfluorocarbon emulsions and hemoglobin-based substitutes;[97–99] these blood substitutes appear to be safe and well-tolerated.[100] Lanzkron *et al.*[101] successfully administered PolyHeme, a novel human-derived HbOC, to a SCD patient with life-threatening ACS, but could not elucidate the effects directly attributable to this agent.

Iron Overload

Hemosiderosis is an anticipated consequence of intensive and long-term transfusion therapy, requiring aggressive chelation therapy to prevent multiorgan toxicity from iron deposition. Transfusional iron overload in SCD has not been well studied, and most treatment principles have been extrapolated from the experience in thalassemia. Although both SCD and thalassemia patients suffer from iron-induced organ injury, the clinical manifestations of iron overload appear to be milder in transfused SCD patients, suggesting a relative protection from iron-induced organ injury.[102] Studies comparing iron toxicity in these two groups are few, but differences in the pattern of iron-related target organ injury have been observed.

In a recent pilot study, Vichinsky *et al.*[103] demonstrated that SCD patients developed liver disease, cardiac disease, or endocrine dysfunction significantly less frequently than β-thalassemia patients matched for age and duration of transfusion. Another recent comparison of cardiac function in iron-loaded SCD and thalassemia patients found that, despite prolonged exposure to similar amounts of iron, none of the SCD patients demonstrated abnormal function or elevated cardiac iron by T2* MRI.[104] In contrast, 80% of the thalassemia patients demonstrated cardiac iron overload by this technique. Why SCD patients fail to exhibit the same degree of cardiac or endocrine organ injury is not clear. The biology of SCD and its secondary inflammatory state, or differences in iron transport and storage proteins, may be protective factors.[105,106] The possible influence of splenic tissue, ineffective erythropoiesis, gastrointestinal iron metabolism and urinary iron loss has not been studied.

Serum ferritin is commonly used to indirectly estimate body iron stores, but it reflects only 1% of the total iron storage pool, and the level is influenced by a variety of conditions. In SCD, the serum ferritin fluctuates widely with illness and inflammation, limiting its value in the assessment of iron stores.[107] Liver biopsy is considered the most accurate and sensitive method for measuring an individual's iron burden; however, this is an invasive technique, and hepatic iron measurements from a liver biopsy specimen may be confounded by hepatic fibrosis and uneven tissue distribution of iron.[108] More sensitive methods to non-invasively measure iron are now being applied in the clinical setting, with the most promising of these based on measurement of hepatic magnetic susceptibility, either using superconducting quantum interference device susceptometry (SQUID) or magnetic resonance susceptometry (MRS).[109–111]

Biomagnetic susceptometry is presently the most reliable non-invasive method for measurement of tissue iron stores.[112,113] Although it has been validated in many clinical studies, its limited availability has restricted widespread clinical use. The utilization of special imaging software with MRI has resulted in reliable non-invasive liver iron results, making this an attractive modality for widespread use.[114] Cardiac iron overload is the most frequent cause of death associated with chronic transfusion therapy; growing evidence from studies using cardiac magnetic resonance (CMR) to assess cardiac iron overload indicates that liver iron measurements or ferritin levels do not accurately reflect myocardial iron stores.[115] In a cohort of thalassemia patients evaluated with T2* magnetic resonance techniques, myocardial iron measurements did not correlate with liver iron measurements or with serum ferritin levels (Anderson LJ and Eur Heart J, 2001). Wood *et al.*[104] confirmed these observations, and found that the rate of cardiac iron clearance was substantially lower than hepatic iron clearance in thalassemia patients who were aggressively treated with Desferrioxamine (DFO) for heart failure. Thus, monitoring chelation treatment using liver iron alone may be misleading in guiding the risk from myocardial iron loading.

DFO is presently considered the most effective iron chelator, but requires daily subcutaneous or intravenous infusion. Recent studies have shown that the duration of infusion, rather than the dose, is the critical factor in preventing organ injury.[116] Unfortunately, up to 32% of patients receiving DFO may develop side effects that affect compliance or preclude continued use.[117] Poor compliance and the high cost of DFO have prompted intensive research in the development of effective oral iron chelators, and several new oral chelators are either licensed in some countries or are presently under investigation in clinical trials, including deferiprone (DFP, or L1) and ICL-670.[118] Deferiprone, a relatively small compound, appears to be effective in entering myocardial cells and removing iron,[119] observations which indicate that deferiprone may be particularly beneficial in patients with cardiac disease due to iron overload. ICL-670, a chelator completing Phase III trials, is another promising agent that requires only single daily dosing.[120] Chelation therapy with more than one agent offers the possibility of more effective removal of iron without compromising safety or compliance.

Erythrocytapheresis

Automated red blood cell exchange (erythrocytapheresis) is an effective method of controlling HbS levels and limiting or preventing iron load in chronically transfused SCD patients.[121–123] Net RBC load, a measure of transfusional iron loading, has been shown to be significantly reduced after conversion to erythrocytapheresis[124] (Kim and Blood, 1998). Moreover, iron chelation therapy and its associated adverse effects could be avoided by early initiation of erythrocytapheresis. Erythrocytapheresis has also been found to be cost-effective when considering the costs of chelation therapy combined with a regular transfusion regimen.[121,125]

Although erythrocytapheresis increases exposures to blood units by approximately 50%, transfusion-related complications may be minimized by using leuko-reduced, phenotypically-matched, and virally-screened units. Lack of venous access presents a major barrier to the widespread use of erythrocytapheresis. The placement of semi-permanent intravenous catheters has partially overcome this problem, but this in itself carries the risks of infection and thrombotic occlusion.[126]

Summary

The STOP trial represents the first randomized clinical trial in SCD to address the question of whether stroke prevention is possible. This trial not only identified a means of predicting and preventing stroke, but made it possible to address questions concerning several other clinical outcomes in SCD. The addition of new neuroimaging techniques to better define stroke phenotypes and to augment the predictive value of TCD in the diagnosis of stroke have permitted subsequent studies looking at genetic modifiers of stroke. Transfusion data from the STOP study also contributed to clinical practice guidelines on the optimal management of alloimmunization and iron overload in SCD. With the development of new diagnostic tools to identify patients at risk for severe complications, the use of transfusions in SCD will undoubtedly increase.

Molecular testing for blood-borne infections and the use of pathogen-inactivated blood products have dramatically improved blood safety. The use of phenotypically matched red cells has reduced alloimmunization rates in SCD. Erythrocytapheresis has become a first-line therapeutic option to prevent iron overload, and oral iron chelators are presently in phase III trials. But despite these advances, the widespread application of transfusion therapy in SCD raises many new questions, and whether an individual's genetic profile alters the response to this highly effective treatment in the setting of SCD still remains to be answered. Along with continued efforts to prevent side effects of transfusion, further investigations on the mechanisms by which transfusions ameliorate SCD are needed to develop selectively targeted therapies.

References

1. Arena JM (1935). Vascular accident and acute hemiplegia in sickle cell anemia. *Am J Dis Child* **49**:722–723.
2. Stockman JA, Nigro MA, Mishkin MM, Oski FA (1972). Occlusion of large cerebral vessels in sickle-cell anemia. *N Engl J Med* **287**:846–849.

3. Prengler M, Pavlakis SG, Prohovnik I, Adams RJ (2002). Sickle cell disease: the neurological complications. *Ann Neurol* **51**:543–552.
4. Earley CJ, Kittner SJ, Feeser BR, Gardner J, Epstein A, Wozniak MA, Wityk R, Stern BJ, Price TR, Macko RF, Johnson C, Sloan MA, Buchholz D (1998). Stroke in children and sickle-cell disease: Baltimore–Washington Cooperative Young Stroke Study. *Neurology* **51**:169–176.
5. Ohene-Frempong K, Weiner SJ, Sleeper LA, Miller ST, Embury S, Moohr JW, Wethers DL, Pegelow CH, Gill FM (1998). Cerebrovascular accidents in sickle cell disease: rates and risk factors. *Blood* **91**:288–294.
6. Powars D, Wilson B, Imbus C, Pegelow C, Allen J (1978). The natural history of stroke in sickle cell disease. *Am J Med* **65**:461–471.
7. Armstrong FD, Thompson RJ Jr, Wang W, Zimmerman R, Pegelow CH, Miller S, Moser F, Bello J, Hurtig A, Vass K (1996). Cognitive functioning and brain magnetic resonance imaging in children with sickle cell disease. Neuropsychology committee of the Cooperative Study of Sickle Cell Disease. *Pediatrics* **97**:864–870.
8. Steen RG, Xiong X, Mulhern RK, Langston JW, Wang WC (1999). Subtle brain abnormalities in children with sickle cell disease: relationship to blood hematocrit. *Ann Neurol* **45**:279–286.
9. Steen RG, Reddick WE, Mulhern RK, Langston JW, Ogg RJ, Bieberich AA, Kingsley PB, Wang WC (1998). Quantitative MRI of the brain in children with sickle cell disease reveals abnormalities unseen by conventional MRI. *J Magn Reson Imaging* **8**:535–543.
10. Kandeel AY, Zimmerman RA, Ohene-Frempong K (1996). Comparison of magnetic resonance angiography and conventional angiography in sickle cell disease: clinical significance and reliability. *Neuroradiology* **38**:409–416.
11. Rodgers GP, Clark CM, Larson SM, Rapoport SI, Nienhuis AW, Schechter AN (1988). Brain glucose metabolism in neurologically normal patients with sickle cell disease. Regional alterations. *Arch Neurol* **45**:78–82.
12. Reed W, Jagust W, Al-Mateen M, Vichinsky E (1999). Role of positron emission tomography in determining the extent of CNS ischemia in patients with sickle cell disease. *Am J Hematol* **60**:268–272.
13. Powars DR, Conti PS, Wong WY, Groncy P, Hyman C, Smith E, Ewing N, Keenan RN, Zee CS, Harold Y, Hiti AL, Teng EL, Chan LS (1999). Cerebral vasculopathy in sickle cell anemia: diagnostic contribution of positron emission tomography. *Blood* **93**:71–79.
14. Kirkham FJ, Calamante F, Bynevelt M, Gadian DG, Evans JP, Cox TC, Connelly A (2001). Perfusion magnetic resonance abnormalities in patients with sickle cell disease. *Ann Neurol* **49**:477–485.
15. Pegelow CH, Macklin EA, Moser FG, Wang WC, Bello JA, Miller ST, Vichinsky EP, DeBaun MR, Guarini L, Zimmerman RA, Younkin DP, Gallagher DM, Kinney TR (2002). Longitudinal changes in brain magnetic resonance imaging findings in children with sickle cell disease. *Blood* **99**:3014–3018.
16. Miller ST, Macklin EA, Pegelow CH, Kinney TR, Sleeper LA, Bello JA, DeWitt LD, Gallagher DM, Guarini L, Moser FG, Ohene-Frempong K, Sanchez N, Vichinsky EP, Wang WC, Wethers DL, Younkin DP, Zimmerman RA, DeBaun MR (2001). Silent infarction as a risk factor for overt stroke in children with sickle cell anemia: a report from the Cooperative Study of Sickle Cell Disease. *J Pediatr* **139**:385–390.
17. Manfre L, Giarratano E, Maggio A, Banco A, Vaccaro G, Lagalla R (1999). MR imaging of the brain: findings in asymptomatic patients with thalassemia intermedia and sickle cell-thalassemia disease. *AJR Am J Roentgenol* **173**:1477–1480.
18. Powars DR (1990). Sickle cell anemia and major organ failure. *Hemoglobin* **14**:573–598.
19. Koshy M, Thomas C, Goodwin J (1990). Vascular lesions in the central nervous system in sickle cell disease (neuropathology). *J Am Assoc Minority Physicians* **1**:71–78.
20. Manci EA, Culberson DE, Yang YM, Gardner TM, Powell R, Haynes J Jr, Shah AK, Mankad VN (2003). Causes of death in sickle cell disease: an autopsy study. *Br J Haematol* **123**:359–365.
21. Rothman SM, Fulling KH, Nelson JS (1986). Sickle cell anemia and central nervous system infarction: a neuropathological study. *Ann Neurol* **20**:684–690.
22. Bernaudin F, Verlhac S, Freard F, Roudot-Thoraval F, Benkerrou M, Thuret I, Mardini R, Vannier JP, Ploix E, Romero M, Casse-Perrot C, Helly M, Gillard E, Sebag G, Kchouk H, Pracros JP, Finck B, Dacher JN, Ickowicz V, Raybaud C, Poncet M, Lesprit E, Reinert PH, Brugieres P (2000). Multicenter prospective study of children with sickle cell disease: radiographic and psychometric correlation. *J Child Neurol* **15**:333–343.

23. Wang W, Riester K, Gallagher D, *et al.* (1999). Progressive neurocognitive deficits occur in school-age children with sickle cell disease (SCD) in the absence of clinical or MRI evidence of infarction [abstract]. In: *23rd Annual Meeting of the National Sickle Cell Disease Program*, San Francisco.
24. Noll RB, Stith L, Gartstein MA, Ris MD, Grueneich R, Vannatta K, Kalinyak K (2001). Neuropsychological functioning of youths with sickle cell disease: comparison with non-chronically ill peers. *J Pediatr Psychol* **26**:69–78.
25. Watkins KE, Hewes DK, Connelly A, Kendall BE, Kingsley DP, Evans JE, Gadian DG, Vargha-Khadem F, Kirkham FJ (1998). Cognitive deficits associated with frontal-lobe infarction in children with sickle cell disease. *Dev Med Child Neurol* **40**:536–543.
26. Powars DR, Wong WY, Vachon LA (2001). Incomplete cerebral infarctions are not silent. *J Pediatr Hematol Oncol* **23**:79–83.
27. Pegelow CH, Wang W, Granger S, Hsu LL, Vichinsky E, Moser FG, Bello J, Zimmerman RA, Adams RJ, Brambilla D (2001). Silent infarcts in children with sickle cell anemia and abnormal cerebral artery velocity. *Arch Neurol* **58**:2017–2021.
28. Oyesiku NM, Barrow DL, Eckman JR, Tindall SC, Colohan AR (1991). Intracranial aneurysms in sickle-cell anemia: clinical features and pathogenesis. *J Neurosurg* **75**:356–363.
29. Vernet O, Montes JL, O'Gorman AM, Baruchel S, Farmer JP (1996). Encephaloduroarterio-synangiosis in a child with sickle cell anemia and moyamoya disease. *Pediatr Neurol* **14**:226–230.
30. Adams RJ, Ohene-Frempong K, Wang W (2001). Sickle cell and the brain. *Hematology* (*Am Soc Hematol Educ Program*) **43**:31–46.
31. Pavlakis SG, Prohovnik I, Piomelli S, DeVivo DC (1989). Neurologic complications of sickle cell disease. *Adv Pediatr* **36**:247–276.
32. Dobson SR, Holden KR, Nietert PJ, Cure JK, Laver JH, Disco D, Abboud MR (2002). Moyamoya syndrome in childhood sickle cell disease: a predictive factor for recurrent cerebrovascular events. *Blood* **99**:3144–3150.
33. Pavlakis SG (1989). Neurologic complications of sickle cell disease. *Adv Pediatr* **36**:247–276.
34. Francis RB (1991). Large-vessel occlusion in sickle cell disease: pathogenesis, clinical consequences, and therapeutic implications. *Med Hypotheses* **35**:88–95.
35. Hebbel RP, Vercellotti GM (1997). The endothelial biology of sickle cell disease. *J Lab Clin Med* **129**:288–293.
36. Solovey A, Lin Y, Browne P, Choong S, Wayner E, Hebbel RP (1997). Circulating activated endothelial cells in sickle cell anemia [see comments]. *N Engl J Med* **337**:1584–1590.
37. Phelan MW, Faller DV (1996). Hypoxia decreases constitutive nitric oxide synthase transcript and protein in cultured endothelial cells. *J Cell Physiol* **167**:469–476.
38. Shreeniwas R, Koga S, Karakurum M, Pinsky D, Kaiser E, Brett J, Wolitzky BA, Norton C, Plocinski J, Benjamin W, Burns DK, Goldstein A, Stern D (1992). Hypoxia-mediated induction of endothelial cell interleukin-1 alpha. An autocrine mechanism promoting expression of leukocyte adhesion molecules on the vessel surface. *J Clin Invest* **90**:2333–2339.
39. Uchida K, Rackoff WR, Ohene-Frempong K, Kim HC, Reilly MP, Asakura T (1998). Effect of erythrocytapheresis on arterial oxygen saturation and hemoglobin oxygen affinity in patients with sickle cell disease. *Am J Hematol* **59**:5–8.
40. Powars DR, Meiselman HJ, Fisher TC, Hiti A, Johnson C (1994). Beta-S gene cluster haplotypes modulate hematologic and hemorheologic expression in sickle cell anemia. Use in predicting clinical severity. *Am J Pediatr Hematol Oncol* **16**:55–61.
41. Hsu LL, Miller ST, Wright E, Kutlar A, McKie V, Wang W, Pegelow CH, Driscoll C, Hurlet A, Woods G, Elsas L, Embury S, Adams RJ (2003). Alpha thalassemia is associated with decreased risk of abnormal transcranial Doppler ultrasonography in children with sickle cell anemia. *J Pediatr Hematol Oncol* **25**:622–628.
42. Nagel RL, Steinberg MH (2001). Role of epistatic (modifier) genes in the modulation of the phenotypic diversity of sickle cell anemia. *Pediatr Pathol Mol Med* **20**:123–136.
43. Kahn MJ, Scher C, Rozans M, Michaels RK, Leissinger C, Krause J (1997). Factor V Leiden is not responsible for stroke in patients with sickling disorders and is uncommon in African Americans with sickle cell disease. *Am J Hematol* **54**:12–15.

44. Cumming AM, Olujohungbe A, Keeney S, Singh H, Hay CR, Serjeant GR (1999). The methylenetetrahydrofolate reductase gene C677T polymorphism in patients with homozygous sickle cell disease and stroke. *Br J Haematol* **107**:569–571.
45. Andrade F, Annichino-Bizzacchi J, Saad S, Costa F, Arruda V (1998). Prothrombin mutant, factor V Leiden, and thermolabile variant of methylenetetrahydrofolate reductase among patients with sickle cell disease in Brazil. *Am J Hematol* **59**:46–50.
46. Zimmerman SA, Ware RE (1998). Inherited DNA mutations contributing to thrombotic complications in patients with sickle cell disease. *Am J Hematol* **59**:267–272.
47. Carroll JE, McKie V, Kutlar A (1998). Are sickle cell disease patients with stroke genetically predisposed to the event by inheriting a tendency to high tumor necrosis factor levels? *Am J Hematol* **58**:250.
48. Taylor JG, Tang D, Foster CB, Serjeant GR, Rodgers GP, Chanock SJ (2002). Patterns of low-affinity immunoglobulin receptor polymorphisms in stroke and homozygous sickle cell disease. *Am J Hematol* **69**:109–114.
49. Tang DC, Prauner R, Liu W, Kim KH, Hirsch RP, Driscoll MC, Rodgers GP (2001). Polymorphisms within the angiotensinogen gene (GT-repeat) and the risk of stroke in pediatric patients with sickle cell disease: a case-control study. *Am J Hematol* **68**:164–169.
50. Tam DA (1997). Protein C and protein S activity in sickle cell disease and stroke. *J Child Neurol* **12**:19–21.
51. Houston PE, Rana S, Sekhsaria S, Perlin E, Kim KS, Castro OL (1997). Homocysteine in sickle cell disease: relationship to stroke. *Am J Med* **103**:192–196.
52. Balasa VV, Gruppo RA, Gartside PS, Kalinyak KA (1999). Correlation of the C677T MTHFR genotype with homocysteine levels in children with sickle cell disease. *J Pediatr Hematol Oncol* **21**:397–400.
53. Hoppe C, Klitz W, Noble J, Vigil L, Vichinsky E, Styles L (2003). Distinct HLA associations by stroke subtype in children with sickle cell anemia. *Blood* **101**:2865–2869.
54. Hoppe C, Klitz W, Cheng S, Apple R, Steiner L, Robles L, Girard T, Vichinsky E, Styles L (2004). Gene interactions and stroke risk in children with sickle cell anemia. *Blood* **103**:2391–2396.
55. Taylor JG, Tang DC, Savage SA, Leitman SF, Heller SI, Serjeant GR, Rodgers GP, Chanock SJ (2002). Variants in the VCAM1 gene and risk for symptomatic stroke in sickle cell disease. *Blood* **100**:4303–4309.
56. Sebastiani P, Ramoni MF, Nolan V, Baldwin CT, Steinberg MH (2005). Genetic dissection and prognostic modeling of overt stroke in sickle cell anemia. *Nat Genet* **37**:435–440.
57. Adams R, McKie V, Nichols F, Carl E, Zhang DL, McKie K, Figueroa R, Litaker M, Thompson W, Hess D (1992). The use of transcranial ultrasonography to predict stroke in sickle cell disease. *N Engl J Med* **326**:605–610.
58. Adams RJ, McKie VC, Carl EM, Nichols FT, Perry R, Brock K, McKie K, Figueroa R, Litaker M, Weiner S, Brambilla D (1997). Long-term stroke risk in children with sickle cell disease screened with transcranial Doppler. *Ann Neurol* **42**:699–704.
59. Nichols FT, Jones AM, Adams RJ (2001). Stroke prevention in sickle cell disease (STOP) study guidelines for transcranial Doppler testing. *J Neuroimaging* **11**:354–362.
60. Adams RJ, Brambilla DJ, McKie VC, Hsu L, Files B, Vichinsky E, Pegelow C, Abboud M, Woods G, Olivieri N, Driscoll C, Miller S, Wang W, Hurlet A, Scher C, Berman B, Carl EM, Nichols FT, Roach S, Sutlar A, Wright E, Zimmerman RA, Gallager D, Waclawiw M, Bonds D (1998). Transfusion prevents first stroke in children with sickle cell disease: the "STOP" study. *N Engl J Med* **339**:5–11.
61. Adams R, McKie V, Hsu L, Files B, Vichinsky E, Pegelow C, Abboud M, Gallagher D, Kutlar A, Nichols F, Bonds D, Brambilla D (1998). Prevention of a first stroke by transfusions in children with sickle cell anemia and abnormal results on transcranial doppler ultrasonagraphy. *N Engl J Med* **339**:5–11.
62. Adams RJ (2000). Lessons from the Stroke Prevention Trial in sickle cell anemia (STOP) study. *J Child Neurol* **15**:344–349.
63. Abboud MR, Cure J, Granger S, Gallagher D, Hsu L, Wang W, Woods G, Berman B, Brambilla D, Pegelow C, Lewin J, Zimmermann RA, Adams RJ (2004). Magnetic resonance angiography in children with sickle cell disease and abnormal transcranial Doppler ultrasonography findings enrolled in the STOP study. *Blood* **103**:2822–2826.
64. Miller ST, Wright E, Abboud M, Berman B, Files B, Scher CD, Styles L, Adams RJ (2001). Impact of chronic transfusion on incidence of pain and acute chest syndrome during the Stroke Prevention Trial (STOP) in sickle-cell anemia. *J Pediatr* **139**:785–789.

65. Vichinsky EP, Luban NL, Wright E, Olivieri N, Driscoll C, Pegelow CH, Adams RJ (2001). Prospective RBC phenotype matching in a stroke-prevention trial in sickle cell anemia: a multicenter transfusion trial. *Transfusion* **41**:1086–1092.
66. Clinical Alert (2004). *NHLBI Stops Study Testing How Long Children with Sickle Cell Anemia Should Have Blood Transfusions to Prevent Stroke*. Bethesda, MD: National Institutes of Health, Health and Human Services.
67. Powars D (2000). Management of cerebral vasculopathy in children with sickle cell anaemia. *Br J Haematol* **108**:666–678.
68. Pegelow CH, Adams RJ, McKie V, Abboud M, Berman B, Miller ST, Olivieri N, Vichinsky E, Wang W, Brambilla D (1995). Risk of recurrent stroke in patients with sickle cell disease treated with erythrocyte transfusions. *J Pediatr* **126**:896–899.
69. Scothorn DJ, Price C, Schwartz D, Terrill C, Buchanan GR, Shurney W, Sarniak I, Fallon R, Chu JY, Pegelow CH, Wang W, Casella JF, Resar LS, Berman B, Adamkiewicz T, Hsu LL, Ohene-Frempong K, Smith-Whitley K, Mahoney D, Scott JP, Woods GM, Watanabe M, Debaun MR (2002). Risk of recurrent stroke in children with sickle cell disease receiving blood transfusion therapy for at least five years after initial stroke. *J Pediatr* **140**:348–354.
70. Cohen AR, Martin MB, Silber JH, Kim HC, Ohene-Frempong K, Schwartz E (1992). A modified transfusion program for prevention of stroke in sickle cell disease. *Blood* **79**:1657–1661.
71. Ware RE, Steinberg MH, Kinney TR (1995). Hydroxyurea: an alternative to transfusion therapy for stroke in sickle cell anemia. *Am J Hematol* **50**:140–143.
72. Adams HP Jr AR, Brott T, del Zoppo GJ, Furlan A, Goldstein LB, Grubb RL, Higashida R, Kidwell C, Kwiatkowski TG, Marler JR, Hademenos GJ (2003). Guidelines for the early management of patients with ischemic stroke: a scientific statement from the Stroke Council of the American Stroke Association. *Stroke* **34**:1056–1083.
73. McQuaker IG, Jaspan T, McConachie NS, Dolan G (1999). Coil embolization of cerebral aneurysms in patients with sickling disorders. *Br J Haematol* **106**:388–390.
74. Anson JA, Koshy M, Ferguson L, Crowell RM (1991). Subarachnoid hemorrhage in sickle-cell disease. *J Neurosurg* **75**:552–558.
75. Firth PG, Peterfreund RA (2000). Management of multiple intracranial aneurysms: neuroanesthetic considerations of sickle cell disease. *J Neurosurg Anesth* **12**:366–371.
76. Fryer RH, Anderson RC, Chiriboga CA, Feldstein NA (2003). Sickle cell anemia with moyamoya disease: outcomes after EDAS procedure. *Pediatr Neurol* **29**:124–130.
77. Charache S, Terrin ML, Moore RD, Dover GJ, Barton FB, Eckert SV, McMahon RP, Bonds DR (1995). Effect of hydroxyurea on the frequency of painful crises in sickle cell anemia. *N Engl J Med* **332**: 1317–1322.
78. Steinberg MH, Barton F, Castro O, Pegelow CH, Ballas SK, Kutlar A, Orringer E, Bellevue R, Olivieri N, Eckman J, Varma M, Ramirez G, Adler B, Smith W, Carlos T, Ataga K, DeCastro L, Bigelow C, Saunthararajah Y, Telfer M, Vichinsky E, Claster S, Shurin S, Bridges K, Waclawiw M, Bonds D, Terrin M (2003). Effect of hydroxyurea on mortality and morbidity in adult sickle cell anemia: risks and benefits up to 9 years of treatment. *JAMA* **289**:1645–1651.
79. Wang W, Wynn LW, Rogers ZR, *et al.* (2001). A two-year pilot trial of hydroxyurea in very young children with sickle cell anemia. *J Pediatr* **139**:790–796.
80. Sumoza A, de Bisotti R, Sumoza D, Fairbanks V (2002). Hydroxyurea (HU) for prevention of recurrent stroke in sickle cell anemia (SCA). *Am J Hematol* **71**:161–165.
81. Bernaudin F, Verlhac S, Coic L, Lesprit E, Brugieres P, Reinert P (2005). Long-term follow-up of pediatric sickle cell disease patients with abnormal high velocities on transcranial Doppler. *Pediatr Radiol* **35**:242–248.
82. Ware RE, Zimmerman SA, Sylvestre PB, Mortier NA, Davis JS, Treem WR, Schultz WH (2004). Prevention of secondary stroke and resolution of transfusional iron overload in children with sickle cell anemia using hydroxyurea and phlebotomy. *J Pediatr* **145**:346–352.
83. Walters MC, Patience M, Edwards S, Robertson S, McMurray M, Donfield S, Sullivan K (2004). Hematopoietic cell transplantation for sickle cell disease: updated results of the multicenter trial. *Blood* **104**:105a.

84. Thame JR, Hambleton IR, Serjeant GR (2001). RBC transfusion in sickle cell anemia (HbSS): experience from the Jamaican Cohort Study. *Transfusion* **41**:596–601.
85. Vichinsky EP (2001). Current issues with blood transfusions in sickle cell disease. *Semin Hematol* **38**:14–22.
86. Styles LA, Vichinsky E (1994). Effects of a long-term transfusion regimen on sickle cell-related illnesses. *J Pediatr* **125**:909–911.
87. Fergusson D, Hebert PC, Lee SK, Walker CR, Barrington KJ, Joseph L, Blajchman MA, Shapiro S (2003). Clinical outcomes following institution of universal leukoreduction of blood transfusions for premature infants. *JAMA* **289**:1950–1956.
88. AuBuchon JP, Birkmeyer JD, Busch MP (1997). Safety of the blood supply in the United States: opportunities and controversies. *Ann Intern Med* **127**:904–909.
89. Busch MP (2000). HIV, HBV and HCV: new developments related to transfusion safety. *Vox Sang* **78**(2 Suppl):253–256.
90. Murthy KK, Henrard DR, Eichberg JW, Cobb KE, Busch MP, Allain JP, Alter HJ (1999). Redefining the HIV-infectious window period in the chimpanzee model: evidence to suggest that viral nucleic acid testing can prevent blood-borne transmission. *Transfusion* **39**:688–693.
91. Laperche S, Le Marrec N, Simon N, Bouchardeau F, Defer C, Maniez-Montreuil M, Levayer T, Zappitelli JP, Lefrere JJ (2003). A new HCV core antigen assay based on disassociation of immune complexes: an alternative to molecular biology in the diagnosis of early HCV infection. *Transfusion* **43**:958–962.
92. Aygun B, Padmanabhan S, Paley C, Chandrasekaran V (2002). Clinical significance of RBC alloantibodies and autoantibodies in sickle cell patients who received transfusions. *Transfusion* **42**:37–43.
93. Tahhan HR, Holbrook CT, Braddy LR, Brewer LD, Christie JD (1994). Antigen-matched donor blood in the transfusion management of patients with sickle cell disease. *Transfusion* **34**:562–569.
94. Vichinsky EP, Earles A, Johnson RA, Hoag MS, Williams A, Lubin B (1990). Alloimmunization in sickle cell anemia and transfusion of racially unmatched blood [see comments]. *N Engl J Med* **322**:1617–1621.
95. Rosse WF, Gallagher D, Kinney TR, Castro O, Dosik H, Moohr J, Wang W, Levy PS (1990). Transfusion and alloimmunization in sickle cell disease. The Cooperative Study of Sickle Cell Disease. *Blood* **76**:1431–1437.
96. Fisher TC (2000). PEG-coated red blood cells-simplifying blood transfusion in the new millennium? *Immunohematology* **16**:37–48.
97. Habler O, Pape A, Meier J, Zwissler B (2005). Artificial oxygen carriers as an alternative to red blood cell transfusion. *Anaesthesist* **54**(8):741–754.
98. Maevsky E, Ivanitsky G, Bogdanova L, Axenova O, Karmen N, Zhiburt E, Senina R, Pushkin S, Maslennikov I, Orlov A, Marinicheva I (2005). Clinical results of Perftoran application: present and future. *Artif Cells Blood Substit Immobil Biotechnol* **33**:37–46.
99. Lowe KC (2003). Engineering blood: synthetic substitutes from fluorinated compounds. *Tissue Eng* **9**:389–399.
100. Gonzalez P (1997). Nuclear medicine test reports. *Rev Med Chil* **125**:1112–1113.
101. Lanzkron S, Moliterno AR, Norris EJ, Gould SA, Segal J, Nuermberger EL, Ness PM (2002). Polymerized human Hb use in acute chest syndrome: a case report. *Transfusion* **42**:1422–1427.
102. Finch CA (1982). The detection of iron overload. *N Engl J Med* **307**:1702–1704.
103. Harmatz P, Butensky E, Quirolo K, Williams R, Ferrell L, Moyer T, Golden D, Neumayr L, Vichinsky E (2000). Severity of iron overload in patients with sickle cell disease receiving chronic red blood cell transfusion therapy. *Blood* **96**:76–79.
104. Wood JC, Tyszka JM, Carson S, Nelson MD, Coates TD (2004). Myocardial iron loading in transfusion-dependent thalassemia and sickle cell disease. *Blood* **103**:1934–1936.
105. Beard JL, Dawson H, Pinero DJ (1996). Iron metabolism: a comprehensive review. *Nutr Rev* **54**:295–317.
106. Bourantas KL, Dalekos GN, Makis A, Chaidos A, Tsiara S, Mavridis A (1998). Acute phase proteins and interleukins in steady state sickle cell disease. *Eur J Haematol* **61**:49–54.
107. Ballas SK (2001). Iron overload is a determinant of morbidity and mortality in adult patients with sickle cell disease. *Semin Hematol* **38**:30–36.
108. Bonkovsky HL, Slaker DP, Bills EB, Wolf DC (1990). Usefulness and limitations of laboratory and hepatic imaging studies in iron-storage disease. *Gastroenterology* **99**:1079–1091.

109. Brittenham GM, Badman DG (2003). Noninvasive measurement of iron: report of an NIDDK workshop. *Blood* **101**:15–19.
110. Fischer R, Longo F, Nielsen P, Engelhardt R, Hider RC, Piga A (2003). Monitoring long-term efficacy of iron chelation therapy by deferiprone and desferrioxamine in patients with beta-thalassaemia major: application of SQUID biomagnetic liver susceptometry. *Br J Haematol* **121**:938–948.
111. Carneiro AA, Fernandes JP, de Araujo DB, Elias J Jr, Martinelli AL, Covas DT, Zago MA, Angulo IL, St Pierre TG, Baffa O (2005). Liver iron concentration evaluated by two magnetic methods: magnetic resonance imaging and magnetic susceptometry. *Magn Reson Med* **54**:122–128.
112. Fischer R, Tiemann CD, Engelhardt R, Nielsen P, Durken M, Gabbe EE, Janka GE (1999). Assessment of iron stores in children with transfusion siderosis by biomagnetic liver susceptometry. *Am J Hematol* **60**:289–299.
113. Nielsen P, Engelhardt R, Duerken M, Janka GE, Fischer R (2000). Using SQUID biomagnetic liver susceptometry in the treatment of thalassemia and other iron loading diseases. *Transfus Sci* **23**:257–258.
114. St Pierre TG, Clark PR, Chua-anusorn W, Fleming AJ, Jeffrey GP, Olynyk JK, Pootrakul P, Robins E, Lindeman R (2005). Noninvasive measurement and imaging of liver iron concentrations using proton magnetic resonance. *Blood* **105**:855–861.
115. Pennell DJ, Sechtem UP, Higgins CB, Manning WJ, Pohost GM, Rademakers FE, van Rossum AC, Shaw LJ, Yucel EK (2004). Clinical indications for cardiovascular magnetic resonance (CMR): Consensus Panel report. *Eur Heart J* **25**:1940–1965.
116. Davis BA, Porter JB (2000). Long-term outcome of continuous 24-hour deferoxamine infusion via indwelling intravenous catheters in high-risk beta-thalassemia. *Blood* **95**:1229–1236.
117. Borgna-Pignatti C, Rugolotto S, De Stefano P, Zhao H, Cappellini MD, Del Vecchio GC, Romeo MA, Forni GL, Gamberini MR, Ghilardi R, Piga A, Cnaan A (2004). Survival and complications in patients with thalassemia major treated with transfusion and deferoxamine. *Haematologica* **89**:1187–1193.
118. Beutler E, Hoffbrand AV, Cook JD (2003). Iron deficiency and overload. *Hematology (Am Soc Hematol Educ Program)* **45**:40–61.
119. Hershko C, Cappellini MD, Galanello R, Piga A, Tognoni G, Masera G (2004). Purging iron from the heart. *Br J Haematol* **125**:545–551.
120. Galanello R, Piga A, Alberti D, Rouan MC, Bigler H, Sechaud R (2003). Safety, tolerability, and pharmacokinetics of ICL670, a new orally active iron-chelating agent in patients with transfusion-dependent iron overload due to beta-thalassemia. *J Clin Pharmacol* **43**:565–572.
121. Adams DM, Schultz WH, Ware RE, Kinney TR (1996). Erythrocytapheresis can reduce iron overload and prevent the need for chelation therapy in chronically transfused pediatric patients. *J Pediatr Hematol Oncol* **18**:46–50.
122. Marques Junior JF, Saad ST, Costa FF, Souza CA (1995). Automated erythrocytapheresis in sickle cell anaemia. *Int J Artif Organs* **18**:345–346.
123. Kim HC, Dugan NP, Silber JH, Martin MB, Schwartz E, Ohene-Frempong K, Cohen AR (1994). Erythrocytapheresis therapy to reduce iron overload in chronically transfused patients with sickle cell disease. *Blood* **83**:1136–1142.
124. Singer ST, Quirolo K, Nishi K, Hackney-Stephens E, Evans C, Vichinsky EP (1999). Erythrocytapheresis for chronically transfused children with sickle cell disease: an effective method for maintaining a low hemoglobin S level and reducing iron overload. *J Clin Apheresis* **14**:122–125.
125. Hilliard LM, Williams BF, Lounsbury AE, Howard TH (1998). Erythrocytapheresis limits iron accumulation in chronically transfused sickle cell patients. *Am J Hematol* **59**:28–35.
126. Raj A, Bertolone S, Bond S, Burnett D, Denker A (2005). Cathlink 20: a subcutaneous implanted central venous access device used in children with sickle cell disease on long-term erythrocytapheresis — a report of low complication rates. *Pediatr Blood Cancer* **44**:669–672.

9

Hemoglobin S Polymerization, Just the Beginning

by Frank A. Ferrone

Introduction

This chapter surveys the molecular events that cause individual hemoglobin tetramers to become a stiff polymer mass with devastating consequences. It outlines the principles of aggregation, so that therapeutic approaches may be developed in the appropriate context of the underlying basic science. Our understanding of hemoglobin polymerization is sufficiently sophisticated that there are now reliable measurements to study many aspects of this phenomenon. Further evolution of our understanding of this subject is inevitable; however the basic measures that form a quantitative counterpart to the conceptual framework have been well established.

Equilibrium

Hemoglobin Function

Hemoglobin has evolved to be an efficient oxygen carrier. Human hemoglobin A is a tetramer of two α and two β chains, each with an oxygen-carrying heme group. The molecule in its oxygenated form readily dissociates into pairs of $\alpha\beta$ dimers at low concentrations, but even at high concentrations this propensity to separate permits subunit exchange. In the typical red cell, the average hemoglobin concentration is 32 g/dl, but can be as high as 45 g/dl. At 32 g/dl, 27% of the cell volume is taken up with hemoglobin; at 45 g/dl that number rises to 38%. At 27% occupancy, the molecules are closer to each other than their average diameter. This will turn out to have profound effects on polymerization. Despite the high concentration, and a variety of charged and hydrophobic surface groups, molecules of HbA slip past one another with the ease of ball-bearings.[1] Given the need of the erythrocyte to deform as it traverses capillaries, it is apparent that adhesion of Hb molecules to one another would represent a serious problem. Erythrocyte deformation, which facilitates intimate contact with the capillary, is another optimization that promotes O_2 unloading. Indeed, if erythrocytes were not in contact with the capillary then O_2 unloading through the plasma would be highly inefficient. After the first or second O_2 is released, the hemoglobin dimers twist about their axis and change their intermolecular packing to the T-structure as part of an intricate mechanism of inter-subunit communication that permits cooperative oxygen binding. Sickle Hemoglobin (HbS) differs only in one surface mutation on both β chains, replacing a glu by val at $\beta 6$. In dilute solutions, its behavior is virtually indistinguishable from HbA.[2]

Polymer Structure

In concentrated solutions, deoxy HbS forms long, multi-stranded fibers comprised of seven double-strands[3] (Fig. 1). Crystals also form double-strands, albeit linear, and the intermolecular contacts of the double strand are shown to appear in the fiber as well.[4,5] There are no x-ray structures of the fiber, but only of the crystal double strands.[6–9] What is known of contacts is deduced from perturbing the double strands, which is a challenging exercise in model building.[10,11]

The double strands form a half-staggered arrangement, with contacts made parallel to the double strand axis (the axial contacts) and zig-zag contacts that hold the double strands together (the lateral contacts) (Fig. 1). The β^S mutation is found at the lateral contacts in the double strand. Each molecule has a donor site ($\beta 6$ — the hydrophobic val) and an acceptor site (a hydrophobic pocket in the tetramer's other β chain) into which the $\beta 6$ from another Hb will dock. Transfer energies of valine from aqueous to hydrophobic environments are about 1.5 kcal/mol.[12] The receptor region contributes another 1.5 kcal/mol. Hence the donor and receptor in a given molecule provide about 3 kcal/mol of stability. The salt bridge contribution is difficult to estimate since ionized histidines can also make ionic contacts in solution. Although the salt bridges in the polymer involve the α chains, most lateral and axial contacts are made between the β subunits, making them a preferred target for genetic modification.

In the axial region, the contacts are less specific, principally hydrophobic in nature, and involving α and β subunits.[7] They are dominated by contact between *Pro*114 and *Ala*115 (both on α chains) with *His*116, *His*117 and *Phe*118 on the adjacent β chain. Above the pK of these histidines, there is no good interaction for the charges; thus below the pK the hydrophobic interactions will be stronger. Therefore the histidines confer pH dependence to the stability of fibers. As a rough scale for the strength of the

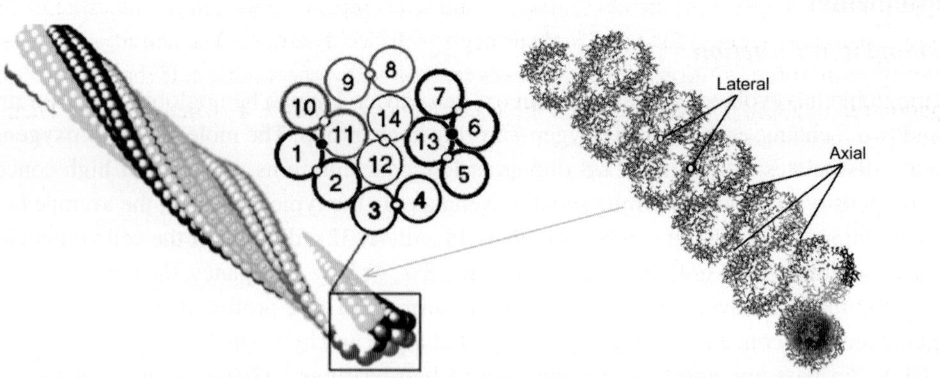

Fig. 1. Molecular structure of hemoglobin S fibers. On the left is the sickle hemoglobin polymer. Seven double strands wrap together, forming a fiber of undulating width. A helical section is shown, with lateral contacts between fibers indicated by small dots. White dots are known from the double strands; the black dots are proposed from model building and thermodynamic measurements. Linear double strands are shown on the right. These are crystallographically determined.[7] A spherical core has been excised from each Hb molecule to highlight contact regions. One end molecule shows the core as a filled sphere. $\beta 6$*Val* is in the lateral contacts, again highlighted by white dots.

axial contacts, the burial of *Ala* is worth about 0.5 kcal/mol[12]; burying the corresponding pocket then gives a total of 1 kcal/mol. It is interesting to ask how much energy keeps HbA from making polymers. If *β6Glu* were neutralized, there would be no penalty for placing it in a hydrophobic environment. From its pK, one infers 4 kcal/mol, also seen in the difference in hydrophobicity between *Glu* and *Val*.[12] A 4 kcal/mole penalty implies that in a mixture of HbS and HbA tetramers, about 1 in every 800 molecules in a polymer would be HbA.

Only the T structure can make axial and lateral contacts simultaneously. There are two R structures[13] and in neither is it possible to make these contacts simultaneously, even though the same amino acids and generally local tertiary structures are present in both R and T structures.[8] The necessity of the T structure was confirmed by studies on an HbS mutant which remains partially in the R-structure on deoxygenation, and required twice the solubility for polymerization,[14] and by the polymerization of methemoglobin S (metHbS) and nitrosylated HbS (NOHbS), using inositol hexa phosphate (IHP) to switch their structures from R to T.[15,16]

What holds the seven double strands together? While some contacts between double strands have been proposed,[11] none have been found that are as strong as the axial or lateral contacts, with one exception. In two of the seven double-strand pairs within the fiber (cf. Fig. 1, between strands 1–11 and 13–16) a second *β6* site forms another donor-acceptor pair.[17] This result, obtained from model building, is supported by measurements of co-polymerization of HbS with HbA.

Polymer Stability

The stability of the polymer is measured by its solubility, typically determined by centrifugation.[18] The supernatant is composed of monomers at a fixed concentration — the solubility (also called c_{SAT}).

Solution phase monomers at concentration c_m have a chemical potential, μ_m that has a "unitary" part due to rotation and translation of the molecule through solution, and a "cratic" part given by RT ln $\gamma_m c_m$. The variable γ is an activity coefficient required to account for the high concentrations of the solution. These high concentrations create a discrepancy between the measure of interaction based on center-to-center distances and surface-to-surface distances. The activity, γc is the effective concentration of the solution, and γ increases rapidly with concentration. If the specific volume of a monomer is v (about 0.75 cm^3/g), then the fractional occupancy is ϕ, where $\phi = vc$. Then

$$\ln \gamma = 8\phi/(1-\phi)^2 \tag{1}$$

Solutions of 32 g/dl act as if they were ~35 times more concentrated ($\gamma = 34.5$). For 45 g/dl, γ is an amazing 812.[1]

At solubility, the monomer chemical potential μ_m is equal to the chemical potential of a molecule in a polymer, μ_p. The polymers are treated as a crystal phase, i.e., with unitary but no cratic terms. A molecule in the polymer has a chemical potential contribution μ_{PC} arising from the intermolecular interactions including axial, lateral, and inter-strand contacts. This chemical potential will include entropic terms such as those found in hydrophobic interactions. In addition to this contact energy the monomer in the polymer has another, unique energy arising from vibrational entropy. The vibrations in question are those of the entire molecule about its equilibrium position in a polymer. This is *not* vibrational motion within the hemoglobin molecule, but of the molecule as a whole, jiggling around its average position in the polymer. This gives rise to the well-known heat capacity of pure solids in crystals.[19] The vibrational entropy chemical potential depends on the specific environment within the polymer or aggregate while the contact chemical potential is determined by interactions in model compounds. Finally, there is a term, $-RT \ln 2$, to account for the two energetically equivalent ways of placing a monomer in a polymer.

$$\mu_P = \mu_{PC} + \mu_{PV} - RT \ln 2 \quad (2)$$

Since at solubility, $\mu_m = \mu_p$ the energy of association related to solubility is given as

$$\Delta G_s = \mu_{PC} + \mu_{PV} - RT \ln 2 - \mu_{TR} = RT \ln \gamma_S c_s \quad (3)$$

ΔG_s is a measure of the overall stability of the hemoglobin polymers and the more positive this value is, the less stable are the polymers. Figure 2 shows the temperature dependent solubility of HbS.[20] At 37°C the solubility is 16.3 g/dl and the energy of association ΔG_s is 1.4 kcal/mol (measured against a 1 mM standard state). This is small relative to the contact energies produced by burying $\beta 6$*Val*. There is extensive cancellation of larger terms since $\mu_{TR} = -34$ kcal/mol.[19] μ_{PC} is about -7.4 kcal/mol by a kinetic analysis[21] and $RT \ln 2 = 0.4$ kcal/mol. The value of μ_{PV} deduced from this equation is -27 kcal/mol. This is substantial because removing a monomer from solution does not delete its motional freedom but transforms it into vibrational motion about its equilibrium position in the polymer.

Why is this separation of chemical potential terms important? It is well known how to determine contact energies from extensive work on calculation and calibration of hydrophobic energies. This is the energy altered if a contact site amino acid is changed or a drug competes for the binding site. The width of the potential well in which a molecule sits when bound to the polymer is determined by direct contact and its neighborhood. Consequently, amino acids that restrict motion of the hemoglobin

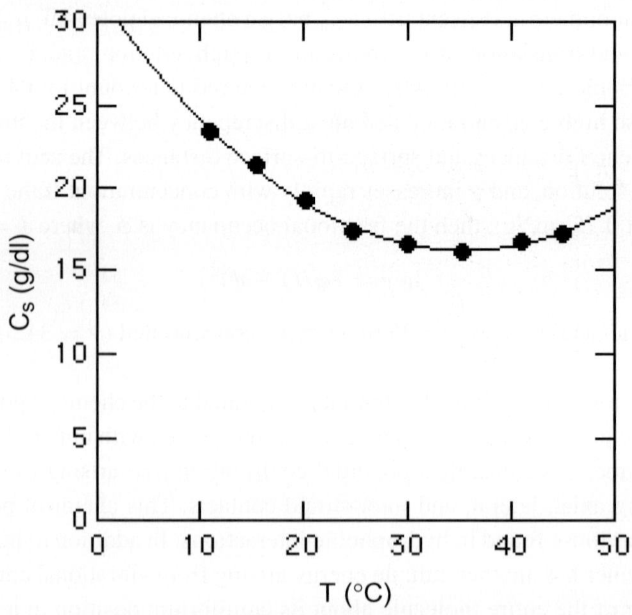

Fig. 2. Solubility of HbS as a function of temperature in 0.15 M phosphate buffer, pH 7.35.[20] The elevation of solubility at low temperatures allows initiation of the reaction by starting at a low temperature and increasing until the temperature crosses the solubility line.

molecules in the polymer limit the vibrational chemical potential. It is possible to destabilize a sickle hemoglobin aggregate by making the docking arrangement more restrictive.

pH and Ionic Dependences

The solubility changes with pH and ionic strength. Figure 3 illustrates the increase in solubility at alkaline pH. A simple model (providing the solid curve) gives a pK of 7.24 for two identical titratable groups, consistent with known pK's for Histidines. The ionic strength of the solution also plays an important role in stability. Molar concentrations of inorganic phosphates reduce the solubility drastically.[22]

Hemoglobin Mixtures

Hemoglobin polymerization *in vivo* involves multiple species distinguished by oxygenation state and globin type. There are two conceptual issues in understanding mixtures. The first is that non-polymerizing hemoglobins affect the reaction by crowding because the monomer activity coefficient is a function of the total occupied volume ϕ, regardless of species. For example, as a sample deoxygenates and the concentration of polymerization-competent deoxyHbS rises, the activity coefficient stays constant since the total Hb does not change. The second issue is the behavior of the hybrid or intermediate species. For example, does singly liganded Hb polymerize? Or what is the fate of hybrid species of HbS and HbA? Analysis requires knowing the fraction of the species involved, and how their interactions within the polymer differ from pure HbS.

For mixtures of different types of Hb, the ability of the molecule to split along its dimer axis generates hybrids as well as pure Hb of both types, in ratios that follow binomial statistics so long as there is no cooperative interaction in dimerization.[23] If X is the fraction of the added Hb (such

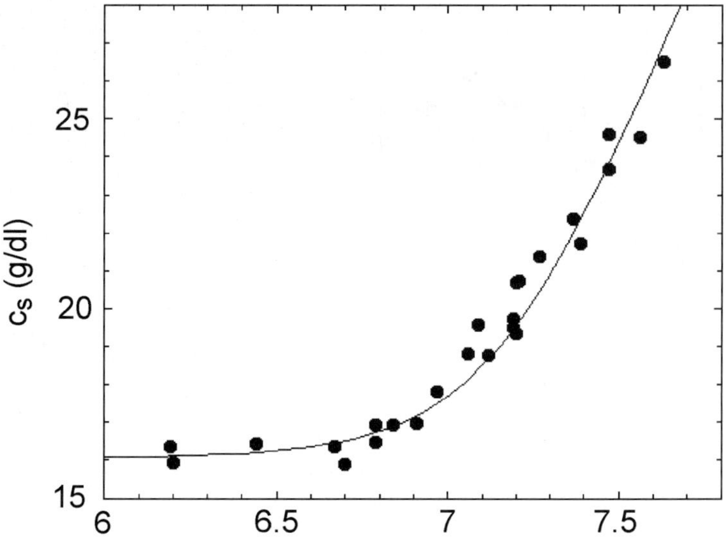

Fig. 3. The effects of pH on HbS solubility.[53] The curve assumes two identical titratable groups per monomer with pK of 7.24 supporting their identification as histidines.

as HbA or HbF) then, X^2 is the fraction of pure species, $(1 - X)^2$ is the fraction of HbS, and the hybrid fraction is $2X(1 - X)$. The preparation of a given experiment (i.e mixing as oxy or deoxy) determines if the hybrids are present or not because oxygenated species repartition much faster than deoxygenated species.

The energy of interaction gives rise to a term e_i designated as the copolymerization probability for a given species i. In terms of association energy (Eq. (3)),

$$e_i = \exp[-(\Delta G_i - \Delta G_s)/RT] \tag{4}$$

where ΔG_i is the association energy of particular species i. The simplest case is HbF. The γ chains possess at least two amino acids that interfere with polymerization.[24,25] Thus incorporating γ chains is so costly (the energy ΔG_i is so large) that neither the HbF nor its S/F hybrids will enter the polymer in any significant amount. Therefore $e \cong 0$ for all species except HbS, for which (by definition) $e = 1$. Figure 4a shows the solubility of HbF compared with theory.

By contrast, HbA/S hybrids do copolymerize (cf. (2)). In the case of the double strands alone, an AS hybrid can polymerize in one orientation (with the $\beta 6Val$ in the acceptor pocket) with no energetic penalty, but the reverse orientation (with $\beta 6glu$ placed in the acceptor pocket) carries enough of an energetic penalty as to make it highly unlikely. Thus for the double strands, the energy of association ΔG (cf. Eq. (3)) for a hybrid would only differ from ΔG for HbS by the absence of the RT ln 2 term. That would make $e = 0.5$. However, the fiber has more interactions than the double strands. Although it was thought that in the fiber only one $\beta 6$ was used (as in the double strand), extensive data collected by various groups consistently found a lower value of e.[17] This was subsequently reconciled by the proposal that some Hb molecules used both $\beta 6$ sites[17] so that the observed e is an average over the complex polymer. The probability of a hybrid entering the polymer is thus 0.375, a number on which observations and modeling agree. Figure 4a shows the dependence of the solubility on total fraction of HbA compared with theoretical predictions.

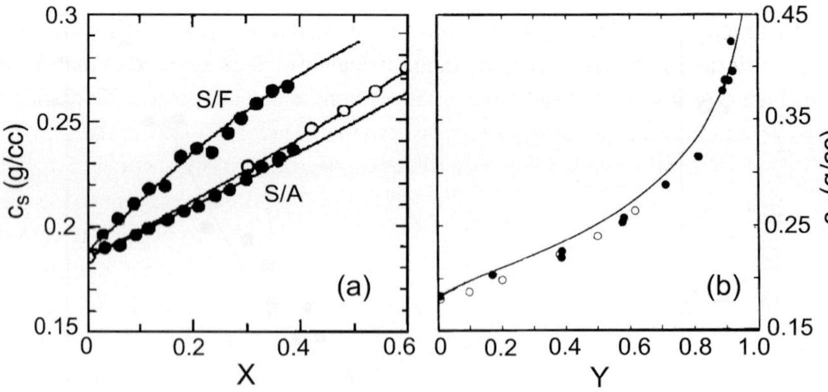

Fig. 4. Solubility of mixtures of hemoglobin. (a) The solubility of HbS mixed with HbF or HbA as a function of fraction (X) of the non-S Hb. HbS/A mixtures are more soluble than HbS/F because S/A hybrids polymerize and S/F hybrids do not. Open and filled circles for S/A data denote different total concentrations. Note that the y-axis does not begin at zero. (b) Solubility of HbS as a function of the fractional saturation, Y. Saturation with O_2 is taken from Sunshine[26] and shown as filled symbols; saturation with CO is from Hofrichter[52] and shown as open symbols. The solid curve shows the solubility inferred by using independent binding curve data.[2] The scales on the right graph are different from that of the left. If superimposed, the solubility curve versus Y would resemble that of S/A mixtures over the range shown.

The coefficients X_i are general designations of fractional species, and can be used in more than one context. The same expressions would apply to oxygenated species if the X_i were the fractions of molecules with 0, 1, 2, 3 and 4 ligands. Analysis of the solubility as a function of oxygen (coupled with linear dichroism studies on partially liganded polymers) reveals that even within the T-structure, the presence of ligands inhibit polymer formation by about a factor of 3 for the first ligand.[26] Figure 4b shows the dependence of solubility on fractional saturation Y. Note the similarity in the initial part of the curves to Fig. 4a.

Fiber and Gel Rigidity

The rigidity of the individual fibers is fundamental to occlusion but has only recently been determined by observing spontaneous bending fluctuations of polymers in solution.[27] The bending modulus (Y_b) is about 10^8 Pa, comparable to other protein fibers such as microtubules. Because polymer formation inevitably generates polymer domains as a result of the interplay between homogeneous and heterogeneous nucleation, rheology will depend on more than just the total concentration of polymerized hemoglobin.

Bulk rheological measurements demonstrate far less stiffness than fiber measurements: a mere 50 Pa shear modulus for a gel.[28] This is likely the result of bulk measurements being strongly influenced by polymer domains moving past one another, a phenomenon that does not appear *in vivo* if cells possess only one domain. The problem of characterizing the rheological properties of a given array of polymers is yet unresolved and only much simpler geometries are tractable. While mechanical occlusion is the most direct result of altered rheology, the ability of polymer-laden cells to induce responses such as inflammation also involves a rheological link that might be exploited for therapeutic intervention.

Kinetics

Mechanism of Hemoglobin S Polymerization

The kinetics of HbS polymerization is an essential part of the pathophysiology. Polymers form by two pathways.[29] In a solution devoid of polymers, random chance creates aggregates that are small pieces of polymers that are unstable with a likely fate of dissolution rather than growth. The random dynamics of the solution eventually produces an aggregate that crosses the stability barrier (by simply being large enough), and grows into a full-fledged polymer. This process is called homogeneous nucleation. Once polymers are present, a second option exists for making polymers, viz. nucleation on the surface of other polymers, called heterogeneous nucleation. This process involves random formation of polymer pieces, and is initially biased against growth, but to a lesser degree than homogeneous nucleation. Once nuclei form polymers by either mechanism the polymers grow linearly and are subsequently indistinguishable. This mechanism, shown in Fig. 5, is known as the double nucleation model. This was proposed on the basis of kinetic experiments,[30] and has subsequently been confirmed directly by Differential Interference Contrast (DIC) microscopy.[31] Since heterogeneous nucleation assists the nucleation process by adding an external contact, it might be supposed that surfaces other than that of the polymer could assist in nucleating polymers. Recent experiments show that sickle membranes (but not HbA membranes) enhance nucleation.[32]

The time course of the initial phase of polymerization in the double nucleation mechanism can be described by

$$\Delta = A[\cosh Bt - 1] \cong (A/2)\exp(Bt) \tag{5}$$

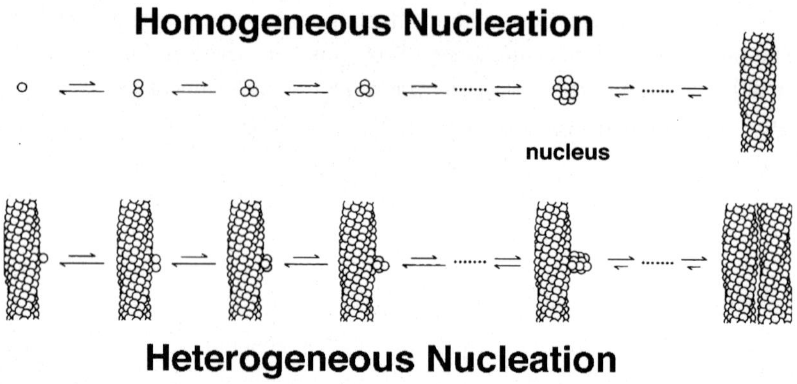

Fig. 5. Double nucleation model of Ferrone, Hofrichter and Eaton.[29] In homogeneous nucleation, a nucleus forms from solution, while in heterogeneous nucleation, a nucleus forms on the surface of another polymer. Nuclei are always unfavorable, so that the equilibrium arrows point more strongly away from nuclei than toward them.

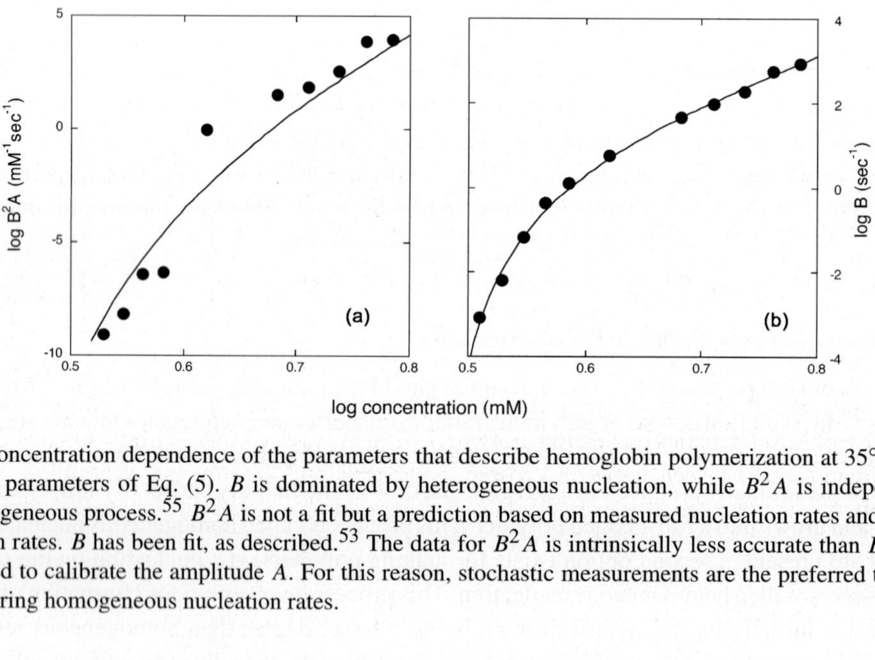

Fig. 6. Concentration dependence of the parameters that describe hemoglobin polymerization at 35°C. B and A are the parameters of Eq. (5). B is dominated by heterogeneous nucleation, while B^2A is independent of the heterogeneous process.[55] B^2A is not a fit but a prediction based on measured nucleation rates and polymer elongation rates. B has been fit, as described.[53] The data for B^2A is intrinsically less accurate than B because of the need to calibrate the amplitude A. For this reason, stochastic measurements are the preferred technique for measuring homogeneous nucleation rates.

in which Δ is the concentration of polymerized monomers, and A and B are constants that are dependent on the specifics of the reaction including homogeneous nucleation rate, f, and heterogeneous nucleation rate, g.[33] Both constants (A and B) are highly concentration dependent. B^2A does not depend on heterogeneous nucleation making data analysis simpler by analyzing the product rather than A alone. Values of B and B^2A are shown in Fig. 6 over a wide range of concentrations, along with the current best theoretical description based on the thermodynamic expression of the double nucleation model.[1] Equation (5) is only accurate when Δ is a small fraction of the initial monomer concentration, and the exponential approximation works well for larger values of Bt. For example, at $Bt = 4$, the difference between the cosh and the exponential is about 1%. Equation (5) illustrates one of the best known features of HbS polymerization, viz.

the appearance of a delay time as a result of exponential growth. For example, at an initial concentration of 5 mM (32 g/dl) and 35°C, one-tenth of the reaction occurs when $\Delta = 0.25$ mM. Under these conditions, B is 60 s^{-1} and A is 8×10^{-4} mM. When $t = 0.11$ s, $\Delta = 0.29$ mM, already exceeding one-tenth, whereas a mere 0.03 s earlier at $t = 0.08$ s, Δ is just 2% of the reaction (= 0.05 mM). In this time interval from 0.08 s to 0.11 s, nothing has changed in the mechanism, but rather the starkness of exponential growth has taken the amount of polymerized monomers from minute amounts to measurable levels. Because the mechanism involves nucleation and growth, it is tempting to equate the delay time with nucleus formation, followed by the time during which polymers grow, but this is fundamentally wrong. The above equation describes the growth of polymers beginning from time zero, not from the end of the delay time. The delay time provides a handy benchmark for reaction kinetics, and also has physiological significance. Since very little polymer forms during the delay, the corresponding occlusive potential is minimal, and the ability of cells to traverse the microcirculation is essentially unimpaired.

At equilibrium, about 5% of sickle red cells would contain polymers at arterial oxygen pressures observed in homozygous sickle cell patients and 85% of red cells would contain polymers at mixed venous pressures.[34] Kinetic experiments agree with the equilibrium result at the arteries, but disagree with the venous results, showing little additional polymerization.[34] Enumeration of deformed cells in clinical studies gives numbers close to the kinetic results, i.e., around 10% sickled cells at arterial pressures, and 20% in veins. The cells have thus *not* reached their equilibrium state, which would entail far more polymerized cells than observed. In short, kinetics have kept the majority of the cells from sickling.

When a small amount of polymer, denoted by Δ_o, is present and gelation is begun abruptly then the concentration of polymerized hemoglobin becomes

$$\Delta = (A + \Delta_o)\cosh Bt \cong [(A + \Delta_o)/2]\exp(Bt) \tag{6}$$

Generally $A < 0.01$ mM,[30] therefore, when incomplete depolymerization is followed by repolymerization, $\Delta_o \gg A$. Equation (6) thus resembles Eq. (5), but with a shorter delay time because of the larger pre-factor. If Δ_o is large enough, there may be no delay at all.[34]

Because of heterogeneous nucleation, polymers form in attached arrays called polymer domains. These domains can be seen between crossed polarizers, and are roughly spherulitic, although close examination revealed that a two-lobed structure forms first (Fig. 7). In DIC microscopy these domains look like wheat-sheaf bundles.[35] Simulations of domain formation based on the double nucleation model successfully reproduce the observed features.[36] Domains become spherulitic because fibers can bend as they grow. Every domain begins from a single homogeneous nucleation event, and the number of domains has been used as a probe for the nucleation rate.[37–39]

The rate of homogeneous nucleation is small (Table 2), and it is possible to execute experiments on a volume in which not even one nucleus would be expected at the end of the delay time. Once a nucleus does form, the process of heterogeneous nucleation provides the same type of exponential growth described above. In such experiments, a new type of delay appears that is inherently random in nature due to the stochastic process of forming a nucleus (cf. Fig. 8). Szabo has derived an expression for the distribution of such delay times, and related it to the homogeneous nucleation rate.[40] Once a nucleus forms then the delay will be followed by the usual instrumental delay time. Table 1 summarizes these and two other types of delays encountered in the kinetics of HbS polymerization.

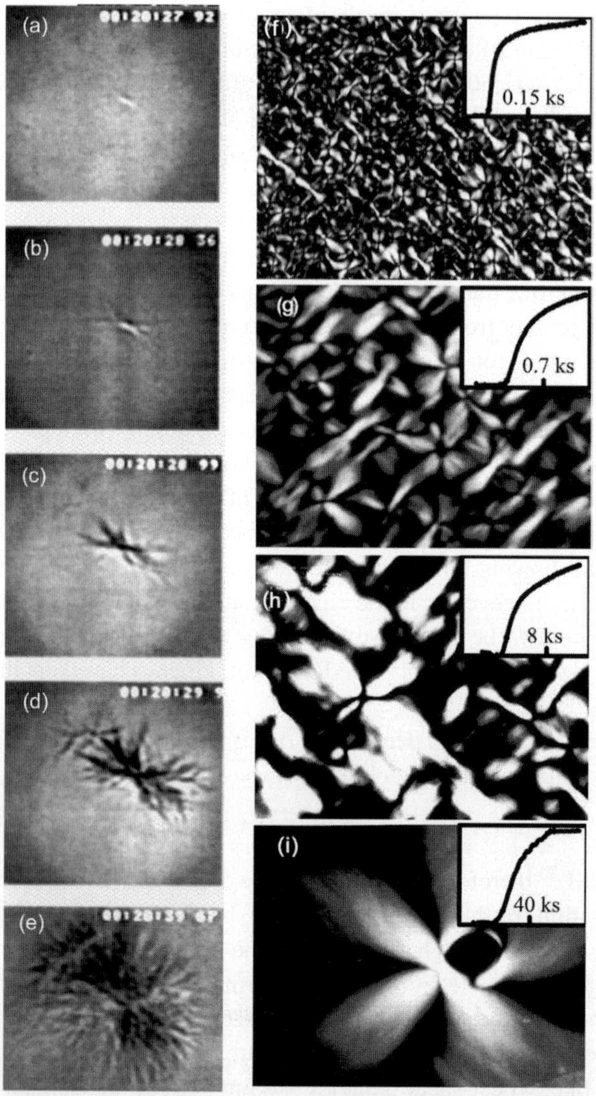

Fig. 7. Heterogeneous nucleation of polymer domains. Left panels, top to bottom (a–e): DIC pictures of a growing domain.[35] Because of heterogeneous nucleation, polymers propagate from the first, homogeneous nucleation event in arrays, called domains. As the polymers lengthen they bend and the wheat sheaf or bow tie pattern closes. Right panels (f–i): Optical birefringence micrographs of multi-domain gels and their kinetic progress curves.[39] Images are 1.16 mm × 0.93 mm. Insets show the fraction polymerized vs time after a jump from 3°C to 30°C. The delay times and initial HbS concentrations, from top to bottom, are (38 s, 26.3 g/dl), (360 s, 25.7 g/dl), (3700 s, 23.0 g/dl), and (24,000 s, 22.6 g/dl), respectively. Domain number depends on the number of homogeneous nuclei, which is highly concentration dependent. Used with permission from Briehl R.[35]

The Thermodynamic Control of Nucleation

The mechanism shown in Fig. 5 gives rise to Eq. (5) and leads to the observed exponential growth of polymers with deoxygenated HbS. The other striking feature of HbS polymerization is the high concentration dependence of the reaction. This is rationalized by the simple assumption that the

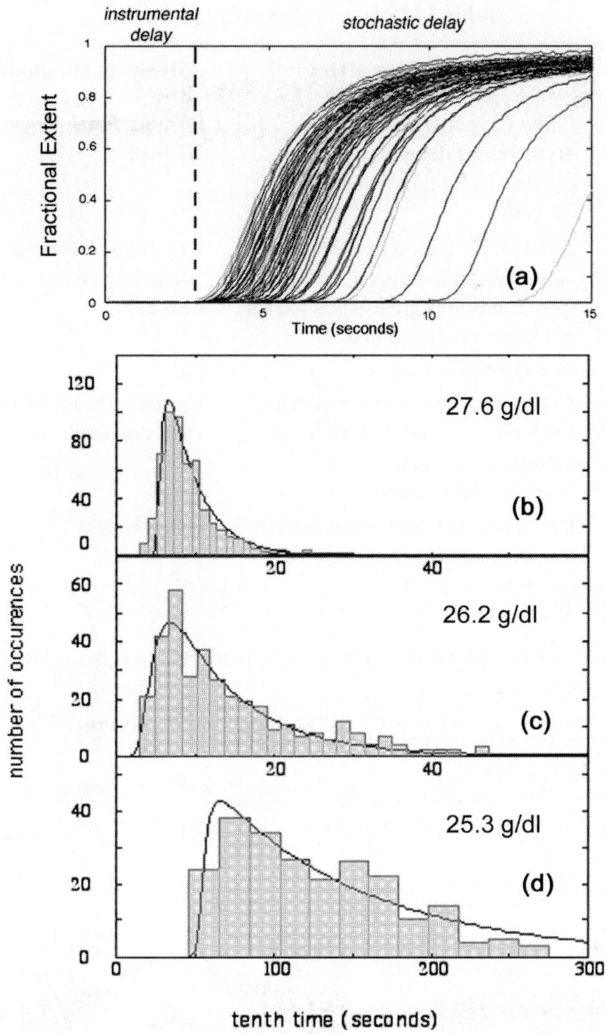

Fig. 8. Stochastic behavior of delay time. The experiments shown are done in parallel on the same sample, using a laser focused in many small spots. Panel (a) shows a typical distribution of progress curves. The shortest delay time arises from the exponential progress curves, and corresponds to nuclei that formed immediately. The other, stochastically distributed delay times, have an added delay because some nuclei will form at later times. (b–d) Distributions of delay times for three different HbS concentrations; fits are to Szabo's equation. The exponential tail arises from the homogeneous nucleation rate.

nucleus consists of a significant number of monomers in equilibrium with the polymer. Because the nucleus is primarily the result of thermodynamic interplay rather than a structural singularity, it can and does vary with initial concentration. Table 2 illustrates typical nucleus sizes. The principal driving force in generating nuclei as well as setting their size is supersaturation, defined as $\gamma c/\gamma_S c_S$ (Crowding also affects the kinetic equations. This effect is large, but fully and accurately prescribed by the theory).[1] The connection between thermodynamics and kinetics can obscure the importance of the latter in the pathophysiology. All thermodynamic variables that are modified have an impact on the nucleation rates, but the amplification that occurs is a feature of the kinetics. Thus, ironically,

Table 1. Delays in HbS polymerization.

Type	Origin of the effect	Means to eliminate the delay
Solubility	Finite oxygen delivery by erythrocytes delays earliest polymerization until solubility is exceeded.	Instantaneous deoxygenation *in vitro*.
Stochastic	Random formation of first nucleus causes variable interval between time conditions are established for polymerization and first observation.	Use of large volume so that nucleation occurs in volume at time zero.
Instrumental	Exponential growth of polymer mass which makes it hard to see earliest accumulation (the "classic" delay time).	Use of sensitive techniques to view polymers after nucleation.
Rheological	Rheological impact is not directly proportional to polymer mass.	Unknown.

Table 2. Selected values characterizing hemoglobin polymerization.[1]

	20 g/dl	25 g/dl	35 g/dl	45 g/dl
Homogeneous nucleus size, i^*	21	9	4	2
Heterogeneous nucleus size, j^*	33	5	2	1
Concentration of homogeneous nuclei (mM)	2.3×10^{-23}	9.3×10^{-16}	7.7×10^{-11}	2.3×10^{-8}
Homogeneous nucleation rate, f_o (mM/s)	1.6×10^{-16}	1.3×10^{-8}	2.5×10^{-3}	0.5
Delay time (s)	5.1×10^{5}	11	0.03	7.6×10^{-4}

*Nucleus sizes have been rounded to the nearest integer.
[1]All values are for 37°C.

while polymer formation in most red cells does not proceed to equilibrium in the microcirculation, the kinetics is still governed by equilibrium properties of the solution.

Kinetics of HbF Mixtures

Mixtures of HbS and HbF are accurately described by the double-nucleation model assuming negligible polymerization of any γ chains (S/F hybrids or HbF).[41] Table 3 summarizes changes in the polymerization process when 20% and 40% of the HbS is replaced by HbF. An unexpected result is the degree to which the crowding effects offset the dilution of the initial concentration by HbF.[41] Because this crowding effect is concentration dependent, the effect on the kinetics of replacing Hb by HbF is concentration dependent as shown in Fig. 9.

Augmenting HbF expression is a major therapeutic goal in sickle cell disease that has been partially realized with the administration of hydroxyurea[42] and several other pharmacological agents.[43] It is therefore important to understand how the kinetic model fits with what is observed in patients.

Table 3. Selected values characterizing polymerization of hemoglobin S/F mixtures.[1]

	0% HbF	**20% HbF**	**40% HbF**
Homogeneous nucleus size, i^*	4	4	5
Heterogeneous nucleus size, j^*	2	2	3
Concentration of homogeneous nuclei (mM)	7.7×10^{-11}	4.3×10^{-12}	3.0×10^{-14}
Homogeneous nucleation rate f_o (mM/s)	2.5×10^{-3}	6.3×10^{-5}	1.6×10^{-7}
Delay time (s)	0.03	0.14	1.3

*Nucleus sizes have been rounded to the nearest integer.
[1]All values are for 37°C; 35 g/dl total Hb.

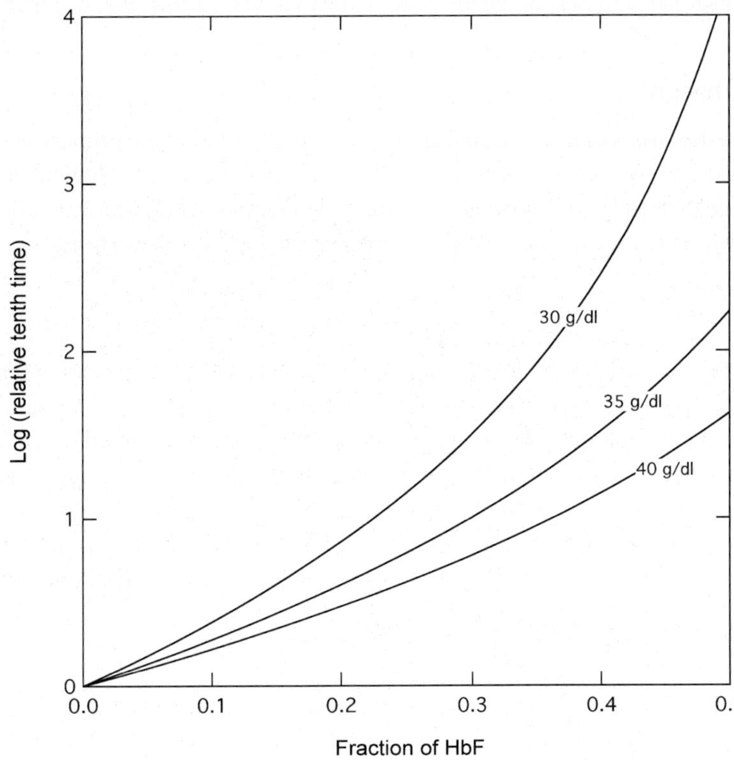

Fig. 9. The anti-polymerization effects of HbF. Log of predicted relative tenth times as a function of X, the fraction of HbF, at 35°C. HbF and HbS are assumed to hybridize, and only pure HbS species are capable of polymerization, such that the polymerizable fraction is $(1-X)^2$. Curves are calculated for total Hb concentrations of 30, 35 and 40 g/dl, labeled as shown. Experimental data confirms these predictions.[41] Tenth times are all deterministic, and shown relative to the delay time for the same concentration at $X = 0$. The same graphs would apply to a drug that rendered a certain fraction X of HbS unable to polymerize, regardless of whether it acted on the α or β chain.

Bridges *et al.*[44] demonstrated that Hydroxyurea has a substantial effect on delay times in patients, and slowing the rate of HbS polymerization correlated with positive clinical outcome. Unexpectedly, the observed effects of hydroxyurea on polymerization in red cells exceeded the predicted effects,[41] suggesting the drug has some other molecular effects on HbS polymerization.

While mixtures of HbF and HbS also change the solubility (cf. Fig. 4a), the effects of kinetics (e.g. Fig. 9) dwarf the effects on the solubility. For example, a cell with 32 g/dl total hemoglobin concentration would form about 13 g/dl of polymer with no HbF, and about 8 g/dl with 20% HbF. Thus 20% HbF in place of HbS changes the *amount* of polymerized hemoglobin by about 40%, but as Fig. 9 shows, it changes the delay time by about 1000%.

Depolymerization

Depolymerization can occur by the loss of polymerized HbS from the ends of the polymers or from their surface.[35,45–47] Side depolymerization is cooperative, and only significant at high concentrations of ligands,[45] while end depolymerization is dominant when the ligand concentration is low. Depolymerization at the high pO_2 of the lungs could be much faster than that occurring in the tissues.

Inhibition by Design

Finally we turn to the issues of how rational approaches might address the primary events of sickling. There are higher order events that will not be addressed here, such as rheological consequences, or membrane damage that lead to deleterious consequences. Our purpose is simply to ask what strategies might be employed to reduce the mass of polymers and increase the delay time.

Drugs

To date, no satisfactory drugs have been found. One approach is to design drugs that will compete efficiently for the contact sites on the hemoglobin surface, thus thwarting polymerization. As seen with HbF, dilution of polymerization competent species by 20% could be quite effective. This has been a strong impetus for discovering contact sites within polymers. In addition, drugs that bind peripheral to a contact site could be effective if they reduce the motional freedom of hemoglobin in the polymer and reduce the amount μ_{PV} contributes to stability. This is the underlying reason that HbC^{Harlem}, a crystal-forming double mutant, polymerizes almost 11 orders of magnitude slower than HbS.[48]

Hemoglobin

In addition to administration of drugs that would interact directly, there is intense interest in inducing HbF and genetically engineering inhibitory recombinant hemoglobin subunits. A useful target is to make HbX in which the β_X subunits effectively prohibit polymerization as is observed with the naturally occurring γ chain. Another avenue is to develop recombinant α-chains (e.g. α_X) that disrupt polymer formation. Sub-unit assembly in erythroid cells expressing this variant, as well as endogenous α-globin will yield a dimer population consisting of $\alpha_X\beta_S$, $\alpha\beta_X$, $\alpha_X\beta_X$, and $\alpha\beta_S$. If only the $(\alpha\beta_S)_2$ species is capable of polymerizing then the propensity of the erythroid cells to sickle will be greatly diminished. This is the motivation behind the HbS Chiapas mutant and variants, all of which have the anti-sickling substitution $\alpha 114 Pro \rightarrow Arg$.[49] It is important that recombinant α-chains retain the capacity to form tetramers, and with enhanced subunit association, their anti-sickling property will increase, as has been reported for a mutant β chain recombinant.[50] A third design strategy is to alter oxygen affinity of mutant hemoglobins. For instance, recombinant mutant hemoglobin that will deoxygenate at lower pO_2 than HbS will yield proportionately less deoxyHbS that is available for polymerization for a given pO_2. Interestingly, this is a potential limitation of HbF since γ chains are not as responsive to allosteric effectors such as 2–3 diphosphoglycerate and thus have slightly higher O_2 affinity than HbS.[51]

References

1. Ferrone FA, Rotter MA (2004). Crowding and the polymerization of sickle hemoglobin. *J Mol Recognition* **17**:497–504.
2. Eaton WA, Hofrichter J (1990). Sickle cell hemoglobin polymerization. *Adv Protein Chem* **40**:63–280.
3. Dykes GW, Crepeau RH, Edelstein SJ (1979). Three-dimensional reconstruction of 14-filament fibers of hemoglobin S. *J Mol Biol* **130**:451–472.
4. Nagel RL, Johnson J, Bookchin RM *et al.* (1980). β-Chain contact sites in the haemoglobin S polymer. *Nature* **283**:832–834.
5. Nagel RL, Bookchin RM (1978). Areas of interaction in the HbS polymer. In: *Biochemical and Clinical Aspects of Hemoglobin Abnormalities*, New York: Academic Press.
6. Wishner BC, Ward KB, Lattman EE, Love WE (1975). Crystal structure of sickle-cell deoxyhemoglobin at 5 Å resolution. *J Mol Biol* **98**:179–194.
7. Harrington DL, Adachi K, Royer Jr WE (1997). The high resolution crystal structure of deoxyhemoglobin S. *J Mol Biol* **272**:398–407.
8. Padlan EA, Love WE (1985). Refined crystal structure of deoxyhemoglobin S. II. Molecular interactions in the crystal. *J Biol Chem* **260**:8280–8291.
9. Padlan EA, Love WE (1985). Refined crystal structure of deoxyhemoglobin S. I. Restrained lest squares refinement at 3.0 Å resolution. *J Biol Chem* **260**: 8272–8279.
10. Watowich SJ, Gross LJ, Josephs R (1993). Analysis of the intermolecular contacts within sickle hemoglobin fibers: effect of site-specific substitutions, fiber pitch and double-strand disorder. *J Struct Biol* **111**:161–179.
11. Watowich SJ, Gross LJ, Josephs RJ (1989). Intramolecular contacts within sickle hemoglobin fibers. *J Mol Biol* **209**:821–828.
12. Creighton TE (1993). *Proteins: Structures and Molecular Properties*. New York: W. H. Freeman and Company.
13. Silva MM, Rogers PH, Arnone A (1992). A third quaternary structure of human hemoglobin A at 1.7-Å resolution. *J Biol Chem* **267**:17248–17256.
14. Adachi K, Sabnekar P, Adachi M, Reddy LR, Pang J, Reddy KS, Surrey S (1995). Polymerization of recombinant Hb S-Kempsey (deoxy-R state) and Hb S-Kansas (oxy-T state). *J Biol Chem* **270**: 26857–26862.
15. Briehl RW, Ewert SA (1974). Gelation of sickle cell hemoglobin. II. Methaemoglobin. *J Mol Biol* **89**: 759–766.
16. Briehl RW, Salhany JM (1975). Gelation of sickle hemoglobin. III. Nitrosyl hemoglobin. *J Mol Biol* **96**: 733–743.
17. Roufberg A, Ferrone FA (2000). A model for the sickle hemoglobin fiber using both mutation sites. *Protein Sci* **9**:1031–1034.
18. Hofrichter J, Ross PD, Eaton WA (1976). Supersaturation in sickle cell hemoglobin solutions. *Proc Natl Acad Sci USA* **73**:3035–3039.
19. Hill TL (1986). *An Introduction to Statistical Thermodynamics*. New York: Dover Publications.
20. Ross PD, Hofrichter J, Eaton WA (1977). Thermodynamics of gelation of sickle cell deoxyhemoglobin. *J Mol Biol* **115**:111–134.
21. Cao Z, Ferrone FA (1997). Homogeneous nucleation in sickle hemoglobin. Stochastic measurements with a parallel method. *Biophys J* **72**:343–372.
22. Adachi K, Asakura T (1979). Nucleation-controlled aggregation of deoxyhemoglobin S. Possible difference in the size of nuclei in different phosphate concentrations. *J Biol Chem* **254**:7765–7771.
23. Perella M, Benazzi L, Shea MA, Ackers GK (1990). Subunit hybridization studies of partially ligated cyanomethemoglobins using a cryogenic method. *Biophys Chem* **35**:97–103.
24. Adachi K, Pang J, Konitzer P, Surrey S (1996). Polymerization of recombinant hemoglobin F γE6V and hemoglobin F γE6V, γQ87T alone, and in mixtures with hemoglobin S. *Blood* **87**:1617–1624.

25. Nagel RL, Bookchin RM, Johnson J, Labie D, Wajcman H, Isaac-Sodeye WA, Honig GR, Schiliro G, Crookston JH, Matsutomo K (1979). Structural bases of the inhibitory effects of hemoglobin F and hemoglobin A2 on the polymerization of hemoglobin S. *Proc Natl Acad Sci USA* **76**:670–672.
26. Sunshine HR, Hofrichter J, Ferrone FA, Eaton WA (1982). Oxygen binding by sickle cell hemoglobin polymers. *J Mol Biol* **158**:251–273.
27. Wang JC, Turner MS, Agarwal G, Kwong S, Josephs R, Ferrone FA, Briehl RW (2002). Micromechanics of isolated sickle hemoglobin fibers: bending moduli and persistence lengths. *J Mol Biol* **315**:6001–6012.
28. Gabriel DA, Smith LA, Johnson CS (1981). Elastic properties of deoxy hemoglobin S gels. *Arch Biochem Biophys* **211**:774–776.
29. Ferrone FA, Hofrichter J, Eaton WA (1985). Kinetics of sickle hemoglobin polymerization II: a double nucleation mechanism. *J Mol Biol* **183**:611–631.
30. Ferrone FA, Hofrichter J, Eaton WA (1985). Kinetics of sickle hemoglobin polymerization I: studies using temperature-jump and laser photolysis techniques. *J Mol Biol* **183**:591–610.
31. Samuel RE, Salmon ED, Briehl RW (1990). Nucleation and growth of fibres and gel formation in sickle cell haemoglobin. *Nature* **345**:833–835.
32. Aprelev A, Rotter MA, Etzion Z, Bookchin RM, Briehl RW, Ferrone FA (2005). The effects of erythrocyte membranes on the nucleation of sickle hemoglobin. *Biophys J* **88**(4):2815–2822.
33. Bishop MF, Ferrone FA (1984). Kinetics of nucleation controlled polymerization: a perturbation treatment for use with a secondary pathway. *Biophys J* **46**:631–644.
34. Mozzarelli A, Hofrichter J, Eaton WA (1987). Delay time of hemoglobin S polymerization prevents most cells from sickling. *Science* **237**:500–506.
35. Briehl R (1995). Nucleation, fiber growth and melting, and domain formation and structure in sickle cell hemoglobin gels. *J Mol Biol* **245**:710–723.
36. Dou Q, Ferrone FA (1993). Simulated formation of polymer domains in sickle hemoglobin. *Biophys J* **65**:2068–2077.
37. Hofrichter J (1986). Kinetics of sickle hemoglobin polymerization III. Nucleation rates determined from stochastic fluctuations in polymerization progress curves. *J Mol Biol* **189**:553–571.
38. Galkin O, Vekilov PG (2004). Mechanisms of homogeneous nucleation of polymers of sickle cell anemia hemoglobin in deoxy state. *J Mol Biol* **336**:43–59.
39. Christoph GW, Hofrichter J, Eaton WA (2005). Understanding the shape of sickled red cells. *Biophys J* **88**:1371–1376.
40. Szabo A (1988). Fluctuations in the polymerization of sickle hemoglobin: a simple analytical model. *J Mol Biol* **199**:539–542.
41. Rotter M, Aprelev A, Adachi K, Ferrone FA (2005). Molecular crowding limits the role of fetal hemoglobin in therapy for sickle cell disease. *J Mol Biol* **347**:1015–1023.
42. Charache S, Terrin ML, Moore RD, Dover GJ, Barton FB, Eckert SV, McMahon RP, Bonds DR (1995). Effect of hydroxyurea on the frequency of painful crises in sickle cell anemia. *N Engl J Med* **332**:1317–1322.
43. Atweh GF, Sutton M, Nassif I, Boosalis V, Dover GJ, Wallenstein S, Wright E, McMahon L, Stamatoyannopoulos G, Faller DV, Perrine SP (1999). Sustained induction of fetal hemoglobin by pulse butyrate therapy in sickle cell disease. *Blood* **93**:1790–1797.
44. Bridges KR, Barabino GD, Brugnara C, Cho MR, Christoph GW, Dover G, Ewenstein BM, Golan DE, Guttmann CR, Hofrichter J, Mulkern RV, Zhang B, Eaton WA (1996). A multiparameter analysis of sickle erythrocytes in patients undergoing hydroxyurea therapy. *Blood* **88**:4701–4710.
45. Agarwal G, Wang JC, Kwong S, Cohen SM, Ferrone FA, Josephs R, Briehl RW (2002). Sickle hemoglobin fibers: mechanisms of depolymerization. *J Mol Biol* **322**:395–412.
46. Louderback JG, Aroutiounian SK, Kerr WC, Ballas SK, Kim-Shapiro DB (1999). Temperature and domain size dependence of sickle cell hemoglobin polymer melting in high concentration phosphate buffer. *Biophys Chem* **80**:21–30.
47. Louderback JG, Ballas SK, Kim-Shapiro DB (1999). Sickle hemoglobin polymer melting in high concentration phosphate buffer. *Biophys J* **76**:2216–2222.
48. Ivanova M, Jasuja R, Krasnosselskaia L, Josephs R, Wang Z, Ding M, Horiuchi K, Adachi K, Ferrone FA (2001). Flexibility and nucleation in sickle hemoglobin. *J Mol Biol* **314**:851–861.

49. Ho C, Willis BF, Shen TJ, Dazhen NT, Sun DP, Tam MF, Suzuka SM, Fabry ME, Nagel RL (1996). Roles of α114 and β87 amino acid residues in the polymerization of hemoglobin S: implications for gene therapy. *J Mol Biol* **263**:475–485.
50. Levasseur DN, Ryan TM, Reilly MP, McCune SL, Asakura T, Townes TM (2004). A recombinant human hemoglobin with anti-sickling properties greater than fetal hemoglobin. *J Biol Chem* **279**:27518–27524.
51. Bunn HF, Briehl RW (1970). The interaction of 2,3-diphosphoglycerate with various human hemoglobins. *J Clin Invest* **49**:1088–1095.
52. Hofrichter J (1979). Ligand binding and the gelation of sickle cell hemoglobin. *J Mol Biol* **128**:335–369.
53. Ferrone FA, Ivanova M, Jasuja R (2002). Heterogeneous nucleation and crowding in sickle hemoglobin: an analytic approach. *Biophys J* **82**:399–406.

10

Damage to the Red Blood Cell Membrane in Sickle Cell Disease

by Steven R. Goodman and Clinton Joiner

Introduction

The red blood cell (RBC) membrane consists of a well-studied lipid bilayer and numerous specialized proteins. The external leaflet of the bilayer contains primarily choline-containing phospholipids, with aminophospholipids (especially phosphatidyl serine) located in the inner leaflet. Cholesterol intercalates between the fatty acyl chains within both leaflets, but is also asymmetrically distributed. The membrane protein complement includes surface-exposed blood group antigens, hormone and cytokine receptors and various effector partners, as well as adhesion molecules; transporters and channels and their regulatory molecules; and the membrane skeleton on the cytoplasmic surface that controls shape, elasticity and flexibility. Some proteins traverse or are embedded in the membrane (integral membrane proteins) and others are bound via interactions with lipids or other proteins.

Mechanical and chemical interactions with hemoglobin S (HbS) damage the plasma membrane of sickle RBCs. Polymerization of deoxygenated HbS leads to dissociation of the cytoskeleton from the lipid bilayer and there is loss of membrane upon reoxygenation. Oxidative damage to proteins and lipids as a result of the altered redox status of sickle cell RBCs leads to pathological consequences. As first pointed out by Robert Hebbel,[1] the sickle RBC is in double jeopardy with increased levels of oxygen radicals and diminished levels of reduced glutathione (GSH). Indeed, Goodman and colleagues[2] have demonstrated an inverse relationship between sickle RBC density and GSH levels, with the highest density RBCs having GSH levels that are below detection. As a result, key functional proteins are damaged, leading to altered adhesion, loss of phospholipid asymmetry, abnormal cation transport and dysfunctional cell volume regulation, and a locked membrane skeleton in the irreversibly sickled cell (ISC). These membrane pathologies perturb many RBC functions and contribute to the hemolytic and vaso-occlusive manifestations of sickle cell disease (SCD). They therefore impact both the acute painful episodes, as well as the chronic manifestations such as organ damage and possibly even premature death.

In this chapter we review the interaction of sickle RBCs with the external environment, exposure of phosphatidyl serine on the external leaflet of the RBC bilayer, altered membrane transport and cell volume regulation, and formation of the locked membrane skeleton leading to the ISC. We will conclude with discussions of the early attempts using proteomics to unravel pathological alterations in the sickle RBC membrane.

The RBC Surface and Vaso-Occlusion

The fundamental basis of the pathology of SCD is the polymerization of deoxygenated HbS. However, the abnormal function of the RBC membrane induced by HbS, as well as pathological responses of other cells to sickle cells, clearly contributes to the complex pathophysiology of the disease. The process of vaso-occlusion involves multiple processes and cellular interactions. Factors that contribute to vaso-occlusion include: endothelial activation and damage; adhesion of sickle cells, especially stress reticulocytes in venules; adherence of leukocytes to activated and damaged endothelium; and platelet and coagulation factor activation. These factors promote entrapment of dense non-deformable ISCs and increase sickling of hydrated sickle cells in capillaries and post-capillary venules.

Increased erythropoietic stress in homozygous SS (HbSS) patients, resulting in increased erythropoietin levels, is associated with increased production of placenta growth factor (PIGF) by $CD34^+$erythroid progenitor cells. PIGF is an angiogenic growth factor belonging to the vascular endothelial growth factor (VEGF) family. Circulating PIGF levels correlate with sickle cell severity and stimulate monocyte chemotaxis. Increased PIGF causes monocytes to increase expression of proinflammatory cytokines and chemokines (IL-1β, IL-8, and MCP-1 and TNF-α), which activate endothelial cells.[3] Adhesion of monocytes to the vascular endothelium is triggered by MCP-1 and IL-8.[4] Markedly more IL-1β and TNF-α is produced by monocytes in patients with SCD (sickle monocytes) than monocytes found in individuals with a normal β-globin genotype. Consequently, there is increased translocation of NF-κB to the nucleus, causing increased expression of adhesion molecules including P- and E-selectin on the surface of endothelial cells, which are virtually in a constant activation state in individuals with SCD. Moreover, leukocytes adhere to venular endothelium expressing P- and E-selectin, and become potential docking targets for sickle RBCs, as elegantly demonstrated by intravital microscopy in mice expressing human sickle hemoglobin.[5] Thus one potential mechanism by which sickle RBCs become attached to the vessel wall is through adhesion to leukocytes.

There is a consensus that sickle cells attach directly to activated endothelium.[6–9] Low density reversibly sickled cells (RSCs) are more adherent than high density ISCs.[10–13] Furthermore, Hebbel *et al.*[6] demonstrated that vaso-occlusive severity in patients with HbSS correlated with *in vitro* HbSS RBC adhesivity. What are the relevant adhesion molecules that cause increased adhesion? On the RBC surface $\alpha_4\beta_1$ integrin, CD36, band 3, sulfated glycolipid, and basal cell adhesion molecule/Lutheran protein are involved. Proteomic studies, discussed later in this chapter, provide a powerful approach for identification of changes in expression and post-translational modifications of specific adhesion molecules, which are critical in HbSS RBC adhesion to the endothelium. Candidate molecules expressed on the endothelial surface include P-selectin, vascular cell adhesion molecule-1, glycoprotein 1-b and CD-36. Plasma proteins that may play a bridging role in increased adhesion include von Willebrand factor and thrombospondin. Thirdly, damage to the endothelium exposes subendothelial matrix molecules such as laminin and fibronectin for adhesion with sickle RBCs.[14–17] Regarding sickle RBCs, there is a consensus that low density reticulocytes and RSCs preferentially adhere to the endothelial wall[6] and to leukocytes.[5] However, non-deformable ISCs[18] become entrapped in venules because of the narrow lumen in these vessels and the reduced pliability of ISCs. These combined events contribute to vaso-occlusion and sickle cell painful episodes.

Altered Phospholipid Asymmetry in the RBC Membrane

As shown in Fig. 1, the normal RBC membrane has phosphatidyl choline (PC) and sphingomyelin (SM) primarily in the outer leaflet, phosphatidyl ethanolamine (PE) primarily in the inner leaflet and phosphatidyl serine (PS) exclusively in the inner leaflet of the lipid bilayer. However, PS becomes exposed on the outer leaflet[19,20] in a subpopulation (varying from 5–10%) of sickle RBCs. PS^+ cells are enriched in both the lightest density fractions (probably reticulocytes) and the densest fractions (ISCs) which may contain as many as one third PS^+ cells.[21,22]

There are two potential mechanisms involved in increased exposure of PS on the surface of HbSS RBCs: 1. Calcium dependent scramblase; and 2. Decreased activity of the ATP driven aminophospholipid translocase or flipase.[23] Flipase is known to be inactivated by oxidative stress.[24] The role of scramblase in PS exposure is less clear and alternative scrambling mechanisms involving lipid rafts and/or a PKC dependent mechanisms have been proposed.[25,26]

Phosphatidyl serine positive RBCs activate coagulation factors and may contribute to altered hemostasis and inflammation in sickle cell patients.[27] In addition PS exposure increases HbSS RBC adhesion to the endothelium and monocytes.[28,29] Enhanced binding to phagocytes and the procoagulant activity likely contributes to the reduced life span of PS^+ RBCs.[21] However, rabbit RBCs exposed to elevated intracellular Ca (which causes both PS externalization and severe dehydration) had nearly normal survival *in vivo*.[30] Studies in HbSS patients demonstrated that more than half of the PS^+ sickle cells are transferrin receptor-positive reticulocytes, which normally exhibit low levels of PS externalization.[22] *In vivo* survival studies of biotin labeled RBCs in SCD patients reveal that most of the rapid clearance of PS^+ sickle cells could be explained by reticulocyte maturation.[22] This may be particularly germane in studies of transgenic mice which have extremely high reticulocyte levels and overall RBC survival of 2–3 days, similar to reticulocyte maturation times.[32,33] Furthermore, in humans, the percentage of PS^+ cells among a biotin labeled population of dense cells remained

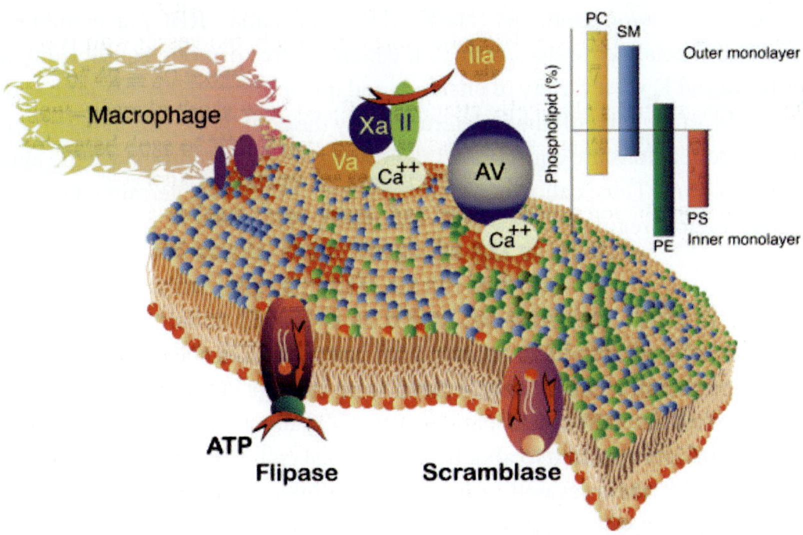

Fig. 1. Exposure of phosphatidyl serine on the outer leaflet of the RBC membrane bilayer leads to calcium dependent binding of clotting factors and shortened life span for the erythrocyte. Used with permission from Kuypers FA and De Jong K (2004). *Cell Mol Biol* **50**:147–158.

relatively constant for several days *in vivo*, which is inconsistent with extremely rapid clearance of these cells.[34] Thus the pathophysiological consequences of PS externalization in sickle cells remain to be clarified.

Membrane Transport and Volume Regulation in Sickle RBC Membrane Transport

An array of membrane proteins facilitate and control the flow of materials across biological barriers, thus providing polar pathways for hydrophilic molecules through the hydrophobic bilayer. Some transporters harness the cell's energy to move materials up an electrochemical gradient. Those that derive energy from cellular chemical sources (ATP) are referred to as pumps. Others couple the uphill movement of one substance against its electrochemical gradient to the down hill flow of another. They are designated co-transporters when the substrates move in the same direction and counter-transporters (or exchangers) when the directions are opposite. Proteins which regulate the movement of substances down their electrochemical gradients are facilitated-diffusion carriers or channels if free diffusion occurs.

Transporters are involved in many cellular processes: intake and egress of substrates, products and wastes; movement of solutes and water across epithelia; signal transduction; propagation of action potentials; maintenance of ion gradients and cellular pH; regulation of water content and cell volume. We will focus on the transport of salt as a mechanism for maintenance of cell volume, as sickle RBCs exhibit dysfunctional volume regulation that contributes to cellular pathology.

Cell Volume Regulation

The paradigm of cell volume control via regulation of cation content in RBCs was developed by Tosteson and Hoffman in 1960.[35] This theoretical framework has supported the accumulation of a remarkable body of detailed information on the molecular nature of the cation transporters involved, and their regulation by cellular processes. Key features of this model include an energy-coupled mechanism to maintain cation gradients (Na/K ATPase in human RBC), a membrane freely permeable to water (via specialized water channels), which ensures osmotic equilibrium, and regulated permeabilities to Na and K via a variety of channels and transporters. Given osmotic equilibrium, cell water content and therefore cell volume is determined by the content of osmolytes, most of which are salt ions. In the red cell, anions are equilibrated rapidly across the membrane via the anion exchanger (AEP1 or Band 3 protein). Thus, cation content represents the major determinant of cell volume that is subject to physiological regulation.

Consequences of Sickle RBC Dehydration

The presence of cells in sickle blood with high cellular hemoglobin concentration (CHC), and resultant high density, has been appreciated for many years. The percentage of this cell population varies among patients and at different times in the same patient. Typically, about 10% of cells in HbSS patients have CHC exceeding 41 gm/dl (compared to about 1% of normal cells) and this value reaches a maximum 50 gm/dl in some cells.[36] This elevated CHC reflects cellular dehydration (reduced water content) and cation depletion. Dense sickle cells have markedly reduced K content, with slightly elevated Na levels, representing dysfunctional volume regulation in sickle cells.

The consequences of cellular dehydration are manifold. Membrane rigidity and cellular viscosity and fragility are all increased in dehydrated cells, as are markers of membrane damage such as membrane skeletal and lipid oxidation. The dense cell population is enriched in ISCs, but all ISCs are not dense and vice versa, so the formation of the two cell types likely involves separate processes. Dense SCD-SS RBCs have markedly reduced life span *in vivo*,[34] and are selectively removed during vaso-occlusive episodes.[37] Hemolysis correlates with dense cell numbers *in vivo*. *Ex vivo* perfusion studies indicate selective trapping of dense cells in the microcirculation, which is potentiated by adhesion of reticulocytes to the venular endothelium.[11]

While the affinity of hemoglobin for oxygen is decreased in dehydrated cells, the most important consequence of cellular dehydration is on the kinetics of HbS polymerization. The delay time for sickling (time from deoxygenation to morphological sickling) is inversely proportional to the 30th power of HbS concentration, such that very small differences in concentration can have profound effects on the polymerization rate.[38] A well hydrated sickle cell (CHC 33 gm/dl) exhibits a delay time exceeding 10 sec, well above the time required to traverse the capillary circulation (about 1 sec) and return to the lungs for reoxygenation.[39] In contrast, a dehydrated cell with CHC of 45 gm/dl polymerizes almost instantaneously upon deoxygenation. Thus, cellular dehydration promotes vaso-occlusion.

The compound heterozygote state of hemoglobin S and C disease (HbSC) illustrates the importance of cellular dehydration in sickling syndromes. With respect to supporting HbS polymerization, HbC and HbA are similar, so that patients with HbSC might be expected to have a phenotype similar to sickle cell trait (HbAS). However, by stimulating KCl co-transport (see below), HbC fosters cellular dehydration, which renders HbSC RBCs capable of sickling under physiological conditions.[40]

Mechanisms of Sickle RBC Dehydration

Sickle cell dehydration represents a volume regulatory pathology derived from abnormal cation transport. Three transport pathways have been implicated in this pathology: A sickling-induced cation leak, a Ca-activated K channel (the Gardos pathway), and the KCl cotransporter. All appear to be active or abnormally regulated in sickle cells, and likely function in concert to produce cation depletion.

Sickling-Induced Cation Pathway

The increased cation permeability of deoxygenated sickle cells was first noted by Tosteson in 1952.[41] This sickling-induced pathway permits passive, down-hill movements of Na and K, which in cells with near-normal cation concentrations, results in balanced Na influx and K efflux. The resultant increase in Na concentration stimulates the Na pump, which because of its unbalanced stoichiometry (3 Na out/2 K in) produces a net cation efflux.[42] This relatively slow process may contribute to some cation loss, but may not explain the extreme cation depletion observed in the densest RBCs. The sickling-induced pathway also mediates increased Ca uptake and Mg efflux from HbSS RBCs,[43–45] which may result in secondary activation of the Ca-dependent K channel and KCl cotransporter, respectively (see below).

The sickling-induced pathway has been characterized as electrogenic, non-saturating, pH sensitive, non-selective among alkali metal cations,[46] and permeable to divalent cations such as Mg[45] and Ca.[43,44] It is difficult to determine whether anions are able to traverse the sickling-induced pathway, since the basal anion flux through the RBC membrane is several orders of magnitude higher than the sickling-induced flux. The pathway is insensitive to standard cation transport inhibitors, but is partially blocked by several compounds which inhibit (non-selectively) anion exchange, including the stilbene disulfonates (DIDS), phloretin, and dipyridamole.[43,47] The magnitude of the sickling-induced cation flux is proportional to the degree of morphological sickling, or more specifically the degree of spicule formation.[48] Spicules are formed under conditions of slow deoxygenation or high pH, which favor single polymer domains, and involve dissociation of the lipid bilayer from the underlying membrane skeleton.[49] These characteristics suggest this pathway represents a non-selective ion channel activated by membrane perturbations associated with sickling. A similar cation leak can be elicited in sickle and normal RBC by membrane shear, and parallels have been drawn to stretch-activated channels in other cells, although the pharmacological profiles of these channels are not similar to that of the sickling-induced leak.[50] Molecular identification of this pathway would help to illuminate this membrane pathology of sickle cells and might provide opportunities for additional therapeutic interventions.

Gardos Pathway

The rapid K efflux from ATP-depleted RBCs, originally described by Gardos in 1958,[51] is now known to be mediated by the intermediate-conductance Ca-activated K channel, hIK1, a product of the Kcnn4 gene. Although the number of channels on a mature RBC may be less than one hundred,[52] activation of this high conductance pathway *in vitro* by treatment of RBC with a Ca ionophore produces rapid K efflux and extreme dehydration of RBCs within minutes. *In vivo* activation of the Gardos pathway in HbSS RBCs is posited to occur as a consequence of Ca influx via the sickling-induced leak pathway. A vigorous Ca-ATPase pump defends against marked elevations of cellular Ca, so long as permeabilization is transient and energy stores are not depleted, conditions which obtain physiologically.[53] Indeed, the high Ca levels measured in HbSS RBCs have been shown to result from Ca sequestration in intracellular vesicles, and cytoplasmic free-Ca levels in the oxygenated state appear to be normal.[54]

Recent evidence indicates, however, that upon deoxygenation of HbSS RBCs, the increase in ionized Ca reaches the threshold to activate the Gardos channel, at least in a fraction of the cell population. Lew and Bookchin have demonstrated that this is a heterogeneous process, in that only a portion of cells experience Gardos channel activation upon deoxygenation and subsequent reoxygenation appears to activate the channel in a similar proportion of additional cells.[53] This behavior reflects the stochastic activation of the sickling-induced pathway to permit Ca influx, rather than differences in the number or activation of K channels in individual cells. A key determinant of whether the Gardos pathway is activated in a given cell is the level of ionized Ca achieved upon sickling, which is a balance between influx mediated by the sickling-induced pathway and efflux mediated by the Ca-pump. In theory this balance could be altered by inhibition of the Ca-pump (e.g., by oxidative damage or proteolysis) or by reduction in the sickling-induced influx (by pharmacological inhibitors or reduction in sickling). The best evidence supporting this pathway of dehydration *in vivo* is the demonstration that inhibitors of the Gardos pathway improve hydration of HbSS RBCs in animal models and in patients with SCD (see below).

The number of Ca-activated K channels on HbSS RBCs appears to be remarkably uniform and similar to that in normal erythrocytes.[55] The kinetics of activation is subject to regulation by receptor-ligand interactions mediated by signal transduction mechanisms. Both the apparent V_{max} of the channel and the $K_{1/2}$ for intracellular Ca activation are altered by a variety of inflammatory cytokines, such as interleukin (IL)-10, platelet activator factor and endothelin-1.[56] Differences between HbSS and HbAA RBCs in Gardos channel activity and kinetics may result from differences in the cytokine milieu *in vivo*. Gardos channel activation and concomitant dehydration occurs when RBC glutathione is decreased to very low levels by exposure to 1-chloro-2,4-dinitrobenzene (CDNB),[2,57] though it is not clear whether this is associated with oxidative damage to the channel itself or to alteration of calcium homeostasis.

KCl Cotransport

Lauf first described the NEM-stimulated Cl-dependent K fluxes mediated by the KCl cotransporter (KCC) in sheep RBCs.[58] When intracellular K concentration is high, the coupled, electroneutral bidirectional fluxes of K and Cl mediated by KCC produce net KCl efflux with consequent water loss and volume reduction.[59] Highly expressed in reticulocytes and young RBCs, KCC probably mediates the reduction in cell volume and resultant increase in CHC associated with normal reticulocyte maturation.[60] Brugnara first demonstrated the striking acid-stimulated K efflux mediated by KCC in cells containing HbS or HbC.[61,62]

KCC activation is associated with a net dephosphorylation event, and serine/threonine (ST) protein phosphatase inhibitors block activation by most stimuli.[63,64] A ST kinase, as yet unidentified, is thought to keep the transporter in a phosphorylated, inactive state. Kinetic studies suggest that swelling inhibits this kinase, thereby shifting the equilibrium to the active, dephosphorylated state. Under some conditions of activation, the phosphatase may be stimulated. Though phosphorylation of the transport protein has not been directly demonstrated, this seems likely, as the KCC transporters exhibit multiple potential phosphorylation sites, and the closely related NKCC transporter is phosphorylated (although inactivated) by stimuli similar to those that activate KCC. Tyrosine kinases also appear to regulate KCC activity, perhaps by modulating the activity of ST kinases and/or phosphatases.[65,66] Parker proposed that the basis of volume sensitivity of KCC (and other volume regulatory transporters) was the modulation of activity of these regulators by macromolecular crowding of intracellular hemoglobin.[67] Changes in CHC associated with swelling or shrinkage produces large changes in the activity coefficients of transport regulators due to the markedly non-ideal behavior of concentrated intracellular protein solutions.

KCC activity is high in HbSS, HbSC, and HbCC blood samples, regardless of the stimulus for activation.[61,62,68] High reticulocyte counts, especially in HbSS samples, complicate interpretation of these findings, since KCC activity is high in reticulocytes and diminishes with maturation. However, the level of KCC in HbCC RBCs is disproportionately high for the modest reticulocytosis of these samples,[62] and heterozygous HbAS and HbAC cells demonstrate small but detectably increased KCC activity despite normal reticulocyte counts.[69] In addition, resealing HbSS into HbAA RBC ghosts elicits abnormal KCC activity.[70] These finding suggest that HbS interacts with KCC or its regulators to increase activity or the response to activators.

KCC is activated *in vitro* by cell swelling,[60,59] acid pH,[61] low cellular Mg levels,[71] urea,[72–74] and sulfhydryl oxidation or alkylation (NEM),[58,68] although the stimuli for activation *in vivo* are less well defined. Cell swelling, induced by incubation in hypotonic solutions or by altering cation content with nystatin, results in progressive activation of KCC as CHC falls from 34 to 26 gm/dl.[75] This response appears to be qualitatively normal in HbSS RBCs; i.e., the proportionate response (percent of maximal activity) of KCC to changes in CHC is similar in HbSS and HbAA cells, even though the maximal activity is greater in sickle RBCs. KCC activation by acid pH shows a maximum between pH 6.7 and 6.9[61] and the proportionate activation is exaggerated in sickle RBCs compared to normal RBCs.[75] Urea is a powerful activator of KCC at concentrations found in the renal medulla (200–800 mM),[72,73] and the response of sickle RBCs is exaggerated compared to normal RBCs.[72,76] Reduction of cellular Mg levels stimulates KCC activity, presumably due to the requirement of Mg for protein kinase activity.[59,71] This may have pathophysiological significance in SCD since RBCs are relatively Mg depleted, especially in the dense cell fraction, perhaps as a consequence of Mg leakage through the sickling-induced pathway.[45] Sulfhydryl alkylation by NEM and oxidation by a variety of agents including hydrogen peroxide, diamide, and chlorodinitrobenzene also activate KCC.[58,59,68] The exaggerated response of KCC to activation by acid pH and urea can be mitigated by exposure of sickle RBCs to sulfhydryl reducing agents, suggesting the increased oxidative stress that occurs *in vivo* may be at least in part responsible for this abnormal KCC regulation.[75]

Activation of KCC results in a reduction in cation and water content and an increase in CHC. When triggered by cell swelling, this response represents a regulatory volume decrease (RVD). Recent studies have examined this RVD by techniques which permit measurement of cell volume and/or CHC in HbSS and HbAA reticulocyte populations, thereby overcoming the problem of varying reticulocyte counts. Upon incubation in isotonic solutions at pH 7.4, swollen reticulocytes (CHC 22–25 gm/dl) rapidly shrink over 10–20 minutes and assume a new steady CHC within an hour; this process is entirely Cl-dependent, indicating its mediation by KCC.[75] Even though the activation of KCC by cell swelling is normal in sickle RBCs, the RVD triggered by swelling results in a higher CHC in sickle reticulocytes compared to normal reticulocytes. Activation by acid pH produce a similarly exaggerated response in sickle versus normal reticulocytes, and this abnormal RVD is partially corrected by treatment with DTT, suggesting it originates from sulfhydryl oxidation.[75] Subtle differences between KCC activity (measured as the initial flux rate after activation) and the RVD and resultant final CHC suggest that physiologically important differences between HbAA and HbSS RBCs may reside in the factors which determine the inactivation of the system at high CHC and therefore control the CHC that results from KCC activation.

Cation Uptake Pathways

It is theoretically possible that inadequate Na uptake, especially in young HbSS RBCs with high transmembrane ion traffic, could contribute to dehydration. Several Na influx pathways have been identified in RBCs, including Na/H exchange,[77] Na/KCl cotransport,[60] Na/Mg exchange,[78] but assessment of their activities and contribution to net Na influx has been variable. Earlier estimates of very high Na/H exchange rates in HbSS RBCs[77] have not been reproducible,[79] and there are no data to suggest that reduction in a Na influx pathway contributes to dehydration in any sickle RBC population.

Interaction of Cation Transport Pathways and the Kinetics of Cellular Dehydration

It has long been appreciated that a substantial number of HbSS reticulocytes are dehydrated. Bookchin and Lew proposed that the dense HbSS RBC population was produced from a subset of reticulocytes which dehydrated rapidly due to permeabilization to Ca (via the sickling-induced pathway) and activation of the Gardos pathway, amplified by KCC activation.[80] Franco *et al.* showed that dense HbSS reticulocytes exhibited exaggerated KCC activity compared to reticulocytes that had not become dense *in vivo*.[81] Using biotin-labeling techniques to track the behavior of cells aging *in vivo*, they also showed that most of the increase in cell density occurred within 2–3 days, consistent with dehydration of young HbSS RBCs.[82] Interaction among cation transport pathways is likely to occur and to amplify their individual dehydrating effects, as shown in Fig. 2.

Thus a vicious cycle may be established in which activation of one or all of the pathological pathways fosters polymerization and thereby potentiates further transporter activation. The presence

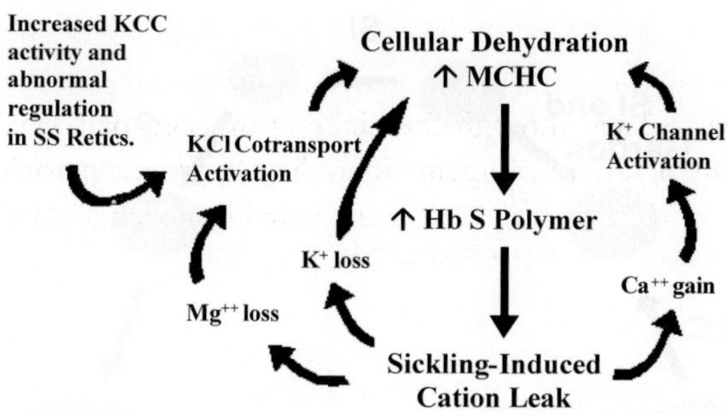

Fig. 2. Interaction of abnormal cation transport pathways in sickle RBCs to produce cellular dehydration.

of dehydrated sickle reticulocytes and their abnormal KCC-mediated RVD implicate KCC as an initiating event in this cycle. On the other hand, sickle reticulocytes have been shown to sickle readily, even those with relatively low CHC, so that the sickling-induced pathway may be readily activated in these cells, with consequent activation of the Gardos channel.

Beyond Dehydration–Pathological Rehydration in Sickle RBCs

The conventional paradigm has depicted dehydrated HbSS RBCs as end-stage cells that are both rigid and are fragile and prone to vascular occlusion and hemolysis, and are rapidly cleared from the body. This model however requires modification to account for the existence of significant numbers of low density, K-depleted, and Na-loaded HbSS RBCs, first reported by Bookchin *et al.*[83] These low density cells are resistant to dehydration by *in vitro* treatment with the K ionophore, valinomycin, are not reticulocytes, and their over-hydration cannot be explained by failure of the Na-pump. Franco and colleagues showed that low density valinomycin-resistant cells represented an older cell population derived from dense cells, and that they had a very short survival (~24 hours) *in vivo*.[34] Such low density RBCs arose spontaneously upon *in vitro* incubation of dense HbSS RBCs under oxygenated condition, and this process was accelerated by cyclic deoxygenation.[84] The steady state *in vivo* levels of valinomycin-resistant cells in HbSS blood (3–10%), together with their short survival, suggest that a significant proportion of sickle RBCs pass through this phase of Na loading and over-hydration prior to their destruction.

A new model of the sickle RBC hydration cycle thus would include pathological rehydration following dehydration as described in Fig. 3. The K loss involved in the initial dense cell production deprives the RBC of the ability to offset cation uptake driven by Donnan forces. Progressive Na loading would then ensue, particularly if dehydration triggers an increase in membrane permeability, as has been shown under other conditions.[85] Provided that the combination of Na influx and K efflux exceeded the capacity for compensation by the Na pump, the cell would be destined to swell to the point of osmotic lysis. This process of osmotic volume regulatory failure may contribute to intravascular hemolysis in SCD, now appreciated as an important aspect of the pathophysiology in light of the perturbations in nitric oxide metabolism caused by free plasma hemoglobin.

Pharmacological Intervention in Dehydrated Sickle RBCs

New treatments for SCD based on pharmacological inhibition of dehydrating pathways are currently being explored. *In vivo* inhibition of sickling-induced Ca influx might be expected to efficiently

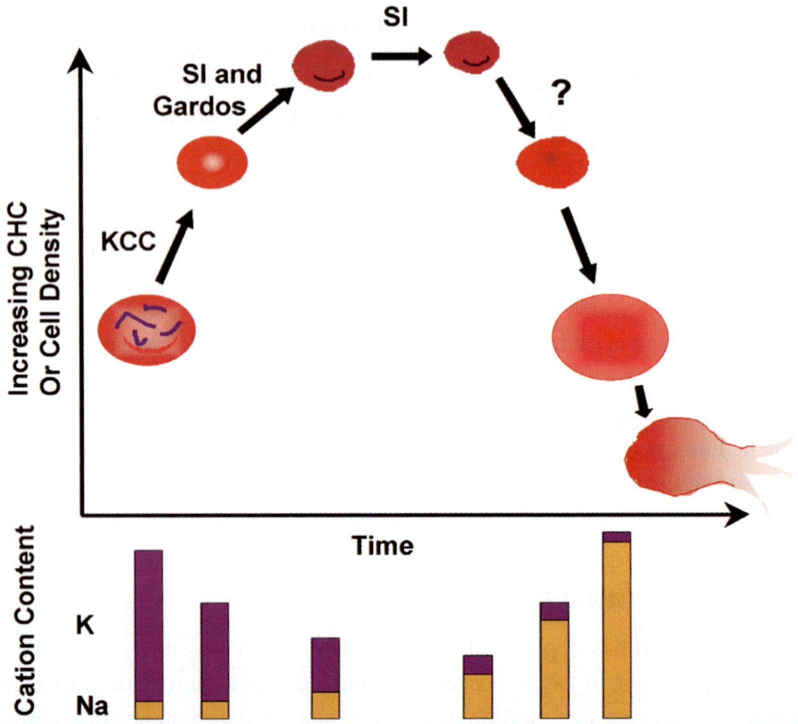

Fig. 3. Sickle RBC hydration cycle. Upon release from the marrow, the sickle reticulocyte loses K content and undergoes initial dehydration via abnormally regulated KCl co-transport (KCC). Activation of the sickling-induced (SI) pathway and consequently the Gardos channel produce further K depletion to produce dense cells. Continued SI pathway activation results in Na loading, initially balance by further K depletion so that total cation content remains low and cells remain dense. However, once the K gradient is depleted, persistence of SIP activity or activation of another Na entry pathway results in progressive Na gain and cell swelling. RBC volume increases (density decreases) until critical hemolytic volume is reached and cell lysis occurs.

reduce Gardos pathway activation, for which ionized Ca levels appear to be the rate-limiting factor (see above). Joiner *et al.*[43] showed that dipyridamole inhibits the sickling-induced leak of Na, K, and Ca. Clinical trials to determine if dipyridamole improves RBC hydration in SCD patients are in progress.

The Gardos channel inhibitor, clotrimazole, has been shown in a sickle cell mouse model as well as humans with SCD to reduce the number of dense sickle cells and to decrease hemolysis.[86,87] Because of limiting side effects of this drug, another compound (ICA17043) was developed, which is specific for Gardos channel inhibition, exhibits a long half life *in vivo*, and has a favorable toxicity profile.[88] In the SAD sickle cell mouse model, the drug prevented cellular dehydration.[88] The preliminary report of randomized, placebo-controlled phase II study in 88 SCD patients confirmed the reduction in dense cell numbers, associated with an increase in hemoglobin levels and reduction in reticulocyte counts, bilirubin, and other markers of hemolysis.[89] This trial had insufficient numbers of subjects to detect a difference in vaso-occlusive episodes. Thus, although there is solid evidence that Gardos channel inhibition improves HbSS RBC hydration status and decreases hemolysis, it remains to be seen whether these effects mitigate vaso-occlusive pathologies and improve clinical outcomes. A large phase III clinical trial of ICA17043 is currently underway to address this question.

A second approach to reduce K depletion via the Gardos channel involves the use of anion permeability inhibitors. K efflux via Ca-activated K channels is limited by accompanying anion movements and reduction in anion permeability, which retards K and water loss. Two related aryl urea compounds that inhibit Cl conductance block K loss via the Gardos channel from human RBCs and improve RBC hydration when given orally to mice with SCD (SAD model).[90] Studies of these compounds in humans have not been reported.

The use of Mg supplementation to inhibit KCC activity in HbSS RBCs was originally proposed by Bookchin,[91] based on the observations that sickle RBCs are relatively Mg depleted *in vivo* and that Mg depletion *in vitro* stimulated KCC activity. In phase II pilot trials, oral Mg pidolate increased RBC Mg levels, decreased KCC activity and reduced dense cell numbers in patients with SCD.[92,93] Hemoglobin levels and measures of hemolysis did not appear to be significantly improved. Diarrhea is a major side effect of Mg supplementation in patients. Larger studies are underway. A trial of Mg in HbSC disease will be particularly interesting, in light of the primary role KCC appears to play in this disorder (see above). There are, as yet, no other inhibitors of KCC with suitable therapeutic indices, but such a compound would potentially be useful.

Anti-oxidants may also have a role in preventing sickle RBC dehydration, given that oxidant stress may affect Gardos channel activation and calcium homeostasis and KCC regulation. Gibson *et al.* demonstrated that the formation of dense cell via oxy/deoxy cycling *in vitro* was substantially diminished in the presence of the sulfhydryl reducing agent N-acetyl cysteine (NAC).[94] Pace and colleagues showed that NAC given orally to SCD patients reduced dense cell numbers *in vivo*.[95] Further clinical trials of anti-oxidants, alone and in combination with transport inhibitors, are warranted.

The Membrane Skeleton and ISC Formation

Figure 4 shows a two step model, described by Goodman, for ISC formation.[96,97] The first step involving activation of the Gardos Channel may be potentiated by oxidative damage. Oxidant stress, as well as direct interaction with HbS, may also contribute to abnormal KCC activity and regulation in sickle RBCs. The resultant excessive K^+ efflux and water loss leads to increased intracellular HbS concentration, with multiple pathological consequences, including an increased number of sickled-shape RBCs.

Dense ISC formation requires sickle RBC dehydration, and thus can be viewed as a consequence of the volume regulation abnormalities engendered by HbS. However, elevated CHC is not sufficient for the generation of the ISC, and the process requires a second step that locks the membrane skeleton as represented in the two step model depicted in Fig. 5. This second step is probably also required for low density ISC formation.[98] To understand this second step, we must first discuss the normal membrane skeleton.

The membrane skeleton viewed by negative staining and electron microscopy is primarily a hexagonal lattice[99] with actin protofilaments at the center and six corners of the hexagons, interconnected by spectrin tetramers. Spectrin is composed of two large subunits of 280 KDa (α spectrin) and 246 KDa (β spectrin) molecular weight.[100,101] There are 22 repeat units of $\sim$106 amino acids within erythrocyte α spectrin[100] and 17 repeat units within β spectrin.[101] The repeats are numbered consecutively beginning at the most N-terminal repeat and the spectrin subunits are arranged in anti-parallel fashion so that the $\alpha 21$ repeat is in contact with the $\beta 1$ repeat and $\alpha 20$ with $\beta 2$ in a zippering mechanism.[102] Spectrin tetramers are formed by head-to-head interaction of heterodimers.

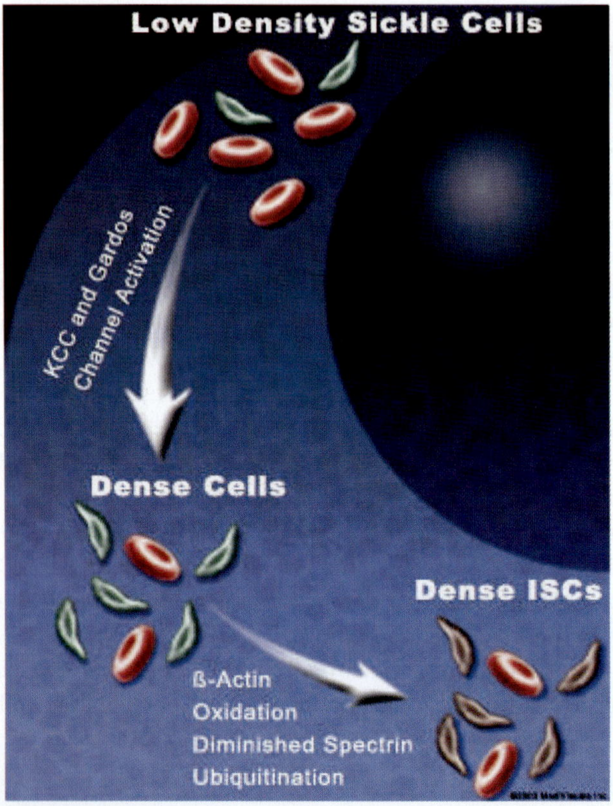

Fig. 4. Two step model for ISC formation. Details are given in the text. Used with permission from Goodman S (2004). *Mol Biol* **50**:53–58.

Spectrin tetramer tails bind actin filaments, thereby cross-linking F-actin protofilaments.[103,104] The actin binding domain is formed by residues 47–186 of β spectrin.[105] The spectrin-actin interaction is strengthened by the binding of protein 4.1 to the tails of spectrin[106,107] and adducin to both spectrin and F-actin.[108,109] β-spectrin contains the binding site for protein 4.1 (residues 207–445) comprising a portion of the N-terminal non-helical region, all of β spectrin repeat 1 (βSp1) and part of repeat 2 (βSp2).[97] The adducin binding site is contained within residues 1–528 of β spectrin, which represents the N-terminal domain plus repeats 1 and 2.[110] The membrane skeleton is attached to the membrane bilayer in two ways. Protein 4.1 binds to the transmembrane protein glycophorin C,[111,112] and ankyrin binds to β spectrin and the transmembrane protein band 3 which serves also as the anion transport channel.[113,114]

The first indication that the membrane skeleton was involved in the etiology of the ISC was the demonstration by Lux *et al.*[115] that most erythrocyte membranes (ghosts) isolated from ISCs remained sickled; and that triton skeletons prepared from ISC ghosts also remained sickled. Almost 20 years later, Goodman and colleagues described defects within the membrane skeleton which cause the ISC to be locked in the sickled shape (Fig. 5). They demonstrated that triton skeletons isolated from ISCs dissociate much more slowly than their counterparts from RSCs or normal RBCs.[116,117] Furthermore, Goodman and colleagues demonstrated that spectrin-4.1-actin ternary complexes dissociate more slowly at 37°C when the individual proteins are isolated from ISCs rather than control RBCs.[116]

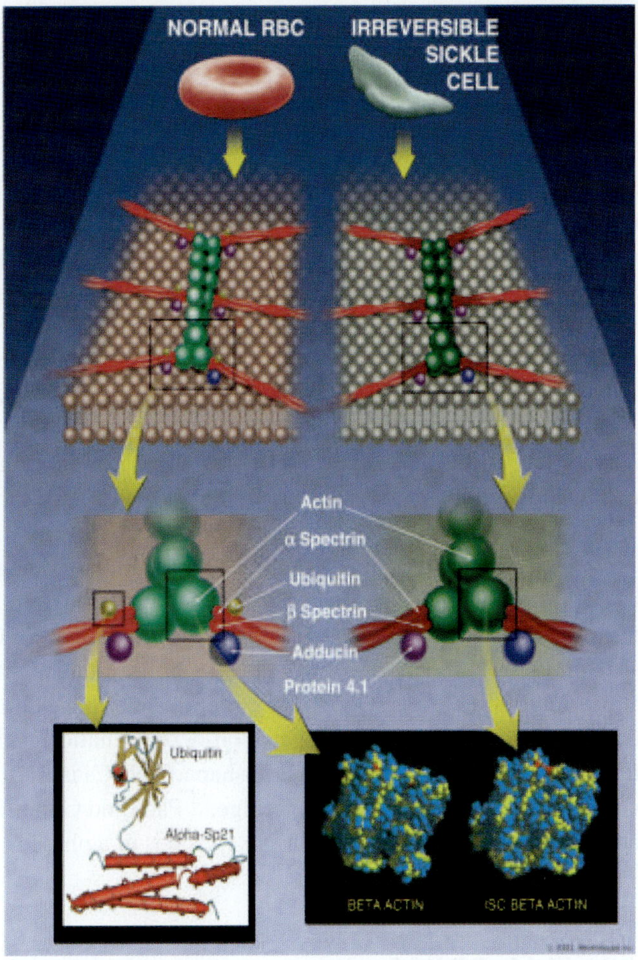

Fig. 5. The molecular basis of the ISC. Details are described in the text. Used with permission from Goodman S (2004). *Mol Biol* **50**:53–58.

This ternary complex dissociation assay allowed them to determine that β actin and spectrin, but not protein 4.1, contain defects that lead to this slow dissociation.[116]

The defect in ISC β-actin is a disulfide bridge formed between Cys284 and Cys373 which is found only at low levels in RSC and control β-actin.[116,118] This single posttranslation modification in β-actin caused it to polymerize and depolymerize more slowly than RSC and control β-actin, but had no effect on its ability to bind spectrin.[117] ISC F-actin at 37°C depolymerizes very slowly and a portion of the filaments do not depolymerize at all.[117] Therefore, part of the mechanism for the slow dissociation was understood. ISC actin monomers are tightly associated within actin protofilaments which disassemble very slowly at physiological temperatures (Fig. 5).

Goodman and associates have demonstrated that normal RBC spectrin has an E2/E3 ubiquitin conjugating and ligating activity which is capable of ubiquitinating itself[119] as well as other target proteins.[120] The ubiquitin-spectrin E2/E3 thio-ester linkages through cysteine residues are found in α-spectrin repeat 20, while the target site is found in a Sp21 repeat.[121] The other target proteins for spectrin's E2/E3 activity are ankyrin, band 3 protein 4.1, protein 4.2 and a protein of unknown function (gi13278939).[119] Sickle RBC α spectrin has vastly diminished ubiquitination (50–90% reduced).[122,123] This is due to the very high GSSG/GSH ratio, especially in the highest density sickle RBCs, which leads to glutathiolation of cysteines including those within spectrin (Shah and Goodman, unpublished data).

Non-ubiquitinated spectrin creates more tightly associated spectrin-4.1-actin and spectrin–adducin–actin ternary complexes than ubiquitinated spectrin. As a result, the rate of dissociation of the sickle RBC ternary complex at 37°C is much slower than the control ternary complex.[122,124] This is a reasonable result as the spectrin ubiquitination sites in repeats $\alpha 20$ and $\alpha 21$ are in direct contact with repeats $\beta 2$ and $\beta 1$ which contain the protein 4.1 and adducin binding domains.[102]

Therefore, the molecular basis of the formation of the ISC is a membrane skeleton that reassembles and disassembles slowly leading to a cell locked into the sickled shape. The slow dissociation of the ISC membrane skeleton is due to: (1) the β-actin disulfide bridge which leads to actin protofilaments that disassemble slowly or not at all; and (2) diminished α spectrin ubiquitination which creates a spectrin-4.1-actin and specrin–adducin–actin ternary complex which disassemble slowly at 37°C (Fig. 5).

In summary, the second step in the two step model for the formation of dense ISCs is the locking of the RBC membrane (Fig. 4). This occurs due to the oxidative damage to β-actin and lack of ubiquitination of spectrin both of which are caused by the diminished levels of GSH, and extremely high GSSG/GSH ratio. Both defects should be reversed by *n*-acetyl-cysteine (NAC) which is an anti-oxidant that raises intracellular GSH levels. Goodman and colleagues have demonstrated that NAC blocks the formation of ISCs formed *in vitro* by oxygenation-deoxygenation cycling.[94] Furthermore, NAC converts ISCs formed *in vivo* back to the biconcave shape.[94] NACs inhibition of ISC formation correlated with reduction of the ISC β-actin disulfide bridge.[94] Pace and Goodman demonstrated the efficacy of NAC in substantially lowering vaso-occlusive episodes in a phase II trial.[95]

The Proteomics of the Sickle RBC Membrane

Our understanding of changes in the content and modification of specific proteins within the HbSS RBC membrane will require sophisticated protein profiling proteomic methods. Goodman and colleagues have made this possible by performing the first comprehensive study of the normal human erythrocyte proteome.[125] A recent protein profiling study, by two dimensional difference gel electrophoresis (2D DIGE) followed by tandem mass spectrometry, has demonstrated that a systems biology approach to understanding global changes occurring in the HbSS RBC membrane can be useful.[126] The results obtained, in this preliminary proof of concept study, have demonstrated an increase in proteins (chaperonins, heat shock proteins, proteasomal subunits, catalase, and peroxyredoxin) involved in launching an adaptive response against the extreme intracellular oxidative stress in HbSS RBCs.[126] A decrease in flotillin and stomatin sickle RBC membranes suggests a loss of vesicles enriched in lipid rafts.[126] This preliminary study will need to be expanded to include samples from large numbers of HbSS versus control subjects, reinvestigated with reticulocyte free RBC preparations (which can now be prepared by magnetic bead and cell sorting protocols), and the impact of cell age in the circulation will need to be evaluated.

Despite the fact that much work remains to be done the protein profiling approach, described above, holds great promise for understanding changes in sickle versus normal RBC membranes.

It will also allow comparisons of protein profiles for patients with mild versus severe phenotypic expression of SCD.

References

1. Hebbel RP (1990). The sickle erythrocyte in double jeopardy: autooxidation and iron decompartmentalization. *Semin Hematol* **27**:51–69.
2. Shartava A, Mclntyre J, Shah AK, Goodman SR (2000). The Gardos channel is responsible for CDNB-induced dense sickle cell formation. *Am J Hematol* **64**(3):184–189.
3. Perelman N, Selvaraj SK, Batra S, Luck LR, Erdreich-Epstein A, Coates TD, Kalra VK, Malik P (2003). Placenta growth factor activates monocytes and correlates with sickle cell disease severity. *Blood* **102**: 1506–1514.
4. Gerszten RE, Garcia-Zepeda EA, Lim YC, Yoshida M, Ding HA, Gimbrone MA Jr, Luster AD, Luscinskas FW, Rosenzweig A (1999). MCP-1 and IL-8 trigger firm adhesion of monocytes to vascular endothelium under flow conditions. *Nature* **398**:718–723.
5. Turhan A, Weiss LA, Mohandas N, Coller BS, Frenette PS (2002). Primary role for adherent leukocytes in sickle cell vascular occlusion: a new paradigm. *Proc Natl Acad Sci USA* **99**:3047–3051.
6. Hebbel RP, Boogaerts MA, Eaton JW, Steinberg MH (1980). Erythrocyte adherence to endothelium in sickle cell anemia. A possible determinant of disease severity. *N Engl J Med* **302**(18):992–995.
7. Hebbel RP, Yamada O, Moldow CF, Jacob HS, White JG, Eaton JW (1980). Abnormal adherence of sickle erythrocytes to cultured vascular endothelium: possible mechanism for microvascular occlusion in sickle cell disease. *J Clin Invest* **65**:154–160.
8. Hines PC, Zen Q, Burney SN, Shea DA, Ataga KI, Orringer EP, Telen MJ, Parise LV (2003). Novel epinephrine and cyclic AMP-mediated activation of BCAM/Lu-dependent sickle (SS) RBC adhesion. *Blood* **101**:3281.
9. Hoover R, Rubin R, Wise G, Warren R (1979). Adhesion of normal and sickle erythrocytes to endothelial monolayer cultures. *Blood* **54**:872–876.
10. Fabry ME, Fine E, Rajanayagam V, Factor SM, Gore J, Sylla M, Nagel RL (1992). Demonstration of endothelial adhesion of sickle cells in vivo: a distinct role for deformable sickle cell discocytes. *Blood* **79**:1602–1611.
11. Kaul DK, Chen D, Zhan J (1994). Adhesion of sickle cells to vascular endothelium is critically dependent on changes in density and shape of the cells. *Blood* **83**(10):3006–3017.
12. Kaul DK, Fabry ME, Nagel RL (1986). Vaso-occlusion by sickle cells: evidence for selective trapping of dense red cells. *Blood* **68**:1162–1169.
13. Kaul DK, Fabry ME, Nagel RL (1989). Microvascular sites and characteristics of sickle cell adhesion to vascular endothelium in sheer flow conditions: pathophysiological implications. *Proc Natl Acad Sci USA* **86**:3356–3362.
14. Frenette PS (2002). Sickle Cell vaso-occlusion: multistep and multicellular paradigm. *Curr Opin Hematol* **9**:101–106.
15. Harlan JM (2000). Introduction: anti-adhesion therapy in sickle cell disease. *Blood* **95**:365–367.
16. Hebbel RP (1997). Adhesive interactions of sickle erythrocytes with endothelium. *J Clin Invest* **99**:2561–2564.
17. Wick TM, Eckman JR (1996). Molecular basis of sickle cell-endothelial cell interactions. *Curr Opin Hematol* **3**:118–124.
18. Bertles JF, Milner PF (1968). Irreversibly sickled erythrocytes: a consequence of the heterogenous distribution of hemoglobin types in sickle cell anemia. *J Clin Invest* **47**:1731–1740.
19. De Jong K, Larkin SK, Styles LA, Bookchin RM, Kuypers FA (2001). Characterization of the phosphatidylserine-exposing subpopulation of sickle cells. *Blood* **98**:860–867.
20. Kuypers FA, Lewis RA, Hua M, Schott MA, Discher D, Ernst JD, Lubin BH (1996). Detection of altered membrane phospholipid asymmetry in subpopulations of human red blood cells using fluorescently labeled annexin V. *Blood* **87**:1179–1187.

21. Kuypers FA, De Jong K (2004). The role of phosphatidylserine in recognition and removal of erythrocytes. *Cell Mol Biol* **50**(2):147–158.
22. Yasin Z, Witting S, Palascak MB, Joiner CH, Rucknagel DL, Franco RS (2003). Phosphatidylserine externalization in sickle red blood cells: associations with cell age, density, and hemoglobin F. *Blood* **102**:365–370.
23. Devaux PF, Zachowski A (1994). Maintenance and consequences of membrane phospholipid asymmetry. *Chem Phys Lipids* **73**:107–120.
24. De Jong K, Geldwerth D, Kuypers FA (1997). Oxidative damage does not alter membrane phospholipid asymmetry in human erythrocytes. *Biochem* **36**:6768–6776.
25. De Jong K, Rettig MP, Low PS, Kuypers FA (2002). Protein kinase C activation induces phosphatidylserine exposure on red blood cells. *Biochem* **41**:12562–12567.
26. Kunzelmann-Marche C, Freyssinet JM, Martinez MC (2002). Loss of plasma membrane phospholipid asymmetry requires raft integrity: role of TRP channels and ERK pathway. *J Biol Chem* **277**: 19876–19881.
27. Chiu D, Lubin B, Roelofsen B, van Deenen LL (1981). Sickled erythrocytes accelerate clotting in vitro: an effect of abnormal membrane lipid asymmetry. *Blood* **58**:398–401.
28. Setty BN, Kulkarni S, Rao AK, Stuart MJ (2000). Fetal hemoglobin in sickle cell disease: relationship to erythrocyte phosphatidylserine exposure and coagulation activation. *Blood* **96**:1119–1124.
29. Setty BN, Kulkarni S, Stuart MJ (2002). Role of erythrocyte phosphatidylserine in sickle red cell-endothelial adhesion. *Blood* **99**:1564–1571.
30. Bookchin KM, Roth EF Jr, Lew VL (1985). Near-normal circulatory survival of rabbit red cells exposed to high levels of Ca and ionophore in vitro. *Blood* **66**(l):220–223.
31. Wood BL, Gibson DF, Tait JF (1996). Increased erythrocyte phosphatidylserine exposure in sickle cell disease: flow-cytometric measurement and clinical associations. *Blood* **88**(5):1873–1880.
32. De Jong K, Emerson RK, Butler J, Bastacky J, Mohandas N, Kuypers FA (2001). Short survival of phosphatidylserine-exposing red blood cells in murine sickle cell anemia. *Blood* **98**(5):1577–1584.
33. Kean LS, Brown LE, Nichols JW, Mohandas N, Archer DR, Hsu LL (2002). Comparison of mechanisms of anemia in mice with sickle cell disease and beta-thalassemia: peripheral destruction, ineffective erythropoiesis, and phospholipid scramblase-mediated phosphatidylserine exposure. *Exp Hematol* **30**(5):394–402.
34. Franco RS, Yasin Z, Lohmann JM, Palascak MB, Nemeth TA, Weiner M, Joiner CH, Rucknagel DL (2000). The survival characteristics of dense sickle cells. *Blood* **96**(10):3610–3617.
35. Tosteson DC, Hoffman JF (1960). Regulation of cell volume by active cation transport in high and low potassium sheep red cells. *J Gen Physiol* **44**:169–194.
36. Fabry ME, Mears JG, Patel P, Schaefer-Rego K, Carmichael LD, Martinez G, Nagel RL (1984). Dense cells in sickle cell anemia: the effects of gene interaction. *Blood* **64**(5):1042–1046.
37. Fabry ME, Benjamin L, Lawrence C, Nagel RL (1984). An objective sign in painful crisis in sickle cell anemia: the concomitant reduction of high density red cells. *Blood* **64**(2):559–563.
38. Ferrone FA (1993). The polymerization of sickle hemoglobin in solutions and cells. *Experientia* **49**(2): 110–117.
39. Coletta M, Hofrichter J, Ferrone FA, Eaton WA (1982). Kinetics of sickle haemoglobin polymerization in single red cells. *Nature* **300**(5888):194–197.
40. Nagel RL, Fabry ME, Steinberg MH (2003). The paradox of hemoglobin SC disease. *Blood Reviews* **17**(3):167–178.
41. Tosteson DC, Shea E, Darling RC (1952). Potassium and sodium in red blood cells in sickle cell anemia. *J Clin Invest* **48**:406–411.
42. Joiner CH, Platt OS, Lux SE 4th (1986). Cation depletion by the sodium pump in red cells with pathologic cation leaks. Sickle cells and xerocytes. *J Clin Invest* **78**(6):1487–1496.
43. Joiner CH, Jiang M, Claussen WJ, Roszell NJ, Yasin Z, Franco RS (2001). Dipyridamole inhibits sickling-induced cation fluxes in sickle red blood cells. *Blood* **97**(12):3976–3983.
44. Etzion Z, Tiffert T, Bookchin RM, Lew VL (1993). Effects of deoxygenation on active and passive Ca2+ transport and on the cytoplasmic Ca2+ levels of sickle cell anemia red cells. *J Clin Invest* **92**(5):2489–2498.

45. Ortiz OE, Lew VL, Bookchin RM (1990). Deoxygenation permeabilizes sickle cell anaemia red cells to magnesium and reverses its gradient in the dense cells. *J Physio* **427**:211–226.
46. Joiner CH, Morris CL, Cooper ES (1993). Deoxygenation-induced cation fluxes in sickle cells. III. Cation selectivity and response to pH and membrane potential. *Am J Physiol* **264**(3 Pt 1):C734–744.
47. Joiner CH (1990). Deoxygenation-induced cation fluxes in sickle cells: II. Inhibition by stilbene disulfonates. *Blood* **76**(1):212–220.
48. Mohandas N, Rossi ME, Clark MR (1986). Association between morphologic distortion of sickle cells and deoxygenation-induced cation permeability increase. *Blood* **68**(2):450–454.
49. Mohandas N, Rossi M, Bernstein S, Ballas S, Ravindranath Y, Wyatt J, Mentzer W (1985). The structural organization of skeletal proteins influences lipid translocation across erythrocyte membrane. *J Bio Chem* **260**(26):14264–14268.
50. Hebbel RP, Mohandas N (1991). Reversible deformation-dependent erythrocyte cation leak. Extreme sensitivity conferred by minimal peroxidation. *Biophys J* **60**(3):712–715.
51. Gardos G (1958) The function of calcium in the potassium permeability of human erythrocytes. *Biochim Biophys Acta* **30**:653–654.
52. Brugnara C, De Franceschi L, Alper SL (1993). Ca(2+)-activated K+ transport in erythrocytes. Comparison of binding and transport inhibition by scorpion toxins. *J Bio Chem* **268**(12):8760–8768.
53. Lew VL, Etzion Z, Bookchin RM (2002). Dehydration response of sickle cells to sickling-induced Ca(++) permeabilization. *Blood* **99**(7):2578–2585.
54. Lew VL, Hockaday A, Sepulveda MI, Somlyo AP, Somlyo AV, Ortiz OE, Bookchin RM (1985). Compartmentalization of sickle-cell calcium in endocytic inside-out vesicles. *Nature* **315**(6020):586–589.
55. Lew VL, Tiffert T, Etzion Z, Perdomo D, Daw N, Macdonald L, Bookchin RM (2005). Distribution of dehydration rates generated by maximal Gardos-channel activation in normal and sickle red blood cells. *Blood* **105**(1):361–367.
56. Rivera A, Jarolim P, Brugnara C (2002). Modulation of Gardos channel activity by cytokines in sickle erythrocytes. *Blood* **99**(1):357–603.
57. Shartava A, Shah AK, Goodman SR (1999). N-acetylcysteine and clotrimazole inhibit sickle erythrocyte dehydration induced by l-choloro-2, 4-dinitrobenzene. *Am J Hematol* **62**:19–24.
58. Lauf PK, Theg BE (1980). A chloride dependent K+ flux induced by N-ethylmaleimide in genetically low K+ sheep and goat erythrocytes. *Bioche & Biophys Res Commun* **92**(4):1422–1428.
59. Lauf PK, Adragna NC (2000). K-C1 cotransport: properties and molecular mechanism, *Cellular Physiology & Biochemistry* **10**(5–6):341–354.
60. Canessa M, Fabry ME, Blumenfeld N, Nagel RL (1987). Volume-stimulated, Cl(−)- dependent K+ efflux is highly expressed in young human red cells containing normal hemoglobin or HbS. *J Memb Biol* **97**(2): 97–105.
61. Brugnara C, Bunn HF, Tosteson DC (1986). Regulation of erythrocyte cation and water content in sickle cell anemia. *Science* **232**(4748):388–390.
62. Brugnara C, Kopin AS, Bunn HF, Tosteson DC (1985). Regulation of cation content and cell volume in hemoglobin erythrocytes from patients with homozygous hemoglobin C disease. *J Clin Invest* **75**(5): 1608–1617.
63. Jennings ML, Schulz RK (1991). Okadaic acid inhibition of KC1 cotransport. Evidence that protein dephosphorylation is necessary for activation of transport by either cell swelling or N-ethylmaleimide. *J Gen Physiol* **97**(4):799–817.
64. Kaji DM, Tsukitani Y (1991). Role of protein phosphatase in activation of KC1 cotransport in human erythrocytes. *Am J Physiol* **260**(1 Pt 1):C176–180.
65. Bize I, Dunham PB (1994). Staurosporine, a protein kinase inhibitor, activates K-C1 cotransport in LK sheep erythrocytes. *Am J Physiol* **266**(3 Pt 1):C759–770.
66. De Franceschi L, Fumagalli L, Olivieri O, Corrocher R, Lowell CA, Berton G (1997). Deficiency of Src family kinases Fgr and Hck results in activation of erythrocyte K/C1 cotransport. *J Clin Invest* **99**(2): 220–227.
67. Parker JC, Colclasure GC (1992). Macromolecular crowding and volume perception in dog red cells. *Molecular & Cellular Biochemistry* **114**(1–2):9–11.

68. Canessa M, Spalvins A, Nagel RL (1986). Volume-dependent and NEM-stimulated K+, C1− transport is elevated in oxygenated SS, SC and CC human red cells. *FEBS Lett* **200**(1):197–202.
69. Olivieri O, Vitoux D, Galacteros F, Bachir D, Blouquit Y, Beuzard Y, Brugnara C (1992). Hemoglobin variants and activity of the (K+C1−) cotransport system in human erythrocytes. *Blood* **79**(3):793–797.
70. Vitoux D, Beuzard Y, Brugnara C (1999). The effect of hemoglobin A and S on the volume- and pH-dependence of K-C1 cotransport in human erythrocyte ghosts. *J Memb Biol* **167**(3):233–240.
71. Brugnara C, Tosteson DC (1987). Inhibition of K transport by divalent cations in sickle erythrocytes. *Blood* **70**(6):1810–1815.
72. Bize I, Taher S, Brugnara C (2003). Regulation of K-Cl cotransport during reticulocyte maturation and erythrocyte aging in normal and sickle erythrozytes. *Am J Physiol Cell Physiol* **285**:C31–38.
73. Culliford SJ, Ellory JC, Gibson JS, Speake PF (1998). Effects of urea and oxygen tension on K flux in sickle cells. *Pflugers Archiv — Eur J Physiol* **435**(5):740–742.
74. Kaji DM, Gasson C (1995). Urea activation of K-C1 transport in human erythrocytes. *Am J Physiol* **268**(4 Pt 1):Cl018–1025.
75. Joiner CH, Rettig RK, Jiang M, Franco RS (2004). KC1 cotransport mediates abnormal sulfhydryl-dependent volume regulation in sickle reticulocytes. *Blood* **104**(9):2954–2960.
76. Joiner CH, Rettig RK, Jiang M, Franco RS (2004). Activation of KCI cotransport by urea induces dehydration in both sickle and normal reticulocytes. *Blood* **104**:3591a.
77. Canessa M, Fabry ME, Suzuka SM, Morgan K, Nagel RL (1990). Na+/H+ exchange is increased in sickle cell anemia and young normal red cells. *J Memb Biol* **116**(2):107–115.
78. Rivera A, Ferreira A, Bertoni D, Romero JR, Brugnara C (2005). Abnormal regulation of Mg2+ transport via Na/Mg exchanger in sickle erythrocytes. *Blood* **105**(1):382–386.
79. Joiner CH, Jiang M, Fathallah H, Giraud F, Franco RS (1998). Deoxygenation of sickle red blood cells stimulates KC1 cotransport without affecting Na+/H+ exchange. *Am J Physiol* **274**(6 Pt 1):C1466–1475.
80. Bookchin RM, Ortiz OE, Lew VL (1991). Evidence for a direct reticulocyte origin of dense red cells in sickle cell anemia. *J Clin Invest* **87**(1):113–124.
81. Franco RS, Palascak M, Thompson H, Joiner CH (1995). KC1 cotransport activity in light versus dense transferrin receptor-positive sickle reticulocytes. *J Clin Invest* **95**(6):2573–2580.
82. Franco RS, Lohmann J, Silberstein EB, Mayfield-Pratt G, Palascak M, Nemeth TA, Joiner CH, Weiner M, Rucknagel DL (1998). Time-dependent changes in the density and hemoglobin F content of biotin-labeled sickle cells. *J Clin Invest* **101**(12): 2730–2740.
83. Bookchin RM, Etzion Z, Sorette M, Mohandas N, Skepper JN, Lew VL (2000). Identification and characterization of a newly recognized population of high-Na+, low-K+, low-density sickle and normal red cells. *Proc Nat Acad Sci USA* **97**(14): 8045–8050.
84. Holtzclaw JD, Jiang M, Yasin Z, Joiner CH, Franco RS (2002). Rehydration of high-density sickle erythrocytes in vitro. *Blood* **100**(8):3017–3025.
85. Crespo LM, Novak TS, Freedman JC (1987). Calcium, cell shrinkage, and prolytic state of human red blood cells. *Am J Physiol* **252**(2 Pt 1):C138–152.
86. Brugnara C, Gee B, Armsby CC, Kurth S, Sakamoto M, Rifai N, Alper SL, Platt OS (1996). Therapy with oral clotrimazole induces inhibition of the Gardos channel and reduction of erythrocyte dehydration in patients with sickle cell disease. *J Clin Invest* **97**(5):1227–1234.
87. De Franceschi L, Brugnara C, Rouyer-Fessard P, Jouault H, Beuzard Y (1999). Formation of dense erythrocytes in SAD mice exposed to chronic hypoxia: evaluation of different therapeutic regimens and of a combination of oral clotrimazole and magnesium therapies. *Blood* **94**(12):4307–4313,
88. Stocker JW, De Franceschi L, McNaughton-Smith GA, Corrocher R, Beuzard Y, Brugnara C (2003). 1C A-17043, a novel Gardos channel blocker, prevents sickled red blood cell dehydration in vitro and in vivo in SAD mice. *Blood* **101**(6): 2412–2418.
89. Ataga KI, DeCastro LM, Swerdlow P, Saunthararajah Y, Smith W, and the ICA-17043 Study Investigators (2004). Efficacy and safety of the Gardos channel inhibitor, ICA-17043, in patients with sickle cell anemia. *Blood* **104**:103a.
90. Bennekou P, De Franceschi L, Pedersen O, Lian L, Asakura T, Evans G, Brugnara C, Christophersen P (2001). Treatment with NS3623, a novel Cl-conductance blocker, ameliorates erythrocyte dehydration in transgenic SAD mice: a possible new therapeutic approach for sickle cell disease. *Blood* **97**(5):1451–1457.

91. Bookchin R, Tiffert T *et al.* (1995). Magnesium therapy for sickle cell anemia: a new rationale. Sickle Cell Disease and Thalassemaemias: New Trends in Therapy. Y. Beuzard, B. Lubin, J. Rosa. London, John Libbey and Co. Ltd. **234**:545–546.
92. De Franceschi L, Bachir D, Galacteros F, Tchernia G, Cynober T, Alper S, Platt O, Beuzard Y, Brugnara C (1997). Oral magnesium supplements reduce erythrocyte dehydration in patients with sickle cell disease. *J Clin Invest* **100**(7):1847–1852.
93. De Franceschi L, Beuzard Y, Jouault H, Brugnara C (1996). Modulation of erythrocyte potassium chloride cotransport, potassium content, and density by dietary magnesium intake in transgenic SAD mouse. *Blood* **88**(7):2738–2744.
94. Gibson XA, Shartava A, Mclntyre J, Monteiro CA, Zhang Y, Shah A, Campbell NF, Goodman SR (1998). The efficacy of reducing agents or antioxidants in blocking the formation of dense cells and irreversibly sickled cells in vitro. *Blood* **91**(11): 4373–4378.
95. Pace BS, Shartava A, Pack-Mabien A, Mulekar M, Ardia A, Goodman SR (2003). Effects of N-acetylcysteine on dense cell formation in sickle cell disease. *Am J Hematol* **73**(1):26–32.
96. Goodman SR, Pace BS, Shartava A (1998). New therapeutic approaches to sickle cell disease: targeting rbc membrane oxidative damage. *Cell Mol Biol Lett* **3**:403–411.
97. Goodman SR (2004). The irreversibly sickled sell: a perspective. *Cell Molec Biol* **50**:53–58.
98. Schwartz RS, Shartava A, *et al.* (1995). Slow dissociation rates of membrane cytoskeletons from hydrated irreversibly sickled cells. *Natl Sickle Cell Prog Mtg Abstract 60.*
99. Liu SC, Derick LH, Palek J (1987). Visualization of the hexagonal lattice in the erythrocyte membrane skeleton. *J Cell Biol* **104**:527–536.
100. Sahr KE, Laurila P, Kotula L, Scarpa AL, Coupal ETL, Linnenbach AJ, Winkelmann JC, Speicher DW, Marchesi VT, Curtis PJ, Forget BG (1990). The complete cDNA and polypeptide sequences of human erythroid a spectrin. *J Biol Chem* **265**:4434–4434.
101. Winkelman JC, Chang JG, Tse WT, Scarpa AL, Marchesi VT, Forget BG (1990). Full-length sequence of the cDNA for human erythroid β-spectrin. *J Biol Chem* **265**:11827–11832.
102. Speicher DW, Weglarz L, DeSilva TM (1992). Properties of human red cell spectrin heterodimer (side-to-side) assembly and identification of an essential nucleation site. *J Biol Chem* **267**:14775–14782.
103. Cohen CM, Tyler JM, Branton D (1980). Spectrin-actin association studied by electron microscopy of shadowed preparations. *Cell* **21**:875–883.
104. Shen BW, Josephs R, Steck TL (1986). Ultrastructure of the intact skeleton of the human erythrocyte membrane. *J Cell Biol* **102**:997–1006.
105. Karinch AM, Zimmer WE, Goodman SR (1990). The identification and sequence of the actin-binding domain of human red blood cell β spectrin. *J Biol Chem* **265**:11833–11840.
106. Tyler JM, Hargreaves WR, Branton D (1979). Purification of two spectrin-binding proteins: biochemical and electron microscopic evidence for site specific reassociation between spectrin and bands 2.1 and 4.1. *Proc Natl Acad Sci USA* **76**:5192–5196.
107. Ungewickell E, Bennett PM, Calvert R, Ohanian V, Gratzer WB (1979). In vitro formation of a complex between cytoskeletal proteins of the human erythrocyte. *Nature* **280**:811–814.
108. Gardner K, Bennett V (1987). Modulation of spectrin-actin assembly by erythrocyte adducin. *Nature* **328**:359–362.
109. Mische SM, Mooseker MS, Morrow JS (1987). Erythrocyte adducin: a calmodulin- regulated actin-bundling protein that stimulates spectrin-actin binding. *J Cell Biol* **105**:2837–2845.
110. Li X, Bennett V (1996). Identification of the spectrin subunit and domains required for formation of spectrin/adducin/actin complexes. *J Biol Chem* **271**: 15695–15702.
111. Mueller TJ, Morrison M (1981). Glycoconnectin (PAS2) a membrane attachment site for the human erythrocyte cytoskeleton. In Erythrocyte membranes 2: recent clinical and experimental advances. Kruckeberg WC, Eaton JW, Brewer GJ (eds.). Alan R. Liss, Inc. New York:95–112.
112. Shiffer KA, Goodman SR (1984). Protein 4.1: its association with the human erythrocyte membrane. *Proc Natl Acad Sci USA* **81**:4404–4408.
113. Wallin R, Gulp EN, Coleman DB, Goodman SR (1984). A structural model of human erythrocyte band 2.1: alignment of chemical and functional domains. *Proc Natl Acad Sci USA* **81**:4095–4099.

114. Yu J, Goodman SR (1979). Syndeins: the spectrin-binding protein(s) of the human erythrocyte membrane. *Proc Natl Acad Sci USA* **76**:2340–2344.
115. Lux SE, John KM, Karnovsky MJ (1976). Irreversible deformation of the spectrin-actin lattice in irreversibly sickled cells. *J Clin Invest* **58**:955–963.
116. Shartava A, Montiero CA, Bencsath FA, Schneider K, Chait BT, Gussio R, Casoria-Scott LA, Shah AK, Heuerman CA, Goodman SR (1995). A posttranslational modification of β-actin contributes to the slow dissociation of the spectrin-protein 4.1-actin complex of irreversibly sickled cells. *J Cell Biol* **128**:805–818.
117. Shartava A, Korn W, Shah AK, Goodman SR (1997). Irreversibly sickled cell β-actin: defective filament formation. *Am J Hematol* **55**:97–103.
118. Bencsath FA, Shartava A, Monteiro CA, Goodman SR (1996). Identification of the disulfide-linked peptide in irreversibly sickled cell β-actin. *Biochem* **35**:4403–4408.
119. Kakhniashvili DG, Chaudhary T, Zimmer WE, Bencsath FA, Jardine I, Goodman SR (2001). Erythrocyte spectrin is an E2 ubiquitin conjugating enzyme. *Biochem* **40**:11630–11642.
120. Chang TL, Cubillos FF, Kakhniashvili DG, Goodman SR (2004). Ankyrin is a target of spectrin's E2/E3 ubiquitin-conjugating/litigating activity. *Cell Mol Biol* **50**:59–66.
121. Hsu YJ, Zimmer WE, Goodman SR (2005). Erythrocyte Spectrin's Chimeric E2/E3 Ubiquitin Conjugating/Ligating Activity. *Cell Mol Biol.*
122. Ghatpande SS, Goodman SR (2004). Ubiquitination of spectrin regulates the erythrocyte spectrin-protein 4.1-actin ternary complex dissociation: implications for the sickle cell membrane skeleton. *Cell Molec Biol* **50**:67–74.
123. Montiero CA, Gibson X, Shartava A, Goodman SR (1998). Preliminary characterization of a structural defect in homozygons sickled cell alpha spectrin demonstrated by a rabbit autoantibody. *Am J Hematol* **58**:200–205.
124. Mishra R, Goodman SR (2004). Ubiquitination of erythrocyte spectrin regulates the dissociation of the spectrin-adducin-f-Actin ternary complex in vitro. *Cell Mol Biol* **50**:75–80.
125. Kakhniashvili DG, Bulla LA Jr, Goodman SR (2004). The Human Erythrocyte Proteome: Analysis by Ion Trap Tandem Mass Spectrometry. *Molec & Cell Proteomics* **3**(5):501–509. Epub 2004 Feb 12.
126. Kakhniashvili DG, Griko NB, Bulla LA Jr, Goodman SR (2005). The Proteomics of Sickle Cell Disease: profiling of erythrocyte membrane proteins by 2d-dige and tandem mass spectrometry. *Exper Biol Med* **230**:787–792.

11

Fetal Hemoglobin for What Ails Sickle Hemoglobin

by Solomon F. Ofori-Acquah and Betty S. Pace

Introduction

Fetal hemoglobin (HbF; $\alpha_2\gamma_2$) dominantly influences the manifestation and phenotype of sickle cell disease (SCD). It is widely acknowledged that the study of this molecule has shaped the management and research of SCD more than any other molecule and it continues to do so in the genome era. The expression of the γ-globin genes is controlled by a developmentally regulated transcriptional program that is widely studied and recapitulated in several experimental systems. Naturally occurring mutations in the γ-genes cause persistent expression of HbF, which is associated with amelioration of symptoms in SCD. Thus, the anti-sickling effect of HbF is primarily due to formation of an asymmetric hemoglobin hybrid containing γ- and β^S-globin that reduces the amount of hemoglobin S (HbS) incorporated into the HbS polymer. Several clinical trials have shown that pharmacological augmentation of HbF ameliorates disease severity in SCD. These positive findings underpin the rationale for studies aimed at understanding transcriptional regulation of the γ-globin genes. Mechanistic insights into HbF-inducing drugs have identified a wealth of signal transduction pathways and transcriptional factors that can be developed to produce safe and efficacious inducers of γ-globin. The task of discovering novel HbF inducers will become more feasible as our understanding of the human genome improves. International collaborations engendered by spin-off projects of the Human Genome Project will be necessary to accomplish this goal.

Developmental Expression and Cellular Distribution of HbF

HbF comprises 50% of total hemoglobin in embryonic blood at seven weeks of gestation, and increases to nearly 90% by nine weeks. It remains the major hemoglobin produced in erythrocytes until 28–34 weeks of gestation when adult hemoglobin predominates. Longitudinal studies have shown that HbF percentage decreases from 80 $\pm$ 4.0% at birth to 9.2 $\pm$ 2.9% at 24 months in infants with SCD compared to normal infants in whom levels of less than one percent are reached by 10 months.[1]

As the level of HbF declines its expression is restricted to a subset of erythrocytes termed F-cells. Consequently, cytochemical analysis of HbF in adult blood smears reveals a patchy or heterocellular distribution of F-cells in most individuals. Family studies show that F-cell numbers are genetically controlled but the genes involved in this process are poorly understood. Nonetheless, the F-cell percentage is a highly reproducible index of HbF production and is therefore used in several studies as the phenotypic measure of γ-gene expression.

Composition of Fetal Hemoglobin

HbF is normally a heterogeneous mixture of globin tetramers distinguished by unique γ-globin polypeptide chains containing either glycine ($^{G}\gamma$) or alanine ($^{A}\gamma$) at residue γ136.[2] At birth HbF contains predominantly $^{G}\gamma$-chains however this fetal pattern progresses gradually to an adult pattern during the first 10 months of life when $^{A}\gamma$ constitutes the majority of γ-globin chains. The $^{G}\gamma \rightarrow {}^{A}\gamma$ switch does not occur in a subset of patients with SCD. This developmental arrest is not unique to individuals with the β^{S} mutation, but is found in association with a polymorphism (C $\rightarrow$ T) at -158 of the $^{G}\gamma$ globin gene.[3] The $^{G}\gamma$:$^{A}\gamma$ ratio has emerged as an important facet of HbF phenotypes in SCD. Highly specialized techniques such as polyacrylamide gel electrophoresis (PAGE),[4] reverse-phase high performance liquid chromatography (HPLC)[5] and more recently mass spectrometry (MS)[6] can be used to quantify globin chains, and the $^{G}\gamma$:$^{A}\gamma$ ratio can be calculated from the relative abundance of the individual γ-globin chains.

While PAGE and reverse-phase HPLC require relatively large sample volumes and reference samples for identification of globin polypeptides, relatively small sample volumes can be used in MS analysis. Moreover MS relies on protein mass for identification and is therefore independent of reference samples (Table 1). For instance, the two γ-globin chains ($^{G}\gamma$ and $^{A}\gamma$) differ by a single methylene [CH_2] group that results in a mass difference of 14 Da, sufficiently large for the two to be resolved by MS. Recent studies have applied MS to examine expression of globin chains in early development and to evaluate the proportions of $^{G}\gamma$ and $^{A}\gamma$ globin in patients with SCD (Fig. 1). The proportion of $^{G}\gamma$-globin in the majority of patients lies within a defined range of 40–50%.

Interactions between Hemoglobin F and S

The protective effect of HbF in SCD was first realized in a landmark observation made by Jane Watson and her colleagues in 1948.[7] It is well established that the beneficial effect of HbF is due

Table 1. Mass of globin polypeptides.

Globin chain	α	ς	$^{G}\gamma$	$^{A}\gamma$	ε	β^{A}	β^{S}
Calculated mass (Da)	15126.4	15547.9	15995.3	16009.3	16071.5	15867.2	15837.3
Mean measured mass (Da)	15126.5	15548.0	15995.5	16009.8	16071.1	15867.1	15837.2
SD	0.2	0.2	0.2	0.4	1.2	0.5	0.2
Number of samples	60	13	60	60	16	25	40

From Ofori-Acquah SF (2000) *Cis Regulation of Fetal Hemoglobin Expression in Sickle Cell Disease*, PhD Thesis, University of London.

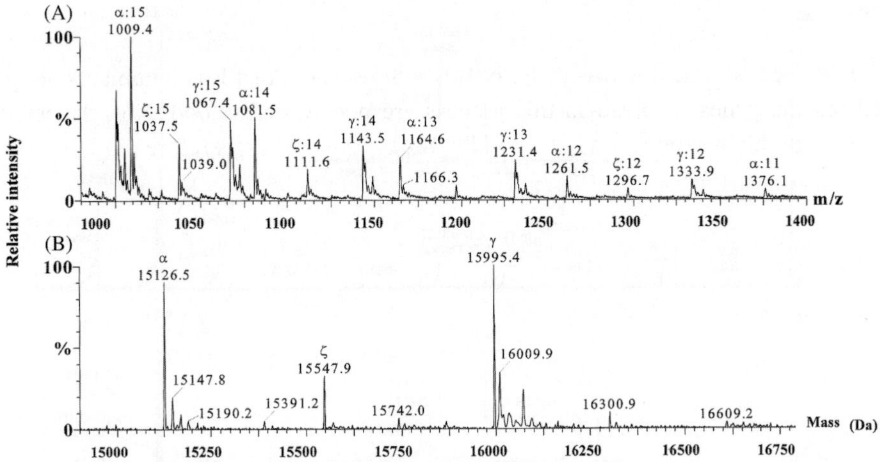

Fig. 1. (A) Electrospray ionization mass spectrometry of fetal blood lysate. MS spectra was scanned over the 980–1400 m/z range (upper panel). Multiply charged globin ions are indicated with the number of positive ions. (B) Deconvolution of the m/z spectra using a maximum entropy software results in identification of individual globin chains and resolution of $^{G}\gamma$ (mass = 15995.4 Da) from $^{A}\gamma$ (mass = 16009.9 Da). Used with permission from Ofori-Acquah SF (2000) *Cis Regulation of Fetal Hemoglobin Expression in Sickle Cell Disease*, PhD Thesis, University of London.

to its interaction with the HbS polypeptide. The anti-sickling effect of HbF reflects both a dilution of intracellular HbS concentration to a threshold below that required for polymerization and to directly influence the stability of the sickle hemoglobin polymer. Experimental evidence from mixing HbS and HbF solutions indicates that the latter mechanism involves formation of an asymmetric hybrid hemoglobin referred to as HbF/S (α_2 $\gamma\beta^S$).[8,9] The issue of HbF/S in patients with SCD was addressed by Ofori-Acquah and associates[10] using electrospray ionization mass spectrometry (ESI-MS), which can, with some limitations, directly reflect the relative abundance of HbF/S complexes in solutions. MS spectra studies by this group demonstrated HbS as the sole species in patients in whom conventional separation techniques revealed the virtual absence of HbF (Fig. 2B). The asymmetric hybrid HbF/S was clearly evident in blood samples from a sickle cell patient with raised HbF (Figs. 2B iii) (see Chapter 9).

Globin Gene Regulation: The Locus Control Region

Regulatory DNA *cis*-active elements that directly control expression of the globin genes are clustered in the promoters and in silencer and enhancer domains located proximal and distant to the individual genes. A major regulator of globin gene expression is the locus control region (LCR), which lies 6.1 kb upstream of the ε-globin gene. The LCR is composed of five DNase I hypersensitive sites (HSs) of which HS1 to HS4 are erythroid-specific[11,12] with the capacity to confer integration position-independent expression of a linked globin gene.[13,14] One of four current models suggests the LCR orchestrates developmental stage-specific globin gene expression by gene competition (Fig. 3). This model is based on the temporal expression of stage-specific transcription factors that medicate interactions between the LCR and target globin gene promoters through a looping mechanism.[15]

The hematopoietic restricted transcription factors GATA-1 and NF-E2[16] and numerous ubiquitous proteins interact with *cis*-elements in the regulatory regions of the globin genes. NF-E2 forms

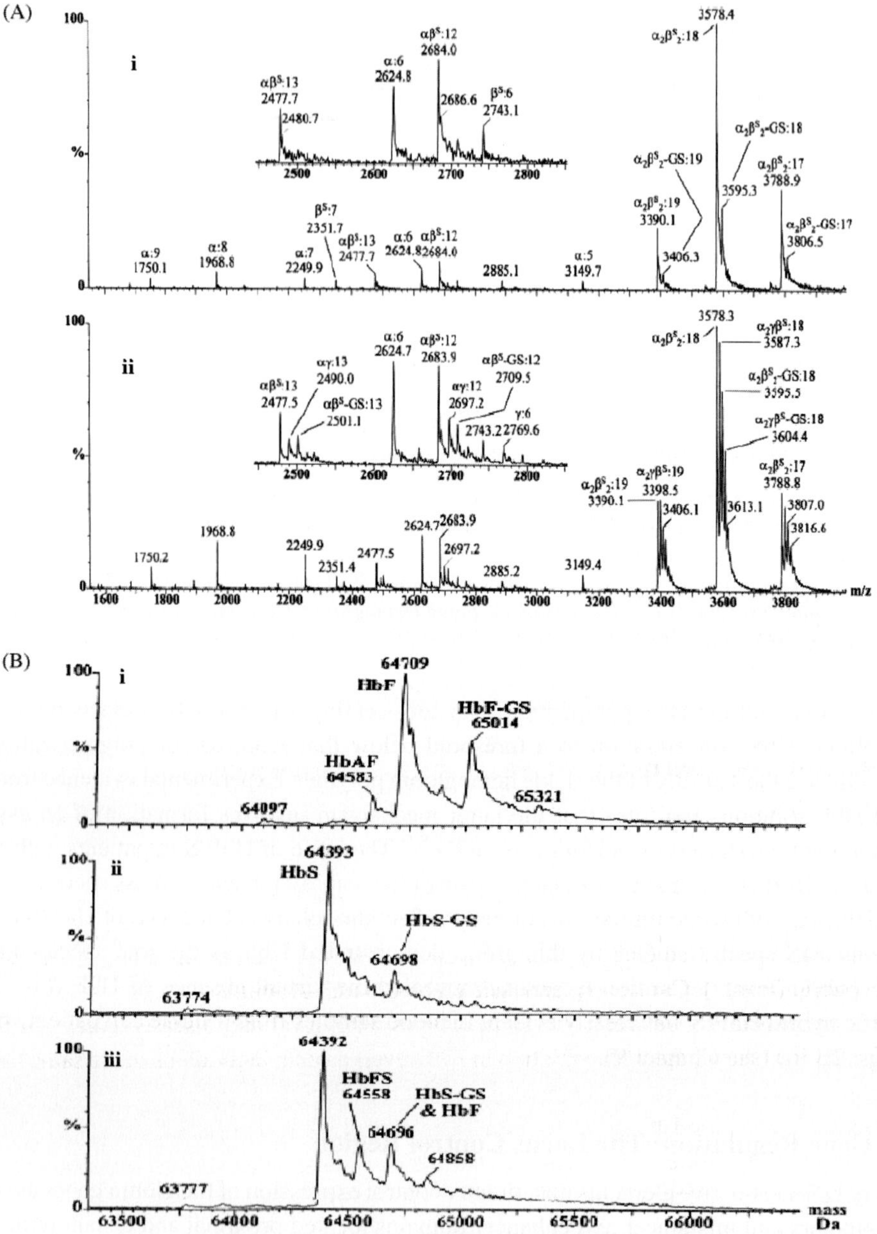

Fig. 2. (A) ESI m/z spectra of blood samples from homozygous sickle cell disease patients with (a) low (1.4%) and (b) high (29.2%) HbF as determined by HPLC. Samples were analyzed under non-denaturing conditions. The insets show m/z region containing Hb dimers on an expanded scale. Each species is associated with the expected number of heme groups, i.e. 1 heme per globin chain. The numbers after the colon indicate the number of positive charges on the ion. (B) Hemoglobin tetramer mass spectra produced by deconvoluting data similar to those shown in Fig. 2A over the m/z range 3250–4000 from (a) fetal blood and samples from patients with HbF levels determined by HPLC of (b) 1.4% and (c) 10.3%. HbFS masses were calculated by assuming there is a binomial distribution of HbF and HbS; (c) 64,558.5. Minor peaks of mass 64,097 and 63,775 correspond to tetramers containing only three heme groups possibly due to loss of a heme in the gas phase. Used with permission from Ofori-Acquah *et al.* (2001) *Anal Biochem* **298**:76–82.

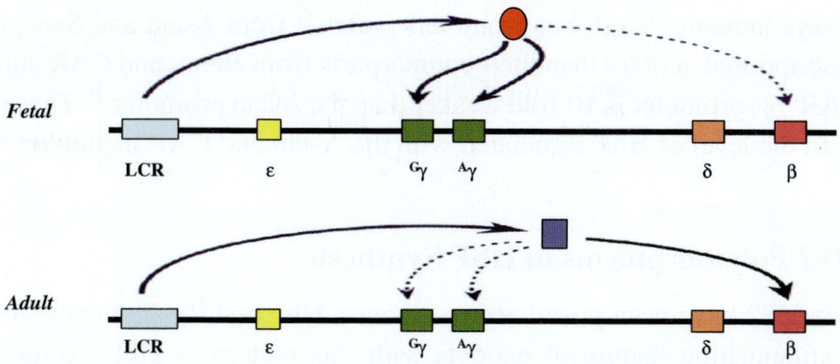

Fig. 3. Competitive model of globin gene switching. The LCR interacts with each globin promoter during the different stages of development by a looping mechanism. Used with permission from Stamatoyannopoulos.[15]

a heterodimer with members of the Maf family of ubiquitously expressed 18 kDa proteins and bind in HS2.[17] The NF-E2/AP-1 binding site in HS2 is largely responsible for its enhancing activity.[18] GATA-1 is found in erythroid cells as well as in megakaryocytes and mast cells.[19] The *trans*-activation potential of GATA-1 is regulated by another factor friend of GATA-1 (FOG), which is co-expressed with GATA-1 during development to promote erythroid and megakaryocytic differentiation.[20] Interactions between multiple factors result in assembly of large transcriptional complexes to drive globin genes expression (see Chapter 13).

β-Globin Cluster Restriction Fragment Length Polymorphisms

The most common variation in DNA is the single nucleotide polymorphism (SNP). A haplotype consists of blocks of SNPs inherited as a group. Haplotypes consisting of eight SNPs distinguishable by restriction fragment length polymorphisms (RFLPs), spanning the β-globin cluster, have been studied extensively to determine the ancestral origin of the β^S mutation in individuals from different parts of Africa and Asia (see Chapter 21). Five common haplotypes exist, Senegal, Benin, Central African Republic (Bantu), Cameroon, and Asia (Indian/Saudi-Arabia).

There is an association between RFLP haplotypes, HbF level and disease-severity in SCD that remains contentious due to the wide variation in HbF levels among individuals of the same haplotype. Despite the controversy, there is acceptance that individuals with the Senegal haplotype generally have higher HbF compared to those with the Benin haplotype who have lower HbF levels.[21,22] Patients heterozygous for the Asian and Benin haplotypes have milder clinical expression and raised HbF compared to their counterparts with homozygosity for the Benin haplotype.[23] Evidence from longitudinal studies indicate the Senegal haplotype confers clinical benefit, which is thought to be mediated in part through raised HbF levels and that the CAR haplotype is associated with a worst outcome. Association studies have shown that the CAR haplotype increases the risk of end-organ or tissue damage; the risk of developing these complications is lowered significantly by inheriting the Senegal haplotype.[24–26] The deleterious effect of the CAR haplotype has now been confirmed in sickle cell populations in South America and Cuba.[27,28]

The level of HbF associated with specific β-locus haplotypes likely reflects linkage disequilibrium with polymorphic *cis*-acting elements in the β-globin locus. While several mutations in the $^G\gamma$ and $^A\gamma$-globin promoters cause hereditary persistence of fetal hemoglobin (HPFH) none have hitherto been found to be in linkage disequilibrium with the Asian or Senegal haplotype. Nonetheless, *in vitro* reporter gene assays indicate $^{G}\gamma$-globin promoters isolated from Asian and Senegal chromosomes exert higher transcriptional activity than their counterparts from Benin and CAR chromosomes.[29] In particular, the CAR $^{G}\gamma$-promoter is 10-fold weaker than the Asian promoter.[29] These *in vitro* findings are consistent with the level of HbF associated with the Asian and CAR haplotype.[29,30]

The Role of HS2 Polymorphisms in HbF Synthesis

Polymorphisms in HS2 have been postulated to increase HbF and $^{G}\gamma$-globin expression in patients with SCD. A distinguishing feature of patients with this pattern of HbF expression is that each has at least one copy of an unusual hybrid chromosome containing both Benin and Senegal elements. The hybrid chromosome is characterized by markers associated with the Senegal haplotype at HS2 and markers typical of the Benin haplotype in the β-globin cluster. A study by Ofori-Acquah and associates examined the relationship between polymorphisms in HS2 and HbF level in two large cohorts from the United Kingdom and Jamaica and found no association between HS2 alleles and HbF levels. They concluded that polymorphism in the γ-globin promoter exerted a dominant influence on HbF levels in SCD.[30] Ironically, reporter gene studies by the same group showed that an HS2 polymorphism associated with the Benin haplotype conferred higher enhancer activity than that associated with the Asian haplotype. The significance of the variable activities of HS2 alleles on different sickle chromosomes is currently not fully understood.

Regulation of γ-Globin Gene Expression

Numerous transcription factors bind a homologous 200-bp segment in the promoters of both γ-genes (Fig. 4). As expected sequence variation in this region alters HbF levels, and accounts for many mutations associated with HPFH (see below). Sp1 and the stage selector protein (SSP)[31] bind a CACC box located at -53 in the γ-gene promoters. SSP binds the stage selector element (SSE) and provides a competitive advantage for γ-globin expression over β-globin.[31] SSP is a heterodimeric protein composed of the ubiquitous protein CP2[32] and a γ-globin activator NF-E4, which alters histone acetylation levels in an erythroid-specific manner.[33] Binding of Sp1 and SSP to the SSE is mutually exclusive and is influenced by methylation of CpG dinucleotides at -55 and -50 in the γ-promoter that may be important in γ-gene silencing.[34] CpG methylation increases Sp1 binding by 10-fold indicating Sp1 acts as a repressor whereas SSP *trans*-activates the γ-promoter at this site.[31]

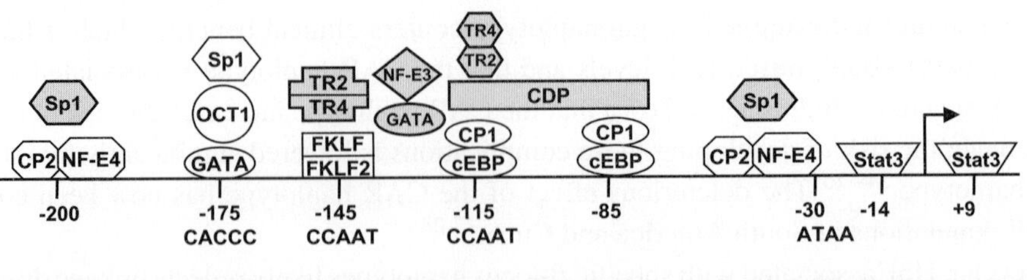

Fig. 4. Transcription factor binding in the γ-globin promoter. Shown are the DNA-binding proteins which have been demonstrated to bind in the minimal γ-globin promoter either as monomers or as homodimers or heterodimers. Shown are ubiquitous and hematopoietic specific transcription factors. The promoter is not drawn to scale.

FKLF-1[35] and FKLF-2[36] bind a CACCC site in this region and activate the γ-promoter *in vitro*. Although a physiological role for the FKLFs remains to be established, deletion of the CACCC reduces γ-gene expression in definitive erythroid cells.[37] EKLF binds to the homologous sequence (CCACACCCT) in the β-globin promoter[38] and a similar motif in HS3,[39] is involved in activating the γ- to β-globin switch.[40] EKLF may therefore provide a competitive advantage for LCR-β-globin interactions over γ-globin to achieve hemoglobin switching.

Recently Douglas Engel and associates identified a direct repeat erythroid-definitive (DRED) repressor complex that binds a DR1 motif[41] adjacent to the ε-globin and γ-globin CCAAT boxes. DRED consists of two nuclear orphan receptors TR2 and TR4, which likely recruit co-repressors to silence the γ-genes. In addition, the γ-globin genes contain a silencer domain located between -382 and -730 that was originally identified in transgenic mice,[35] as well as an enhancer 750-bp downstream of the Aγ-gene[42] with binding motifs for GATA-1 and a nuclear matrix-associated protein SABT1.[43] Each of these elements serve as a potential target to modulate γ-gene expression.

Hereditary Persistence of Fetal Hemoglobin

HPFH is characterized by sustained synthesis of one or both γ-globin chains in adults. This phenotype is normally due to deletions of DNA sequences downstream of the γ-globin genes or mutations within the promoters of the γ genes. When co-inherited with the β^S mutation, the HFPH phenotype ameliorates the severity of SCD.

Deletional HPFH

At least six different deletions in the β-globin cluster have been described that are associated with HPFH phenotypes. HPFH-1 and HPFH-2 found in people of Africa origin involve deletions of approximately 105 kb of DNA, with breakpoints staggered by about 6 kb.[44] In common with other deletion-type HPFH, the molecular basis for persistent γ-globin expression in HPFH-1 and 2, is that DNA sequences at the 3′ deletion breakpoint juxtaposes regulatory elements downstream of the γ-globin genes that prevent developmental silencing. Therefore, HbF in individuals with these deletions contain both $^G\gamma$-globin and $^A\gamma$-globin chains.[45] The distribution of HbF in erythrocytes in HPFH-1 and HPFH-2 is uniform, and is commonly described as pancellular. There is normally no hematological deficiency associated with either HPFH mutation so affected individuals have normal red cell indices. The DNA deletions associated with HPFH-3, which is commonly found in Asian Indians and HPFH-4 first described in southern Italian kindred, are less extensive, encompassing approximately 50 kb and 40 kb respectively.[46] The 3′ breakpoints of these two deletions are separated by 2 kb, and are located about 30 kb downstream the β-globin gene. HPFH-6 deletion first identified in Thais is similar to HPFH-1 and 2 but is shifted in the 5′ direction and involves deletion of the $^A\gamma$-globin gene.[47,48]

Non-deletional HPFH

Non-deletional HPFH is due to mutations of *cis*-acting elements in the $^G\gamma$ or $^A\gamma$-gene promoters, which result in selective up-regulation of the corresponding globin polypeptide. Mutations are clustered in three regions in the promoters (Table 2). At least six different mutations are located approximately 200-bp from the cap site of both genes. This region of the promoter is identical in both

Table 2. Non-deletion HPFH Mutants.

Type and Ethnic Group	Mutation	HbF in Heterozygotes (%)	References
G_γ HPFH			
Japanese	G_γ-114 C to T	11–14	135
Australian	G_γ-114 C to G	8.6	136
Black/Sardinian	G_γ-175 T to C	17–30	137, 138
Tunician	G_γ-200 + C	18–49	139
Black	G_γ-202 C to G	15–25	140
A_γ HPFH			
Georgian	A_γ-114 C to T	3–6.5	141
Black	A_γ-114 to -102 Deleted	30–32	142
Greek	A_γ-117 G to A	10–20	43, 143, 144
Cretan	A_γ-158 C to T	2.9–5.1	145
Black	A_γ-175 T to C	36–41	146
Brazilian	A_γ-195 C to G	4.5–7	147
Chinese/Italian	A_γ-196 C to T	14–21	148–150
British	A_γ-198 To to C	3.5–10	151
Georgian	A_γ-202 C to T	1.6–3.9	152

From Stamtoyannopoulos G, Grosveld F.[48]

γ-globin genes, is GC rich and contains a binding site for the transcription factor Sp1. Mutations in the -198 T $\rightarrow$ C and -202 C $\rightarrow$ $^G\gamma$ region enhance or create new binding sites for Sp1[49] and SSP.[31]

A second region containing non-deletion HPFH mutations is located at -175 where a T $\rightarrow$ C substitution originally found in a patient with raised HbF[50] was later shown to alter GATA-1 and octamer 1 binding.[51] The third region of HPFH mutation is found in the proximal CCAAT box. This motif interacts with CP1, a ubiquitous *trans*-activator[52] which competes with the repressor protein CCAAT displacement protein for binding.[51,53] The C $\rightarrow$ T mutation at base -114 in the $^G\gamma$-promoter abolishes repressor binding to the CCAAT motif and increases γ-globin expression. Naturally occurring mutations in the CAAT box that alter binding of NF-E3 and GATA-1 are associated with HPFH,[54] however, experimentally induced mutations in the CCAAT box have not produced an HPFH phenotype.[55] A summary of non-deletional HPFH mutations is shown in Table 2.

Nucleotide substitutions in the $^A\gamma$-globin enhancer located 750-bp downstream [$+2285^A\gamma$T $\rightarrow$ C, $+2460^A\gamma$C $\rightarrow$ A, and $+2676^A\gamma$A $\rightarrow$ G] are associated with raised HbF in SCD patients of African origin, and was originally termed Seattle HPFH, however further studies revealed this variation is a common polymorphism.[56] An unrelated mutation (C $\rightarrow$ T) at +2401 $^A\gamma$ is responsible for raised HbF in an unusual type of $\delta\beta$-thalassemia in a Chinese family.[57] The $^A\gamma$-globin enhancer may therefore be a potential site for modulating HbF levels in individuals with SCD, although the enhancer function of this region has not been confirmed *in vivo*.[58]

Emerging Cell Signaling Mechanisms for Drug-Mediated HbF Induction

Numerous pharmacological agents induce HbF synthesis. The mechanisms responsible for this effect are not clearly understood however emerging data have implicated several signal transduction pathways in this phenomenon. For example, infants with severe anemia and congenital cyanotic heart disease have elevated HbF levels and a delay in the switch from HbF to HbA synthesis that is associated with oxygen-related signaling.[59] Indeed, evidence that erythropoietin (Epo) signaling influences the switch from primitive to definitive erythropoiesis is highly relevant to raised HbF associated with severe anemia.[60] Stem cell factor (SCF) and transforming growth factor-β (TGF-β) are produced in the placenta, and have been implicated in the γ to β switch as well. There is a delay in the γ to β switch in infants of diabetic mothers due to increased serum levels of α-amino butyric acid.[61,62] These observations have provided a conceptual model for several studies that have uncovered a role for diverse signaling pathways in γ-globin gene regulation.

Mitogen Activated Protein Kinase Signaling Pathway

A variety of pharmacological agents including histone deacetylase (HDAC) inhibitors,[63,65] hydroxyurea[66,67] and decitabine[68] converge at the p38 mitogen activated protein kinase (MAPK) pathway to induce HbF expression. Four MAPK pathways have been identified: ERK1/2, ERK5/BMK1, cJun amino-terminal signal kinases 1, 2, 3 and p38 MAPKs.[69] Studies using different experimental systems including primary erythroid progenitors,[69] knockout mice[70] and K562 stable lines[64] demonstrated a role for p38 MAPK in globin gene regulation. Indeed, enforced p38 MAPK expression in K562 stable lines increases γ-globin synthesis.[64] Several factors that are downstream effectors of p38 MAPK such as MAPK-activated protein kinases 1 and 2,[71] PRAK,[72] ATF-1-4, CREB and CREM[73] are therefore candidate regulators of the γ-globin gene. The Pace group have identified a functional consensus ATF-2/CREB motif in the Gγ-globin promoter that binds ATF-2 and CREB1. Interestingly, HDAC inhibitors alter nuclear protein binding to this motif to modulate γ-gene expression (personal communication, Betty Pace).

Signal Transducer and Activators of Transcription (STAT) Signaling Pathways

The Janus kinase (JAK) family consists of the ubiquitous kinases Jak1, Jak2 and Tyk2 and the hematopoietic cell specific Jak3 kinase. The STAT family of tyrosine kinases is the most common downstream signaling molecules activated by Jaks. There are six major members, Stat1–Stat6 which exists as latent cytoplasmic proteins.[74] Two Stat3 variants (Stat3α and Stat3β) have identical cognate DNA-binding motifs, the Interleukin 6 (IL-6) response elements.[75] Pace and colleagues have demonstrated that Stat3β, the truncated dominant negative isoform, silences γ-gene expression via the IL-6 response element at base +9 in the γ-gene 5′-untranslated region.[76,77]

Nitric Oxide and Cyclic Guanosine Monophosphate (cGMP) Signaling Pathways

Nitric oxide (NO) is a second messenger signaling molecule with autocrine and paracrine properties.[78] NO is generated from arginine by the action of nitric oxide synthase (NOS). In the vasculature, NO reacts with iron in the active site of the enzyme guanylyl cyclase (GC) to stimulate cGMP signaling,[79] and moreover NO activates MAPK thus broadening its role in γ-globin regulation.[80] High soluble GC levels have been observed in erythroid cells expressing a γ-globin program compared to those expressing β-globin.[81]

Two research groups showed that hydroxyurea (HU) increase γ-globin expression in K562 cells and human erythroid progenitors[82] via cGMP signaling. Moreover, hemin and butyrate activate

γ-globin via cGMP activation in K562 cells.[81] NO/cGMP signaling targets several transcription factors including c-fos and c-jun[79] which have become candidate regulators of the γ-gene. Indeed, AP-1 binds HS2 and modulates LCR function. In addition, NO increases binding of Sp-1 to the CACCC box[83,84] and may therefore increase expression of the γ-globin gene by this mechanism.

Pharmacological Fetal Hemoglobin Induction

Pharmacological approaches to increase HbF levels rely on both direct and indirect mechanisms to activate the γ-genes. Experience from hematological diseases revealed that perturbations in the kinetics of erythropoiesis increase HbF synthesis. Three major cell signaling pathways have been implicated in activation of the γ-gene. More direct mechanisms of γ-gene activation involve strategies that alter chromatin structure by histone acetylation, DNA methylation, and/or increased transcription factor binding.

Cytokines and Growth Factors

Numerous growth factors and cytokines influence γ-globin expression. The function of Epo is primarily expansion of late erythroid progenitors and mobilization of burst-forming unit-erythroid (BFU-E) colonies. High dose Epo increases HbF synthesis in primates[85] and is required for survival, proliferation and differentiation of erythroid progenitors and globin gene expression through Jak2/Stat5 and MAPK cell signaling.[86] There is an increase in the number of HbF-programmed progenitors that accounts for the increased F-cell numbers associated with Epo therapy. Reversible inhibition of p38 MAPK using SB203580, blocks Epo-dependent accumulation of mouse globin mRNA[86] and p38α-/- knockout mice lack α-globin expression.[87] In addition, p38 MAPK is required for the stability of Epo mRNA and hemoglobin synthesis.[88] These findings support a role for p38 MAPK in Epo-induced HbF augmentation.

Stem cell factor (SCF) increases proliferation of erythroid progenitors and HbF synthesis through MAPK signaling.[89] In addition, SCF-mediated signaling increases expression of TAl-1[90] and FKLF,[36,37] which suggests that these two transcription factors may contribute to HbF induction by SCF. TGF-β markedly augments Epo and SCF mediated induction of HbF; recent studies correlating plasma TGF-β levels with HbF level in patients with SCD[91] lend *in vivo* support for this mechanism of γ-gene reactivation.

Cytotoxic Agents

S-stage-specific drugs including hydroxyurea (HU),[92,93] cytosine arabinoside,[94] myleran,[95] and vinblastine[96] induce HbF production in primates and humans. These agents induce HbF through cellular growth perturbations secondary to alterations in erythroid-regeneration kinetics after cytotoxic events.[92] However, mechanisms are unclear by which γ-gene activation occurs under conditions of rapid erythropoiesis and how erythroid progenitors with an active HbF program are selectively recruited.

A large scale randomized multicenter clinical trial, the Multicenter Study of Hydroxyurea (MSH) showed HU reduced the incidence of vaso-occlusive episodes, and increased HbF levels in two-thirds of adult subjects.[97] These encouraging results led to the approval of HU treatment in adults with recurrent symptoms and complications of SCD, by the Food and Drug Administration in 1997. Recent studies in children indicate HU is efficacious in raising HbF levels in this patient population.[98,99] Hankins and associates[100] found improvement in splenic function with no indication of toxicity in infants treated up to four years with HU. The long-term effect of treating infants with HU however remains a major socio-medical concern that needs to be addressed in the future.

The goal of understanding how HU reduces the severity of SCD serves both a scientific and important socio-medical need. In one study the number of BFU-E colonies was inversely correlated with HbF levels in the peripheral blood of patients treated with HU[101] secondary to cytotoxicity against late erythroid precursors. Our current understanding of HU-mediated induction of HbF synthesis is summarized in Fig. 5.[78] HU act as an NO donor[82] however, an analogue of HU, zileuton induces HbF in an NO-independent mechanism.[102] Nonetheless, the National Institutes of Health[103] has initiated

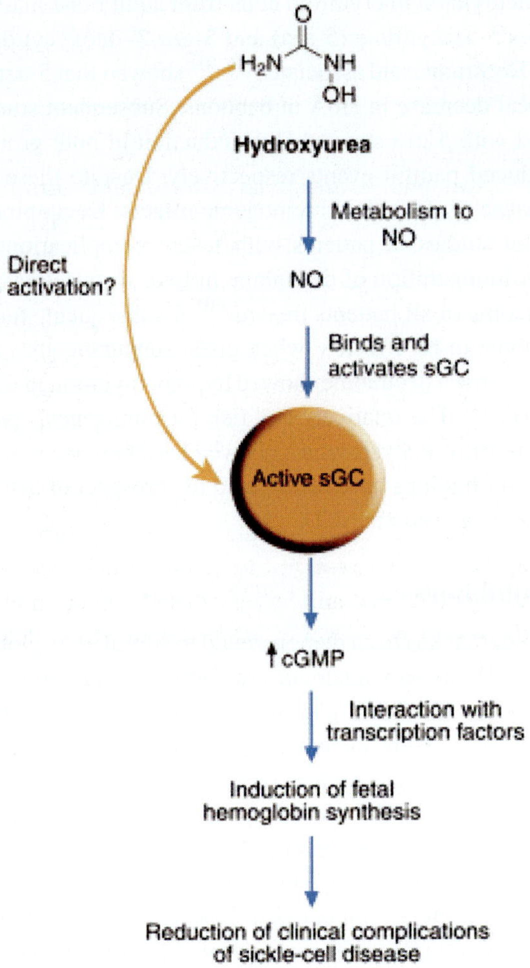

Fig. 5. Mechanism of fetal hemoglobin induction by hydroxyurea. Hydroxyurea is metabolized by an unknown pathway to produce nitric oxide (NO) that binds and activates sGC (soluble guanylyl cyclase). Alternatively, hydroxyurea may be able to directly activate sGC. Activation of sGC increases production of cGMP, which probably influences some transcription factors leading to increased fetal hemoglobin synthesis. Used with permission from King BS.[78]

a phase I clinical trial in sickle cell patients to determine whether HU and the NO donors L-arginine and Sildenafil (Viagra) synergistically increase HbF synthesis.

DNA Methyltransferase Inhibitors

Epigenetic modifications such as DNA methylation and histone acetylation alter chromatin structure and are therefore intimately involved in transcriptional regulation.[104,105] It is widely accepted that gene silencing is accompanied by histone deacetylation and DNA methylation at CpG dinucleotides mediated by DNA methyltransferase (DNMT) enzymes. Methylation patterns in the β-globin cluster were first studied by southern blot analysis of genomic DNA digested with methylation-sensitive restriction enzymes. These studies showed that CpG dinucleotides in the γ-globin gene promoter are unmethylated in fetal liver erythroid and K562 cells, which actively express the γ-gene. However, these sites were heavily methylated in erythroid cells from adult bone marrow.[105]

The cytidine analogues 5-azacytidine (5-aza) and 5-aza-2′-deoxycytidine (decitabine) bind and deplete DNMT enzymes. DeSimone and associates[106,107] showed that 5-aza enhanced HbF synthesis up to 80% with a reciprocal decrease in HbA in baboons. Subsequent studies in β-thalassemia and sickle cell patients treated with 5-aza showed HbF induction in both groups[108] and an increase in total hemoglobin and reduced painful events respectively. Despite these promising results, 5-aza is not used clinically because of potential carcinogenic effects. Decitabine is less carcinogenic,[109] and has been used in pilot studies in patients with severe complications of SCD for whom HU therapy was ineffective. Administration of decitabine in two, six-week cycles increased HbF levels and improved disease outcome in all patients treated.[110] A subsequent study by Saunthararajah and colleagues showed decitabine to be effective when given subcutaneously.[111] Analysis of erythroid cells in patients and baboons given decitabine showed hypomethylation in the five CpG residues in the proximal γ-globin promoter.[112] The relatively low risk for mutagenesis predicted for decitabine[113] has been bolstered by microarray analysis which showed less than one percent activation of genes in tumor cell lines. These recent findings have resurrected the prospect of using DNA hypomethylating agents to reactivate γ-gene expression in SCD.

Histone Deacetylase Inhibitors

Butyrate is a HDAC inhibitor that has been demonstrated to stimulate γ-globin expression in a variety of experimental systems.[114,115] Several analogues including sodium butyrate, arginine butyrate and sodium 4-phenylbutyrate delay the γ to β-globin switch and/or enhance HbF synthesis in cultured cells, primates and hemoglobinopathy patients.[116,117] The increase in HbF is however lost after extended periods of butyrate therapy, which is presumably due to the myelotoxic effect of these agents. This interpretation prompted the use of intermittent butyrate dosing in humans, which is associated with sustained increases in HbF levels.[118]

Butyrate can induce HbF synthesis in the absence of erythropoietic stress and therefore exerts a direct influence on the activity of the γ-globin transcriptome.[119,120] Mechanistically, these inhibitors bind to a central zinc atom in HDACs to block enzymatic deacetylation of histone H3 and H4. The resulting hyperacetylated histones have a lower affinity for DNA and are therefore easily displaced from *cis*-active butyrate response elements (BREs) by transcription factors.[121] Pace and colleagues have proposed a dual mechanism for HDAC inhibitor action (Fig. 6). This model involves displacement of hyperacetylated histones as a primary event postulated to "open" euchromatin, and a second event involving up-regulation of specific transcription factors to activate γ-globin expression.[122]

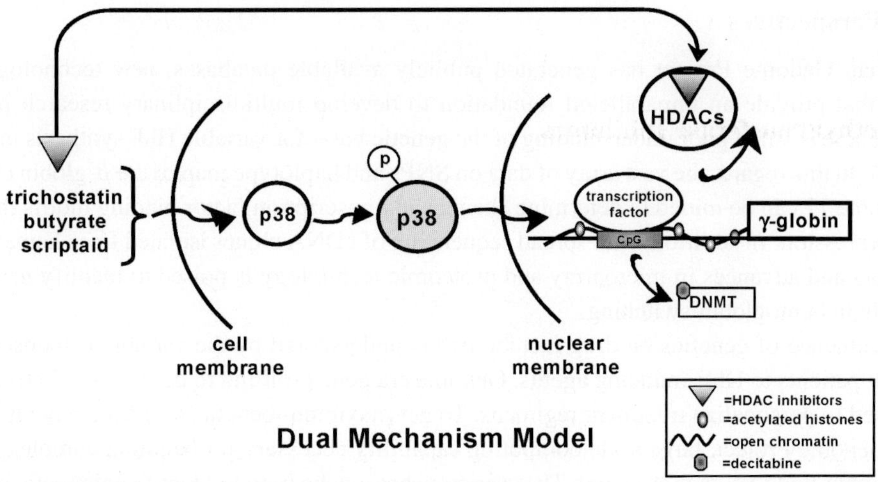

Fig. 6. Experimental model for γ-globin induction by HDAC inhibitors. A schematic is shown of the molecular events believed to occur during HDAC inhibitor-mediated γ-globin activation. The HDAC inhibitors butyrate, trichostatin and scriptaid mediate histone hyperacetylation and open chromatin domains. All three agents stimulate p38 MAPK phoshorylation followed by p-p38 translocation and transcription factor activation which bind *cis*-regulatory elements to augment γ-gene expression. Combination therapy with DNA methyltransferase (DNMT) inhibitors such as decitabine (5-aza-2′-deoxycytidine) to produce hypomethylation of CpG islands should enhance HbF induction by HDAC inhibitors. Used with permission from McElveen.[122]

There are several BREs in the proximal and distal γ-globin promoter,[120,123,124] including the distal CCAAT box[120] and the SSE.[123] *In vivo* studies in transgenic mice have demonstrated an additional BRE at −822[124] in the γ-promoter, which was later shown by Pace and associates to bind a transcriptional complex upon stimulation with butyrate.[125] Recently, another mechanism for γ-gene activation by butyrate has been proposed, which involves increased efficiency of γ-globin mRNA translation.[126]

Several other HDAC inhibitors including trichostatin A,[64,120] M275[127] and M344,[128] and hydroxamic acid[127] induce HbF *in vitro*; valproic acid-induced HbF *in vivo*[129] when given orally. A major challenge in SCD research is to rigorously determine, which of the currently available HDAC inhibitors is the most efficacious for augmenting HbF expression in SCD.

Short-Chain Fatty Acids

Several short-chain fatty acid (SCFA) derivatives increase HbF synthesis in humans and baboons. Phenyl acetic and phenylalkyl acids induce HbF production in cultured cells and in primates.[130] Studies with the three carbon fatty acid propionate showed HbF induction in transgenic mice and baboons.[131] Using computer modeling, Perrine and associates[132] tested several SCFAs to deduce the functional group(s) required for γ-globin reactivation. Compounds such as α-methylhydrocinnamic acid, phenoxyacetic acid, 2,2-dimethylbutyric acid, and butyrate-induced Stat-5 cell signaling and the growth-related immediate early genes *c-myc* and *c-myb* during HbF induction.[133] These synthetic SCFA derivatives do not inhibit HDACs therefore γ-globin induction occurred independent of hyperacetylated histones. Subsequent studies by Pace and associates[125] showed the ability of the SCFA derivatives to induce HbF in β-YAC mice and baboons demonstrating an *in vivo* effect. These agents are being developed for human trials.

Future Perspectives

The Human Genome Project has generated publicly available databases, new technologies, and reagents that provide an unparalleled foundation to develop multidisciplinary research programs in SCD. There is still a poor understanding of the genetic basis for variable HbF synthesis in patients with SCD. In this regard, the vast array of data on SNPs and haplotype map of the β-globin cluster on chromosome 11 can be mined to determine changes in transcription factor binding motifs that effect γ-gene expression. In addition, widespread sequencing of cDNA clones isolated from hematopoietic progenitors and advances in microarray and proteomic technology is poised to identify new factors with a role in hemoglobin switching.

The influence of genetics on drug metabolism is underscored by the variable responses among sickle cell patients to HbF-inducing agents. Genome era gene profiling to develop predictive models can be used to personalize treatment regimens. To get maximum benefit from data generated by the Human Genome Project, large scale computing capability necessary for handling complex analyses using genome-wide SNPs is essential. These approaches can be used to identify epigenetic modifiers of SCD. The importance of a multidisciplinary approach cannot be overemphasized, exemplified by the four-part structure of this timely text book.

References

1. Maier-Redelsperger M, Noguchi de CT, Montalembert M, Rodgers GP, Schechter AN, Gourbil A, Blanchard D, Jais JP, Ducrocq R, Peltier JY (1994). Variation in fetal hemoglobin parameters and predicted hemoglobin S polymerization in sickle cell children in the first two years of life: parisian prospective study on sickle cell disease. *Blood* **84**:3182–3188.
2. Schroeder WA, Huisman TH, Shelton JR, Shelton JB, Kleihauer EF, Dozy AM, Roberson B (1968). Evidence for multiple structural genes for the gamma chain of human fetal hemoglobin. *Proc Natl Acad Sci USA* **60**:537–544.
3. Sampietro M, Thein SL, Contreras M, Pazmany L (1992). Variation of HbF and F-cell number with the G-gamma Xmn I (C-T) polymorphism in normal individuals. *Blood* **79**:832–833.
4. Alter BP (1981). Gel electrophoretic separation of globin chains. *Prog Clin Biol Res* **60**:157–175.
5. Schroeder WA, Shelton JB, Shelton JR (1981). High performance liquid chromatography in the identification of human hemoglobin variants. *Prog Clin Biol Res* **60**:1–22.
6. Ofori-Acquah SF, Green BN, Wild BJ, Lalloz MRA, Layton DM (1998). Quantification of Gc- and Acglobins by electrospray ionisation mass spectrometry. *Int J Mol Med* **2**:451–453.
7. Watson J, Stahman AW, Bilello FP (1948). The significance of the paucity of sickle cells in newborn Negro infants. *Am J Med Sci* **215**:419–423.
8. Bookchin RM, Balazs T, Nagel RL, Tellez I (1977). Polymerisation of haemoglobin SA hybrid tetramers. *Nature* **269**:526–527.
9. Nagel RL, Bookchin RM, Johnson J, Johnson J, Labie D, Wajcman H, Isaac-Sodeye WA, Honig GR, Schiliro G, Crookston JH, Matsutomo K (1979). Structural bases of the inhibitory effects of hemoglobin F and hemoglobin A2 on the polymerization of hemoglobin S. *Proc Natl Acad Sci USA* **76**:670–672.
10. Ofori-Acquah SF, Green BN, Davies SC, Nicolaides KH, Serjeant GR, Layton DM (2001). Mass spectral analysis of asymmetric hemoglobin hybrids: demonstration of Hb FS (alpha(2)gamma beta(s)) in sickle cell disease. *Anal Biochem* **298**:76–82.
11. Forrester WC, Takegawa S, Papayannopoulou T, Stamatoyannopoulos G, Groudine M (1987). Evidence for a locus activation region: the formation of developmentally stable hypersensitive sites in globin-expressing hybrids. *Nucleic Acids Res* **15**:10159–10177.
12. Tuan D, Abeliovich A, Lee-Oldham M, Lee D (1987). Identification of regulatory elements of human beta-like globin genes. *Prog Clin Biol Res* **251**:211–220.

13. Forrester WC, Epner E, Driscoll MC Driscoll, MC Enver T, Brice M, Papayannopoulou T, Groudine M (1990). A deletion of the human beta-globin locus activation region causes a major alteration in chromatin structure and replication across the entire beta-globin locus. *Genes Dev* **4**:1637–1649.
14. Fraser P, Gribnau J, Trimborn T (1998). Mechanisms of developmental regulation in globin loci. *Curr Opin Hematol* **5**:139–144.
15. Stamatoyannopoulos G (2005). Control of globin gene expression during development and erythroid differentiation. *Exp Hematol* **33**:259–271.
16. Bresnick EH, Felsenfeld G (1993). Evidence that the transcription factor USF is a component of the human beta-globin locus control region heteromeric protein complex. *J Biol Chem* **268**:18824–18834.
17. Andrews NC, Kotkow KJ, Ney PA, Erdjument-Bromage H, Tempst P, Orkin SH (1993). The ubiquitous subunit of erythroid transcription factor NF-E2 is a small basic-leucine zipper protein related to the v-maf oncogene. *Proc Natl Acad Sci USA* **90**:11488–11492.
18. Ney PA, Sorrentino BP, McDonagh KT, Nienhuis AW (1990). Tandem AP-1-binding sites within the human beta-globin dominant control region function as an inducible enhancer in erythroid cells. *Genes Dev* **4**:993–1006.
19. Ito E, Toki T, Ishihara H, Ohtani H, Gu L, Yokoyama M, Engel JD, Yamamoto M (1993). Erythroid transcription factor GATA-1 is abundantly transcribed in mouse testis. *Nature* **362**:466–468.
20. Tsang AP, Fujiwara Y, Hom DB, Orkin SH (1998). Failure of megakaryopoiesis and arrested erythropoiesis in mice lacking the GATA-1 transcriptional cofactor FOG. *Genes Dev* **12**:1176–1188.
21. Nagel RL, Erlingsson S, Fabry ME, Fabry ME, Croizat H, Susuka SM, Lachman H, Sutton M, Driscoll C, Bouhassira E, Billett HH (1991). The Senegal DNA haplotype is associated with the amelioration of anemia in African-American sickle cell anemia patients. *Blood* **77**:1371–1375.
22. Nagel RL, Fabry ME, Pagnier J, Zohoun I, Wajcman H, Baudin V, Labie D (1985). Hematologically and genetically distinct forms of sickle cell anemia in Africa. The Senegal type and the Benin type. *N Engl J Med* **312**:880–884.
23. Padmos MA, Roberts GT, Sackey K, Kulozik A, Bail S, Morris JS, Serjeant BE, Serjeant GR (1991). Two different forms of homozygous sickle cell disease occur in Saudi Arabia. *Br J Haematol* **79**:93–98.
24. Powars DR, Meiselman HJ, Fisher TC, Hiti A, Johnson C (1994). Beta-S gene cluster haplotypes modulate hematologic and hemorheologic expression in sickle cell anemia. Use in predicting clinical severity. *Am J Pediatr Hematol Oncol* **16**:55–61.
25. Powars DR, Elliott-Mills DD, Chan L, Niland J, Hiti AL, Opas LM, Johnson C (1991). Chronic renal failure in sickle cell disease: risk factors, clinical course, and mortality. *Ann Intern Med* **115**:614–620.
26. Powars DR (1991). Beta s-gene-cluster haplotypes in sickle cell anemia. Clinical and hematologic features. *Hematol Oncol Clin North Am* **5**:475–493.
27. Muniz A, Corral L, Alaez C, Svarch E, Espinosa E, Carbonell N, di Leo R, Felicetti L, Nagel RL, Martinez G (1995). Sickle cell anemia and beta-gene cluster haplotypes in Cuba. *Am J Hematol* **49**:163–164.
28. Figueiredo MS, Kerbauy J, Goncalves MS, Arruda VR, Saad ST, Sonati MF, Stoming T, Costa FF (1996). Effect of alpha-thalassemia and beta-globin gene cluster haplotypes on the hematological and clinical features of sickle-cell anemia in Brazil. *Am J Hematol* **53**:72–76.
29. Ofori-Acquah SF, Lalloz MRA, Layton DM (2001). Nucleotide variation regulates the level of enhancement by hypersensitive site 2 of the beta-globin locus control region. *Blood Cells Mol Dis* **27**:803–811.
30. Ofori-Acquah SF, Lalloz MRA, Serjeant G, Layton DM (2004). Dominant influence of gamma-globin promoter polymorphisms on fetal haemoglobin expression in sickle cell disease. *Cell Mol Biol* **50**:35–42.
31. Jane SM, Gumucio DL, Ney PA, Cunningham JM, Nienhuis AW (1993). Methylation-enhanced binding of Sp1 to the stage selector element of the human gamma-globin gene promoter may regulate development specificity of expression. *Mol Cell Biol* **13**:3272–3281.
32. Jane SM, Amrolia P, Cunningham JM (1995). Developmental regulation of the human beta-globin cluster. *N Engl J Med* **25**:865–869.
33. Zhao Q, Cumming H, Cerruti L, Cunningham JM, Jane SM (2004). Site-specific acetylation of the fetal globin activator NF-E4 prevents its ubiquitination and regulates its interaction with the histone deacetylase, HDAC1. *J Biol Chem* **279**:41477–41486.
34. Mavilio F, Giampaolo A, Care A, Migliaccio G, Calandrini M, Russo G, Pagliardi GL, Mastroberardino G, Marinucci M, Peschle C (1983). Molecular mechanisms of human hemoglobin switching: selective

undermethylation and expression of globin genes in embryonic, fetal, and adult erythroblasts. *Proc Natl Acad Sci USA* **80**:6907–6911.
35. Asano H, Li XS, Stamatoyannopoulos G (1999). FKLF, a novel Kruppel-like factor that activates human embryonic and fetal beta-like globin genes. *Mol Cell Biol* **19**:3571–3579.
36. Asano H, Li XS, Stamatoyannopoulos G (2000). FKLF-2: a novel Kruppel-like transcriptional factor that activates globin and other erythroid lineage genes. *Blood* **95**:3578–3584.
37. Stamatoyannopoulos G, Josephson B, Zhang JW, Li Q (1993). Developmental regulation of human gamma-globin genes in transgenic mice. *Mol Cell Biol* **13**:7636–7644.
38. Miller IJ, Bieker JJ (1993). A novel, erythroid cell-specific murine transcription factor that binds to the CACCC element and is related to the Kruppel family of nuclear proteins. *Mol Cell Biol* **13**:2776–2786.
39. Gillemans N, Tewari R, Lindeboom F, Rottier R, de Wit T, Wijgerde M, Grosveld F, Philipsen S (1998). Altered DNA-binding specificity mutants of EKLF and Sp1 show that EKLF is an activator of the beta-globin locus control region *in vivo*. *Genes Dev* **12**:2863–2873.
40. Ellis J, Tan-Un KC, Harper A, Michalovich D, Yannoutsos N, Philipsen S, Grosveld F (1996). A dominant chromatin-opening activity in 5′ hypersensitive site 3 of the human beta-globin locus control region. *EMBO J* **15**:562–568.
41. Tanabe O, Katsuoka F, Campbell AD, Song W, Yamamoto M, Tanimoto K, Engel JD (2002). An embryonic/fetal beta-type globin gene repressor contains a nuclear receptor TR2/TR4 heterodimer. *EMBO J* **21**:3434–3442.
42. Bodine DM, Ley TJ (1987). An enhancer element lies 3′ to the human A gamma globin gene. *EMBO J* **6**:2997–3004.
43. Wen J, Huang S, Rogers H, Dickinson LA, Kohwi-Shigematsu T, Noguchi CT (2005). SATB1 family protein expressed during early erythroid differentiation modifies globin gene expression. *Blood* **105**: 3330–3339.
44. Elder JT, Forrester WC, Thompson C, Mager D, Henthorn P, Peretz M, Papayannopoulou T, Groudine M (1990). Translocation of an erythroid-specific hypersensitive site in deletion-type hereditary persistence of fetal hemoglobin. *Mol Cell Biol* **10**:1382–1389.
45. Camaschella C, Serra A, Saglio G, Baiget M, Malgaretti N, Mantovani R, Ottolenghi S (1987). The 3′ ends of the deletions of Spanish delta beta zero-thalassemia and black HPFH 1 and 2 lie within 17 kilobases. *Blood* **70**:593–596.
46. Anagnou NP, Papayannopoulou T, Nienhuis AW, Stamatoyannopoulos G (1988). Molecular characterization of a novel form of (A gamma delta beta) zero-thalassemia deletion with a 3′ breakpoint close to those of HPFH-3 and HPFH-4: insights for a common regulatory mechanism. *Nucleic Acids Res* **16**:6057–6066.
47. Kosteas T, Palena A, Anagnou NP (1997). Molecular cloning of the breakpoints of the hereditary persistence of fetal hemoglobin Type-6 (HPFH-6) deletion and sequence analysis of the novel juxtaposed region from the 3′ end of the ß-globin gene cluster. *Hum Genet* **100**:441–445.
48. Stamatoyannopoulos G, Grosveld F (2001). Hemoglobin switching. In: Stamatoyannopoulos G, Majerus PW, Perlmutter RM, Varmus H, editors. *The Molecular Basis of Blood Disease*, Vol. 3. Philadelphia: Saunders.
49. Ronchi A, Nicolis S, Santoro C, Ottolenghi S (1998). Increased Sp1 binding mediates erythroid-specific overexpression of a mutated (HPFH) gamma-globulin promoter. *Nucleic Acids Res* **17**:10231–10241.
50. Stoming TA, Stoming GS, Lanclos KD, Lanclos KD, Fei YJ, Altay C, Kutlar F, Huisman TH (1989). An A gamma type of nondeletional hereditary persistence of fetal hemoglobin with a T----C mutation at position −175 to the cap site of the A gamma globin gene. *Blood* **73**:329–333.
51. McDonagh KT, Lin HJ, Lowrey CH, Bodine DM, Nienhuis AW (1991). The upstream region of the human gamma-globin gene promoter. Identification and functional analysis of nuclear protein binding sites. *J Biol Chem* **266**:11965–11974.
52. Fucharoen S, Shimizu K, Fukumaki Y (1990). A novel C-T transition within the distal CCAAT motif of the G gamma-globin gene in the Japanese HPFH: implication of factor binding in elevated fetal globin expression. *Nucleic Acids Res* **18**:5245–5253.
53. Skalnik DG, Strauss EC, Orkin SH (1991). CCAAT displacement protein as a repressor of the myelomonocytic-specific gp91-phox gene promoter. *J Biol Chem* **266**:16736–16744.

54. Ronchi AE, Bottardi S, Mazzucchelli C, Ottolenghi S, Santoro C (1995). Differential binding of the NFE3 and CP1/NFY transcription factors to the human gamma- and epsilon-globin CCAAT boxes. *J Biol Chem* **270**:21934–21941.
55. Ronchi A, Berry M, Raguz S, Raguz S, Imam A, Yannoutsos N, Ottolenghi S, Grosveld F, Dillon N (1996). Role of the duplicated CCAAT box region in gamma-globin gene regulation and hereditary persistence of fetal haemoglobin. *EMBO J* **15**:143–149.
56. Bouhassira EE, Krishnamoorthy R, Ragusa A, Driscoll C, Labie D, Nagel RL (1989). The enhancer-like sequence 3′ to the A gamma gene is polymorphic in human populations. *Blood* **73**:1050–1053.
57. Balta G, Brickner HE, Takegawa S, Kazazian HH Jr, Papayannopoulou T, Forget BG, Atweh GF (1994). Increased expression of the G gamma and A gamma globin genes associated with a mutation in the A gamma enhancer. *Blood* **83**:3727–3737.
58. Liu Q, Tanimoto K, Bungert J, Engel JD (1998). The A gamma-globin 3′ element provides no unique function(s) for human beta-globin locus gene regulation. *Proc Natl Acad Sci USA* **95**:9944–9949.
59. Bard H, Lachance C, Widness JA, Gagnon C (1994). The reactivation of fetal hemoglobin synthesis during anemia of prematurity. *Pediatr Res* **36**:253–256.
60. Hisakawa H, Sugiyama D, Nishijima I, Xu M, Wu H, Nakao K, Watanabe S, Katsuki M, Asano S, Arai K, Nakahata T, Tsugi K (2001). Epo primitive to definitive erythropoiesis. *Blood* **98**:3618–3625.
61. Perrine SP, Greene MF, Faller DV (1985). Delay in the fetal globin switch in infants of diabetic mothers. *N Engl J Med* **312**:334–338.
62. Bard H, Prosmanne J (1985). Relative rates of fetal hemoglobin and adult hemoglobin synthesis in cord blood of infants of insulin-dependent diabetic mothers. *Pediatrics* **75**:1143–1147.
63. Witt O, Sand K, Pekrun A (2000). Butyrate-induced erythroid differentiation of human K562 leukemia cells involves inhibition of ERK and activation of p38 MAP kinase pathways. *Blood* **95**:2391–2396.
64. Pace BS, Qian XH, Sangerman J, Ofori-Acquah SF, Baliga BS, Han J, Critz SD (2003). p38 MAP kinase activation mediates gamma-globin gene induction in erythroid progenitors. *Exp Hematol* **31**:1089–1096.
65. Witt O, Monkemeyer S, Kanbach K, Pekrun A (2002). Induction of fetal hemoglobin synthesis by valproate: modulation of MAP kinase pathways. *Am J Hematol* **71**:45–47.
66. Park JI, Jeong JS, Han JY, Kim DI, Gao YH, Park SC, Rodgers GP, Kim IH (2000). Hydroxyurea induces a senescence-like change of K562 human erythroleukemia cell. *J Cancer Res Clin Oncol* **126**:455–460.
67. Park JI, Choi HS, Jeong JS, Han JY, Kim IH (2001). Involvement of p38 kinase in hydroxyurea-induced differentiation of K562 cells. *Cell Growth Differ* **12**:481–486.
68. Lavelle D, DeSimone J, Hankewych M, Kousnetzova T, Chen YH (2003). Decitabine induces cell cycle arrest at the G1 phase via p21(WAF1) and the G2/M phase via the p38 MAP kinase pathway. *Leuk Res* **27**:999–1007.
69. Uddin S, Ah-Kang J, Ulaszek J, Mahmud D, Wickrema A (2004). Differentiation stage-specific activation of p38 mitogen-activated protein kinase isoforms in primary human erythroid cells. *Proc Natl Acad Sci USA* **101**:147–152.
70. Tamura K, Sudo T, Senftleben U, Dadak AM, Johnson R, Karin M (2000). Requirement for p38alpha in erythropoietin expression: a role for stress kinases in erythropoiesis. *Cell* **102**:221–231.
71. Hazzalin CA, Cano E, Cuenda A, Barratt MJ, Cohen P, Mahadevan LC (1996). p38/RK is essential for stress-induced nuclear responses: JNK/SAPKs and c-Jun/ATF-2 phosphorylation are insufficient. *Curr Biol* **6**:1028–1031.
72. New L, Jiang Y, Zhao M, Liu, K, Zhu W, Flood LJ, Kato Y, Parry GC, Han J (1998). PRAK, a novel protein kinase regulated by the p38 MAP kinase. *EMBO J* **17**:3372–3384.
73. Gustin JA, Pincheira R, Mayo LD, Ozes ON, Kessler KM, Baerwald MR, Korgaonkar CK, Donner DB (2004). Tumor necrosis factor activates CRE-binding protein through a p38 MAPK/MSK1 signaling pathway in endothelial cells. *Am J Physiol Cell Physiol* **286**:C547–555.
74. Darnell JE Jr. (1997). STATs and gene regulation. *Science* **277**:1630–1635.
75. Schaefer TS, Sanders LK, Nathans D (1995). Cooperative transcriptional activity of Jun and Stat3 Beta, a short form of Stat3. *Proc Natl Acad Sci USA* **92**:9097–9101.
76. Foley HA, Ofori-Acquah SF, Yoshimura A, Critz S, Baliga BS, Pace BS (2002). tat3 Beta inhibits gamma-globin gene expression in erythroid cells. *J Biol Chem* **277**:16211–16219.

77. Ferry AE, Baliga SB, Monteiro C, Pace BS (1997). Globin gene silencing in primary erythroid cultures. An inhibitory role for interleukin-6. *J Biol Chem* **272**:20030–20037.
78. King BS (2003). A role for nitric oxide in hydroxyurea-mediated fetal hemoglobin induction. *J Clin Invest* **111**:171–172.
79. Haby C, Lisovoski F, Aunis D, Zwiller J (1994). Stimulation of the cyclic GMP pathway by NO induces expression of the immediate early genes c-fos and junB in PC12 cells. *J Neurochem* **62**:496–501.
80. Akashi T, Sawada Y, Shimada N, Sakurai N, Aoki T, Ayabe S (2003). cDNA cloning and biochemical characterization of S-adenosyl-L-methionine: 2,7,4′-trihydroxyisoflavanone 4′-O-methyltransferase, a critical enzyme of the legume isoflavonoid phytoalexin pathway. *Plant Cell Physiol* **44**:103–112.
81. Ikuta T, Ausenda S, Cappellini MD (2001). Mechanism for fetal globin gene expression: role of the soluble guanylate cyclase-cGMP-dependent protein kinase pathway. *Proc Natl Acad Sci USA* **98**:1847–1852.
82. Cokic VP, Smith RD, Beleslin-Cokic BB, Njoroge JM, Miller JL, Gladwin MT, Schechter AN (2003). Hydroxyurea induces fetal hemoglobin by the nitric oxide-dependent activation of soluble guanylyl cyclase. *J Clin Invest* **111**:231–239.
83. Suhasini M, Boss GR, Pascual FE, Pilz RB (1995). Nitric oxide-releasing agents and cGMP analogues inhibit murine erythroleukemia cell differentiation and suppress erythroid-specific gene expression: correlation with decreased DNA binding of NF-E2 and altered c-myb mRNA expression. *Cell Growth Differ* **6**:1559–1566.
84. Fischer KD, Haese A, Nowock J (1993). Cooperation of GATA-1 and Sp1 can result in synergistic transcriptional activation or interference. *J Biol Chem* **268**:23915–23923.
85. Al-Khatti A, Veith RW, Papayannopoulou T, Fritsch EF, Goldwasser E, Stamatoyannopoulos G (1987). Stimulation of fetal hemoglobin synthesis by erythropoietin in baboons. *N Engl J Med* **317**:415–420.
86. Nagata Y, Todokoro K (1999). Requirement of activation of JNK and p38 for environmental stress-induced erythroid differentiation and apoptosis and of inhibition of ERK for apoptosis. *Blood* **94**:853–863.
87. Tamura K, Sudo T, Senftleben U, Dadak AM, Johnson R, Karin M (2000). Requirement for p38alpha in erythropoietin expression: a role for stress kinases in erythropoiesis. *Cell* **102**:221–231.
88. Dean JL, Brook M, Clark AR, Saklatvala J (1999). p38 mitogen-activated protein kinase regulates cyclooxygenase-2 mRNA stability and transcription in lipopolysaccharide-treated human monocytes. *J Biol Chem* **274**:264–269.
89. Wojda U, Leigh KR, Njoroge JM, Jackson KA, Natarajan B, Stitely M, Miller JL (2003). Fetal hemoglobin modulation during human erythropoiesis: stem cell factor has "late" effects related to the expression pattern of CD117. *Blood* **101**:492–497.
90. Miller BA, Floros J, Cheung JY, Wojchowski DM, Bell L, Begley CG, Elwood NJ, Kreider J, Christian C (1994). Steel factor affects SCL expression during normal erythroid differentiation. *Blood* **84**:2971–2976.
91. Bhanu NV, Trice TA, Lee YT, Gantt NM, Oneal P, Schwartz JD, Noel P, Miller JL (2005). A sustained and pancellular reversal of gamma-globin gene silencing in adult human erythroid precursor cells. *Blood* **105**:387–393.
92. Galanello R, Stamatoyannopoulos G, Papayannopoulou T (1988). Mechanism of HbF stimulation by S-stage compounds. *In vitro* studies with bone marrow cells exposed to 5-azacytidine, Ara-C, or hydroxyurea. *J Clin Invest* **81**:1209–1216.
93. Letvin NL, Linch DC, Beardsley GP, McIntyre KW, Nathan DG (1984). Augmentation of fetal-hemoglobin production in anemic monkeys by hydroxyurea. *N Engl J Med* **310**:869–873.
94. Papayannopoulou T, Torrealba de Ron A, Veith R, Knitter G, Stamatoyannopoulos G (1984). Arabinosylcytosine induces fetal hemoglobin in baboons by perturbing erythroid cell differentiation kinetics. *Science* **224**:617–619.
95. Liu DP, Liang CC, Ao ZH, Jia PC, Chen SS, Wang RX, Liu LJ, Jin HQ, Zha DY, Huang YW (1990). Treatment of severe beta-thalassemia (patients) with myleran. *Am J Hematol* **33**:50–55.
96. Veith R, Papayannopoulou T, Kurachi S, Stamatoyannopoulos G (1985). Treatment of baboon with vinblastine: insights into the mechanisms of pharmacologic stimulation of HbF in the adult. *Blood* **66**:456–459.
97. Charache S, Terrin ML, Moore RD, Dover GJ, Barton FB, Eckert SV, McMahon RP, Bonds DR (1995). Effect of hydroxyurea on the frequency of painful crises in sickle cell anemia. Investigators of the multicenter study of hydroxyurea in sickle cell anemia. *N Engl J Med* **332**:1317–1322.

98. Wang WC, Helms RW, Lynn HS, Redding-Lallinger R, Gee BE, Ohene-Frempong K, Smith-Whitley K, Waclawiw MA, Vichinsky EP, Styles LA, Ware RE, Kinney TR (2002). Effect of hydroxyurea on growth in children with sickle cell anemia: results of the HUG-KIDS Study. *J Pediatr* **140**:225–229.
99. Rogers ZR, Buchanan GR (2004). Expanding the role of hydroxyurea in children with sickle cell disease. *J Pediatr* **145**:287–288.
100. Hankins JS, Ware RE, Rogers ZR *et al.* (2005). Long-term hydroxyurea therapy for infants with sickle cell anemia — the HUSOFT extension study. *Blood* **106**:2269–2275.
101. Yang YM, Pace B, Kitchens D, Ballas SK, Shah A, Baliga BS (1997). BFU-E colony growth in response to hydroxyurea: correlation between *in vitro* and *in vivo* fetal hemoglobin induction. *Am J Hematol* **56**: 252–258.
102. Haynes J Jr, Baliga BS, Obiako B, Ofori-Acquah S, Pace B (2004). Zileuton induces hemoglobin F synthesis in erythroid progenitors: role of the L-arginine-nitric oxide signaling pathway. *Blood* **103**:3945–3950.
103. National Institutes of Health Government Identifier (2005). NCT00056433.
104. Belinsky SA, Klinge DM, Stidley CA, Issa JP, Herman JG, March TH, Baylin SB (2003). Inhibition of DNA methylation and histone deacetylation prevents murine lung cancer. *Cancer Res* **63**:7089–7093.
105. van der Ploeg LH, Flavell RA (1980). DNA methylation in the human gamma delta beta-globin locus in erythroid and nonerythroid tissues. *Cell* **19**:947–958.
106. DeSimone J, Heller P, Hall L, Zwiers D (1982). 5-Azacytidine stimulates fetal hemoglobin synthesis in anemic baboons. *Proc Natl Acad Sci USA* **79**:4428–4431.
107. Lavelle D, DeSimone J, Heller P (1993). Fetal hemoglobin reactivation in baboon and man: a short perspective. *Am J Hematol* **42**:91–95.
108. Ley TJ, DeSimone J, Noguchi CT, Turner PH, Schechter AN, Heller P, Nienhuis AW (1983). 5-Azacytidine increases gamma-globin synthesis and reduces the proportion of dense cells in patients with sickle cell anemia. *Blood* **62**:370–380.
109. Cortvrindt R, Bernheim J, Buyssens N, Robol K (1987). 5-Azacytidine and 5-aza-2′-deoxycytidine behave as different antineoplastic agents in B16 melanoma. *Br J Cancer* **56**:261–265.
110. Koshy M, Dorn L, Bressler L, Molokie R, Lavelle D, Talischy N, Hoffman R, van Overveld W, DeSimone J (2000). 2-deoxy 5-azacytidine and fetal hemoglobin induction in sickle cell anemia. *Blood* **96**:2379–2384.
111. Saunthararajah Y, Lavelle D, DeSimone J (2004). DNA hypo-methylating agents and sickle cell disease. *Br J Haematol* **126**:629–636.
112. Karpf AR, Lasek AW, Ririe TO, Hanks AN, Grossman D, Jones DA (2004). Limited gene activation in tumor and normal epithelial cells treated with the DNA methyltransferase inhibitor 5-aza-2′-deoxycytidine. *Mol Pharmacol* **65**:18–27.
113. Jackson-Grusby L, Laird PW, Magge SN, Moeller BJ, Jaenisch R (1997). Mutagenicity of 5-aza-2′-deoxycytidine is mediated by the mammalian DNA methyltransferase. *Proc Natl Acad Sci USA* **94**: 4681–4685.
114. Perrine SP, Rudolph A, Faller DV, Roman C, Cohen RA, Chen SJ, Kan YW (1988). Butyrate infusions in the ovine fetus delay the biologic clock for globin gene switching. *Proc Natl Acad Sci USA* **85**:8540–8542.
115. Constantoulakis P, Knitter G, Stamatoyannopoulos G (1989). On the induction of fetal hemoglobin by butyrates: *in vivo* and *in vitro* studies with sodium butyrate and comparison of combination treatments with 5-AzaC and AraC. *Blood* **74**:1963–1971.
116. Sher GD, Olivieri NF (1994). Rapid healing of chronic leg ulcers during arginine butyrate therapy in patients with sickle cell disease and thalassemia. *Blood* **84**:2378–2380.
117. Perrine SP, Faller DV (1993). Butyrate-induced reactivation of the fetal globin genes: a molecular treatment for the beta-hemoglobinopathies. *Experientia* **49**:133–137.
118. Atweh GF, Sutton M, Nassif I, Dover GJ, Wallenstein S, Wright E, McMahon L, Stamatoyannopoulos G, Faller DV, Perrine SP (1999). Sustained induction of fetal hemoglobin by pulse butyrate therapy in sickle cell disease. *Blood* **93**:1790–1797.
119. Candido EP, Reeves R, Davie JR (1978). Sodium butyrate inhibits histone deacetylation in cultured cells. *Cell* **14**:105–113.
120. McCaffrey PG, Newsome DA, Fibach E, Yoshida M, Su MS (1997). Induction of gamma-globin by histone deacetylase inhibitors. *Blood* **90**:2075–2083.

121. Norton VG, Imai BS, Yau P, Bradbury EM (1989). Histone acetylation reduces nucleosome core particle linking number change. *Cell* **57**:449–457.
122. McElveen RL, Lou TF, Reese K, Xia S, Baliga BS, Pace BS (2005). Erk pathway inhibitor U0126 induces gamma-globin expression in erythroid cells. *Cell Mol Biol* **51**:215–227.
123. Ikuta T, Kan YW, Swerdlow PS, Faller DV, Perrine SP (1998). Alterations in protein-DNA interactions in the gamma-globin gene promoter in response to butyrate therapy. *Blood* **92**:2924–2933.
124. Pace BS, Li Q, Stamatoyannopoulos G (1996). *In vivo* search for butyrate responsive sequences using transgenic mice carrying A gamma gene promoter mutants. *Blood* **88**:1079–1083.
125. Pace BS, White GL, Dover GJ, Boosalis MS, Faller DV, Perrine SP (2002). Short-chain fatty acid derivatives induce fetal globin expression and erythropoiesis *in vivo*. *Blood* **100**:4640–4648.
126. Weinberg RS, Ji X, Sutton M, Perrine S, Galperin Y, Li Q, Liebhaber SA, Stamatoyannopoulos G, Atweh GF (2005). Butyrate increases the efficiency of translation of gamma-globin mRNA. *Blood* **105**: 1807–1809.
127. Cao H, Stamatoyannopoulos G, Jung M (2004). Induction of human gamma globin gene expression by histone deacetylase inhibitors. *Blood* **103**:701–709.
128. Skarpidi E, Cao H, Heltweg B, White BF, Marhenke RL, Jung M, Stamatoyannopoulos G (2003). Hydroxamide derivatives of short-chain fatty acids are potent inducers of human fetal globin gene expression. *Exp Hematol* **131**:197–203.
129. Collins AF, Dover GJ, Luban NL (1994). Increased fetal hemoglobin production in patients receiving valproic acid for epilepsy. *Blood* **84**:1690–1691.
130. Torkelson S, White B, Faller DV, Phipps K, Pantazis C, Perrine SP (1996). Erythroid progenitor proliferation is stimulated by phenoxyacetic and phenylalkyl acids. *Blood Cells Mol Dis* **22**:150–158.
131. Liakopoulou E, Blau CA, Li Q, Josephson B, Wolf JA, Fournarakis B, Raisys V, Dover G, Papayannopoulou T, Stamatoyannopoulos G (1995). Stimulation of fetal hemoglobin production by short chain fatty acids. *Blood* **86**:3227–3235.
132. Boosalis MS, Ikuta T, Pace BS, da Fonseca S, White GL, Faller DV, Perrine SP (1997). Abrogation of IL-3 requirements and stimulation of hematopoietic cell proliferation *in vitro* and *in vivo* by carboxylic acids. *Blood Cells Mol Dis* **23**:434–442.
133. Boosalis MS, Bandyopadhyay R, Bresnick EH, Pace BS, Van DeMark K, Zhang B, Faller DV, Perrine SP (2001). Short-chain fatty acid derivatives stimulate cell proliferation and induce STAT-5 activation. *Blood* **97**:3259–3267.

12

Genetic Modulation of Sickle Cell Disease

by Martin H. Steinberg and Swee Lay Thein

Introduction

Sickle cell anemia is a prototypical Mendelian single gene disorder, caused by homozygosity for a mutation affecting the β-globin gene (*HBB*; β^6 GAG $\rightarrow$ GTG; glu^6 $\rightarrow$ val^6) to produce HbS. In addition to homozygosity for HbS (SCD-SS), sickle cell disease (SCD) can result from compound heterozygosity of HbS and HbC (SCD-SC), or β-thalassemia variants, to produce a spectrum of sickle cell/β-thalassemia (SCD-Sβthal) conditions. Generally the compound heterozygotes are less severe than SCD-SS, but HbSS, despite its genetic simplicity, displays a remarkable clinical heterogeneity.[1]

As with all monogenic diseases, the clinical heterogeneity is the result of multiple genes interacting with each other and with the environment. Twin studies, based on the concordance and disconcordance of disease complications, have traditionally been used to assess the relative contributions of genetic and environmental factors in complex disorders such as schizophrenia and diabetes. Remarkably, there are only two reports of this kind, both limited to single pairs of identical twins, one with SCD-SS and α-thalassemia,[2] and the other with SCD-Sβthal.[3] Both case studies report that, despite identical α and β genotypes, and similarities in their laboratory and hematological parameters, the twins have quite different frequency and severity of painful crises. These limited data suggest that environmental factors may be of considerable importance in determining the clinical course of SCD. Such environmental factors include the nutritional state, access to social support, and medical care, all of which influence risk factors such as infections.

Primary genetic determinants are the underlying genotypes, SCD-SS generally being more severe than SCD-SC. Factors that influence the primary event of polymerization of deoxyHbS include the co-inheritance of α-thalassemia and the levels of fetal hemoglobin (HbF, $\alpha_2\gamma_2$), itself a quantitative trait that is predominantly genetically controlled, involving multiple loci (see below).[4] Roles for many other modifying factors and genes have been proposed based on the pathophysiology of vasoocclusion which follows the primary event of HbS polymerization.

Due to the complex interplay of a multitude of factors, a genetic approach might be the most efficacious way of dissecting the molecular mechanisms and identifying the modifier genes. The elucidation of genetic variants and mechanisms responsible for the phenotypic variability of SCD will have important clinical implications for genetic counseling and clinical management; identificationx of these factors may improve the accuracy of prediction of the disease severity and facilitate risk stratification. High risk patients might then be followed more closely and intensively, and preventative and experimental therapies (which carry risks of their own) could be targeted.

Variation in the Human Genome — Similarities and Differences

With the exception of identical twins, every individual has a unique genome. It is generally accepted that any two genomes are roughly 99.9% identical, but that still leaves several million differences among the $2 \times 3 \times 10^9$ base pairs. Recent genome-wide studies also provide evidence for large-scale variation of up to several hundred kilobases in size in the human genome.[5,6] The most common types of variant are single nucleotide polymorphisms (SNPs),[7] which by definition exist at a frequency of more than 1% in at least one population. SNPs are distributed throughout the human genome at an overall frequency of one every 1330 bp,[7] but the distribution is highly variable within the genome, and only a small proportion lie within coding regions.

These SNPs are estimated to result in 100,000 amino acid differences between the human proteomes that provide the underlying basis for differences in the physiology and genetic background between individuals. Although it is intuitively apparent that the amino acid differences that result from the coding SNPs can change the function of a gene, gene expression can also be altered by SNPs positioned in critical regulatory sequences.

It is not difficult to envisage how multiple genetic variants contribute to the heterogeneity of the so-called "simple Mendelian" disorders such as SCD-SS. They influence disease-related quantitative traits such as levels of HbF, in the same way as they confer susceptibility to common diseases such as diabetes and hypertension. The significance of a functional SNP should be interpreted in the context of the study population. Small differences in the expression of key genes can amplify or dampen a biological pathway — for example, in the co-ordination of responses following the polymerization of deoxyHbS — ultimately leading to vasoocclusion.

Identification of Genetic Variants, Haplotypes and Linkage Disequilibrium

Identification of SNPs has been accelerated by three developments: availability of sequence data from the Human Genome Project (HGP)[8,9]; improved bioinformatics tools for mining the sequence; and high-throughput genotyping platforms.[10] Approximately 10 million SNPs are believed to exist in humans, the vast majority of which have been discovered through public effort using sequence data from the public HGP. As part of the draft sequence, 1.42 million SNPs have been reported in db-SNP (http://www.ncbi.nlm.nih.gov/SNP/), and public databases now contain more than 3.7 million mapped human SNPs.[11] The majority of the SNPs in db-SNP have been discovered in-silico, using bioinformatics algorithms to search for single nucleotide differences between aligned sequence reads and available genomic sequence. Approximately 5% of the SNPs have been discovered deliberately by resequencing gene(s) or genetic regions.

Combinations of SNPs are inherited together on the same DNA strand as haplotypes.[12,13] The genome is thus portrayed as a series of high linkage disequilibrium (LD) regions ("haplotype blocks") separated by short discrete segments of very low LD described as "recombination hotspots." Strong LD within a genomic region implies that most of the variation within that region can be captured by just a few informative SNPs, referred to as tag SNPs, thereby streamlining and reducing genotyping efforts by avoiding the genotyping of redundant SNPs. The International HapMap Project (The International

HapMap Consortium[a]), a multi-country effort, was initiated in 2003 to identify and catalog genetic similarities and differences in humans, based on identifying common haplotypes composed of SNPs in four populations from different parts of the world.[14]

The HapMap Project aims to identify the tagging SNPs that capture those haplotypes (http://www.hapmap.org); thus, without prior knowledge of the functional variant or causative SNP, genomic regions can be evaluated for association with disease outcomes using these tag SNPs. But the causal variant that is responsible may be anywhere within the haplotype block, and it can be a considerable challenge to identify it, so the choice of tagging SNPs that capture most of the variation over a genomic region is the key to association studies. It is estimated that the number of tag SNPs that contain most of the information about the patterns of genetic variation is between 300,000 and 600,000. Since non-African populations generally have higher levels of LD than African populations, capturing variation in African populations will almost certainly require a substantially larger number of tagging SNPs.[15]

A Genetic Approach Towards the Study of SCD

Generally, two genetic approaches have been used to locate genetic variants in human disease: linkage analysis and association studies.[16] Linkage mapping follows the segregation of chromosomal regions marked by genetic variants in affected families, in search of regions that co-segregate with the disease or trait. Although extremely successful in highly penetrative single gene disorders, linkage analysis has had limited success in the common diseases; this lack of success has been attributed to low heritability of most complex traits, genetic heterogeneity, imprecise definition of phenotypes, and inadequately powered study designs.

Association studies look for differences in the frequencies of genetic variants between ethnically matched cases and controls to find variants that are strongly associated with the disease. If a variant is more common in cases than controls, the association can be inferred. Such studies require large sample numbers and until recently have not been feasible due to the cost and labor of genotyping. SNPs identified in pilot studies should always be validated in additional, independent populations.

To date, the genetic association studies in SCD-SS have been based on candidate genes and genotyping SNPs within and flanking these genes. Many of these studies are poorly designed, with inadequate sample sizes. Single or a few SNPs in the candidates were selected for genotyping with no consideration of the LD structure. Regardless of population genetic reasons, the SNPs selected for genotyping in these candidate gene studies may not be disease-causing, and also not in LD with the causal SNP. Hence it is not too surprising that linkage with the putative culprits was not replicated in independent studies. Candidate genes were selected based on our understanding of the sickle cell pathophysiology underpinned by the two central mechanisms of deoxyHbS polymerization and vasoocclusion.

It has often been argued that association studies in complex diseases should be genome-wide and hypothesis-free, without having to guess the identity of the causal genes, as inevitably there will be some aspects of the multi-faceted disease etiology that cannot be interpreted correctly. Association studies based on hypothesis-driven candidate genes could well leave out some of the key genetic players.[16]

---

[a]The International HapMap Consortium, Altshuler D, Brooks LD, *et al.* (2005). A haplotype map of the human genome. *Nature* **437**:1299–1320.

Established Predictors of Complications in Sickle Cell Anemia

Fetal Hemoglobin (HbF)

HbF, a quantitative trait in healthy adults, is the best known modulator of symptoms in sickle cell anemia; it has its effect by inhibiting polymerization of deoxyHbS. Studies have shown that even allowing for the expected age and gender differences, HbF levels vary considerably from less than 1% to more than 20% among patients with SCD-SS.[17]

The beneficial effect of HbF has led to the use of pharmacologic agents such as hydroxyurea (HU)[18] and high doses of intermittent butyrate to augment its production in SCD-SS.[19,20] Although not universally effective, a pertinent observation was the variable response. Strong correlation in HbF levels was present in siblings before and during treatment with HU,[21] and individuals with higher baseline levels of HbF also have a higher HbF response to intravenous butyrate.[20] These observations suggest that the HbF response is modulated by underlying genetic determinants. Therefore, identifying potential regulatory elements may help us to devise better therapeutics to augment HbF production in the treatment of SCD-SS. But how do we find these genetic factors?

Genetic Regulation of Fetal Hemoglobin

HbF expression is regulated by complex interactions involving chromosome remodeling activities, transcription factors, cytokines modulating erythropoiesis, controlled by elements linked and unlinked to the β-globin gene cluster, providing ample opportunity for genetic modulation.[22]

HbF levels vary considerably not only in patients with SCD-SS, but also in normal adults. Every study of the levels of HbF or F cells (a sub-set of erythrocytes that contain measurable HbF) in normal adults has shown a continuous distribution that is substantially positively skewed, with the levels varying by more than 20-fold.[23] Twin studies have shown that 89% of the quantitative variation in HbF and F cell (FC) levels in normal adults is genetically controlled, but the genetic etiology is complex.[4]

Measured genotype analyses have shown the trait to be influenced by several factors including age, sex, and a common SNP (C to T) at position -158 upstream of the $^{G}\gamma$-globin gene, referred to as the *Xmn*1-$^{G}\gamma$ polymorphism.[24,25] The *Xmn*1-$^{G}\gamma$ polymorphism accounts for about one third of the FC variance in normal adults, but more than 50% of the variation in FC levels is due to factor(s) not linked to the β-globin complex.

The "T" variant that creates a cutting site for the *Xmn*1 restriction enzyme is common in all population groups and is present at a frequency of 0.32 to 0.35. Unlike the rare mutations in the γ-globin promoter that are consistently associated with large discrete effects of increased HbF levels of 10–35% in heterozygotes, the so-called non-deletional pancellular HPFHs, the SNP in $^{G}\gamma$-158 does not always raise the HbF levels in otherwise normal individuals.[26] Nonetheless, although it has little effect in normal individuals, clinical studies have shown that under conditions of erythropoietic stress — for example, in homozygous β-thalassemia[27] and SCD-SS[28] — presence of the *Xmn*1-$^{G}\gamma$ site favors a higher HbF response. Presence of the *Xmn*1-$^{G}\gamma$ site on the Senegal and Arab-Indian β^{S} haplotypes[29,30] may explain the higher HbF levels in SCD-SS patients with such haplotypes, compared to those with the Bantu or Benin β^{S} haplotypes (see below). However, the HbF response associated with the *Xmn*1-$^{G}\gamma$ site is usually moderate, and may not be sufficient to explain the wide difference in phenotype observed in many cases. Family studies by Kulozik and co-workers indicate the presence of an independent hereditary persistence of fetal hemoglobin (HPFH) determinant active in both HbS heterozygotes (AS), and SCD-SS, and which segregates independently of the Arab-Indian

β^S haplotype.[31] This is in keeping with genetic studies which showed that >50% of the variation in FC levels in the general population is accounted for by trans-acting factors[25] or quantitative trait loci (QTLs).

Intensive linkage studies have mapped QTLs controlling FC levels to three regions of the genome–chromosomes 6q23, Xp22 and 8q11.

The putative F cell production (FCP) locus on chromosome Xp22.2 was mapped in a group of normal individuals and individuals with SCD[32]; the QTL on chromosome 6q23 was mapped in an extensive Asian-Indian kindred with β-thalassemia and heterocellular HPFH.[33,34] Further analysis reduced the candidate region to 1.5 Mb and showed that it contains five protein coding genes, including a very large uncharacterized gene with 28 exons that spanned 215 kb (AHI1), and several genes that do not appear to be protein coding.[35] In the same Asian-Indian family with β-thalassemia and heterocellular HPFH, after adjusting the FC levels for the effects of the *Xmn*1-$^G\gamma$ site and 6q QTL, a repeat linkage analysis led to the localization of another QTL on 8q11.[36]

However, unlike the 6q QTL, the effects of the 8q locus are conditional on the *Xmn*1-$^G\gamma$ site, suggesting a genetic interaction between the site and the 8q QTL. The 8q QTL defines another class of genetic determinants of FC levels, one that is conditional on cis-acting sequences of the β-globin complex. This could explain some of the inconsistent associations of high HbF levels with the *Xmn*1-$^G\gamma$ site that has been observed even within families. Linkage to chromosome 8q has recently been replicated in a different ethnic population; the linkage was also conditional on the *Xmn*1-$^G\gamma$ site in the European population.[37,38] In another recent study, 180 SNPs in 38 candidate genes that might modulate HbF levels were studied in 280 patients with SCD-SS. Strong associations with HbF were found with SNPs in phosphodiesterase 7 *(PDE7B)*, microtubule-associated protein 7 *(MAP7)*, mitogen-activated protein kinase kinase kinase 5 (*MAP3K5*) and peroxisomal biogenesis factor 7 (*PEX7*), genes that abut 6q 23.2.[39] *PDE7B*, *MAP3K5* and *PEX7* all have putative roles in globin synthesis but *PDE7B* is an especially interesting candidate given the inhibitory effect of cAMP on γ-globin gene expression.

Despite these QTLs, studies indicate that chromosomal locations of other QTLs contributing to HbF levels in adults have yet to be determined;[40] thus heterocellular HPFH appears to be genetically heterogeneous. As the genetic basis for the propensity to produce HbF is unraveled, it is becoming clear that the conglomeration of the *Xmn*1-$^G\gamma$ polymorphism, the QTLs on 6q, Xp and 8q and others, linked and unlinked to the β-globin complex, contribute to the quantitative trait of HbF that constitutes the loosely defined syndrome of heterocellular HPFH. These QTLs presumably play an important role in the fine tuning of γ-globin production in normal adults, in response to "erythropoietic stress" and possibly in an individual's capacity to respond to pharmacologic inducers of HbF synthesis.

β-globin Gene Cluster Haplotypes

The HbS gene is found on a genetic background of four major β-globin gene cluster haplotypes. Carriers of the HbS gene on the Senegal or Arab-Indian β^S haplotype usually have the highest HbF level and PCV, and the mildest clinical course. Individuals with the HbS gene on a Bantu (Central African Republic) haplotype have the lowest HbF level and PCV, and the most severe clinical course. Carriers of the Benin haplotype have intermediate features.[41] These differences are accounted for in large part by an association of the *Xmn*1-$^G\gamma$ site with the Senegal and Arab-Indian β^S haplotypes.[28,29] However, the prognostic value of knowing the β^S haplotype in an individual remains to be determined.[42]

α-Thalassemia

About a third of patients with SCD-SS have coincidental α-thalassemia.[43] These individuals have less hemolysis, higher packed cell volume (PCV), lower mean corpuscular volume (MCV), and lower reticulocyte counts.[44,45] Coinheritance of α-thalassemia results in relatively longer erythrocyte lifespan because of the reduction of dense and rigid sickle red cells. The resulting increased PCV and blood viscosity may increase the incidence of certain vasoocclusive complications such as painful episodes, acute chest syndrome, and osteonecrosis. In Jamaicans, the absence of α-thalassemia coupled with a high HbF appears to predict a benign disease.[46] When phenotypes of SCD-SS patients were clustered into two or three major phenotype groups, neither α-thalassemia nor the β-globin gene cluster haplotype appeared to influence the clinical events defining the groups.[47]

Genetic Variants as Predictors of Disease Severity

The diversity of SCD-SS cannot be explained by the levels of HbF alone. Factors that potentially affect the pathogenesis and modulate the phenotype of disease include mediators of inflammation, oxidant injury, nitric oxide biology, vaso-regulation, cell–cell interaction, blood coagulation, and hemostasis. These factors have modifying effects independent of modulating HbS polymerization, and all are likely to be complex traits contributed by several QTLs. Based on this pathophysiology, candidate genes that plausibly affect the phenotype of SCD-SS have been selected and SNPs identified through association studies (Table 1).

Painful Episodes

Acute painful episodes, the major clinical event in SCD, are probably the complications most influenced by environmental factors. They are a measure of disease severity and a predictor of early death in adults. The rate of painful episodes varies widely among patients: highest pain rates are found in patients with high PCV and low HbF.[48] Besides HbF concentration and a possible role of α-thalassemia, very little is known about the genetic basis for the variability of painful episodes among patients.

Stroke

A familial predisposition to stroke was first suggested by increased prevalence of this complication among siblings with SCD-SS.[49] A recent study of 42 sibships with SCD-SS, in which at least one sibling had a stroke, showed that a greater than expected number of families had at least two children with stroke.[50] Stroke in SCD-SS, has been associated with variants in genes such as vascular adhesion molecule-1 (*VCAM1*), interleukin 4 receptor gene (*IL4R*), tumor necrosis factor α-gene (*TNFAα*), β-adrenergic receptor 2 (*ADRB2*), and low density lipoprotein receptor (*LDLR*).[51,52]

Given the association of human leukocyte antigen (HLA) with moya-moya disease in Japanese children, HLA locus was considered as a potential modifier for stroke in SCD-SS.[53] Complete HLA genotyping was performed in 230 SCD-SS children previously enrolled in the Cooperative Study Group of Sickle Cell Disease (CSSCD).[54] Cerebral infarction on magnetic resonance imaging (MRI) was documented in 71 patients, and 160 patients with negative MRI served as controls. Of the 71 MRI positive patients, 36 were classified as having large vessel (LV) stroke, while 35 had small vessel (SV) stroke. Although distinct HLA alleles were identified as risk factors for LV and SV stroke indicating

Table 1. Genetic polymorphisms affecting sickle cell anemia.

Sub-phenotype	Gene/SNP marker*	Effect	Reference(s)
Stroke	*VCAM1*/G1238C	Protective	Taylor JG, Tang DC, Savage SA *et al.* (2002)[51]
	VCAM1/T1594C	Permissive	Hoppe C, Klitz W, Cheng S *et al.* (2004)[52]
	IL4R/S503P	Permissive	Hoppe C, Klitz W, Cheng S *et al.* (2004)[52]
	TNFA/G-308S	Protective	Hoppe C, Klitz W, Cheng S *et al.* (2004)[52]
	LDLR/Ncol +/−	Protective	Hoppe C, Klitz W, Cheng S *et al.* (2004)[52]
	ADRB2/Q/27E	Protective	Hoppe C, Klitz W, Cheng S *et al.* (2004)[52]
	AGT/AG repeats	Permissive	Tang DC, Prauner R, Liu W *et al.* (2001)[70]
	HLA genes	Protective; Permissive	See text
Osteonecrosis	*MTHFR*/ C677T	Permissive, but questionable	Kutlar F, Tural C, Park D *et al.* (1998), Zimmerman SA and Ware RE (1998)[57,58,71,72]
Acute chest syndrome	*NOS3*/ T-786C	Permissive	Sharan K, Surrey S, Ballas S *et al.* (2004)[61]
	NOS1/AAT repeats	Permissive	Sullivan KJ, Kissoon N, Duckworth LJ *et al.* (2001)[62]
Cholelithiasis	*UGT1A*/promoter repeats	7/7 Permissive	Passon RG, Howard TA, Zimmerman SA *et al.* (2001) Fertrin KY, Melo MB, Assis AM *et al.* (2003)[64,65]
Priapism	*KL*	Permissive	Nolan VG, Baldwin C, Ma Q *et al.* (2005)[56]

Genetic polymorphisms affecting the sub-phenotypes of sickle cell anemia (NCBI gene abbreviations are used). Only genetic variants with a protective or permissive association with a phenotype and that have been published as complete papers are listed.
*Indicates the nucleotide or amino acid substituted.

separate pathogenetic mechanisms, a continuum must exist between these two stroke sub-types. As in all the candidate gene association studies in SCD, these findings need confirmation in independent studies.

To examine the interaction among genes and their variant SNPs, and to develop a prognostic model for stroke in SCD-SS, a Bayesian network was developed to analyze 235 SNPs in 80 candidate genes in 1398 unrelated subjects. SNPs on 11 genes and four clinical variables, including α-thalassemia and HbF, interacted in a complex network of dependency to modulate the risk of stroke. This network of interactions included three genes (*BMP6, TGFBR2, TGFBR3)* with a functional role in the TGF-β pathway and one gene (*SELP*) associated with stroke in the general population. The model was validated in a different population by predicting the occurrence of stroke in 114 unrelated individuals with 98.2% accuracy: the correct outcome for all seven stroke subjects, and for 105 of 107 non-stroke subjects. This gave a 100% true positive rate, 98.14% true negative rate, and an overall predictive accuracy of 98.2%.[55]

As traditional analytical methods are often inadequate for the discovery of the genetic basis of complex traits in large association studies, Bayesian networks are a promising approach. The predictive accuracy of this stroke model is a step toward the development of prognostic tests better able to identify patients at risk for stroke. The presence of genes known to be associated with stroke in the general population, such as genes in the TGF-β pathway and *SELP*, suggests that some genetic factors predisposing to stroke may be shared by both SCD-SS patients and stroke victims in general.

Priapism

SNPs in 44 candidate genes were examined for their association with priapism in 148 patients who had priapism compared with 529 controls who had not developed priapism. Polymorphisms in KLOTHO (*KL*) showed an association with priapism by genotypic and haplotype analyses.[56] *KL*, located on chromosome 13q12, encodes a membrane protein (Klotho) that regulates many vascular functions, including vascular endothelial growth factor expression and endothelial nitric oxide release. The results again need to be validated in an independent set of patients.

Osteonecrosis

Osteonecrosis (avascular necrosis of bone) can be found in nearly half of all adults with SCD-SS; its prevalence is highest in individuals with SCD-SS disease and α-thalassemia. Association studies which examined polymorphisms in *MTHFR*, platelet glycoprotein IIIa (*ITPB3*) and plasminogen activator inhibitor-1 (*PAI1*) in small numbers of patients were non-conclusive.[57,58]

In another study,[59] 442 HbSS subjects with osteonecrosis were compared with 455 controls. Individuals with osteonecrosis had a higher prevalence of α-thalassemia, and there was no difference in PCV or HbF levels. In the initial screen, three to five SNPS in 66 candidate genes were genotyped, and significant associations were observed with seven SNPS in seven genes (*BMP6, TGFBR2, TGFBR3, EDN1, ERG, KL, ECE1*). Additional SNPs were typed in all seven genes, and a significant association with many SNPs in *KL* and *BMP6* was found. SNPs in *ANXA2* were also typed because of a previous finding of association between this gene and stroke among patients with SCD (Fig. 1).

The three genes identified are important in bone metabolism. KL encodes a glycosyl hydrolase that participates in a negative regulatory network of the vitamin D endocrine system. Bone morphogenic proteins (BMP), including *BMP6*, are pleiotropic secreted proteins structurally related to transforming growth factor (TGF)-β and activins. It is involved in inflammatory processes and is important for bone formation. *ANXA2* is a member of the calcium-dependent phospholipid-binding protein family; it regulates cell growth and is involved in signal transduction pathways. It is also involved in osteoblast mineralization; lipid rafts containing annexin 2 appear to be important for alkaline phosphatase activity in bone and the neuronal response to hypoxia.

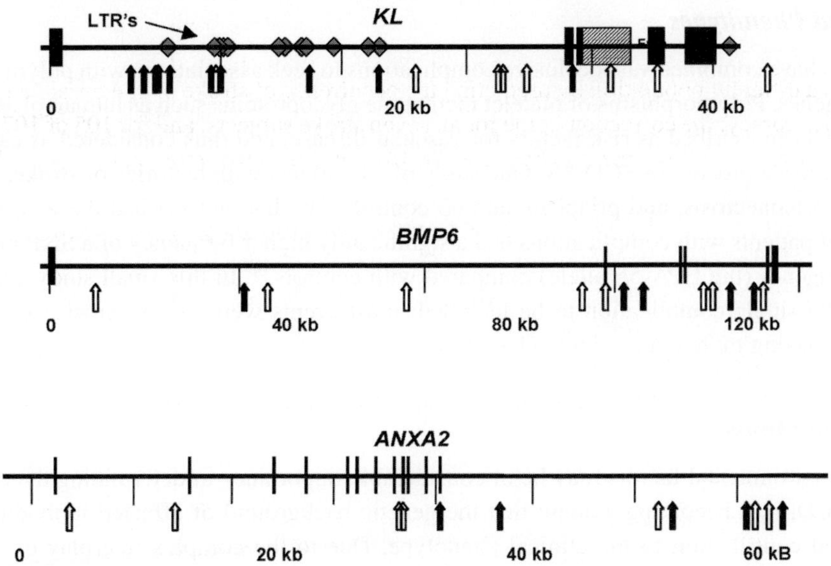

Fig. 1. Gene structure and location of SNPs in the *KL*, *BMP6* and *ANXA2* genes. Exons are indicated as solid boxes within the gene. The hashed box in *KL* shows the location of an alternative exon and stop codon. Solid arrows indicate the location of SNPs that are associated with osteonecrosis. Open arrows show the location of SNPs studied that were not associated with osteonecrosis. The gray diamonds indicate the location of retroviral LTRs (long terminal repeats) within the *KL* gene.

Pulmonary Disease

Pulmonary hypertension or pulmonary vascular disease has emerged as an important risk factor for premature death in patients with SCD-SS.[60] To date, data on the genetics of pulmonary hypertension in SCD-SS are limited.

Increased susceptibility to sickle acute chest syndrome in females only was associated with a T-786C SNP in the endothelial NO synthase gene (*NOS3*).[61] Exhaled NO levels were reduced in patients with acute chest syndrome compared with controls, and this was associated with the number of AAT repeats in intron 20 of NOS1.[62]

Gallstones

Promoter polymorphisms in the uridine diphosphate (UDP)-glucuronosyl-transferase 1A (*UGT1A*) gene are associated with unconjugated hyperbilirubinemia and Gilbert syndrome.[63] Children with SCD had a significantly higher mean bilirubin level if they carried the abnormal 7/7 *UGT1A* genotype compared with the 6/6 wild type or 6/7 UGT1A genotypes; patients with the 7/7 genotype were also more likely to have had a cholecystectomy. This suggested that symptomatic cholelithiasis is more common in carriers of this genotype.[64,65] Influence of the *UGT1A* promoter polymorphisms on bilirubin levels become more evident in SCD-SS patients while on hydroxyurea therapy. Children with wild type 6/6 *UGT1A* genotype demonstrated normalized bilirubin levels, but children with 6/7 or 7/7 genotypes did not.[66] Another study suggested that the 7/7 and 7/8 genotype were risk factors for symptomatic gallstones only in older subjects.[67]

Compound Phenotypes

Few studies have combined vasoocclusive complications to seek associations with polymorphisms in candidate genes. Polymorphisms of platelet membrane glycoproteins such as human platelet antigen (HPA) have been ascribed as risk factors for vascular disease, and thus considered as candidates in the vasoocclusive process in SCD-SS. One study of 34 patients with histories of stroke, acute chest syndrome, osteonecrosis, and priapism, and 63 controls who had not yet had these complications, showed that patients with complications had a significantly higher frequency of a SNP in the human platelet antigen-5 (the HPA-5b allele) compared with controls.[68] In this small study, an individual needed but a single complication to be included; most events were osteonecrosis, with only four individuals having more than a single phenotype.

Future Directions

Although environmental factors may be of considerable importance in determining the clinical outcome of SCD, it is becoming evident that the genetic background of affected individuals imparts a substantial contribution to the clinical phenotype. Due to the complex interplay of a multitude of factors, a genetic approach, parallel with functional genomics, might be the most efficacious way to identify these modifier genes. It is timely to take advantage of the opportunities provided by the completion of the human genome sequence, the availability of the millions of SNPs in the public domain, improvements in SNP typing technology, and the HapMap, to "fish" for these modulators.

Such genetic approaches require a large number (thousands) of well-characterized patients, and a critical mass of investigators with expertise in many fields — clinical research, genetics, molecular biology, bioinformatics, and quantitative statistics. These genome-wide approaches, however, are feasible only for well defined phenotypes. As well as relating association to clinical end points (for example, overt stroke) which are relatively rare events, intermediate end-points (such as imaging and physiological markers) that contribute to the phenotype of SCD-SS should also be documented and integrated with the association analyses. As with all association studies, identification of "case controls" is just as important as the cases.

The accumulation of a comprehensive panel of clinical and laboratory markers provide opportunities to use Baysian networks to predict occurrence of certain complications and early death, facilitating risk stratification in the management of SCD.

Summary

It is still not feasible to integrate the many clinical and laboratory abnormalities of SCD-SS to accurately predict the clinical outcome. Miller and colleagues[69] have used the presence of dactylitis in infants, leukocyte count and hemoglobin level to predict severe disease outcomes as defined by death, stroke, frequent pain, and acute chest syndrome in SCD-SS and SCD-S/β^{o} thalassemia.

Numerous association studies with candidate genes for the various complications (subphenotypes) of SCD-SS have been published, but the significant findings of these associations have not been replicated. Critics argue that our current knowledge of the pathophysiology of disease is not sufficient to accurately predict the functional candidate genes and variants. Proponents of the genome-wide approach argue that not only will genome-wide association studies be more cost-effective in the longer term, but the "hypothesis-free" approach could well enlighten us as to the key players controlling the disease-related quantitative traits and the molecular mechanisms underlying the pathophysiology.

References

1. Embury SH, Hebbel RP, Mohandas N, Steinberg MH, eds. (1996). *Sickle Cell Disease: Basic Principles and Clinical Practice*. Philadelphia, NY: Lippincott-Raven Publishers.
2. Amin BR, Bauersachs RM, Meiselman HJ, Mohandas N, Hebbel RP, Bowen PE, Schlegel RA, Williamson P, Westerman MP (1991). Monozygotic twins with sickle cell anemia and discordant clinical courses: clinical and laboratory studies. *Hemoglobin* **15**:247–256.
3. Joishy SK, Griner PF, Rowley PT (1976). Sickle β-thalassemia: identical twins differing in severity implicate nongenetic factors influencing course. *Am J Hematol* **1**:23–33.
4. Garner C, Tatu T, Reittie JE, Littlewood T, Darley J, Cervino S, Farrall M, Kelly P, Spector TD, Thein SL (2000). Genetic influences on F cells and other hematologic variables: a twin heritability study. *Blood* **95**:342–346.
5. Sebat J, Lakshmi B, Troge J, Alexander J, Young J, Lundin P, Maner S, Massa H, Walker M, Chi M, Navin N, Lucito R, Healy J, Hicks J, Ye K, Reiner A, Gilliam TC, Trask B, Patterson N, Zetterberg A, Wigler M (2004). Large-scale copy number polymorphism in the human genome. *Science* **305**:525–528.
6. Iafrate AJ, Feuk L, Rivera MN, Listewnik ML, Donahoe PK, Qi Y, Scherer SW, Lee C (2004). Detection of large-scale variation in the human genome. *Nat Genet* **36**:949–951.
7. Sachidanandam R, Weissman D, Schmidt SC, Kakol JM, Stein LD, Marth G, Sherry S, Mullikin JC, Mortimore BJ, Willey DL, Hunt SE, Cole CG, Coggill PC, Rice CM, Ning Z, Rogers J, Bentley DR, Kwok PY, Mardis ER, Yeh RT, Schultz B, Cook L, Davenport R, Dante M, Fulton L, Hillier L, Waterston RH, McPherson JD, Gilman B, Schaffner S, Van Etten WJ, Reich D, Higgins J, Daly MJ, Blumenstiel B, Baldwin J, Stange-Thomann N, Zody MC, Linton L, Lander ES, Altshuler D (2001). A map of human genome sequence variation containing 1.42 million single nucleotide polymorphisms. The International SNP Map Working Group. *Nature* **409**:928–933.
8. Venter JC, Adams MD, Myers EW *et al.* (2001). The sequence of the human genome. *Science* **291**: 1304–1351.
9. Lander ES, Linton LM, Birren B, Nusbaum C *et al.* (2001). Initial sequencing and analysis of the human genome. *Nature* **409**:860–921.
10. Syvanen AC (2005). Toward genome-wide SNP genotyping. *Nat Genet* **37**(Suppl):S5–10.
11. Collins FS, Morgan M, Patrinos A (2003). The Human Genome Project: lessons from large-scale biology. *Science* **300**:286–290.
12. Daly MJ, Rioux JD, Schaffner SF, Hudson TJ, Lander ES (2001). High-resolution haplotype structure in the human genome. *Nat Genet* **29**:229–232.
13. Gabriel SB, Schaffner SF, Nguyen H, Moore JM, Roy J, Blumenstiel B, Higgins J, DeFelice M, Lochner A, Faggart M, Liu-Cordero SN, Rotimi C, Adeyemo A, Cooper R, Ward R, Lander ES, Daly MJ, Altshuler D (2002). The structure of haplotype blocks in the human genome. *Science* **296**:2225–2229.
14. The International HapMap Project (2003). *Nature* **426**:789–796.
15. Reich DE, Cargill M, Bolk S, Ireland J, Sabeti PC, Richter DJ, Lavery T, Kouyoumjian R, Farhadian SF, Ward R, Lander ES (2001). Linkage disequilibrium in the human genome. *Nature* **411**:199–204.
16. Hirschhorn JN, Daly MJ (2005). Genome-wide association studies for common diseases and complex traits. *Nat Rev Genet* **6**:95–108.
17. Powars DR, Weiss JN, Chan LS, Schroeder WA (1984). Is there a threshold level of fetal hemoglobin that ameliorates morbidity in sickle cell anemia. *Blood* **63**:921–926.
18. Charache S, Terrin ML, Moore RD, Dover GJ, Barton FB, Eckert SV, McMahon RP, Bonds DR (1995). Effect of hydroxyurea on the frequency of painful crises in sickle cell anemia. Investigators of the multicenter study of hydroxyurea in sickle cell anemia. *N Engl J Med* **332**:1317–1322.
19. Perrine SP, Ginder GD, Faller DV, Dover GH, Ikuta T, Witkowska HE, Cai SP, Vichinsky EP, Olivieri NF (1993). A short-term trial of butyrate to stimulate fetal-globin-gene expression in the β globin disorders. *N Engl J Med* **328**:81–86.

20. Atweh GF, Sutton M, Nassif I, Boosalis V, Dover GJ, Wallenstein S, Wright E, McMahon L, Stamatoyannopoulos G, Faller DV, Perrine SP (1999). Sustained induction of fetal hemoglobin by pulse butyrate therapy in sickle cell disease. *Blood* **93**:1790–1797.
21. Steinberg MH, Voskaridou E, Kutlar A, Loukopoulos D, Koshy M, Ballas SK, Castro O, Barton F (2003). Concordant fetal hemoglobin response to hydroxyurea in siblings with sickle cell disease. *Am J Hematol* **72**:121–126.
22. Blobel GA, Weiss MJ (2001). Nuclear factors that regulate erythropoeisis. In: Steinberg MH, Forget BG, Higgs DR, Nagel RL, editors. *Disorders of Hemoglobin: Genetics, Pathophysiology, and Clinical Management*. Cambridge UK: Cambridge University Press, pp. 72–94.
23. Wood WG (1993). Increased HbF in adult life. In: Higgs DR, Weatherall DJ, editors. *Baillière's Clinical Haematology: International Practice and Research*. London: Baillière Tindall, W.B. Saunders, Vol. 6, pp. 177–213.
24. Gilman JG, Huisman THJ (1985). DNA sequence variation associated with elevated fetal $^{G}\gamma$ globin production. *Blood* **66**:783–787.
25. Garner C, Tatu T, Game L *et al.* (2000). A candidate gene study of F cell levels in sibling pairs using a joint linkage and association analysis. *Gene Screen* **1**:9–14.
26. Wood WG (2001). Hereditary persistence of fetal hemoglobin and delta beta thalassemia. In: Steinberg MH, Forget BG, Higgs DR, Nagel RL, editors. *Disorders of Hemoglobin: Genetics, Pathophysiology, and Clinical Management*. Cambridge, UK: Cambridge University Press, pp. 356–388.
27. Thein SL, Hesketh C, Wallace RB, Weatherall DJ (1988). The molecular basis of thalassaemia major and thalassaemia intermedia in Asian Indians: application to prenatal diagnosis. *Br J Haematol* **70**:225–231.
28. Labie D, Pagnier J, Lapoumeroulie C, Rouabhi F, Dunda-Belkhodja O, Chardin P, Beldjord C, Wajcman H, Fabry ME, Nagel RL (1985). Common haplotype dependency of high G gamma-globin gene expression and high HbF levels in beta-thalassemia and sickle cell anemia patients. *Proc Natl Acad Sci USA* **82**:2111–2114.
29. Nagel RL, Fabry ME, Pagnier J, Zohoun I, Wajcman H, Baudin V, Labie D (1985). Hematologically and genetically distinct forms of sickle cell anemia in Africa. The Senegal type and the Benin type. *N Engl J Med* **312**:880–884.
30. Steinberg MH, Hsu H, Nagel RL, Milner PF, Adams JG, Benjamin L, Fryd S, Gillette P, Gilman J, Josifovska O *et al.* (1995). Gender and haplotype effects upon hematological manifestations of adult sickle cell anemia. *Am J Hematol* **48**:175–181.
31. Kulozik AE, Kar BC, Satapathy RK, Serjeant BE, Serjeant GR, Weatherall DJ (1987). Fetal hemoglobin levels and β^{S} globin haplotypes in an Indian population with sickle cell disease. *Blood* **69**:1742–1746.
32. Dover GJ, Smith KD, Chang YC, Milner PF, Adams JG, Benjamin L, Fryd S, Gillette P, Gilman J, Josifovska O *et al.* (1992). Fetal hemoglobin levels in sickle cell disease and normal individuals are partially controlled by an X-linked gene located at Xp22.2. *Blood* **80**:816–824.
33. Thein SL, Sampietro M, Rohde K, Rochette J, Weatherall DJ, Lathrop GM, Demenais F (1994). Detection of a major gene for heterocellular hereditary persistence of fetal hemoglobin after accounting for genetic modifiers. *Am J Hum Genet* **54**:214–228.
34. Craig JE, Rochette J, Fisher CA, Weatherall DJ, Marc S, Lathrop GM, Demenais F, Thein SL (1996). Dissecting the loci controlling fetal haemoglobin production on chromosomes 11p and 6q by the regressive approach. *Nat Genet* **12**:58–64.
35. Close J, Game L, Clark BE, Bergounioux J, Gerovassili A, Thein SL (2004). Genome annotation of a 1.5 Mb region of human chromosome 6q23 encompassing a quantitative trait locus for fetal hemoglobin expression in adults. *BMC Genom* **5**:33.
36. Garner C, Tatu T, Best S, Creary L, Thein SL (2002). Evidence for genetic interaction between the beta-globin complex and chromosome 8q in the expression of fetal hemoglobin. *Am J Hum Genet* **70**:793–799.
37. Garner C, Silver N, Best S, Menzel S, Martin C, Spector TD, Thein SL (2004). Quantitative trait locus on chromosome 8q influences the switch from fetal to adult hemoglobin. *Blood* **104**:2184–2186.
38. Garner C, Menzel S, Martin C, Silver N, Best S, Spector TD, Thein SL (2005). Interaction between two quantitative trait loci affects fetal hemoglobin expression. *Ann Hum Genet* **69**:707–714.
39. Wyszynski DF, Baldwin CT, Cleves MA, Amirault Y, Nolan VG, Farrell JJ, Bisbee A, Kutlar A, Farrer LA, Steinberg MH (2004). Polymorphisms near a chromosome 6q QTL area are associated with modulation of fetal hemoglobin levels in sickle cell anemia. *Cell Mol Biol (Noisy-le-grand)* **50**:23–33.

40. Craig JE, Rochette J, Sampietro M, Wilkie AO, Barnetson R, Hatton CS, Demenais F, Thein SL (1997). Genetic heterogeneity in heterocellular hereditary persistence of fetal hemoglobin. *Blood* **90**:428–434.
41. Nagel RL, Steinberg MH (2001). Genetics of the β^S gene: Origins, genetic epidemiology, and epistasis in sickle cell anemia. In: Steinberg MH, Forget BG, Higgs DR, Nagel RL, editors. *Disorders of Hemoglobin: Genetics, Pathophysiology, and Clinical Management*. Cambridge, UK: Cambridge University Press, pp. 711–755.
42. Powars DR (1991). Sickle cell anemia: beta S-gene-cluster haplotypes as prognostic indicators of vital organ failure. *Semin Hematol* **28**:202–208.
43. Steinberg MH, Embury SH (1986). Alpha-thalassemia in blacks: genetic and clinical aspects and interactions with the sickle hemoglobin gene. *Blood* **68**:985–990.
44. Embury SH, Dozy AM, Miller J, Davis JR Jr, Kleman KM, Preisler H, Vichinsky E, Lande WN, Lubin BH, Kan YW, Mentzer WC (1982). Concurrent sickle-cell anemia and alpha-thalassemia: effect on severity of anemia. *N Engl J Med* **306**:270–274.
45. Steinberg MH, Rosenstock W, Coleman MB *et al.* (1984). Effects of thalassemia and microcytosis on the hematologic and vasoocclusive severity of sickle cell anemia. *Blood* **63**:1353–1360.
46. Thomas PW, Higgs DR, Serjeant GR (1997). Benign clinical course in homozygous sickle cell disease: a search for predictors. *J Clin Epidemiol* **50**:121–126.
47. Alexander N, Higgs D, Dover G, Serjeant GR (2004). Are there clinical phenotypes of homozygous sickle cell disease? *Br J Haematol* **126**:606–611.
48. Platt OS, Thorington BD, Brambilla DJ, Milner PF, Rosse WF, Vichinsky E, Kinney TR (1991). Pain in sickle cell disease: rates and risk factor. *N Engl J Med* **325**:11–16.
49. Russell MO, Goldberg HI, Hodson A, Kim HC, Halus J, Reivich M, Schwartz E (1984). Effect of transfusion therapy on arteriographic abnormalities and on recurrence of stroke in sickle cell disease. *Blood* **63**:162–169.
50. Driscoll MC, Hurlet A, Styles L, McKie V, Files B, Olivieri N, Pegelow C, Berman B, Drachtman R, Patel K, Brambilla D (2003). Stroke risk in siblings with sickle cell anemia. *Blood* **101**:2401–2404.
51. Taylor JG, Tang DC, Savage SA, Leitman SF, Heller SI, Serjeant GR, Rodgers GP, Chanock SJ (2002). Variants in the VCAM1 gene and risk for symptomatic stroke in sickle cell disease. *Blood* **100**:4303–4309.
52. Hoppe C, Klitz W, Cheng S, Apple R, Steiner L, Robles L, Girard T, Vichinsky E, Styles L (2004). Gene interactions and stroke risk in children with sickle cell anemia. *Blood* **103**:2391–2396.
53. Styles L, Hoppe C, Klitz W, Vichinsky E, Lubin B, Trachtenberg E (2000). Evidence for HLA-related susceptibility for stroke in children with sickle cell disease. *Blood* **95**:3562–3567.
54. Hoppe C, Klitz W, Noble J, Vigil L, Vichinsky E, Styles L (2003). Distinct HLA associations by stroke subtype in children with sickle cell anemia. *Blood* **101**:2865–2869.
55. Sebastiani P, Ramoni MF, Nolan V, Baldwin CT, Steinberg MH (2005). Genetic dissection and prognostic modeling of overt stroke in sickle cell anemia. *Nat Genet* **37**:435–440.
56. Nolan VG, Baldwin C, Ma Q, Wyszynski DF, Amirault Y, Farrell JJ, Bisbee A, Embury SH, Farrer LA, Steinberg MH (2005). Association of single nucleotide polymorphisms in klotho with priapism in sickle cell anaemia. *Br J Haematol* **128**:266–272.
57. Kutlar F, Tural C, Park D, Markowitz RB, Woods KF, Kutlar A (1998). MTHFR (5,10 methylenetetrahydrofolate reductase) 677 → T mutation as a candidate risk factor for avascular necrosis (AVN) in patients with sickle cell disease. *Blood* **92**:695a.
58. Zimmerman SA, Ware RE (1998). Inherited DNA mutations contributing to thrombotic complications in patients with sickle cell disease. *Am J Hematol* **59**:267–272.
59. Baldwin C, Nolan VG, Wyszynski DF, Ma QL, Sebastiani P, Embury SH, Bisbee A, Farrell J, Farrer L, Steinberg MH (2005). Association of klotho, bone morphogenic protein 6, and annexin A2 polymorphisms with sickle cell osteonecrosis. *Blood* **106**:372–375.
60. Gladwin MT, Sachdev V, Jison ML, Shizukuda Y, Plehn JF, Minter K, Brown B, Coles WA, Nichols JS, Ernst I, Hunter LA, Blackwelder WC, Schechter AN, Rodgers GP, Castro O, Ognibene FP (2004). Pulmonary hypertension as a risk factor for death in patients with sickle cell disease. *N Engl J Med* **350**:886–895.
61. Sharan K, Surrey S, Ballas S, Borowski M, Devoto M, Wang KF, Sandler E, Keller M (2004). Association of T-786C eNOS gene polymorphism with increased susceptibility to acute chest syndrome in females with sickle cell disease. *Br J Haematol* **124**:240–243.

62. Sullivan KJ, Kissoon N, Duckworth LJ, Sandler E, Freeman B, Bayne E, Sylvester JE, Lima JJ (2001). Low exhaled nitric oxide and a polymorphism in the NOS I gene is associated with acute chest syndrome. *Am J Respir Crit Care Med* **164**:2186–2190.
63. Bosma PJ, Chowdhury JR, Bakker C, Gantla S, de Boer A, Oostra BA, Lindhout D, Tytgat GN, Jansen PL, Oude Elferink RP *et al.* (1995). The genetic basis of the reduced expression of bilirubin UDP-glucuronosyltransferase 1 in Gilbert's syndrome. *N Engl J Med* **333**:1171–1175.
64. Passon RG, Howard TA, Zimmerman SA, Schultz WH, Ware RE (2001). Influence of bilirubin uridine diphosphate-glucuronosyltransferase 1A promoter polymorphisms on serum bilirubin levels and cholelithiasis in children with sickle cell anemia. *Am J Pediatr Hematol Oncol* **23**:448–451.
65. Fertrin KY, Melo MB, Assis AM, Saad ST, Costa FF (2003). UDP-glucuronosyltransferase 1 gene promoter polymorphism is associated with increased serum bilirubin levels and cholecystectomy in patients with sickle cell anemia. *Clin Genet* **64**:160–162.
66. Heeney MM, Howard TA, Zimmerman SA, Ware RE (2003). UGT1A promoter polymorphisms influence bilirubin response to hydroxyurea therapy in sickle cell anemia. *J Lab Clin Med* **141**:279–282.
67. Haverfield EV, McKenzie CA, Forrester T, Bouzekri N, Harding R, Serjeant G, Walker T, Peto TE, Ward R, Weatherall DJ (2004). UGT1A1 variation and gallstone formation in sickle cell disease. *Blood* **105**:968–972.
68. Castro V, Alberto FL, Costa RN, Lepikson-Neto J, Gualandro SF, Figueiredo MS, Annichino-Bizzacchi JM, Saad ST, Costa FF (2004). Polymorphism of the human platelet antigen-5 system is a risk factor for occlusive vascular complications in patients with sickle cell anemia. *Vox Sang* **87**:118–123.
69. Miller ST, Sleeper LA, Pegelow CH, Enos LE, Wang WC, Weiner SJ, Wethers DL, Smith J, Kinney TR (2000). Prediction of adverse outcomes in children with sickle cell disease. *N Engl J Med* **342**:83–89.
70. Tang DC, Prauner R, Liu W, Kim KH, Hirsch RP, Driscoll MC, Rodgers GP (2001). Polymorphisms within the angiotensinogen gene (GT-repeat) and the risk of stroke in pediatric patients with sickle cell disease: a case-control study. *Am J Hematol* **68**:164–169.
71. De Castro L, Rinder HM, Howe JG, Smith BR (1998). Thrombophilic genotypes do not adversely affect the course of sickle cell disease (SCD). *Blood* **92**:161a.
72. Zimmerman SA, Howard TA, Whorton MR, Rosse WF, Ware RE (1998). The A312G polymorphism in alpha-fibrinogen is associated with stroke and avascular necrosis in patients with sickle cell anemia. *Blood* **92**:36b.

13

Molecular Framework of Hemoglobin Switching

by Steven Fiering

Introduction

The assumption that detailed molecular information regarding both normal and pathological biology will provide productive new approaches clinically is a central tenet of biomedical research. This assumption has proven to be correct in a variety of pathological situations. It is clear that γ-globin expression will ameliorate the symptoms of sickle cell disease (SCD) and therefore, understanding the molecular basis of γ-globin expression prenatally and γ silencing postnatally has long been a goal among those seeking a cure for SCD. Despite effort by many researchers over many years and considerable progress, the detailed molecular mechanisms of hemoglobin switching in general and γ-globin regulation in particular are not fully elucidated. Once available, this information will provide molecular targets that can be manipulated pharmacologically to activate γ-globin expression and manifest a cure for SCD. Recent progress has been significant and new techniques that are being applied to the problem will fill in the molecular details and provide a basis for new therapeutic approaches.

Developmental Patterns of β-like Globin Gene Expression

The α-globin and β-globin loci contain multiple related genes that are expressed differentially during development and are assembled into functional hemoglobin molecules. The developmental changes in globin gene expression are collectively called hemoglobin switching. The switch from γ-globin expression to β-globin expression is relevant for discovering new treatment strategies for SCD.

The β-globin switch is regulated transcriptionally, therefore understanding the molecular mechanisms of switching is a problem in transcriptional biology. However, switching also incorporates cell biology processes since erythroblasts derived from the three sites of erythropoiesis during development, each express sequentially different patterns of β-like and α-like globin genes. Shifts in the site of erythropoiesis and the type of globins expressed during development are found in all vertebrates.[1] In humans the ε-gene is expressed in primitive erythroid cells that develop in the embryonic yolk sac

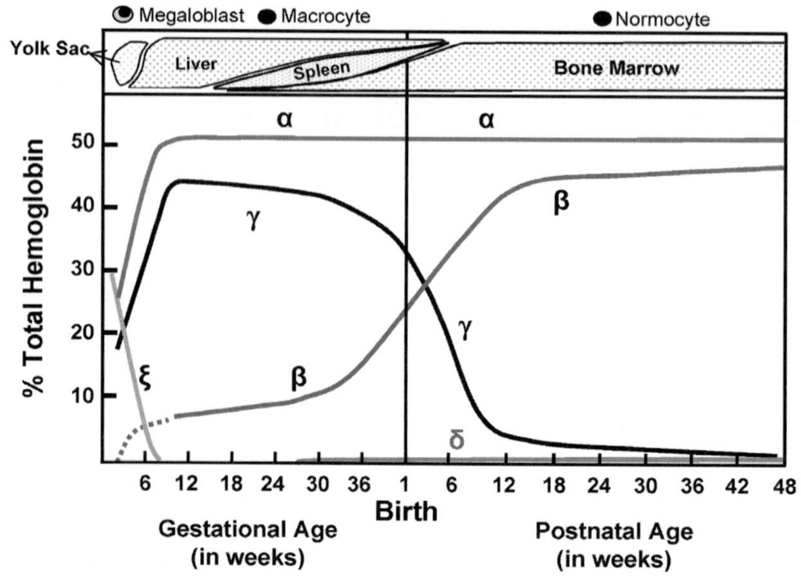

Fig. 1. Changes in globin chain production, sites of hematopoiesis, red cell morphology, and size of erythrocytes during the course of development. Used with permission from Wood WG (1976) *Br Med Bull* **32**:282–287.

up to about 6 weeks of gestation. The site of erythropoiesis shifts to the fetal liver during fetal stage development from 6 weeks to birth, where definitive erythroid cells expressing predominantly Gγ and Aγ β-like globin chains are produced. As the bone marrow forms late in prenatal development, erythropoiesis moves to the bone marrow where β-globin is expressed at high levels and δ-globin and γ-globin chains are produced at low levels in most people postnatally (Fig. 1).

Although the expressed hemoglobin genes are the most obvious molecular difference between erythrocytes produced during different stages of erythropoiesis, there are other variable characteristics. Primitive erythrocytes are nucleated and are the largest red blood cells (RBCs). Fetal liver and bone marrow-derived definitive erythrocytes enucleate, however fetal liver-derived erythrocytes are 50% larger than those produced in bone marrow. In addition to globin genes, there are other genes known to be expressed differentially in fetal liver and bone marrow-derived erythrocytes.[2]

The most relevant aspect of the switch for SCD is γ-globin gene silencing in bone marrow-derived erythrocytes. Virtually all sickle cell patients carry normal Gγ and Aγ genes. After birth, γ-globin expression decreases with a concomitant drop in HbF (hemoglobin containing γ and α) levels to 1–3% of total hemoglobin during the first year of life.[3] This expression is not spread evenly over all erythrocytes but is generally concentrated in 3–7% of RBCs known as F-cells, where HbF levels remain high compared to other erythrocytes. In a benign condition known as hereditary persistence of fetal hemoglobin (HPFH) the amount of γ-globin expressed after birth remains high. Co-inheritance of HPFH in association with the β^S mutation demonstrated HbF to be a powerful anti-sickling molecule and when HbF is greater than 20% of total hemoglobin, the symptoms of SCD are greatly ameliorated.[4] Therefore, to develop novel treatment strategies for SCD we must answer the key question, how is γ-globin silenced after birth? How close are we to understanding that mechanism? Significant progress has been made, moving us closer to an answer but we are not there yet, as outlined in detail below.

Two ideas underlie most mechanistic thinking about the control of hemoglobin switching. First, the binding of stage-specific and/or ubiquitous activating or silencing transcription factors are required to achieve gene expression or repression. The second explanation that *is not* mutually exclusive with the first is that expression of one β-like globin gene suppresses expression of the others in the locus. The most relevant hypothesis is that expression of embryonic and fetal globin genes somehow suppress expression of postnatal genes during embryonic and fetal development. However when the early genes are silenced, adult stage genes are activated since suppression by fetal-stage genes is eliminated. Current thinking regarding γ-globin is that its silencing is intrinsic to the proximal *cis* elements in the promoter regions. This implies that either there is a specific protein complex that silences γ-globin or the loss of γ-globin expression is due to absence of a specific protein(s) in bone marrow-derived erythroid progenitors that is needed to activate γ-globin to high levels. Determining which of these possibilities is relevant is a central question to be answered and it could be that both possibilities are mechanistically involved. It is likely that multiple mechanisms are involved in γ-globin regulation after birth. A variety of pathologies, clinical treatments or genetic changes can alter HbF levels.[1] However, generating HbF levels that approach those of HbA (hemoglobin containing β and α) after birth is rare; it is very difficult to generate "full" γ-gene expression, which likely will involve several modifications to overcome what is possibly a multifaceted mechanism of γ-gene silencing.

One potential mechanism (which mirrors the early genes suppress later genes expectation) is that expression of β-globin and/or δ-globin suppresses γ-globin expression. This concept has been supported by transgenic mouse studies from the laboratories of Frank Grosveld and George Stamatoyannopoulos[5,6] and human studies (reviewed in Ref. 7). The prevailing mechanistic idea is that the β-like genes compete for interaction with a large enhancer element, the locus control region (LCR), located near the ε-globin gene. When one gene is interacting with the LCR, the others are silenced. However, although the LCR-competition model is supported by considerable correlative data, it is still unproven.

Model Systems for Study of Hemoglobin Switching

The most commonly used systems for studying the molecular mechanisms of the γ- to β-globin switch are erythroid cell lines, transgenic mice, primates whose γ-gene is regulated similarly to humans, *ex vivo* erythroid progenitor differentiation, normal human primary erythroid progenitors and humans with mutations causing abnormal globin gene expression. Each system has strengths and weaknesses. In general, the further the system is from primary human cells, the easier it is to manipulate, but also the less relevant the derived information is to normal switching. Much work has been completed in erythroid cell lines because they are easy to manipulate and have a fixed pattern of globin gene expression. If that pattern can be altered by experimental approaches then it is likely that the approach is worthy of further investigation.

Transgenic mice have been heavily utilized for genetic studies of *cis* elements and gene arrangements that influence the β-globin switch. Mice with human β-globin locus transgenes that are either plasmid-based or derived from yeast or bacterial artificial chromosomes,[8–11] and mutation of the murine β-like globin locus using homologous recombination[12–15] have been used to elucidate mechanisms of switching. The obvious strengths of mouse transgenic systems are that *cis* elements are studied in animals during normal development. In addition, mice with mutations of potentially relevant genes can be studied alone or in combination with globin transgenes to evaluate the role of the protein product of the mutant gene in globin regulation. The fact that mice do not have the three-stage switch that higher primates have limits the relevance of transgenic mouse data, but the system has been productive and will continue to be heavily relied upon.

Some non-human primates such as baboons are informative since they have fetal liver-specific γ-gene expression, like humans and other higher primates. The baboon model is expensive, difficult to work with and cannot be genetically manipulated, however baboons can be treated with a variety of drugs to test γ-globin inducibility after birth.[16] The system has been very useful to test a variety of pharmacological HbF inducers and related hypotheses but baboons generally respond to a specific treatment with much stronger γ induction than similarly treated humans.

Another highly productive system has been culturing human erythroid progenitors that form erythroblasts when cultured *ex vivo*.[17] Importantly, this system can be manipulated either pharmacologically or genetically. Recent development of efficient retroviral vectors for forced gene expression and RNAi approaches for gene inhibition make *ex vivo* culture studies more informative. However, the erythroid progenitors differentiate in an artificial environment that impacts the value of derived information. Uncultured primary erythroblasts from humans are available to a limited extent and with these cells, specific aspects of chromatin structure or gene expression can be assayed with minimal manipulation. This provides a very dependable system for questions that are accessible with the type and numbers of cells available.

Due to selection pressure from malaria, mutations in globin expression or globin proteins occur commonly in humans. Human mutations in globin genes have been a rich research resource, despite the limited experimentation that can be done with humans. These mutations provide insights into *cis* elements that influence the β-locus and cause HPFH. Many of the most important advances, such as the discovery of the LCR can be traced to studies of mutant human loci.[18] While there is certainly more to be learned from naturally occurring human mutations, most have been characterized to some extent. Undoubtedly there will be new mutations identified but since the common mutations have been identified and characterized, the pace of such discovery has slowed markedly. A related area of research that likely will be important in the future is the effort to identify *trans* loci that are linked either to HPFH or to variations in clinical phenotypes in sickle cell disease.[19,20]

The LCR and Its Role in Hemoglobin Switching

The LCR is a region of clustered regulatory binding sites for transcription factors spread over 15 kb upstream of the ε gene (for review see Ref. 1). It is a powerful positive regulator of the globin genes, and dependably stimulates expression of linked transgenes in erythroblasts. The functional LCR regions have high interspecies homology and are recognized by erythroid-specific DNaseI hypersensitive sites (HS) in chromatin, of which there are five major HS in the human LCR.

The "looping model" is the most widely accepted potential mechanism for LCR interactions with globin genes. This model postulates physical contact between the promoter of target globin genes and *trans*-factor(s) bound to the LCR with looping out of the intervening DNA. Studies showing that the HSs are in closer physical proximity to the active promoters than to inactive promoters of intervening genes support the existence of this loop.[21,22] Most other studies cited to support the looping model also support other models for initiating LCR/globin gene interaction, such as protein tracking along the chromatin from the LCR to the globin gene promoters.

It has been proposed that the HSs of the LCR are organized into a three dimensional protein/DNA holocomplex.[23] The best data supporting this hypothesis is that deletion of a single HS sensitizes transgenes with the LCR to position effects in which there is little or no globin gene expression and no formation of the remaining HS, demonstrating synergism among the HS.[23–26] Data arguing against existence of a 3D holocomplex comes from studies in which the individual HS of the endogenous murine LCR are deleted.[12–14] In these animals the remaining HS form normally[27] and the reduction in gene expression is additive, rather than synergistic as it appears to be in the human β-locus. This implies that, in the endogenous murine locus, either the hypersensitive sites do not work together or they do not form a specific complex required for functional activity as proposed in humans.

Without the LCR, the globin genes undergo switching at the proper time but are expressed at very low levels. This is true in mice with human transgenes[28] or in the endogenous mouse locus,[29,30] demonstrating that the information required for normal switching is contained in other *cis* elements, such as the proximal promoter. Finding that the globin genes "switch" without the LCR implies that the LCR does not initiate the switch but its function enhances output from already active promoters. The low γ-globin level observed in humans by one year of age mirrors expression observed with globin genes lacking an LCR, supporting the notion that β-globin expression produces γ-globin suppression because β-globin interacts with the LCR and therefore γ-globin cannot. Based on this mechanism, HPFH mutations would cause high HbF levels through increased physical interactions between the γ-gene promoter(s) and the LCR. This could be through altered *trans*-factor binding at the promoters that stabilize or destabilize the LCR-promoter complex. One interpretation of published data is a model in which globin gene activity is primary and the LCR is recruited to the promoter by the active promoter and then boosts gene expression. Experimental data also supports the complementary model where the LCR activates globin genes as the primary event. The differences in the models are subtle but mechanistically relevant and it is still possible that physical proximity of the LCR to promoters through looping is not the key variable that many researchers in the hemoglobin switching field assume it to be.

A recent paper calls into question the relevance of data for physical proximity of promoters and the LCR from another perspective. Osborne *et al.*[31] showed that not only is the LCR in closer physical proximity to expressed globin genes than to silent β-like genes, but an unrelated expressed gene quite distant on the same chromosome is also occasionally in physical proximity to the LCR. This suggests that transcriptionally active genes and the LCR are clustered in "transcription factories" where multiple genes are transcribed. Perhaps the closer physical proximity of the LCR to active β-like gene promoters is because both the active promoters and the LCR are transcribed by RNA polII,[32] whereas the inactive genes are not transcribed. This raises the possibility that physical proximity between the LCR and active promoters is coincidental to being transcribed and not mechanistically relevant to regulating the active promoter.

The actual HSs are characterized not only by DNase hypersensitivity (thought to reflect an area devoid of histones) but also the presence of polII and modifications of histones such as acetylation that are also associated with actively transcribed genes.[33] The complement of *trans*-factors that bind to the LCR-HSs is thought to be very similar for each site, including GATA-1 and NF-E2.[34] Detailed studies of requirements for forming a HS suggest that all transcription factors are required to bind before a HS is observed (Lowrey, personal communication).

Studies utilizing the human locus as transgenes in mice have provided much of our current understanding of LCR function in supporting position independent and copy number dependent expression of linked transgenes. Without the LCR or with an impaired LCR globin transgenes are more susceptible to position effects and are frequently silenced and found in inactive chromatin.[24] These characteristics demonstrate the ability of the LCR to modulate chromatin at transgene integration sites and support models in which chromatin-opening is one of the main functions of the LCR.

The chromatin structure of the murine β-like globin locus has been examined in detail.[33,35] It shows a gradation of acetylated histones with moderate acetylation levels across the locus compared to hyperacetylated histones in the LCR hypersensitive sites and promoters of expressed genes. Similarly the general DNaseI sensitivity across the locus is high in erythroblasts, regardless of the developmental stage. These results suggest that the switch is not due to a major change in chromatin structure in association with silenced genes and that the LCR is mediator of the "active" chromatin structure that is characteristic of the whole locus. However, studies of the murine β-like globin locus using homologous recombination to delete the entire LCR unexpectedly show that the LCR is not required for appropriate chromatin structure across the locus.[29,30] This implies that unlike human transgenes, where lack of a functional LCR often leads to heterochromatin and lack of globin gene expression, the murine LCR is not necessary for generating the open chromatin structure of the endogenous murine locus.

The assumption that the LCR recruits RNA polII to active promoters must be further re-examined in light of the LCR-deletion mice, in which only a two fold reduction of RNA polII at the globin promoters was produced.[36] Instead it appears that the LCR is most important for permitting proper mRNA elongation since the presence of polII in the distal portions of the gene is reduced five fold. Despite good experimentation and some very attractive models, the mechanistic role of the LCR in globin regulation and switching is not yet understood. The more the LCR is studied, the more surprises are revealed and the more complex the previously simple models become.

Activation of γ-globin Expression after Birth

The research thrust to reverse γ-gene silencing to accomplish HbF induction as a viable approach for treating SCD has occurred for the following reasons: HbF has anti-sickling properties that inhibits HbS polymerization;[37] HbF is expressed at low levels in the absence of disease and therefore is not fully "silenced"; a variety of clinical conditions occur with high HbF levels[1] and a variety of drugs cause increases in γ-globin expression.[38] The difficulty is not getting three to four fold γ-gene activation routinely, this can be achieved, but this level does not have a significant beneficial effect. The challenge is getting therapeutic effects in SCD which are optimal with consistently higher than 20% HbF induction in a homogenous pattern in red blood cells.

Data support a temporary increase in γ-gene expression in clinical circumstances where the rate of erythropoiesis is increased such as during rapid erythroid expansion in response to acute anemia (reviewed in Ref. 1). Erythropoietin levels are increased during erythropoietic stress and exogenous administration is associated with increased HbF *in vivo*. However, clinically relevant increases in HbF do not occur during the chronic anemia of SCD or β-thalassemia unless associated with a specific HPFH mutation.

A variety of studies support high γ-globin gene expression during early stages of erythroid differentiation from erythroid progenitors, followed by increased adult β-globin expression as differentiation proceeds.[39] This may be the basis for the heterogeneous nature of HbF distribution in clinical diseases with F cells produced from prematurely committed precursors. However, a recent *ex vivo* study in which erythroid progenitors were cultured synchronously from $CD34^+$ precursors argues against the hypothesis that erythroblasts in the early stages of differentiation have high γ-gene expression.[40]

As noted above, one system for studying hemoglobin switching is *ex vivo* differentiation of erythroid precursors. Results vary with differentiation conditions but in general the levels of γ-globin expression are higher in progenitor cultures from donors producing high HbF in comparison to

γ-globin levels produced from low HbF donors. A variety of conditions influence HbF levels during *ex vivo* differentiation including the age of the donor and experimental variables such as the species source of the serum, or the experimental growth factors added in cultures (reviewed in Ref. 1). These findings support models in which there is a cell signaling component of switching regulation.

Erythroid Progenitor Commitment to γ or β Globin Expression

One current model is that cell–cell interactions in the tissue of origin programs progenitor cells to express either predominantly γ-globin or predominantly β-globin. Fetal liver-derived burst-forming unit-erythroid (BFU-E) colonies express predominantly γ-globin and bone marrow-derived BFU-E colonies express predominantly β-globin when grown *ex-vivo* in identical culture conditions, indicating that the committed precursors are already programmed to express either γ or β globin as predicted by the developmental stage and tissue of origin.[41] In addition, *ex-vivo* erythroid cultures derived from a single BFU-E progenitor showed that a single colony can produce cells with only HbA or a mixture of HbF and HbA. The pattern of HbF expression is "sectored" suggesting that as the precursor cells divide and expand there is a stochastic chance that HbF will be active in a cell and its descendants.[1] This implies that any progenitor programming affects probability of γ expression rather than fully determining expression. This concept is supported by the existence of sporadic F-cells.

Studies of γ-globin and β-globin chromatin structure in multipotent hematopoietic stem cells from human $CD34^{+}$ cells or murine progenitors that carry a human β-locus transgene have been completed.[42] Both γ and β globin gene transcription occurs at low levels in hematopoietic stem cells (HSC) from either fetal liver or bone marrow, implying that both genes are equally poised for expression as cells proceed down the erythroid lineage. However, histone modifications in HSCs from bone marrow indicate that the β-globin gene but not γ-globin gene has increased histone H3 acetylation which is generally associated with high level gene expression, suggesting that bone marrow stem cells are already programmed to preferentially express β.

Trans-factors Involved in Globin Switching

As with all eukaryotic genes, the transcriptional pattern of the β-like genes is primarily determined by the binding of transcription factors to the promoter region and to other regulatory elements that may be quite distant, like the LCR. Most of the factors have a consensus DNA sequence to which they bind. Transcription factors also bind to other transcription factors and recruit them to regulatory sites; they sometimes act to silence rather than activate transcription either directly or due to associated factors they recruit. Differential transcription in various tissues has long been associated with expression of transcription factors that are expressed only in specific tissues suggesting that the simple presence of tissue-specific trans-factors is sufficient to explain tissue-specific gene expression. However, it is clear that the presence of a factor in cells does not equate with binding of that factor to a regulatory site unless the chromatin at the site is appropriate. Perhaps the most difficult task in deciphering transcriptional regulation has been determining what factors are actually bound to a specific site *in vivo*. Recent progress in developing and utilizing chromatin immunoprecipitation (ChIP) assays promises to provide answers to the question of what factors are bound *in vivo*.

There has been considerable success in defining transcription factors involved in γ-globin and β-globin gene regulation, including potential activators or repressors. Some of these proteins have been shown to modify γ-gene expression levels in various cell lines or to delay the γ to β switch in transgenic mice, however none of the experimental manipulations to date are sufficient to mediate substantially higher levels of γ-gene expression permanently. Surprisingly, none of the factors that play a role in γ-gene regulation are specifically expressed in fetal liver or bone marrow erythroblasts. Many are erythroid-specific but none are developmental stage-specific. This implies that either a key erythroid/stage-specific factor exists that has not been isolated or the identified transcription factors influence γ-gene expression through: 1) modest increases or decreases in expression level, 2) posttranslational modification of other factors or 3) differential combinations with other proteins assembled into multi-protein complexes.

GATA-1 is an important regulator of erythroid genes and essential for erythroid development. Like other β-like globin genes, ε-globin transcription is regulated by GATA-1, which can activate or repress gene expression. Mutations of GATA-1 binding sites at -163 and -269 suppress ε-globin but mutating the GATA-1 site at -208 partially inhibits developmental silencing and there is further data suggesting that GATA-1 bound at -208 is associated with the known silencing protein, YY1.[43]

All of the β-like globin genes have a CACCC box in the promoter that plays an important role in gene regulation. Many members of the SP1 family have an affinity for the CACCC sequence and attention has been focused on identifying factors in this family that regulate globins such as erythroid Kruppel-like factor (EKLF).[44] EKLF is the one factor that is unequivocally associated with regulation of a particular β-like globin gene. Although EKLF is expressed at all three stages of erythropoiesis, it is only necessary for β-globin expression. When EKLF knockout mice carry a human β-globin locus transgene, expression of ε and γ is normal but β-globin is not expressed.[45]

Three KLF family factors are under investigation for potential roles in γ-gene regulation, fetal Kruppel-like factors 1 and 2 (FKLF1,[46] FKLF2[47]) and basic Kruppel-like factor (BKLF[48]). Data from *in vitro* transfection experiments supports a role for FKLF1 and FKLF2 in γ-globin activation and BKLF in γ-globin repression. Another protein complex that preferentially interacts with the CACCC boxes of ε and γ globins is the DRED complex[49] that is hypothesized to suppress both γ-genes after birth. This conclusion is supported by transgenic mouse studies using human β-locus yeast artificial chromosome constructs with mutations in the CACCC boxes.[50] Two orphan nuclear hormone receptors, TR2 and TR4 are contained in the DRED complex.[49] The stage selector protein (SSP) which consists of ubiquitously expressed CP2 and ALY and erythroid specific NF-E4 factors is a potential γ-globin regulatory complex as well.[51] This complex activates γ-gene expression in some experimental systems.

New Approaches

Techniques for molecular biology analysis are developing rapidly and being applied to the problem of globin gene transcriptional regulation. It is now possible to analyze transcription levels from all known genes in a tissue. If the problem is reduced to its essence, it can be stated that the key to the γ to β globin switch lies in the differences between fetal liver and bone marrow erythroblasts. One such difference is the differentially transcribed genes in erythroblasts from each site of erythropoiesis. Studies are underway to identify transcriptional differences between normal human fetal liver and bone marrow erythroblasts. Of course the key differences may not be in transcription but rather in protein complexes that form or acquire modifications through signaling pathways. Genomic scale proteomic studies in mammals will someday contribute to solving many biological problems but the technologies are still in their infancy.

When regulatory proteins are bound to chromatin, DNase hypersensitive sites are formed. Assay for these sites has been very productive in globin regulation studies, leading to the identification of the LCR. New approaches are being developed to assay changes in DNase hypersensitive sites across the entire genome.[52] When applied to human fetal liver and bone marrow erythroblasts, this could reveal important information about global changes in gene expression that could be involved in switching.

It is possible to study the binding of known factors to a specific genomic site through chromatin immunoprecipitation (ChIP). This has already been applied to histones and some transcription factors bound in the β-globin locus in mice. There is much to be done to produce a comprehensive comparison of histone modifications and regulatory protein binding patterns in the β-globin locus in human fetal liver versus bone marrow erythroblasts.

Although the β-globin locus was among the first large loci sequenced, the availability of the human genome sequence will have an impact on the study of globin gene regulation. Traditional gene mapping techniques used to identify loci that influence γ-gene silencing and/or reduce the severity of SCD are ongoing. The powerful tools provided by the human genome project will increase productivity of these studies and all other efforts to identify relevant genes in humans.

The availability of the human genome sequence, genomics databases and comparison tools will support new approaches which are often the key to unlocking a problem. Human embryonic stem cells can be differentiated into erythroid progenitors *in vitro* providing a new system for switching studies (reviewed in Ref. 53). Although not yet routine, these cells can be genetically modified which opens new avenues for genetic approaches using human cells.

The globin genes and their protein product hemoglobin were the first mammalian genes to have the techniques of molecular biology applied to them, yielding an immense store of information about gene regulation. Many of those working in this field 20 years ago expected that the switching mechanism would be understood in detail by now, however it is not. Most likely this has been a recalcitrant problem because there are multiple mechanisms that each contribute to achieve normal switching. Yet there remains a continued effort and enthusiasm to solve this problem since the potential for curing sickle cell disease by γ-gene reactivation is so tangible, yet remains elusive.

Acknowledgments

I would like to thank George Stamatoyannopolous and Chris Lowrey for critical reading of the manuscript.

References

1. Stamatoyannopoulos G, Grosveld F (2001). Hemoglobin switching. In: Stamatoyannopoulos G, Majerus PW, Perlmutter RM, Varmus H, editors. *The Molecular Basis of Blood Diseases*. Philadelphia, PA: Saunders, pp. 135–182.
2. Chen S-H, Anderson J, Giblett E, Stamatoyannopolous G (1977). Isozyme patterns in erythrocytes from human fetuses. *Am J Hematol* **2**:23.
3. Wood W, Stamatoyannopolous G, Lim G, Nute P (1975). F-cells in the adult: normal values and levels in individuals with hereditary and acquired elevations of HbF. *Blood* **46**:671.
4. Pembry M, Wood W, Weatherall D, Perrine R (1978). Fetal haemoglobin production and the sickle gene in the oases of eastern Saudi Arabia. *Br J Haematol* **40**:415.
5. Enver T, Raich N, Ebens AJ, Papayannopoulou T, Costantini F, Stamatoyannopoulos G (1990). Developmental regulation of human fetal-to-adult globin gene switching in transgenic mice. *Nature* **344**:309–313.
6. Hanscombe O, Whyatt D, Fraser P, Yannoutsos N, Greaves D, Dillon N, Grosveld F (1991). Importance of globin gene order for correct developmental expression. *Genes Dev* **5**(8):1387–1394.

7. Forget B (1998). Molecular basis of hereditary persistence of fetal hemoglobin. In: Cohen AR, editors. *Annals of the New York Academy of Sciences: Cooleys Anemia Seventh Symposium*, Vol. 850. New York, NY: New York Academy of Sciences, pp. 38–44.
8. Gaensler KM, Kitamura N, Kan YW (1993). Germ-line transmission and developmental regulation of a 150 kb yeast artificial chromosome containing the human β-globin locus in transgenic mice. *Proc Natl Acad Sci USA* **90**:11381–11385.
9. Peterson KR, Clegg CH, Navas PA, Norton EJ, Kimbrough TG, Stamatoyannopoulos G (1996). Effect of deletion of 5′HS3 or 5′HS2 of the human β-globin LCR on the developmental regulation of globin gene expression in β-YAC transgenic mice. *Proc Natl Acad Sci USA* **93**:6605–6609.
10. Alami R, Greally J, Tanimoto K, Hwang S, Feng Y, Engel J, Fiering S, Bouhassira E (2000). β-globin YAC transgenes exhibit uniform expression levels but position effect variegation in mice. *Hum Mol Genet* **9**:631–636.
11. Raich N, Enver T, Nakamoto B, Josephson B, Papayannopoulou T, Stamatoyannopoulos G (1990). Autonomous developmental control of human embryonic globin gene switching in transgenic mice. *Science* **250**:1147–1149.
12. Fiering S, Epner E, Robinson K, Zhuang Y, Telling A, Hu M, Martin DIK, Enver T, Ley T, Groudine M (1995). Targeted deletion of 5′HS2 of the murine β-globin locus reveals that it is not essential for proper regulation of the β-globin locus. *Genes Dev* **9**:2203–2213.
13. Hug B, Wesselschmidt RL, Fiering S, Bender MA, Groudine M, Ley TJ (1996). Analysis of mice containing a targeted deletion of the β-globin locus control region 5′HS-3. *Mol Cell Biol* **16**:2906–2912.
14. Bender M, Roach J, Halow J, Close J, Alami R, Bouhassira E, Groudine M, Fiering S (2001). Targeted deletion of 5′HS1 and 5′HS4 of the β-globin locus control region reveals the additive activity of the DNaseI hypersensitive sites. *Blood* **98**:2022–2027.
15. Hu X, Bulger M, Roach JN, Eszterhas S, Olivier E, Bouhassira EE, Groudine MT, Fiering S (2003). Promoters of the murine embryonic β-like globin genes Ey and βh1 do not compete for interaction with the β-globin locus control region. *Proc Natl Acad Sci USA* **100**:1111–1115.
16. Lavelle D, Molokie R, Ducksworth J, DeSimone J (2001). Effects of hydroxyurea, stem cell factor and erythropoietin in combination on fetal hemoglobin in the baboon. *Exp Hematol* **29**:156–162.
17. Fibach E, Manor D, Oppenheim A, Rachmilewitz E (1969). Proliferation and maturation of human erythroid progenitors in liquid culture. *Blood* **73**:100–103.
18. Curtin P, Pirastu M, Kan YW, Gobert-Jones JA, Stephens AD, Lehmann H (1985). A distant deletion affects β-globin gene function in an atypical gdβ-thalassemia. *J Clin Invest* **76**:1554–1558.
19. Wyszynski DF, Baldwin CT, Cleves MA, Amiranti Y, Nolvan VG, Farrell JJ, Bisbee A, Kullar A, Ferrer LA, Steinberg MH (2004). Polymorphisms near a chromsome 6 QTL area are associated with modulation of fetal hemoglobin in sickle cell anemia. *Cell Mol Biol* **50**:23–33.
20. Garner SC, Silver N, Best S, Menzel S, Martin C, Specter TD, Thein SL (2004). Quantitative trait locus on 8Q influences the switch from fetal to adult hemoglobin. *Blood* **104**:2184–2186.
21. Carter D, Chaklova L, Osborne C, Fraser P (2002). Long-range chromatin regulatory interactions in vivo. *Nat Genet* **32**:623–626.
22. Tolhuis B, Palstra R, Splinter E, Grosveld F, de Laat W (2002). Looping and interaction between hypersensitive sites in the active beta-globin locus. *Mol Cell Biol* **10**:1453–1465.
23. Bungert J, Dave U, Lim K-C, Lieuw KH, Shavit JA, Liu Q, Engel JD (1995). Synergistic regulation of human β-globin gene switching by locus control region elements HS3 and HS4. *Genes Dev* **9**:3083–3096.
24. Milot E, Stroubolis J, Trimborn T, Wijgerde M, de Boer E, Langeveld A, Tan-un K, Vergeer W, Yannoutsos N, Grosveld F, Fraser P (1996). Heterochromatin effects on the frequency and duration of LCR-mediated gene transcription. *Cell* **87**:105–114.
25. Navas P, Peterson K, Li Q, Skarpidi E, Rohde A, Shaw S, Clegg C, Asano H, Stamatoyannopoulos G (1998). Developmental specificity of the interaction between the locus control region and embryonic or fetal globin genes in transgenic mice with an HS3 core deletion. *Mol Cell Biol* **18**:4188–4196.
26. Li G, Lim K, Engel J, Bungert J (1998). Individual LCR hypersensitive sites cooperate to generate an open chromatin domain spanning the human β-globin locus. *Genes Cells* **3**:415–429.
27. Bender M, Mehaffey M, Telling A, Hug B, Ley T, Groudine M, Fiering S (2000). Independent formation of DNaseI hypersensitivity sites in the murine β-globin locus control region. *Blood* **95**:3600–3604.

28. Starck J, Sarkar R, Romana M, Bhargava A, Scarpa AL, Tanaka M, Chamberlain JW, Weissman SM, Forget BG (1994). Developmental regulation of human γ and β-globin genes in the absence of the locus control region. *Blood* **84**:1656–1665.
29. Epner E, Reik A, Cimbora D, Telling A, Bender MA, Fiering S, Enver T, Martin DIK, Kennedy M, Keller G, Groudine M (1998). The β-globin LCR is not necessary for an open chromatin structure or developmentally regulated transcription of the native mouse β-globin locus. *Mol Cell* **2**:447–455.
30. Bender M, Bulger M, Close J, Groudine M (2000). β-globin gene switching and DNaseI sensitivity of the endogenous β-globin locus in mice do not require the locus control region. *Mol Cell* **5**:387–393.
31. Osborne C, Chakalova L, Brown K, Carter D, Horton A, Debrand E, Goyenechea B, Mitchell J, Lopes S, Reik W, Fraser P (2004). Active genes dynamically colocalize to shared sites of ongoing transcription. *Nat Genet* **36**:1065–1071.
32. Johnson K, Grass J, Park C, Im H, Choi K, Bresnick E (2003). Highly restricted localization of RNA polymerase II within a locus control region of a tissue-specific chromatin domain. *Mol Cell Biol* **23**: 6484–6493.
33. Bulger M, Schubeler D, Bender MA, Hamilton J, Farrell CM, Hardison RC, Groudine M (2003). A complex chromatin landscape revealed by patterns of nuclease sensitivity and histone modification within the mouse β-globin locus. *Mol Cell Biol* **23**:5234–5244.
34. Stamatoyannopolous JA, Goodwin A, Joyce T, Lowrey CH (1995). NF-E2 and GATA binding motifs are required for the formation of DNaseI-hypersensitive site 4 of the human β-globin locus. *EMBO J* **14**: 106–116.
35. Forsberg E, Downs K, Christensen H, Im H, Nuzzi N, Bresnick E (2000). Developmentally dynamic histone acetylation pattern of a tissue-sepcific chromatin domain. *Proc Nat Acad Sci USA* **97**:14494–14499.
36. Sawado T, Halow J, Bender MA, Groudine M (2003). The β-globin locus control region (LCR) functions primarily by enhancing the transition from transcription initiation to elongation. *Genes Dev* **17**:1009–1018.
37. Poillon W, Kim B, Rodgers G, Noguchi C, Schechter A (1993). Sparing effect of hemoglobin F and hemoglobin A2 on the polymerization of hemoglobin S at physiologic ligand concentrations. *Proc Natl Acad Sci USA* **90**:5039–5043.
38. Atweh G, Schechter A (2001). Pharmocolgic induction of fetal hemoglobin: raising the therapeutic bar in sickle cell disease. *Curr Opin Hematol* **8**:123–130.
39. Farquhar M, Turner P, Papayannopoulou T, Brice M, Nienhuis AW, Stamatoyannopoulos G (1981). The asynchrony of gamma and beta synthesis during human erythroid cell maturation. *Dev Biol* **85**:403–409.
40. Wojda U, Noel P, Miller J (2002). Fetal and adult hemoglobin production during adult erythropoiesis: coordinate expression correlates with cell proliferation. *Blood* **99**:3005–3013.
41. Papayannopoulou T, Kurachi S, Nakamoto B, Zanjani E, Stamatoyannopoulos G (1982). Hemoglobin switching in culture: evidence for a humoral factor that induces switching in adult and neonatal but not fetal erythroid cells. *Proc Natl Acad Sci USA* **79**:6579.
42. Bottardi S, Aumont A, Grosveld F, Milot E (2003). Developmental stage-specific epigenetic control of human β-globin gene expression is potentiated in hematopoietic progenitor cells prior to their transcriptional activation. *Blood* **102**:3989–3997.
43. Raich N, Clegg CH, Grofti J, Romeo PH, Stamatoyannopoulos G (1995). GATA-1 and YY-1 are developmental repressors or the human epsilon-globin gene. *EMBO J* **14**:801.
44. Miller I, Bieker J (1993). A novel erythroid cell-specific murine transcription factor that binds to the CACCC element and is related to the Kruppel family of nuclear proteins. *Mol Cell Biol* **13**:2776–2786.
45. Perkins A, Gaensler K, Orkin S (1996). Silencing of human fetal globin expression is impaired in the absence of the adult beta-globin gene activator, EKLF. *Proc Natl Acad Sci USA* **93**:12267–12271.
46. Asano H, Li S, Stamatoyannopoulos G (1999). FKLF, a novel Kruppel-like factor that activates human embryonic and fetal β-like globin genes. *Mol Cell Biol* **19**:3571–3579.
47. Asano H, Li S, Stamatoyannopoulos G (2000). FKLF-2: a novel Kruppel-like transcription factor that activates globin and other erythroid lineage genes. *Blood* **95**:3578–3584.
48. Crossley M, Whitelaw E, Perkins A, Williams G, Fujiwara Y, Orkin SH (1996). Isolation and characterization of cDNA encoding BKLF/TEF-2, a major CACCC-box binding protein in erythroid and selected other cell-types. *Mol Cell Biol* **16**:1695–1705.

49. Tanabe O, Katsuoka F, Campbell A, Song W, Yamamoto M, Tanimoto K, Engel J (2002). An embryonic/fetal beta-type globin gene repressor contains nuclear receptor TR2/TR4 heterodimer. *EMBO J* **21**:3434–3442.
50. Tanimoto K, Liu Q, Grosveld F, Bungert J, Engel J (2000). Context-dependent EKLF responsiveness defines the developmental specificity of the human epsilon-globin gene in erythroid cells of YAC mice. *Genes Dev* **14**:2778–2794.
51. Zhou W, Clouston D, Wang X, Cerruti L, Cunningham J, Jane S (2000). Induction of human fetal globin gene expression by a novel erythroid factor, NF-E4. *Mol Cell Biol* **20**:7662–7672.
52. Sabo P, Humbert R, Hawrylycz M, Wallace J, Dorschner M, McArthur M, Stamatoyannopolous J (2004). Genome-wide identification of DNaseI hypersensitive sites using active chromatin sequence libraries. *Proc Natl Acad Sci USA* **101**:4537–4542.
53. Chen D, Lewis R, Kaufman D (2003). Mouse and human embryonic stem cell models of hematopoiesis: past, present and future. *Biotechniques* **35**:1253–1261.

14

Dynamic Nucleoprotein Structure of the β-Globin Locus: Establishing a Rational Molecular Basis for Therapeutic Modulation of Hemoglobin Switching

Emery Bresnick, Hogune Im, Kirby D. Johnson, Meghan E. Boyer and Jeffrey A. Grass

Introduction

Given the essential physiological role of hemoglobin and diseases resulting from deregulated β-globin transcription, elucidating the underlying transcriptional mechanisms has high significance. Progress in understanding these mechanisms can be categorized with respect to defining and dissecting regulatory modules that collectively determine the transcriptional activity of the endogenous β-globin genes during development (Fig. 1). The major goal of this chapter is to provide an overview of the "Factor Occupancy and Recruitment" and "Histone Modification Pattern" modules. Whereas understanding the interrelationships between modules is critical, studies directed towards achieving this goal are in their infancy.

Transcriptional Mechanisms within Endogenous Chromatin Domains: The β-Globin Locus System

Despite the recent surge of publications on chromatin structure/function, and the apparent depth in which this topic has been studied, our knowledge of how mammalian genes are regulated within chromatin domains remains sketchy. It is common for such domains to span 100 kb or more of DNA, with the intragenic and extragenic sequences littered with repetitive elements and many prospective regulatory motifs.[1,2] As disruption of the physical integrity of chromatin domains has been implicated in diverse disease processes,[3] understanding how domains are established and regulated has far-reaching implications.

The murine β-globin locus, consisting of two genes active in embryonic erythroid cells, *Ey* and $\beta H1$, and two genes active in adult erythroid cells, *βmajor* and *βminor*,[4] typifies a complex

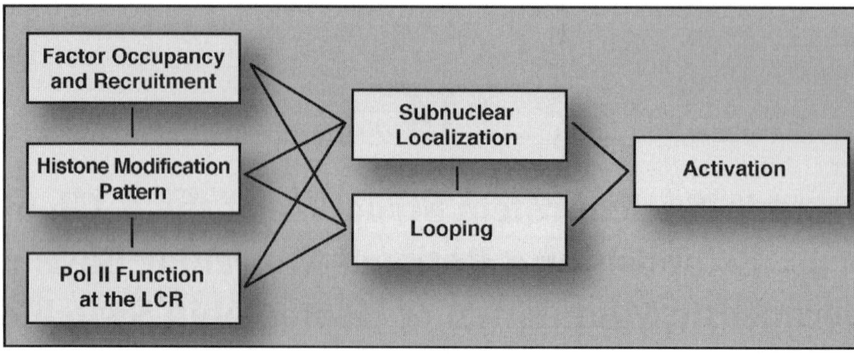

Fig. 1. Regulatory modules that control β-like globin gene transcription. The spatial relationship between the modules has been arbitrarily assigned, as it is likely that multi-directionality exists. For example, factor occupancy is a determinant of looping, and looping will likely impact upon factor occupancy.

chromatin domain organization scenario. These genes reside within a sea of odorant receptor genes and repetitive elements.[4] Similarly, the human β-globin locus has embryonic (ε) and adult (β) genes, but uniquely has two genes (Gγ and Aγ) expressed during the fetal stage of development.[5–7] Upstream of the β-globin genes reside four erythroid-specific DNaseI hypersensitive sites (HSs),[8–10] which are collectively termed the locus control region (LCR).[11,12]

The LCR is defined by its ability to confer copy number-dependent and position-independent expression to linked β-globin transgenes.[12] Based on its large size (greater than 10 kb), and extensive conserved sequences, not surprisingly, the LCR contains a daunting number of conserved motifs that bind erythroid-specific and broadly expressed transcription factors.[13,14] Initial studies of a chromosomal deletion in Hispanic thalassemic patients, which removes the LCR and upstream sequences, indicated that these sequences are required to establish erythroid-specific DNaseI sensitivity throughout the β-globin locus.[15] This important finding was interpreted with respect to a long-range, broad chromatin opening activity of the LCR. More recent studies in which the endogenous murine LCR was knocked out via homologous recombination in ES cells, indicate that erythroid cells derived from mutant ES cells *in vitro* and *in vivo* are highly impaired in β-globin transcription, thus confirming the crucial role of the LCR.[16–18]

Despite the dramatic loss of transcriptional activity upon targeted deletion of the LCR, the broad general DNaseI sensitivity is LCR-independent.[16–18] Although the molecular basis for the ~2-fold enhancement of DNaseI sensitivity (a hallmark of "active" chromatin) is unknown, this parameter has been assumed to reflect an unfolded or "open" chromatin state. Subsequent studies involving targeted deletions of HSs of the LCR provided evidence that high-level transcription of the β-globin genes requires multiple HSs.[19–21] Although the HSs function synergistically at ectopic chromosomal sites[22] and in transfection assays,[23] they function additively at the endogenous mouse locus to establish the fully active state.[19] The targeted deletions confirmed that the LCR functions over a long distance on the chromosome to confer high-level transcription to the β-globin genes, and importantly showed that the LCR has no obvious role in determining the precise developmental timing of the endogenous murine β-globin genes. Lastly, the results suggest that the LCR functions via a long-range mechanism to regulate chromatin structure at the β-globin genes, rather than instigating biochemical events that simultaneously influence the chromatin structure locus-wide.

The targeted deletion studies described above were paradigm-shifting, since many studies with β-globin locus transgenes have implicated LCR HSs in conferring developmental regulation to the β-globin promoters.[24–27] What might explain the different conclusions reached from analysis of the endogenous mouse locus and human β-globin locus transgenes? The unique activity of the LCR to confer developmental specificity of transgenes might be explained by divergent mechanisms that regulate the mouse and human loci. The mouse LCR knockout results suggest that sequences conferring developmental specificity reside outside of the LCR, and evidence exists for such sequences near the β-globin genes.[28–30] Alternatively, physiological mechanisms that confer developmental specificity might breakdown upon transgene integration into ectopic chromosomal sites. It would not be suprising if the strong activation potential of the LCR differentially opposes chromosomal position effects for constructs that assemble distinct nucleoprotein complexes. Lastly, if the endogenous murine LCR indeed confers developmental specificity, despite the insensitivity of hemoglobin switching to deletion of the LCR, compensatory mechanisms might normalize the pattern of β-globin expression upon perturbation of the endogenous LCR. Understanding the basis for the different conclusions requires major additional investigation.

Regardless of the exact mechanism conferring developmental specificity, it is clear that the LCR functions over a long distance on the chromosome to regulate globin promoter transcription. Attempts to address how such long-range activity is conferred need to consider the impact of the LCR on higher-order chromatin organization, perhaps the simplest permutation being "looping".[31] As the name implies, looping involves physical interactions between components bound at regions of a locus that are separated in two-dimensional space, thereby allowing for molecular events that induce Pol II recruitment and/or function. A wealth of results obtained with the β-globin system have been interpreted with respect to "looping",[32–37] but additional conceptually attractive models, including "tracking",[38] "linking"[39] and "long-range polymerase transfer", have been invoked.[40] Looping appears to be common in considerably less complex systems,[31] but considering how entire chromosomal segments "loop" is a daunting biophysical problem. Nevertheless, a clever, modified RNA fluorescence *in situ* hybridization method (RNA Tagging and Recovery of Associated Proteins)[41] and chromosome conformation coupling methodology (3C)[42] have been used to demonstrate looping at the β-globin locus.[41,43] This contact requires the essential hematopoietic transcription factors erythroid Kruppel-like factor (EKLF)[44] and GATA-1.[45]

Besides looping, factors recruited far upstream of the β-like globin genes might processively migrate along the chromosome to a downstream promoter (tracking).[38] The linking model, which incorporates features of looping and tracking models, assumes that LCR-bound factors are physically linked to factors at the promoters via a continuous chain of protein–protein interactions.[46] Both looping and linking models focus on the requirements to recruit and/or activate RNA polymerase II (PolII) at a promoter. As LCR- and promoter-binding factors appear to be the same, and the density of such factors is likely to be higher at the large LCR versus the small promoter, it is reasonable to assume that the LCR recruits Pol II. In fact, high levels of Pol II reside at the LCR even under conditions in which Pol II is undetectable at the promoter.[40,47] This finding led to the proposal of a third model, long-range polymerase transfer (LPT),[40,47] which assumes that Pol II is recruited to the LCR and then re-localizes to the promoters in a regulated fashion. It is impossible to critically assess these models without considering the nonrandom positioning of chromosomal regions in the three-dimensional nuclear milieu.[48] The likely existence of replication and transcription "factories"[49,50] constitutes a further level of regulation and adds an important dimension to the models described above. Importantly, the models are not mutually exclusive. For example, LPT requires looping, and looping could conceivably regulate tracking. Discriminating among models of transcriptional regulation within chromatin domains will be greatly facilitated by the development of cellular systems that allow for facile genetic, molecular and biochemical analyses. Biochemical studies in cell-free systems have not reached a level of sophistication whereby long-range regulation can be studied over chromosomal distances of hundreds of thousands to millions of base pairs. Furthermore, it is not a trivial task to validate the biological relevance of chromatin structures assembled *in vitro*.

Experimental Systems to Analyze Mechanisms of β-like Globin Gene Activation Versus Developmental Regulation

Transcriptional mechanisms involving long-range activation within multi-gene chromatin domains are complex, and no single approach will likely suffice to unravel them. In the context of the β-like globin genes, a major goal has been to tackle the intellectually challenging problem of how hemoglobin switching is regulated. Based on experimental complexities, notably the lack of easily manipulatable systems for studying endogenous β-globin gene switching, and the apparent involvement of multiple factors, progress has been slow. It is important to differentiate between systems optimal for studying activation versus developmental switching, two distinct, but obviously interrelated, events. Large transgenes are powerful tools for conducting developmental studies,[13,27,51–54] but factor-dependent activation of the endogenous adult β-globin genes can be recapitulated via facile complementation assays in cells that mimic primary cells.[55–57] Such cellular systems allow one to analyze complex mechanisms that mediate activation of the endogenous adult β-globin genes and provide an opportunity to establish a strong foundation for elucidating switching mechanisms. Whereas it is possible that the endogenous locus in cultured cells has certain inherent differences versus the locus *in vivo*, the merits of using physiologically-relevant cell systems, coupled with primary cell validation for rigorously analyzing mechanisms, far outweigh potential negatives.

The development of physiologically-relevant murine cell models has allowed for major progress in dissecting the role of chromatin structure in regulating β-globin transcription. As comparable human models have not been developed, this review focuses predominantly on the murine system. Knockouts of hematopoietic transcription factors that regulate β-globin transcription, e.g., GATA-1,[58–60] p45/NF-E2,[61,62] and EKLF,[63,64] and the GATA factor co-regulator Friend of GATA-1 (FOG-1)[65] are lethal, thereby complicating transcriptional analyses, which often require large numbers of primary hematopoietic cells. However, in the case of GATA-1,[55] FOG-1,[56] and EKLF,[66] immortalized cell lines have been derived from ES cells, in which the respective genes are knocked out. These cell lines retain phenotypic properties of normal hematopoietic precursors: proerythroblast-like in the case of the GATA-1-null G1E line, and hematopoietic precursors with dual erythroid- and megakaryocytic-potential in the case of FOG-1-null cells. GATA-1 and FOG-1 expression in the respective null cells rescues β-like globin gene transcription and a normal differentiation program.[55,56,67]

Another important advance *vis-à-vis* dissecting the role of chromatin structure in regulating β-globin locus transcription involved the application of estrogen receptor ligand binding domain fusion technology[68] to generate a conditionally-active derivative of GATA-1 (ER-GATA-1).[69] Since expressing ER-GATA-1 in GATA-1-null G1E cells instigates differentiation, it is crucial to maintain GATA-1 in an inactive state upon expression in cells. Although the ER fusion system does not completely inactivate GATA-1, it strongly suppresses GATA-1 activity, allowing one to rapidly activate ER-GATA-1 upon addition of an estrogen receptor agonist. Thus, stable cell lines expressing wild-type and mutant GATA-1 can be generated. Lastly, the p45/NF-E2-null CB3 cell line was isolated, in which the p45/NF-E2 gene is disrupted via retroviral insertion.[57] As with the GATA-1-null cells, the adult β-globin genes are strongly activated upon p45/NF-E2 expression.[57,70] Although a conditionally-active p45/NF-E2 line has not been described, CB3 cells have allowed for mechanistic studies,[70–72] as the stable expression of p45/NF-E2 does not result in cessation of proliferation and terminal differentiation.

Factor Occupancy/Recruitment Module

Whereas GATA-1, NF-E2, and EKLF have been implicated for years in β-globin transcriptional regulation,[73–77] only recently has progress been made as to where these factors function within the locus. This progress has been made possible by the development of chromatin immunoprecipitation (ChIP) assays, in which factors are crosslinked to their occupancy sites in living cells.[14,78] Although ultraviolet irradiation was used previously to crosslink proteins to their natural occupancy sites in cells,[79,80] formaldehyde has emerged as the preferred reagent for generating reproducible data in diverse contexts. Efforts to define the composition and spatial/temporal regulation of endogenous β-globin locus regulatory complexes are ongoing, and much remains to be learned about how even the known transcription factors select chromatin sites and function post-chromatin occupancy.

GATA-1 binds WGATAR motifs, which are present abundantly throughout the genome.[74,81,82] However, analyses of endogenous GATA-1 and ER-GATA-1 occupancy indicate that only a small fraction of the WGATAR motifs are occupied in cells.[83–85] Quantitative ChIP analysis of endogenous GATA-1 and ER-GATA-1 occupancy of the murine β-globin locus in adult erythroid cells revealed occupancy at HS1, HS2, HS3, HS4, and the *βmajor* promoter.[83] GATA-1 does not occupy all accessible chromatin regions however, as no occupancy was detected at HS5. GATA-1 occupancy was not detected at additional regions containing WGATAR motifs between the LCR and *βmajor*. Since it is commonly assumed that evolutionarily conserved *cis*-elements are more likely to be functional versus nonconserved motifs, careful attention was paid to assessing whether ER-GATA-1 preferentially occupies conserved motifs. Whereas the occupancy sites within the LCR contain conserved WGATAR motifs, multiple regions of the locus containing conserved motifs are not occupied by ER-GATA-1. Importantly, ER-GATA-1 recapitulates the endogenous GATA-1 occupancy pattern,[83] justifying the use of G1E-ER-GATA-1 cells for mechanistic studies.

Since GATA-1 can be crosslinked to multiple HSs of the LCR, an important question is whether GATA-1 directly binds to each HS, or if a GATA-1-containing complex at one HS is crosslinked to other HSs. As the chromatin size used in our ChIP analyses with 0.4% formaldehyde is ~300–500 bp, and the HSs are considerably farther apart, it is likely that ER-GATA-1 directly binds HS1, HS2, HS3, and HS4, but this has not been proven. The chromatin binding specificity of other GATA factors has not been studied in detail, and therefore it is unclear whether the selective utilization of a small subset of WGATAR motifs is a hallmark of all GATA factor family members.

Multiple GATA factors bind a similar, if not identical, WGATAR motif *in vitro*,[81,82] suggesting that such factors, when expressed in the same cell, can have common target genes. As noted above GATA factor chromatin occupancy has largely been studied with GATA-1, but significant progress has been made in defining the GATA-2 chromatin occupancy pattern. Analysis of endogenous GATA-2 occupancy at the β-globin and *Gata*-2 loci revealed a result similar to GATA-1 occupancy; only a subset of the total WGATAR motifs are occupied.[84–86] An important unresolved issue is whether the similar GATA-1 and GATA-2 DNA binding specificities *in vitro* are applicable to chromatin occupancy in cells. Initial ChIP studies indicated that certain common occupancy sites, e.g. the region −2.8 kb upstream of the *Gata*-2 1S promoter, are occupied by both GATA-1 and GATA-2 at distinct developmental stages.[84–86] Our finding that GATA-1 displaces GATA-2 from the −2.8 kb region in a manner that is tightly coupled to *Gata*-2 repression provided the basis for the GATA switch model,[84] which assumes that GATA-1 and GATA-2 occupy certain common chromatin sites at distinct developmental stages and function differentially through these sites.[87] GATA-1-mediated displacement of GATA-2 instigates either repression (e.g., the *Gata*-2 locus)[84,85] or activation (e.g., the *Tac-2* locus),[88] dependent upon the chromosomal context.

By contrast to the GATA-1/GATA-2 occupancy of the −2.8 kb region, GATA-2, but not GATA-1, is readily crosslinked to the −1.8 kb region of the *Gata*-2 locus.[85] This result might reflect the preferential capacity of GATA-2 to bind the −1.8 kb region in cells, or the preferential GATA-2 crosslinking might relate to different conformations and/or complexes assembled at the −1.8 kb region. Comparisons of GATA-1 and GATA-2 occupancy genome-wide will allow one to ascertain the frequency of sites that are preferred by GATA-1 or GATA-2 and to gain insights into mechanisms underlying site discrimination. These comparisons will lay the groundwork for determining whether distinct crosslinking preferences equate to GATA factor-specific chromatin binding preferences. We hypothesize that a GATA Recognition Code (GRC) exists in which site occupancy is determined by intrinsic features of the GATA motif, nearest-neighbor *cis*-elements, and the chromatin environment.[87] Elucidating the GRC has potential to yield fundamental insights into GATA factor function and therefore the control of diverse developmental processes, including hemoglobin switching.

As FOG-1 is required for GATA-1 to access certain chromatin sites,[86,89] we proposed that FOG-1 has an important biological role as a Chromatin Occupancy Facilitator or "COF coregulator".[86] However, it is unlikely that COF activity represents the sole function of FOG-1, since FOG-1 contains an N-terminal repression domain[90] that binds the NURD corepressor complex.[91] It appears that FOG-1 has a dual activity as a COF coregulator and a recruiter of chromatin-modifying coregulators.

Based on the apparently common usage of FOG-1 by GATA-1 to regulate transcription, GATA-1 and FOG-1 would be expected to co-localize at certain chromatin sites. Though studies have only begun to investigate the pattern and determinants of FOG-1 chromatin occupancy,[86] initial work indicated that FOG-1 can be crosslinked to certain chromatin sites that are occupied by GATA-2 or GATA-1.[86] In cases whereby GATA-2 is displaced from chromatin by GATA-1, FOG-1 occupancy persists.

In common with GATA-1, p45/NF-E2 can be crosslinked to HS1, HS2, HS3, and HS4 of the β-globin LCR and the *βmajor* promoter in adult erythroid cells.[83,92] Analysis of p45/NF-E2 occupancy revealed an intriguing differential regulation dependent upon GATA-1. In GATA-1-null G1E cells, p45/NF-E2 occupancy persists at HS2, but is reduced at the other HSs and the *βmajor* promoter.[83] The GATA-1-dependent p45/NF-E2 occupancy cannot be explained by changes in p45/NF-E2 protein concentration, since p45/NF-E2 levels are identical in the presence and absence of GATA-1. Although mechanisms underlying GATA-1-dependent p45/NF-E2 occupancy have not been resolved, the results can be explained by one of the following models. Stable complexes of p45/NF-E2 at HS4, HS3, HS1, and the promoter might require cooperative interactions between GATA-1 and/or GATA-1-associated factors. Alternatively, the GATA-1-dependent p45/NF-E2 crosslinking might reflect crosslinking of HS2-bound GATA-1 to additional regions, as GATA-1 promotes physical interactions between regions of the locus that are separated in two-dimensional space.[45]

Initial studies suggest that EKLF has an occupancy pattern distinct from that of GATA-1 and p45/NF-E2. EKLF occupies only a subset of the HSs of the LCR and the *βmajor* promoter.[93] ER-GATA-1 induces EKLF protein levels in G1E cells, resulting in increased EKLF occupancy of chromatin target sites.[93] As GATA-1 can bind EKLF,[94] such factors might cooperatively assemble at certain chromatin sites. However, this does not appear to be relevant to all sites, based on the distinct GATA-1 and EKLF chromatin occupancy patterns.

GATA-1, NF-E2, and EKLF interact with the histone acetyltransferase CBP/p300, a common transcriptional coactivator.[95–97] Thus, one would predict that CBP/p300 is recruited to all chromatin sites bound by these factors. A comprehensive analysis of CBP/p300 occupancy at the β-globin locus has not been conducted, but CBP/p300 occupancy at HS2 and the *βmajor* promoter is enhanced by ER-GATA-1 activation in G1E cells.[98,99] Since GATA-1, NF-E2, and EKLF do not have obvious redundant functions to regulate the transcription of β-like globin genes, CBP/p300 recruitment by any single factor appears insufficient to activate β-like globin gene transcription. Besides CBP/p300 and FOG-1, other protein–protein interactions of potential relevance include the interaction of NF-E2 with TATA globin binding protein (TBP)-associated factor 130[100] and proteins that contain a protein interaction module flanked by tryptophan residues (WW domains).[71,101,102]

Themes emerging from ChIP analyses of the "Factor Occupancy and Recruitment Module" include: (i) only a subset of the many motifs within the β-globin locus are occupied in cells[83]; (ii) occupancy is developmentally-dynamic[83,93]; (iii) occupancy by a single activator can be differentially regulated at distinct sites[83,103]; and (iv) activators regulate the occupancy of additional activators through protein synthesis-dependent and -independent mechanisms. Collectively, these steps form a regulatory network that constitutes the "Factor Occupancy and Recruitment" module, which mediates assembly of the native nucleoprotein structure of the endogenous β-globin locus.

"Histone Modification Pattern" Module

The organization of DNA into nucleosomes and higher-order chromatin structures represents an essential mechanism underlying transcriptional regulation and the control of all nuclear processes.[104–106] The core histone components of nucleosomes have multiple amino acids, especially lysines, that are subjected to diverse posttranslational modifications including acetylation, methylation, phosphorylation, ubiquitination, and ADP ribosylation.[107] This array of signal-dependent modifications regulates DNA accessibility at the level of single nucleosomes,[108,109] higher-order chromatin folding,[110] and provides docking sites or excludes regulatory factors.[107] The histone modifications and ensuing biochemical consequences are essential components of transcriptional mechanisms. Snapshots of the end product of these signaling events, the dynamic histone modification pattern of a chromatin domain, provide important insights into mechanisms underlying transcriptional regulation.

The β-globin locus assembles into a developmentally-dynamic, erythroid-specific histone modification pattern.[103,111–114] Both the LCR and the region containing the adult β-globin genes are highly enriched in acetylated histones H3 and H4 and histone H3 di-methylated at lysine 4 (H3-meK4) in adult erythroid cells of mouse E13.5 fetal liver and in mouse erythroleukemia (MEL) cells. The histone modification pattern is erythroid cell-specific, as histone acetylation is not enriched in E13.5 fetal brain and in non-erythroid cell lines. Dimethyl sulfoxide-induced maturation of MEL cells induces a $\sim$2-fold increase in histone acetylation at HS2 and the *βmajor* promoter, but the overall histone modification pattern is pre-established in "uninduced" MEL cells. The histone modification pattern of MEL cells resembles that of the locus in E13.5 fetal liver, thus validating the MEL cell system for analyzing determinants and functional consequences of the pattern.

By contrast to the adult murine β-globin genes, the middle of the β-globin locus containing the Ey and βH1 genes, termed the central subdomain, has nearly undetectable acetylated H3 and H3-meK4 and low levels of acetylated H4 relative to other regions.[103,111–113] In embryonic erythroid cells derived from murine embryonic stem cells (EryP colonies), acetylated H3 and H4 and H3-meK4 are high at the LCR, the adult β-globin genes, and the central subdomain.[112] A limited analysis of histone acetylation within the central subdomain and the *βmajor* promoter in mouse yolk sac, which contains embryonic erythroid cells, also provided evidence for broadly enriched acetylation in embryonic erythroid cells.[111] In human K562 erythroleukemia cells, which express embryonic and fetal γ-globin genes, the locus is broadly enriched in histone acetylation and H3-meK4.[114] Hemin-induced erythroid maturation of K562 cells is accompanied by a ~2-fold increase in embryonic ε-globin transcription and a broad ~2-fold increase in histone acetylation throughout the locus, with little or no change in H3-meK4.[114] Collectively, these studies indicate that the histone modification pattern of the locus in embryonic and adult erythroid cells differs, suggesting that the histone modification pattern is an important determinant of β-like globin gene transcriptional activity during development.

An obvious question that arises is when does the histone modification pattern assemble during development and are specific components of the pattern established and/or regulated distinctly from other components? ChIP analyses in p45/NF-E2-null CB3 cells using antibodies generated against di-acetylated histone H3, multi-acetylated H4, and H3-meK4 revealed that p45/NF-E2 is required to establish maximally acetylated H3, and to a lesser degree acetylated H4 at the *βmajor* promoter.[40,93,98,99] A similar regulation is apparent within the *βmajor* coding region. Acetylation of H3 and H4 at the LCR is p45/NF-E2-independent. Establishment of H3-meK4 at the promoter and the coding region, but not at the LCR, is p45/NF-E2-dependent. GATA-1 also modestly increases (~2-fold) acetylated H3 at the promoter, but unlike p45/NF-E2, modestly increases acetylation at the LCR as well. GATA-1 is not required for establishment of H3-meK4 at the promoter, but is required for maximal H3-meK4 in the coding region. GATA-1 has little or no effect on H3-meK4 at the LCR. These results demonstrate that, through common and unique activities, the two activators establish important components of the histone modification pattern of the β-globin locus. Importantly, however, even when GATA-1 or p45/NF-E2 are lacking, the LCR retains erythroid-specific histone modifications. Significant acetylation also exists at *βmajor* without GATA-1 or p45/NF-E2. Thus, these factors are not absolutely required to establish the histone modification pattern at certain sites.

In the studies summarized above, antibodies recognizing histones modified at individual sites (e.g., mono acetylation at single residues) and at various combinations of these sites (e.g., mono-versus di-versus tri-methylation) were not used. As histone acetylation at different lysine residues can convey distinct consequences in certain chromosomal contexts,[107,115] additional work is required to determine whether the initial acetylation and methylation patterns recapitulate the patterns of all potential acetylation and methylation sites on the H3 and H4 N-terminal tails. Furthermore, other functionally-important histone modifications have not been defined at the β-globin locus. For example, histone H3 methylated at lysine 9 is enriched at heterochromatic sites that are transcriptionally-inactive[116–118] and is recognized by heterochromatin protein-1[119–123] that mediates the formation of repressive chromatin. Efforts to measure H3-meK9 at the murine β-globin locus revealed low signals, which might relate to inadequate antibody efficacy or a low density of H3-meK9. Methylation of histone H3 at lysines 27 and 36,[124,125] which are likely to have important roles in regulating chromatin domains, have not been studied at the β-globin locus or in detail at other endogenous mammalian chromatin domains. Thus, even the initial phase of describing the chromatin architecture of the endogenous β-globin locus requires considerably more work. As histone modifications induce functionally-important chromatin structural changes and provide docking sites for regulatory factors,[2] understanding how histone modification patterns are established/modulated will undoubtedly reveal new steps in β-globin locus transcriptional regulation.

Targeting the "Histone Modification Module"

It has been known for almost 30 years that short chain fatty acids, such as butyrate, inhibit histone deacetylases (HDACs)[126,127] and therefore regulate transcription and other nuclear processes. It is not surprising therefore that such chemicals affect β-like globin gene transcription. The finding that butyrate and other HDAC inhibitors can induce γ-globin expression in certain systems[128–134] provides a useful tool for efforts to understand the complex problem of how switching is regulated and importantly raises the possibility that this activity can be exploited for the development of sickle cell disease therapeutics. However, even though multiple HDAC inhibitors have been developed and can be quite specific,[135] HDAC inhibition affects multiple aspects of cell function, given the broad roles of histone acetylation and non-histone acetylation.[136] It will therefore be a complex task to ascertain the mechanism of how HDAC inhibitors influence β-like globin gene transcription.

ChIP analysis was used to identify whether HDAC inhibitors influence the native nucleoprotein structure of the β-globin locus in MEL cells.[111,137] HDAC inhibition could potentially result in unopposed histone acetyltransferase (HAT) function at the locus, thereby leading to broadly increased histone H3 and H4 acetylation. Alternatively, if HAT access is highly regulated, the attenuation of HDAC activity might not suffice to promote HAT recruitment and altered acetylation. The data could not be explained exclusively by either scenario. Whereas butyrate and trichostatin A induced H4 acetylation at the central subdomain, low-level H3 acetylation within this region was unchanged. The HDAC inhibitors did not affect the high-level H3 and H4 acetylation at the LCR and the *βmajor* promoter. These results indicate that HATs responsible for establishing and maintaining acetylated H4, but not acetylated H3, can readily access the locus upon attenuation of HDAC activity. The surprising differential restoration of H4 versus H3 acetylation illustrates the complexities involved in specifically modulating endogenous chromatin structure. First, a component of the histone modification pattern is pre-established (high-level acetylation at the LCR and *βmajor*) and is non-responsive to HDAC inhibition. Second, only one specific modification (acetylated H4) is responsive to HDAC inhibitors at the murine locus, whereas both H3 and H4 will likely need to be restored to yield a transcriptionally-permissive template. Consistent with the incomplete restoration of the active structure, HDAC inhibitors did not induce Pol II recruitment to the *Ey* and $\beta H1$ promoters, which reside in the central subdomain of the locus, nor did they reactivate *Ey* and $\beta H1$ transcription.

It is reasonable to assume that the inability of HDAC inhibitors to restore H3 acetylation at the central subdomain of the murine β-globin locus is related to the absence of signals required for the respective HATs to be recruited and/or activated at the central subdomain. In this regard, phosphorylation of histone H3 at serine 10 can be functionally coupled to H3 acetylation.[138–140] H3 phosphorylation at serine 10 is catalyzed by Rsk2[141] and MSK[142] kinases, which, in turn, are activated via diverse cell signaling mechanisms. It is attractive to propose a strategy to increase both H3 and H4 acetylation, involving the induction of cell signaling to increase serine 10 phosphorylation, coupled with HDAC inhibition. One potential serious issue, however, is that other studies suggest that serine 10 phosphorylation occurs independent of H3 acetylation.[143,144] Nevertheless, the concept that the full histone acetylation profile is not reestablished by HDAC inhibitors due to the absence of requisite cell signals is compelling and deserves intensive investigation. Interestingly, in the presence of erythropoietin and stem cell factor, MEK1/2 signals are required for maximal γ-globin expression.[145]

As novel chemical strategies are developed to target specific components of the chromatin remodeling and modification machinery,[146] it will be important to test whether combinations of such molecules, and importantly including signaling modulators, will reactivate the repressed fetal γ-globin genes. Furthermore, there has been interest in using DNA methyltransferase inhibitors as therapeutic agents to elevate the levels of fetal hemoglobin.[147] Whereas it is conceivable that such drugs will constitute important components of future combination therapies, major uncertainties exist regarding whether efficacy requires complex indirect effects, which would not be ideal.

Whereas the mechanistic studies described above focused on reactivating the repressed murine embryonic β-globin genes, it is formally possible that mechanisms underlying murine embryonic β-globin gene repression and activation differ from those applicable to the human fetal γ-globin genes. However, it is not possible at this time to assign similarities and differences between the systems, since tractable, physiologically-relevant human systems to implement definitive mechanistic experiments do not exist. The development of methods to differentiate human embryonic stem cells into hematopoietic precursors should allow for the derivation of immortalized hematopoietic cell lines and perhaps the generation of unlimited numbers of primary cells. Analogous to G1E and FOG-1-null cells, such lines, and more generally, efficient embryonic stem cell differentiation systems will likely provide important tools for dissecting endogenous human β-like globin gene regulation. Furthermore, the establishment of systems in which hematopoietic development from human embryonic stem cells is supported by physiologically-relevant human stromal lines will allow one to study the impact of signals instigated by cell–cell interactions. We anticipate that such signals will constitute a crucial intercellular signaling module, which will require careful consideration at the systems level.

Summary

In the past two decades, a daunting amount of evidence has accumulated supporting a role for chromatin organization in regulating nuclear processes such as transcription. A paradigm has been established in which chromatin modifying enzymes are recruited to regulatory regions of genes, resulting in targeted histone modifications, as a fundamental step in transcriptional mechanisms. Thus, it is widely accepted that it is not possible to understand transcriptional mechanisms without considering the endogenous nucleoprotein structure of the locus of interest and how this structure is established, maintained, and dynamically regulated. This chapter describes what is known about the native nucleoprotein structure of the endogenous β-globin locus and how this structure is targeted to establish erythroid cell-specific and developmental stage-specific transcription of the β-like globin genes. Furthermore, as chemical inducers of fetal hemoglobin are modifiers of chromatin structure, we also address mechanisms underlying their actions on the β-globin locus.

References

1. Forsberg EC, Bresnick EH (2001). Histone acetylation beyond promoters: long-range acetylation patterns in the chromatin world. *Bioessays* **23**:820–830.
2. Bresnick EH, Im H, Johnson KD (2003). Long-range acetylation patterns in the genome. Encyclopedia of the Human Genome.
3. Kleinjan DA, van Heyningen V (2005). Long-range control of gene expression: emerging mechanisms and disruption in disease. *Am J Hum Genet* **76**:8–32.
4. Bulger M, van Doorninck JH, Saitoh N, Telling A, Farrell C, Bender MA, Felsenfeld G, Axel R, Groudine M, von Doorninck JH (1999). Conservation of sequence and structure flanking the mouse and human beta-globin loci: the beta-globin genes are embedded within an array of odorant receptor genes. *Proc Natl Acad Sci USA* **96**:5129–5134.

5. Collins FS, Weissman SM (1984). The molecular genetics of human hemoglobin. *Prog Nucleic Acid Res Mol Biol* **31**:315–462.
6. Karlsson S, Nienhuis AW (1984). Developmental regulation of human globin genes. *Ann Rev Biochem* **54**:1071–1108.
7. Blau CA, Stamatoyannopoulos G (1994). Hemoglobin switching and its clinical implications. *Curr Opin Hematol* **1**:136–142.
8. Tuan D, London IM (1984). Mapping of DNase I-hypersensitive sites in the upstream DNA of human embryonic epsilon-globin gene in K562 leukemia cells. *Proc Natl Acad Sci USA* **81**:2718–2722.
9. Tuan D, Solomon W, Li Q, London IM (1985). The "beta-like-globin" gene domain in human erythroid cells. *Proc Natl Acad Sci USA* **82**:6384–6388.
10. Forrester WC, Thompson C, Elder JT, Groudine M (1986). A developmentally stable chromatin structure in the human β-globin locus. *Proc Natl Acad Sci USA* **83**:1359–1363.
11. Forrester WC, Takegawa S, Papayannopoulou T, Stamatoyannopoulos G, Groudine M (1987). Evidence for a locus activation region: the formation of developmentally stable hypersensitive sites in globin-expressing hybrids. *Nucleic Acids Res* **15**:10159–10177.
12. Grosveld F, van Assendelft GB, Greaves DR, Kollias G (1987). Position-independent, high-level expression of the human beta-globin gene in transgenic mice. *Cell* **51**:975–985.
13. Li Q, Peterson KR, Fang XR, Stamatoyannopoulos G (2002). Locus control regions. *Blood* **1094–1104**: 3077–3086.
14. Johnson KD, Bresnick EH (2002). Dissecting long-range transcriptional mechanisms by chromatin immunoprecipitation. *Methods* **26**:27–36.
15. Forrester WC, Epner E, Driscoll MC, Enver T, Brice M, Papayannopoulou T, Groudine M (1990). A deletion of the human beta-globin locus activation region causes a major alteration in chromatin structure and replication across the entire beta-globin locus. *Genes Dev* **4**:1637–1649.
16. Epner E, Reik A, Cimbora D, Telling A, Bender MA, Fiering S, Enver T, Martin DI, Kennedy M, Keller G, Groudine M (1998). The beta-globin LCR is not necessary for an open chromatin structure or developmentally regulated transcription of the native mouse beta-globin locus. *Mol Cell* **2**:447–455.
17. Reik A, Telling A, Zitnik G, Cimbora D, Epner E, Groudine M (1998). The locus control region is necessary for gene expression in the human beta-globin locus but not the maintenance of an open chromatin structure in erythroid cells. *Mol Cell Biol* **18**:5992–6000.
18. Bender MA, Bulger M, Close J, Groudine M (2000). Beta-globin gene switching and DNase I sensitivity of the endogenous beta-globin locus in mice do not require the locus control region. *Mol Cell* **5**:387–393.
19. Bender MA, Mehaffey MG, Telling A, Hug B, Ley TJ, Groudine M, Fiering S (2000). Independent formation of DnaseI hypersensitive sites in the murine beta-globin locus control region. *Blood* **95**: 3600–3604.
20. Hug BA, Wesselschmidt RL, Fiering S, Bender MA, Epner E, Groudine M, Ley TJ (1996). Analysis of mice containing a targeted deletion of beta-globin locus control region 5′ hypersensitive site 3. *Mol Cell Biol* **16**:2906–2912.
21. Fiering S, Epner E, Robinson K, Zhuang Y, Telling A, Hu M, Martin DI, Enver T, Ley TJ, Groudine M (1995). Targeted deletion of 5′HS2 of the murine beta-globin LCR reveals that it is not essential for proper regulation of the beta-globin locus. *Genes Dev* **9**:2203–2213.
22. Bresnick EH, Tze L (1997). Synergism between hypersensitive sites confers long-range gene activation by the beta-globin locus control region. *Proc Natl Acad Sci USA* **94**:4566–4571.
23. Jackson JD, Miller W, Hardison RC (1996). Sequences within and flanking hypersensitive sites 3 and 2 of the beta-globin locus control region required for synergistic versus additive interaction with the epsilon-globin gene promoter. *Nucleic Acids Res* **24**:4327–4335.
24. Peterson KR, Clegg CH, Navas PA, Norton EJ, Kimbrough TG, Stamatoyannopoulos G (1996). Effect of deletion of 5′HS3 or 5′HS2 of the human beta-globin locus control region on the developmental regulation of globin gene expression in beta-globin locus yeast artificial chromosome transgenic mice. *Proc Natl Acad Sci USA* **93**:6605–6609.
25. Navas PA, Peterson KR, Li Q, Skarpidi E, Rohde A, Shaw SE, Clegg CH, Asano H, Stamatoyannopoulos G (1998). Developmental specificity of the interaction between the locus control region and embryonic or fetal globin genes in transgenic mice with an HS3 core deletion. *Mol Cell Biol* **18**:4188–4196.

26. Bungert J, Dave U, Lim KC, Lieuw KH, Shavit JA, Liu Q, Engel JD (1995). Synergistic regulation of human beta-globin gene switching by locus control region elements HS3 and HS4. *Genes Dev* **9**:3083–3096.
27. Navas PA, Peterson KR, Li Q, McArthur M, Stamatoyannopoulos G (2001). The 5′HS4 core element of the human beta-globin locus control region is required for high-level globin gene expression in definitive but not in primitive erythropoiesis. *J Mol Biol* **312**:17–26.
28. Filipe A, Li Q, Deveaux S, Godin I, Romeo P-H, Stamatopannopoulos G, Mignotte V (1999). Regulation of embryonic/fetal globin genes by nuclear hormone receptors: a novel perspective on hemoglobin switching. *EMBO J* **18**:687–697.
29. Tanabe O, Katsuoka F, Cambell AD, Song W, Yamamoto M, Tanimoto K, Engel JD (2002). An embryonic/fetal beta-type globin gene repressor contains a nuclear receptor TR2/TR4 heterodimer. *EMBO J* **21**:3434–3442.
30. Omori A, Tanabe O, Engel JD, Fukamizu A, Tanimoto K (2005). Adult stage gamma-globin silencing is mediated by a promoter direct repeat element. *Mol Cell Biol* **25**:3443–3451.
31. Schleif R (1992). DNA looping. *Annu Rev Biochem* **61**:199–223.
32. Engel JD (1993). Developmental regulation of human beta-globin gene transcription: a switch of loyalties? *Trends Genet* **9**:304–309.
33. Wijgerde M, Grosveld F, Fraser P (1995). Transcription complex stability and chromatin dynamics in vivo. *Nature* **377**:209–213.
34. Wijgerde M, Gribnau J, Trimborn T, Nuez B, Philipsen S, Grosveld F, Fraser P (1996). The role of EKLF in human beta-globin gene competition. *Genes Dev* **10**:2894–2902.
35. Dillon N, Trimborn T, Strouboulis J, Fraser P, Grosveld F (1997). The effect of distance on long-range chromatin interactions. *Mol Cell* **1**:131–139.
36. Gribnau J, de Boer E, Trimborn T, Wijgerde M, Milot E, Grosveld F, Fraser P (1998). Chromatin interaction mechanism of transcriptional control in vivo. *EMBO J* **17**:6020–6027.
37. Carter D, Chakalova L, Osborne CS, Dai YF, Fraser P (2002). Long-range chromatin regulatory interactions in vivo. *Nat Genet* **32**:623–626.
38. Ling J, Ainol L, Zhang L, Yu X, Pi W, Tuan D (2004). HS2 enhancer function is blocked by a transcriptional terminator inserted between the enhancer and the promoter. *J Biol Chem* **279**:51704–51713.
39. Bulger M, Sawado T, Schubeler D, Groudine M (2002). ChIPs of the beta-globin locus: unraveling gene regulation within an active domain. *Curr Opin Genet Dev* **12**:170–177.
40. Johnson KD, Christensen HM, Zhao B, Bresnick EH (2001). Distinct mechanisms control RNA polymerase II recruitment to a tissue-specific locus control region and a downstream promoter. *Mol Cell* **8**:465–471.
41. Carter D, Chakalova L, Osborne CS, Dai YF, Fraser P (2002). Long-range chromatin regulatory interactions in vivo. *Nat Genet* **32**:623–626.
42. Dekker J, Rippe K, Dekker M, Kleckner N (2002). Capturing chromosome conformation. *Science* **295**:1306–1311.
43. Tolhuis B, Palstra RJ, Splinter E, Grosveld F, de Laat W (2002). Looping and interaction between hypersensitive sites in the active beta-globin locus. *Mol Cell* **10**:1453–1475.
44. Drissen R, Palstra RJ, Gillemans N, Splinter E, Grosveld F, Philipsen S, de Laat W (2004). The active spatial organization of the beta-globin locus requires the transcription factor EKLF. *Genes Dev* **18**:2485–2490.
45. Vakoc CR, Letting DL, Gheldof N, Sawado T, Bender MA, Groudine M, Weiss MJ, Dekker J, Blobel GA (2005). Proximity among distant regulatory elements at the beta globin locus requires GATA-1 and FOG-1. *Mol Cell* **17**:453–462.
46. Bulger M, Groudine M (1999). Looping versus linking: toward a model for long-distance gene activation. *Genes Dev* **13**:2465–2477.
47. Johnson KD, Grass JA, Im H, Park C, Choi K, Bresnick EH (2003). Highly restricted localization of RNA polymerase II to the hypersensitive site cores of a tissue-specific locus control region. *Mol Cell Biol* **23**:6484–6493.
48. Kosak ST, Groudine M (2002). The undiscovered country: territories and the organization of transcription. *Dev Cell* **2**:690–692.
49. Cook PR (2002). Predicting three-dimensional genome structure from transcriptional activity. *Nat Genet* **32**:347–352.

50. Swedlow JR, Agard DA, Sedat JW (1993). Chromosome structure inside the nucleus. *Curr Opin Cell Biol* **5**:412–416.
51. Tanimoto K, Liu Q, Grosveld F, Bungert J, Engel JD (2000). Context-dependent EKLF responsiveness defines the developmental specificity of the human epsilon-globin gene in erythroid cells of YAC transgenic mice. *Genes Dev* **14**:2778–2794.
52. Navas PA, Li Q, Peterson KR, Swank RA, Rohde A, Roy J, Stamatoyannopoulos G (2002). Activation of the beta-like globin genes in transgenic mice is dependent on the presence of the beta-locus control region. *Hum Mol Genet* **11**:893–903.
53. Peterson KR (2003). Hemoglobin switching: new insights. *Curr Opin Hematol* **10**:123–129.
54. Bungert J, Tanimoto K, Patel S, Liu Q, Fear M, Engel JD (1999). Hypersensitive site 2 specifies a unique function within the human beta-globin locus control region to stimulate globin gene transcription. *Mol Cell Biol* **19**:3062–3072.
55. Weiss MJ, Yu C, Orkin SH (1997). Erythroid-cell-specific properties of transcription factor GATA-1 revealed by phenotypic rescue of a gene-targeted cell line. *Mol Cell Biol* **17**:1642–1651.
56. Cantor AB, Katz SG, Orkin SH (2002). Distinct domains of the GATA-1 cofactor FOG-1 differentially influence erythroid versus megakaryocytic maturation. *Mol Cell Biol* **22**:4268–4279.
57. Lu SJ, Rowan S, Bani MR, Ben-David Y (1994). Retroviral integration within the Fli-2 locus results in inactivation of the erythroid transcription factor NF-E2 in Friend erythroleukemias: evidence that NF-E2 is essential for globin expression. *Proc Natl Acad Sci USA* **91**:8398–8402.
58. Pevny L, Simon MC, Robertson E, Klein WH, Tsai SF, D'Agati V, Orkin SH, Costantini F (1991). Erythroid differentiation in chimaeric mice blocked by a targeted mutation in the gene for transcription factor GATA-1. *Nature* **349**:257–260.
59. Simon MC, Pevny L, Wiles MV, Keller G, Costantini F, Orkin SH (1992). Rescue of erythroid development in gene targeted GATA-1 mouse embryonic stem cells. *Nat Genet* **1**:92–98.
60. Pevny L, Lin CS, D'Agati V, Simon MC, Orkin SH, Costantini F (1995). Development of hematopoietic cells lacking transcription factor GATA-1. *Development* **121**:163–172.
61. Shivdasani RA, Rosenblatt MF, Zucker-Franklin D, Jackson CW, Hunt P, Saris CJ, Orkin SH (1995). Transcription factor NF-E2 is required for platelet formation independent of the actions of thrombopoietin/MGDF in megakaryocyte development. *Cell* **81**:695–704.
62. Shivdasani RA, Orkin SH (1995). Erythropoiesis and globin gene expression in mice lacking the transcription factor NF-E2. *Proc Natl Acad Sci USA* **92**:8690–8694.
63. Nuez B, Michalovich D, Bygrave A, Ploemacher R, Grosveld F (1995). Defective haematopoiesis in fetal liver resulting from inactivation of the EKLF gene. *Nature* **375**:316–318.
64. Perkins AC, Sharpe AH, Orkin SH (1995). Lethal beta-thalassaemia in mice lacking the erythroid CACCC-transcription factor EKLF. *Nature* **375**:318–322.
65. Tsang AP, Visvader JE, Turner CA, Fujuwara Y, Yu C, Weiss MJ, Crossley M, Orkin SH (1997). FOG, a multitype zinc finger protein as a cofactor for transcription factor GATA-1 in erythroid and megakaryocytic differentiation. *Cell* **90**:109–119.
66. Coghill E, Eccleston S, Fox V, Cerruti L, Brown C, Cunningham J, Jane S, Perkins A (2001). Erythroid Kruppel-like factor (EKLF) coordinates erythroid cell proliferation and hemoglobinization in cell lines derived from EKLF null mice. *Blood* **97**:1861–1868.
67. Welch JJ, Watts JA, Vakoc CR, Yao Y, Wang H, Hardison RC, Blobel GA, Chodosh LA, Weiss MJ (2004). Global regulation of erythroid gene expression by transcription factor GATA-1. *Blood* **104**:3136–3147.
68. Mattioni T, Louvion JF, Picard D (1994). Regulation of protein activities by fusion to steroid binding domains. *Methods Cell Biol* **43**:335–352.
69. Gregory T, Yu C, Ma A, Orkin SH, Blobel GA, Weiss MJ (1999). GATA-1 and erythropoietin cooperate to promoter erythroid cell survival by regulating bcl-xl expression. *Blood* **94**:87–96.
70. Kotkow KJ, Orkin SH (1995). Dependence of globin gene expression in mouse erythroleukemia cells on the NF-E2 heterodimer. *Mol Cell Biol* **15**:4640–4647.
71. Mosser EA, Kasanov JD, Forsberg EC, Kay BK, Ney PA, Bresnick EH (1998). Physical and functional interactions between the transactivation domain of the hematopoietic transcription factor NF-E2 and WW domains. *Biochemistry* **37**:13686–13695.

72. Li YJ, Higgins RR, Pak BJ, Shivdasani RA, Ney PA, Archer M, Ben-David Y (2001). p45(NFE2) is a negative regulator of erythroid proliferation which contributes to the progression of friend virus-induced erythroleukemias. *Mol Cell Biol* **21**:73–80.
73. Mignotte V, Eleouet JF, Raich N, Romeo PH (1989). Cis- and trans-acting elements involved in the regulation of the erythroid promoter of the human porphobilinogen deaminase gene. *Proc Natl Acad Sci USA* **86**:6548–6552.
74. Evans T, Reitman M, Felsenfeld G (1988). An erythrocyte-specific DNA-binding factor recognizes a regulatory sequence common to all chicken globin genes. *Proc Natl Acad Sci USA* **85**:5976–5980.
75. Ney PA, Sorrentino BP, Lowrey CH, Nienhuis AW (1990). Inducibility of the HS II enhancer depends on binding of an erythroid specific nuclear protein. *Nucleic Acids Res* **18**:6011–6017.
76. Miller IJ, Bieker JJ (1993). A novel, erythroid cell-specific murine transcription factor that binds to the CACCC element and is related to the Kruppel family of nuclear proteins. *Mol Cell Biol* **13**:2776–2786.
77. Orkin SH (1995). Regulation of globin gene expression in erythroid cells. *Eur J Biochem* **231**:271–281.
78. Orlando V, Strutt H, Paro R (1997). Analysis of chromatin structure by in vivo formaldehyde cross-linking. *Methods* **11**:205–214.
79. Bresnick EH, Bustin M, Marsaud V, Richard-Foy H, Hager GL (1992). The transcriptionally-active MMTV promoter is depleted of histone H1. *Nucleic Acids Res* **20**:273–278.
80. Nacheva GA, Guschin DY, Preobrazhenskaya OV, Karpov VL, Ebralidse KK, Mirzabekov AD (1989). Change in the pattern of histone binding to DNA upon transcriptional activation. *Cell* **58**:27–36.
81. Ko LJ, Engel JD (1993). DNA-binding specificities of the GATA transcription factor family. *Mol Cell Biol* **13**:4011–4022.
82. Merika M, Orkin SH (1993). DNA-binding specificity of GATA family transcription factors. *Mol Cell Biol* **13**:3999–4010.
83. Johnson KD, Grass JD, Boyer ME, Kiekhaefer CM, Blobel GA, Weiss MJ, Bresnick EH (2002). Cooperative activities of hematopoietic regulators recruit RNA polymerase II to a tissue-specific chromatin domain. *Proc Natl Acad Sci USA* **99**:11760–11765.
84. Grass JA, Boyer ME, Paul S, Wu J, Weiss MJ, Bresnick EH (2003). GATA-1-dependent transcriptional repression of GATA-2 via disruption of positive autoregulation and domain-wide chromatin remodeling. *Proc Natl Acad Sci USA* **100**:8811–8816.
85. Martowicz ML, Grass JA, Boyer ME, Guend H, Bresnick EH (2005). Dynamic GATA factor interplay at a multi-component regulatory region of the GATA-2 locus. *J Biol Chem* **280**:1724–1732.
86. Pal S, Cantor AB, Johnson KD, Moran T, Boyer ME, Orkin SH, Bresnick EH (2004). Coregulator-dependent facilitation of chromatin occupancy by GATA-1. *Proc Natl Acad Sci USA* **101**:980–985.
87. Bresnick EH, Martowicz ML, Pal S, Johnson KD (2005). Developmental control via GATA factor interplay at chromatin domains. *J Cell Physiol*, in press.
88. Pal S, Nemeth MJ, Bodine DM, Miller JL, Svaren J, Thein SL, Lowry PJ, Bresnick EH (2004). Neurokinin-B transcription in erythroid cells: direct activation by the hematopoietic transcription factor GATA-1. *J Biol Chem* **279**:31348–31356.
89. Letting DL, Chen YY, Rakowski C, Reedy S, Blobel GA (2004). Context-dependent regulation of GATA-1 by friend of GATA-1. *Proc Natl Acad Sci USA* **101**:476–481.
90. Lin AC, Roche AE, Wilk J, Svensson EC (2004). The N termini of Friend of GATA (FOG) proteins define a novel transcriptional repression motif and a superfamily of transcriptional repressors. *J Biol Chem* **279**:55017–55023.
91. Hong W, Nakazawa M, Chen YY, Kori R, Vakoc CR, Rakowski C, Blobel GA (2005). *EMBO J* **24**: 2367–2378.
92. Forsberg EC, Downs KM, Bresnick EH (2000). Direct interaction of NF-E2 with hypersensitive site 2 of the beta-globin locus control region in living cells. *Blood* **96**:334–339.
93. Im H, Grass JA, Johnson, KD, Kim, SI, Boyer ME, Imbalzano, AN, Bieker, JJ, Bresnick EH (2005). Chromatin domain activation via GATA-1 utilization of a small subset of dispersed GATA motifs within a chromatin domain. *Proc Natl Acad Sci USA*, in press.
94. Merika M, Orkin SH (1995). Functional synergy and physical interactions of the erythroid transcription factor GATA-1 with the Kruppel family proteins Sp1 and EKLF. *Mol Cell Biol* **15**:2437–2447.

95. Blobel GA, Nakajima T, Eckner R, Montminy M, Orkin SH (1998). CREB-binding protein cooperates with transcription factor GATA-1 and is required for erythroid differentiation. *Proc Natl Acad Sci USA* **95**:2061–2066.
96. Forsberg EC, Johnson K, Zaboikina TN, Mosser EA, Bresnick EH (1999). Requirement of an E1A-sensitive coactivator for long-range transactivation by the beta-globin locus control region. *J Biol Chem* **274**:26850–26859.
97. Zhang W, Kadam S, Emerson BM, Bieker JJ (2001). Site-specific acetylation by p300 or CREB binding protein regulates erythroid Kruppel-like factor transcriptional activity via its interaction with the SWI-SNF complex. *Mol Cell Biol* **21**:2413–2422.
98. Letting DL, Rakowski C, Weiss MJ, Blobel GA (2003). Formation of a tissue-specific histone acetylation pattern by the hematopoietic transcription factor GATA-1. *Mol Cell Biol* **23**:1334–1340.
99. Kiekhaefer CM, Grass JD, Bresnick EH (2004). WW domain binding motif within the activation domain of the hematopoietic activator p45/NF-E2 is essential for establishment of a tissue-specific histone modification pattern. *J Biol Chem* **279**:7456–7461.
100. Amrolia PJ, Ramamurthy L, Saluja D, Tanese N, Jane SM, Cunningham JM (1997). The activation domain of the enhancer binding protein p45NF-E2 interacts with TAFII130 and mediates long-range activation of the alpha- and beta-globin gene loci in an erythroid cell line. *Proc Natl Acad Sci USA* **94**:10051–10056.
101. Gavva NR, Gavva R, Ermekova K, Sudol M, Shen JCK (1997). Interaction of WW domains with hematopoietic transcription factor p45/NF-E2 and RNA polymerase II. *J Biol Chem* **272**:24105–24108.
102. Chen X, Wen S, Fukuda MN, Gavva NR, Hsu D, Akamo TO, Yang-Feng T, Shen CK (2001). Human itch is a coregulator of the hematopoietic transcription factor NF-E2. *Genomics* **73**:238–241.
103. Kiekhaefer CM, Grass JA, Johnson KD, Boyer ME, Bresnick EH (2002). Hematopoietic activators establish an overlapping pattern of histone acetylation and methylation within a tissue-specific chromatin domain. *Proc Natl Acad Sci USA* **99**:14309–14314.
104. Wolffe A (1995). *Chromatin*. Academic Press.
105. Hager GL, Archer TK, Fragoso G, Bresnick EH, Tsukagoshi Y, John S, Smith CL (1993). Influence of chromatin structure on the binding of transcription factors to DNA. *Cold Spring Harb Symp Quant Biol* **58**:63–71.
106. Felsenfeld G (1992). Chromatin as an essential part of the transcriptional mechanism. *Nature* **355**: 219–224.
107. Fischle W, Wang Y, Allis CD (2003). Histone and chromatin crosstalk. *Curr Opin Cell Biol* **15**: 172–183.
108. Vettese-Dadey M, Grant PA, Hebbes TR, Crane-Robinson C, Allis CD, Workman JL (1996). Acetylation of histone H4 plays a primary role in enhancing transcription factor binding to nucleosomal DNA in vitro. *EMBO J* **15**:2508–2518.
109. Lee DY, Hayes JJ, Pruss D, Wolffe AP (1993). A positive role for histone acetylation in transcription factor access to nucleosomal DNA. *Cell* **72**:73–84.
110. Tse C, Sera T, Wolffe AP, Hansen JC (1998). Disruption of higher-order folding by core histone acetylation dramatically enhances transcription of nucleosomal arrays by RNA polymerase III. *Mol Cell Biol* **18**: 4629–4638.
111. Forsberg EC, Downs KM, Christensen HM, Im H, Nuzzi PA, Bresnick EH (2000). Developmentally dynamic histone acetylation pattern of a tissue-specific chromatin domain. *Proc Natl Acad Sci USA* **97**:14494–14499.
112. Im H, Park C, Feng Q, Johnson KD, Kiekhaefer CM, Choi K, Zhang Y, Bresnick EH (2003). Dynamic regulation of histone H3 methylated at lysine 79 within a tissue-specific chromatin domain. *J Biol Chem* **278**:18346–18352.
113. Bulger M, Schubeler D, Bender MA, Hamilton J, Farrell CM, Hardison RC, Groudine M (2003). A complex chromatin landscape revealed by patterns of nuclease sensitivity and histone modification within the mouse beta-globin locus. *Mol Cell Biol* **23**:5234–5244.
114. Kim A, Dean A (2004). Developmental stage differences in chromatin subdomains of the beta-globin locus. *Proc Natl Acad Sci USA* **101**:7028–7033.
115. Suka N, Suka Y, Carmen AA, Wu J, Grunstein M (2001). Highly specific antibodies determine histone acetylation site usage in yeast heterochromatin and euchromatin. *Mol Cell* **8**:473–479.

116. Litt MD, Simpson M, Gaszner M, Allis CD, Felsenfeld G (2001). Correlation between histone lysine methylation and developmental changes at the chicken beta-globin locus. *Science* **9**:9.
117. Litt MD, Simpson M, Recillas-Targa F, Prioleau MN, Felsenfeld G (2001). Transitions in histone acetylation reveal boundaries of three separately regulated neighboring loci. *EMBO J* **20**:2224–2235.
118. Noma K, Allis CD, Grewal SI (2001). Transitions in distinct histone H3 methylation patterns at the heterochromatin domain boundaries. *Science* **293**:1150–1155.
119. Eissenberg JC, Elgin SC (2000). The HP1 protein family: getting a grip on chromatin. *Curr Opin Genet Dev* **10**:204–210.
120. Jacobs SA, Taverna SD, Zhang Y, Briggs SD, Li J, Eissenberg JC, Allis CD, Khorasanizadeh S (2001). Specificity of the HP1 chromo domain for the methylated N-terminus of histone H3. *EMBO J* **20**: 5232–5241.
121. Bannister AJ, Zegerman P, Partridge JF, Miska EA, Thomas JO, Allshire RC, Kouzarides T (2001). Selective recognition of methylated lysine 9 on histone H3 by the HP1 chromo domain. *Nature* **410**:120–124.
122. Lachner M, O'Carroll D, Rea S, Mechtler K, Jenuwein T (2001). Methylation of histone H3 lysine 9 creates a binding site for HP1 proteins. *Nature* **410**:116–120.
123. Nakayama J, Rice JC, Strahl BD, Allis CD, Grewal SI (2001). Role of histone H3 lysine 9 methylation in epigenetic control of heterochromatin assembly. *Science* **292**:110–113.
124. Cao R, Wang L, Wang H, Xia L, Erdjument-Bromage H, Tempst P, Jones RS, Zhang Y (2002). Role of histone H3 lysine 27 methylation in Polycomb-group silencing. *Science* **298**:1039–1043.
125. Kizer KO, Phatnani HP, Shibata Y, Hall H, Greenleaf AL, Strahl BD (2005). A novel domain in Set2 mediates RNA polymerase II interaction and couples histone H3 K36 methylation with transcript elongation. *J Biol Chem* **25**:3305–3316.
126. Boffa LC, Vidali G, Mann RS, Allfrey VG (1978). Suppression of histone deacetylation in vivo and in vitro by sodium butyrate. *J Biol Chem* **253**:3364–3366.
127. Vidali G, Boffa LC, Bradbury EM, Allfrey VG (1978). Butyrate suppression of histone deacetylation leads to accumulation of multiacetylated forms of histones H3 and H4 and increased DNase I sensitivity of the associated DNA sequences. *Proc Natl Acad Sci USA* **75**:2239–2243.
128. Alter BP, Goff SC (1977). Globin synthesis in mouse erythroleukemia cells in vitro: a switch in beta chains due to inducing agent. *Blood* **50**:867–876.
129. Kameji R, Obinata M, Natori Y, Ikawa Y (1977). Induction of globin gene expression in cultured erythroleukemic cells by butyric acid. *J Biochem (Tokyo)* **81**:1901–1910.
130. Profous-Juchelka HR, Reuben RC, Marks PA, Rifkind RA (1983). Transcriptional and post-transcriptional regulation of globin gene accumulation in murine erythroleukemia cells. *Mol Cell Biol* **3**:229–232.
131. Ginder GD, Whitters MJ, Pohlman JK (1984). Activation of a chicken embryonic globin gene in adult erythroid cells by 5-azacytidine and sodium butyrate. *Proc Natl Acad Sci USA* **81**:3954–3958.
132. Constantoulakis P, Knitter G, Stamatoyannopoulos G (1989). On the induction of fetal hemoglobin by butyrates: in vivo and in vitro studies with sodium butyrate and comparison of combination treatments with 5-AzaC and AraC. *Blood* **74**:1963–1971.
133. Perrine SP, Miller BA, Papayannopoulou T, Faller DV, Rudolph AM, Vichinsky E, Hurst D, Kan YW (1989). Butyrate analogues modulate globin gene expression in human and ovine erythroid cells. *Prog Clin Biol Res* **316B**:341–350.
134. Pace BS, White GL, Dover GJ, Boosalis MS, Faller DV, Perrine SP (2002). Short-chain fatty acid derivatives induce fetal globin expression and erythropoiesis in vivo. *Blood* **100**:4640–4648.
135. Marks PA, Richon VM, Miller T, Kelly WK (2004). Histone deacetylase inhibitors. *Adv Cancer Res* **91**:137–168.
136. Kouzarides T (2000). Acetylation: a regulatory modification to rival phosphorylation? *EMBO J* **19**: 1176–1179.
137. Im H, Grass JA, Christensen HM, Perkins A, Bresnick EH (2002). Histone deacetylase-dependent establishment and maintenance of broad low-level histone acetylation within a tissue-specific chromatin domain. *Biochem* **41**:15152–15160.
138. Lo WS, Trievel RC, Rojas JR, Duggan L, Hsu JY, Allis CD, Marmorstein R, Berger SL (2000). Phosphorylation of serine 10 in histone H3 is functionally linked in vitro and in vivo to Gcn5-mediated acetylation of lysine 14. *Mol Cell* **5**:917–926.

139. Cheung P, Tanner KG, Cheung WL, Sassone-Corsi P, Denu JM, Allis CD (2000). Synergistic coupling of histone H3 phosphorylation and acetylation in response to epidermal growth factor stimulation. *Mol Cell* **5**:905–915.
140. Nowak SJ, Corces VG (2004). Phosphorylation of histone H3: a balancing act between chromosome condensation and transcriptional activation. *Trends Genet* **20**:214–220.
141. Sassone-Corsi P, Mizzen CA, Cheung P, Crosio C, Monaco L, Jacquot S, Hanauer A, Allis CD (1999). Requirement of Rsk-2 for epidermal growth factor-activated phosphorylation of histone H3. *Science* **285**:886–891.
142. Thomson S, Clayton AL, Hazzalin CA, Rose S, Barratt MJ, Mahadevan LC (1999). The nucleosomal response associated with immediate-early gene induction is mediated via alternative MAP kinase cascades: MSK1 as a potential histone H3/HMG-14 kinase. *EMBO J* **18**:4779–4793.
143. Clayton AL, Rose S, Barratt MJ, Mahadevan LC (2000). Phosphoacetylation of histone H3 on c-fos- and c-jun-associated nucleosomes upon gene activation. *EMBO J* **19**:3714–3726.
144. Mahadevan LC, Clayton AL, Hazzalin CA, Thomson S (2004). Phosphorylation and acetylation of histone H3 at inducible genes: two controversies revisited. *Novartis Found Symp* **259**:102–111.
145. Bhanu NV, Trice TA, Lee YT, Miller JL (2004). A signaling mechanism for growth-related expression of fetal hemoglobin. *Blood* **103**:1929–1933.
146. Zheng Y, Thompson PR, Cebrat M, Wang L, Devlin MK, Alani RM, Cole PA (2004). Selective HAT inhibitors as mechanistic tools for protein acetylation. *Methods Enzymol* **376**:188–199.
147. Ley TJ, DeSimone J, Noguchi CT, Turner PH, Schechter AN, Heller P, Nienhuis AW (1983). 5-Azacytidine increases gamma-globin synthesis and reduces the proportion of dense cells in patients with sickle cell anemia. *Blood* **62**:370–380.

15

Vertebrate Models for Sickle Cell Disease Research

by Barry H. Paw, Seong-Kyu Choe, Flavia C. Costa, Shirin V. Sundar and Kenneth Peterson

Introduction

Transgenic Mice

Animal systems that model human diseases provide invaluable resources to understand the molecular mechanisms underlying the pathologies associated with these maladies, as well as provide experimental vehicles in which to test pharmacologic therapeutics and gene therapies. The common house mouse, *Mus musculus*, has been at the forefront of this approach for several decades. Mice have a physiology similar to humans and offer many positive advantages over other animal models. First, several thousand spontaneous and radiation-induced mutant strains are available from a number of vendors. Mutations in homologous genes often result in disease etiologies similar to that in humans. Second, the derivation of inbred and congenic strains provides a uniform genetic background on which to perform studies free from the effect of modifier genes and polymorphisms. More recently, the advent of transgenic and gene targeting methodologies have made it possible to probe the biochemical and molecular pathways of human gene expression, offering the possibility that understanding these processes might lead to points of intervention in disease etiology. It is precisely these two procedures that made mice the animal of choice for modeling sickle cell disease (SCD) and thalassemia. Transgenesis allows introduction of human globin gene sequences or entire loci into the mouse genome and gene targeting permits genetic studies of the endogenous globin loci, which are arranged and regulated to a large degree like the human loci. Finally, the 21-day parturition period in mice, compacts developmental studies into a useful time frame.

Transgenic mice were first produced between 1980 and 1981 by a number of investigators using simple gene or cDNA sequences driven only by promoters. Most of the transgenes integrated in mice with this approach suffered from position effect variegation (PEV). Various *cis*-acting regulatory elements including enhancers, introns, insulators and poly-adenylation signals were included in transgene constructs in an attempt to overcome PEV, which was achieved only after locus control regions (LCRs) were discovered and incorporated into transgenes.[1] More recently, the use of whole locus transgenes, such as yeast artificial chromosomes (YACs), bacterial artificial chromosomes (BACs) and bacteriophage P1 artificial chromosomes (PACs), has resulted in murine models that more accurately mirror human gene expression.[2–4]

Globin Gene Switching

Development of transgenesis technology created opportunities for studying human gene expression *in vivo* within an animal model. Previous work was limited to analysis of transgenes in established cell lines. Murine transgenesis was particularly attractive, since endogenous globin synthesis during embryogenesis had been well characterized.[5] Developmental studies on hematopoiesis and globin gene switching could be carried out on staged embryos and fetuses during pregnancy with the onset of erythropoiesis at day seven post-conception and post-partum in adult mice. All of the hematopoietic tissues present at a particular stage of development could be assessed, including yolk sac, liver, bone marrow, spleen and peripheral blood.

Early Transgenic Models

Initial experiments involved microinjection of simple transgenes encompassing the human adult β-globin gene or the fetal γ-globin gene (Fig. 1A).[6,7] Expression was affected by position-of-integration of the transgenes within the murine genome and copy number-dependence was not observed, hallmarks of PEV. However, when the transgenes were expressed, correct regulation regarding developmental stage- and tissue-specificity was observed, demonstrating that *cis*-regulatory elements controlling developmental expression were gene-proximal. Discovery of the β-locus LCR in 1987 provided a major breakthrough in globin gene switching.[1] The LCR consists of a collection of DNA regions upstream of the β-globin cluster that are hypersensitive to DNaseI. The LCR opens up the globin domain and makes it available for transcription. It is absolutely required for the expression of all the genes within the cluster. When this element was linked to either a globin gene or a heterologous gene, PEV disappeared; that is, site-of-integration-independent, copy number-dependent expression of the linked transgene was obtained.[1] These two properties eventually defined the functional properties of all LCRs subsequently discovered.

A flurry of transgenic experiments using small transgenes defined and characterized a number of *cis*-regulatory elements and molecular mechanisms operative within the human locus. Generally, all of these transgenes used fragments that encompassed one or two human β-like globin genes coupled to derivative LCR sequences, such as mini- or micro-LCR cassettes (μLCR), individual, or multiple DNaseI-hypersensitive sites (HSs). As transgenic technology advanced, and more DNA could be successfully microinjected into oocytes in an intact state, multiple β-like genes could be linked to μLCRs or the intact LCR could be linked to single globin genes in cosmid constructs.[8] These experiments established the function of individual LCR HSs in globin gene expression. A role for gene proximity to the LCR or spatial gene order in globin gene switching was demarcated, although the trans-acting transcriptional environment was established as the primary determinant of globin gene expression.[9,10] Other studies identified an ε-globin gene silencer and its autonomous function in ε-globin gene repression, as well as provided evidence for γ-globin gene silencing.[11,12] Investigators demonstrated that competition between the γ- and β-globin genes for interaction with the LCR was operative during development.[8,13] Other analyses characterized essential or minimal promoter and enhancer functions for the individual genes, identified various regulatory-promoter and drug-response elements, and identified sites of chromatin remodeling complex interaction (reviewed in Refs. 14, 15). The function of both erythroid-specific and ubiquitous transcription factors was also genetically inferred from many studies (reviewed in Ref. 14).

Studies of α-globin transgenics identified the HS-40 upstream master regulatory region, and *cis*-regulatory elements controlling ζ-globin and α-globin gene regulation,[16,17] including the 3′ untranslated region (UTR), which controls α-globin mRNA stability (reviewed in Ref. 18).

YAC and BAC Transgenesis

The accumulation of large amounts of sequence information from the Human Genome Project has led to the development of new technologies that allow gene function to be assessed *in vivo*. BACs

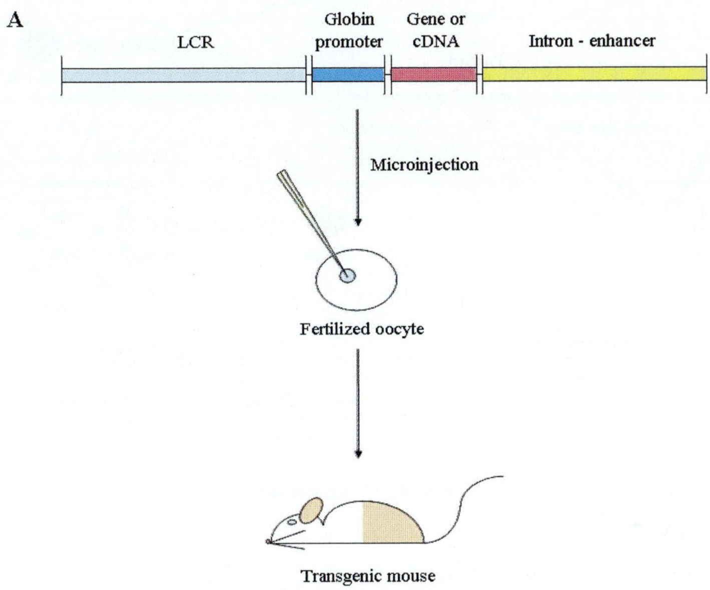

Fig. 1. Transgenesis with small recombinant or whole loci globin constructs. (A) A prototypical recombinant globin transgene is illustrated. Most early transgenes consisted of LCR sequences (mini- or micro-LCRs, one or more 5′HSs, or the entire LCR), a globin promoter linked to its gene sequences, hybrid globin gene sequences, or a cDNA reporter (occasionally a non-globin promoter was linked to globin gene or cDNA cassettes), and a globin or heterologous intron and/or enhancer (or not). These purified transgenes were microinjected into the pronucleus of fertilized mouse oocytes to produce transgenic mice as shown schematically. (B) Diagram of the 213 Kb human β-globin locus yeast artificial chromosome (β-YAC). This β-YAC, or a 150 Kb β-YAC, was used in most published whole loci transgenic studies. The insert is 187 Kb and contains the LCR, the five functional β-like globin genes, 3′HS1 and extensive 5′ and 3′ flanking regions. The LCR 5′HSs and 3′HS1 are indicated by arrows above the line. The genes are shown as colored boxes. Some key restriction enzyme sites are displayed below the line. The YAC arms are represented by rectangles at the ends of the line. Each contains a telomere (not shown) and other chromosome functional elements or selectable markers. ARS1, autonomous replicating sequence; CEN1, centromere; TRP1, tryptophan prototrophy selectable marker; LYS2, lysine prototrophy selectable marker; PGKneo, G418^R selectable marker. (C) Methods used to generate transgenic mice with YACs. The β-YAC may be isolated from the yeast host and microinjected as outlined in (A) or lipofected into embryonic stem (ES) cells, which are then utilized to produce chimeric mice and establish transgenic lines (see Fig. 2). Alternately, the YAC may be transferred to ES cells by yeast spheroplast-ES cell protoplast fusion. Lipofection or fusion must be employed when the YAC exceeds 650–800 Kb in size.

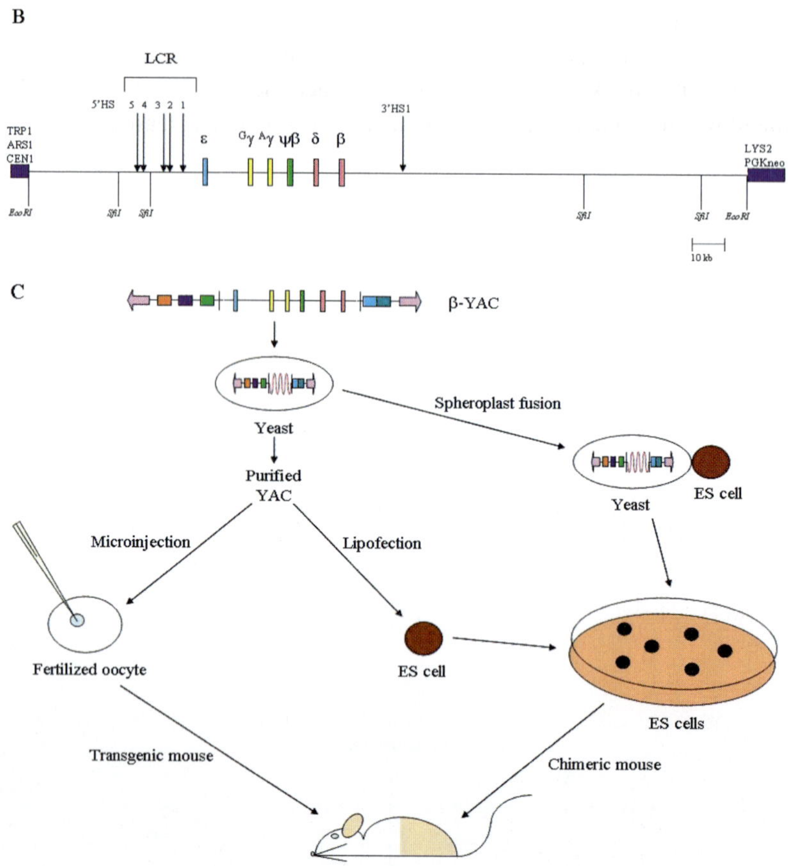

Fig. 1. (*Continued*)

and YACs allow the manipulation of very large stretches of DNA, permitting the analysis of entire gene clusters such as the β-globin locus. The first human β-globin locus yeast artificial chromosome (β-YAC) transgenic mice were produced in 1993,[19,20] ushering in the era of whole locus transgene studies (Fig. 1B). More recently, the use of β-globin locus bacterial artificial chromosomes (β-BACs), coupled with "recombineering" has been employed.[21] Both types of transgenes allow the introduction of alterations within the context of the entire human β-locus using homologous recombination in either yeast or bacteria, respectively, without retention of exogenous DNA sequences. Thus, the effect of mutations can be studied in an intact locus throughout the development of a mouse (Fig. 1C). Initial studies demonstrated that the human globin gene expression pattern was, by and large, recapitulated during ontogeny in the mouse. The concept of the LCR holocomplex or active chromatin hub (ACH) was derived from studies on LCR HS deletions; the function of individual HSs in globin gene switching was ascertained; silencing of the ε- and γ-globin genes was further characterized; the role of gene order and gene-LCR proximity was better assessed; enhancer and chromatin remodeling response elements were tested; and transgenes were utilized to produce hereditary persistence of fetal hemoglobin (HPFH) and sickle cell mouse models (reviewed in Refs. 14, 15).

Insights gained from these transgenic studies opened up many avenues of treatment regimens for SCD and thalassemia, including both pharmacologic and gene therapies. Many of these transgenics were mated with murine knockout mutant lines to produce the mouse models of sickle cell disease and thalassemia now utilized to study the pathophysiology of these diseases.

Knockout Models

Coincidental and in parallel to transgenic studies, were projects analyzing the regulation of the endogenous murine β-like and α-like globin genes. The strategy in this case utilized gene targeting and chimeric mice to produce knockout mutations of genes or globin regulatory sequences (Fig. 2). In general, the deficiencies produced revealed similar mechanisms of action regarding globin gene switching. One important caveat is that mice do not have a fetal globin analogous to the human γ-globin genes. A single switch from embryonic globin synthesis to definitive globin synthesis occurs, whereas two switches exist in humans, from primitive embryonic erythropoiesis to fetal definitive erythropoiesis and later from fetal definitive erythropoiesis to adult definitive erythropoiesis. However, the sites of hematopoiesis during development are conserved between mice and humans. Thus, studies on γ-globin gene expression can only be carried out in mice using the human transgene approach. Knockouts of adult α-globin and β-globin provided the mouse lines that were bred with

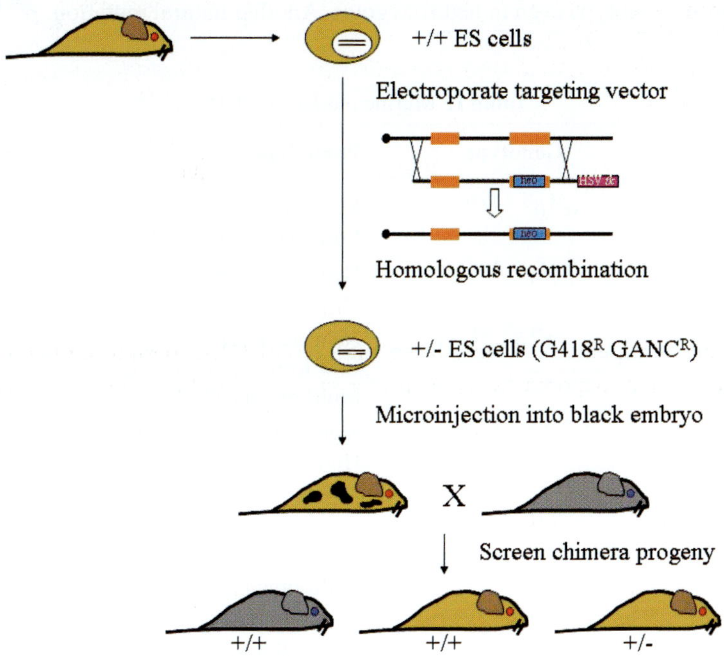

Fig. 2. Generation of targeted mutations in endogenous murine globin loci. ES cells usually derived from 129 mice (brown coat color) are electroporated with a gene targeting construct, a generalized version of which is shown. Following drug selection and screening for proper integration of the targeting vector, the ES cells are microinjected into blastocysts normally derived from C57/BL6 mice (black coat color). Chimeric mice are identified initially based on coat color, which is a varying mixture of brown and black. If the ES cells contributed to the germ line, the targeted mutation will be inherited by the next generation of mice as indicated at the bottom of the figure.

human transgenic lines to produce mouse models of hemoglobinopathies, in which only human globin chains are incorporated into the hemoglobin molecule.

Sickle Cell and Thalassemia Models

A number of mouse models for various hemoglobinopathies have been generated (Table 1). Three sickle cell anemia mouse lines were generated using slightly variant approaches and lines modeling SCD variants and β- or α-thalassemias have been produced. In many instances, the human phenotype for each of these diseases is remarkably conserved in the mouse (reviewed in Ref. 22). These mice offer the possibility for understanding the pathophysiology of these diseases in more detail, as well as serve as vehicles to test emerging pharmaceutical compounds and gene therapy approaches, including isolation and proliferation of stem cells, analysis of gene therapy vectors and corrective constructs, and efficacy of bone marrow transplantation and engraftment of gene-corrected stem cells.

S Antilles and S + S Antilles Models

Hemoglobin S Antilles (HbS Antilles) is a naturally occurring human sickling variant. The $\beta^{S\ Antilles}$-globin gene has two mutations, resulting in a glutamic acid to valine change at codon 6 and a valine to isoleucine change at codon 23. Hemoglobin polymerization is increased in this double mutant, causing a significant disease phenotype even in heterozygotes. Another natural mutation, $\beta^{D\ Punjab}$ (HbD Los

Table 1. Murine models of SCD.

Mouse model	Genotype	Phenotype	Severity of disease
S Antilles	$\alpha^H\beta^{S\ Antilles}$	Mild adult anemia Low irreversible sickling	Mild
S + S Antilles	$\alpha^H\beta^S\beta^{S\ Antilles}$	Neonatal anemia Significant sickling	Mild–moderate
SAD	$\alpha^H\beta^{SAD}$	Neonatal anemia Low irreversible sickling Multi-organ pathology	Moderate
Berkeley model (murine knockout)	$\alpha^H\beta^S\gamma^H\delta^H$	Significant adult anemia High irreversible sickling Chronic multi-organ damage	Moderate–severe
Birmingham model (murine knockout)	$\alpha^H\beta^S\gamma^H$	Severe anemia High irreversible sickling Multi-organ pathology	Moderate–severe
San Francisco model (murine knockout)	$\alpha^H\beta^S\gamma^H\delta^H$ (YAC)	Anemia Irreversible sickling	Moderate–severe
Enhanced γ-globin expression models	$\alpha^H\beta^S$ ($\gamma^{Ha}/\gamma^{Hb}/\gamma^{Hc}$) and $\alpha^H\beta^{SAD}\gamma^H$	Decreased severity of disease with increasing HbF levels	Moderate–severe

α^H, γ^H, δ^H, human α-, γ- and δ-globin genes; β^S, human β-globin gene with HbS mutation; $\beta^{S\ Antilles}$, human β-globin gene with HbS Antilles mutation; β^{SAD}, human β-globin gene with HbS, HbS Antilles and HbD Punjab mutations; $\gamma^{Ha}/\gamma^{Hb}/\gamma^{Hc}$, human γ-globin genes expressing three different levels of HbF. See text for detailed descriptions of each model and references.

Angeles, glutamic acid to glutamine at codon 121) causes pathology only in double heterozygotes for both β^{S} and $\beta^{D\ Punjab}$.

The S Antilles mouse model was created using human α_2- and $\beta^{S\ Antilles}$-globin genes, each linked to the erythroid-specific human LCR 5′HS2, to produce transgenic mice and crossing the resultant animals onto a β-thalassemic background.[23] Approximately 50% of the total β-globin chains in the transgenic animals were $\beta^{S\ Antilles}$. These animals were slightly anemic and had a reduced hematocrit. Measurable induction of irreversible red cell sickling was observed only in response to hypoxia. A modified version of this model exhibited high and balanced synthesis of human α- and $\beta^{S\ Antilles}$-globin genes, erythrocyte sickling and secondary end-organ pathology commonly found in patients with SCD.[24]

The S + S Antilles model sought to increase the severity of disease by combining S Antilles with other murine models. A transgenic mouse line was created by co-integrating the human β^{S}- and α_2-globin genes, each linked to the β-globin LCR. These mice were mated with a mutant mouse line carrying a β^{major}-globin deletion to produce transgenic mice homozygous for the β^{major}-globin deletion.[25,26] Finally, the β^{major}-globin deletion mice bearing the human β^{S}- and α_2-globin transgenes were then crossed with a line expressing human α- and $\beta^{S\ Antilles}$-globins.[27] These doubly transgenic mice had more severe phenotype than either of the parent lines. β^{S}- and $\beta^{S\ Antilles}$-globins together constituted 60–80% of the total β-globin chains in the transgenic animals. Neonatal anemia, significant sickling, enhanced vasoocclusion, RBC destruction and pathology were observed.

SAD Model

Trudel and colleagues synthesized a novel human β-globin gene, β^{SAD}, to obtain increased polymerization of HbS *in vivo*.[28] The triple mutant β^{SAD}-globin was constructed by site directed mutagenesis, and incorporated the β^{S}, $\beta^{S\ Antilles}$ and $\beta^{D\ Punjab}$ mutations ($\beta^{S\text{-}Antilles\text{-}D\ Punjab}$ or β^{SAD}). Human α_2-globin and β^{SAD}-globin were separately linked to the β-globin LCR and co-injected to generate transgenic mice. Hb polymerization was further increased by mating SAD mice with mice homozygous for a β-thalassemia mutation.[28] The β-thal/SAD mice contained 26% HbSAD compared to 19% in SAD mice. This model was characterized by neonatal anemia, a red cell phenotype very similar to human SCD, and death upon exposure to hypoxia.[28,29]

Although these models provided important information about SCD, they only mimicked the sickle cell trait exhibited in human β^{S} heterozygous carriers and not the severe hemolytic anemia that is a hallmark of homozygous sickle cell disease. The mild phenotypes associated with these models were due to the anti-sickling effect of endogenous murine hemoglobin. To eliminate this problem, transgenic mice were generated with deletions of murine adult β-like (β^{major} and β^{minor})[30,31] and α-globin genes.[32] The disease phenotypes of these mice were rescued by introduction of human globin transgenes, leading to the development of improved sickle cell models expressing high levels of HbS.

Berkeley Model

SCD model transgenic mice expressing exclusively human globins are especially relevant for testing anti-sickling agents. The Berkeley transgenic model was developed by Paszty and co-workers[33] to overcome the failure of earlier models to exhibit faithful sickle cell pathology. These mice no longer expressed murine α- and β-globin genes; instead, they expressed exclusively human α-, β^{S}- and γ-globin genes.[33] Three DNA fragments, one containing a human β-globin mini-LCR, the second containing the human α_1-globin gene, and the third containing the human ${}^{G}\gamma$-, ${}^{A}\gamma$-, δ- and β^{S}-globin genes, were used to create transgenics. The ${}^{G}\gamma$- and ${}^{A}\gamma$-globin genes were included because γ-globin has an anti-sickling property that was expected to decrease the likelihood of fetal death in early development prior to activation of the adult globin genes.[33] Transgenic mice thus created were successively bred with knockout mice heterozygous for murine α- and β-globin deletions to generate mice homozygous for the deletions and containing the transgenes. Erythrocytes in the adult Berkeley mice synthesized exclusively human α- and β-globin. These mice exhibited abundant irreversibly sickled red cells, anemia, and chronic multi-organ damage. This model also showed some characteristics of β-thalassemia, including an increased susceptibility to oxidative damage due to free-radical production.

Birmingham Model

Ryan and Townes were among the first investigators to produce transgenic mice that displayed the sickle trait.[34] Subsequent improvements led to the creation of a more comprehensive SCD mouse model in which a DNA fragment containing human ${}^{A}\gamma$- and β^{S}-globin genes were co-injected with a fragment containing an α_1-globin gene (both driven by the human β-globin LCR). The transgenic mice thus produced were bred to mouse α- and β-globin knockout lines to produce transgenic animals homozygous for the knockout alleles.[35] Similar to the Berkeley mice, adult mice did not produce any murine hemoglobin. These mice showed severe hemolytic anemia, a large number of irreversibly sickled cells, a marked reduction in red cell count and Hb concentration, and multiple organ pathology. The course of hemolytic anemia mimicked the onset of anemia in human sickle cell infants.

San Francisco Model

Although co-injection of human globin constructs in the Berkeley and Birmingham models led to the integration of all the genes at a single chromosomal site, normal gene order and spatial organization required for appropriate expression of the members of the β-like globin gene family was disrupted. To remedy this, Chang *et al.*[36] created novel sickle cell mice harboring a β^{S}-globin YAC containing the entire human β-globin gene cluster. The β^{S}-YAC transgenic mice were bred with β^{major}-, β^{minor}-globin knockout mice to eliminate murine β-globin gene expression. Transgenic mice expressing exclusively human α-globin were separately generated by crossing a transgenic line carrying two copies of the human α_2-globin gene linked to a mini-LCR with mice heterozygous for a deletion of the murine α_1- and α_2-globin genes. The β^{S}-YAC mice were mated with the mini-LCRα_2 mice to produce viable animals with hemolytic anemia and irreversibly sickled red blood cells. This transgenic mouse line should be very useful to study globin switching and for evaluating new SCD therapeutics aimed at reactivating fetal γ-globin.

Enhanced γ-globin Expression Models

A new generation of transgenic studies describes the effect of increasing the level of human γ-globin in transgenic mice. Blouin *et al.*[37] mated transgenic mice that express various levels of human ${}^{A}\gamma$-globin with the SAD mouse line and found improved life span, pathology and hematological profile. They were also able to establish the physiological range of HbF that would alleviate the SCD condition in mice. Another set of knockout mice expressing exclusively human globins and three levels of HbF (low, medium and high) also showed a more faithful SCD pathology than the previous models.[38] In this model, knockout mice created by Paszty *et al.*[33] and Yang *et al.*[31] were mated with mice that expressed the co-integrated mini-LCRα_2 and mini-LCRβ^S used to create the S + S Antilles model. Lethality of these mice was rescued by breeding with transgenic mice expressing three different human γ-globin constructs (γL, γM, γH) with increasing postnatal HbF levels. These mice had high reticulocyte counts and anemia, which were generally more severe than the Berkeley mice expressing comparable levels of HbF.[22] These mice also exhibited a more balanced globin chain synthesis. The features presented by this model make it a closer approximation of human SCD and a good model for screening anti-sickling drugs and evaluating gene therapy protocols.

Thalassemia Models

Mouse models with deletions in β^{major}- and α-globin chains show varying severity of disease.[30,32,39] Successful gene therapy in mice using lentiviral vectors has been reported by different groups.[40,41] A more recent study that rescued a severely β-thalassemic mouse model using lentivirus mediated globin gene transfer strongly supports the efficacy of gene therapy in severe hemoglobinopathies.[42]

In Vivo Drug Screening

Increased HbF level is of considerable importance in improving the clinical symptoms of sickle cell patients. Pharmacological agents that are capable of up regulating the γ-globin genes have been studied in mice. Several drugs are available with distinct pathways of action, and a few of them, including hydroxyurea, butyric acid-related chemicals and 5-azacytidine, have been used in clinical trials of thalassemia and sickle cell patients. HbF induction by some of these compounds involves initial rapid regeneration of erythroid precursors followed by a cytoreduction phase. Changes in chromatin structure and transcription factor binding have been identified as the molecular mechanism of action for other pharmacological agents.

Hydroxyurea

Hydroxyurea, a ribonucleotide reductase inhibitor that exhibits an anti-tumor effect by arresting DNA synthesis, is largely used in the chemotherapy of myeloproliferative diseases. This agent has recently been approved for augmenting HbF expression in SCD and other hematological disorders. The administration of hydroxyurea to thalassemic mouse increased the synthesis of the β^{minor}-globin chain and ameliorated the disease phenotype which underscored its efficacy in humans.[43] Transgenic SAD mice that mimic the clinical and pathological features of human SCD treated with hydroxyurea have significantly increased mean corpuscular volume (MCV), hemoglobin and reticulocyte counts.[44] There is also an increase in potassium content, but no enhancement of Gardos channel activity in erythroid cells from treated SAD mice.

Short Chain Fatty Acid Analogs

Butyrate and other short chain fatty acid analogs such as α-aminobutyric acid are compounds that have been reported to enhance the synthesis of HbF in human erythroid progenitor cells of normal, sickle cell and thalassemia patients.[45,46] These reagents enhance HbF levels, in part, by increasing transcription of the γ-globin gene through changes in histone acetylation. Recently, butyrate was shown to activate transcription through the MEK-ERK pathway of signal transduction in K562 cells.[47] Another short chain fatty acid, valproic acid, was reported to induce HbF in K562 cells by modulating the p38 MAPK pathway.[48] These data suggest that the MAPK pathway plays a role in HbF induction (see Chapter 11). The administration of sodium butyrate to β-YAC transgenic mice was shown to increase the expression of γ-globin mRNA levels, but only after pre-treatment of the mice with 5-azacytidine.[49]

Zebrafish

Studying human diseases using animal models has greatly facilitated the understanding of molecular mechanisms governing causes and consequences of the diseases. Zebrafish was originally introduced as a model organism for studying neurobiology and developmental biology. Since its introduction, advantages of its developmental biology combined with genetic screens have made zebrafish a powerful model for studying vertebrate organogenesis, and such studies have contributed to our understandings of basic and pathological biology.

In this part of the chapter, zebrafish models for organogenesis will be introduced with an emphasis on models for hematopoiesis. Available approaches using the zebrafish and its applications for studying human diseases will be discussed.

Zebrafish as a Model System for Organogenesis

Zebrafish has been developed as a genetic model for studying vertebrate organogenesis by providing several advantages during embryonic development. First, organ development in zebrafish is characteristic of vertebrate organisms. Molecular mechanisms occurring during zebrafish organogenesis is largely shared with those found in higher vertebrates. Second, zebrafish provides transparent and externally developing embryos. These embryos can be examined directly under the microscope and readily accessible for many experimental manipulations, such as microinjection, cell-lineage tracing and cell transplantation. Third, zebrafish provides large brood size of embryos (~200 eggs/spawn per week), which rapidly develop to sexual maturity within 3–4 months so that genetic studies can be performed. These characteristics allowed a number of genetic screens that identified novel genes and additional functions for known genes during zebrafish development. Previously performed forward genetic screens in zebrafish have identified hundreds of mutants that cover major areas of organogenesis.[50,51] For hematopoiesis in particular, the vast majority of the genes recovered from the first generation genetic screens represent erythroid-specific lineage defects.[52,53] Recent application of other lineage markers such as *pu.1* and *myeloperoxidase* (*mpo*) for myelopoiesis[54,55] and *rag-1* for lymphocyte development[56] has shown fruitful outcome from forward genetic screens to isolate mutants defective in other hematopoietic lineages.

Zebrafish as a Model for Hematopoiesis

Hematopoiesis is a regulated process that produces the full complement of terminally differentiated blood cell types from hematopoietic progenitors and stem cells. Stem cells originate from the mesodermal layer of the embryo which becomes specified to a hematopoietic fate. In common with other vertebrates, zebrafish blood cell development occurs in two distinct waves: primitive (embryonic) and definitive (adult). Primitive hematopoiesis occurs in the intraembryonic intermediate cell mass (ICM) when blood circulation is initiated and blood cells, largely of the erythroid lineage, are produced.[57] The ventral wall of the developing aorta in zebrafish serves as the site for initiating definitive hematopoiesis, which is complementary to dorsal mesentery (region of aorta, gonad and mesonephros, or AGM in short) in mammals. Subsequently, the kidney serves as the site for definitive hematopoiesis in zebrafish.[57] In mammals, definitive hematopoietic stem cells originated in the AGM are believed to enter the circulation to populate the fetal liver and eventually the bone marrow for life-long production of stem and blood cells.[58] Zebrafish blood contains red cells, granulocytes,[59] lymphocytes[60] and platelet-equivalent hemostatic cells, called thrombocytes.[61] Zebrafish red cells are nucleated as in amphibian, reptilian, and avian organisms. They share similar molecular, functional and morphological characteristics (Figs. 3A, B) with corresponding blood cells in other vertebrates.

Large-scale mutagenesis screens have identified many mutants defective in hematopoiesis. For examples, mutants defective in early patterning molecules (e.g., *bmp* and *chordin*), lineage-specific transcription factors (e.g., *gata-1*), genes required for formation of hemoglobin (e.g., *globin*, *ferroportin-1*) and structural membrane proteins for erythroid cells (e.g., *band3*, *β-spectrin*, *protein 4.1R*) have been isolated.[62] These mutants were categorized in terms of defects during hematopoiesis. Mutations in genes involved in body patterning during early embryogenesis display global defects indirectly producing abnormal hematopoiesis. One example with such broad defects is found in the *swirl* (*swl*) mutant where early dorsoventral patterning is disturbed due to mutations in the bone morphogenetic protein (BMP) required for the formation of ventral structures, such as blood and nephros.[63] Other mutants defective in processes of hematopoiesis display more defined phenotypes

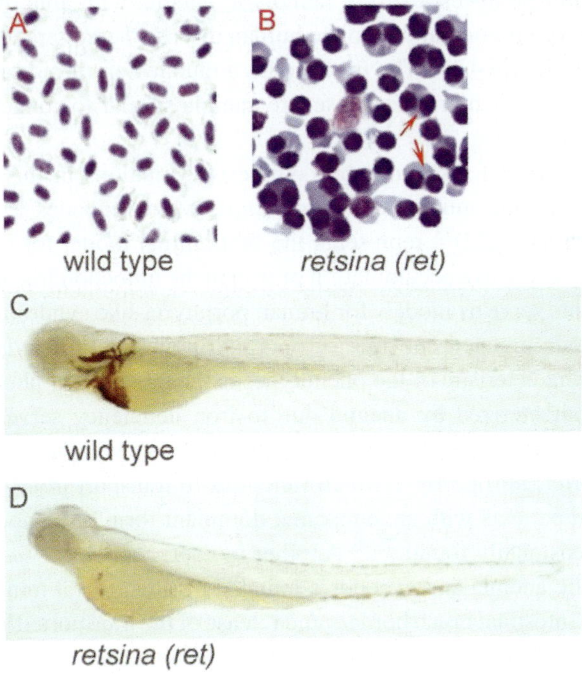

Fig. 3. Blood and gross embryologic morphology in wild type and mutant zebrafish. (A, B) Wild type blood shows normal nucleated erythrocytes (A), while the *ret* mutant contains erythroblasts some with two nuclei (B, arrows). (C, D) Hemoglobinized cells stained with O-dianisidine are noted in wild type embryo at 72 hours post-fertilization (hpf). In contrast to wild type (C), the *ret* mutant is grossly deficient in circulating erythrocytes, shown by lack of O-dianisidine stained cells (D). Used with permission from Paw BW *et al.*[72]

so that they may represent defects in critical steps during hematopoiesis. An example of such mutants is *cloche*, in which both the hematopoietic and endothelial lineages are affected.[64,65] The *cloche* mutant therefore represents a valuable model for studying common progenitors producing the endothelial and blood lineages.

For erythroid lineage development, several mutant models for studying human hematological disorders have been identified based on zebrafish phenotypes (Table 2). During the different stages of erythrocyte development, mutants defective in one of the steps may severely interfere with normal erythropoiesis.

The *vlad tepes* (*vlt*) mutant is characterized by severe anemia due to complete loss of circulating red cells leading to embryonic lethality.[66] This is due to a mutation in *gata-1*, a critical factor for differentiation and survival of erythroid cells.[67,68] Another mutant, *moonshine* (*mon*) is defective in both primitive and definitive erythropoiesis. Recent cloning of the mutated gene has demonstrated that the *mon* gene encodes the zebrafish ortholog of mammalian intermediary factor 1γ (Tif1γ) whose function had not been reported in hematopoiesis.[69] In summary these mutants confirmed gene function in different organisms or identified novel genes involved in erythropoiesis.

Mutations in genes that are structurally important for red cell membrane, including *riesling* (*ris*), *merlot/chablis* (*mot/cha*) and *retsina* (*ret*) cause defects in hematopoietic proliferation and differentiation due to the hemolytic anemia caused by the defective membrane structure (Fig. 3). The *ris* encodes a mutated erythroid specific β-spectrin isoform and is characterized by a phenotype similar to spectrin deficiency in humans and mice.[70] The *mot/cha* encodes a mutated erythroid specific protein 4.1R isoform and produces a severe hemolytic anemia with abnormal red cell membrane structure.[71] This phenotype is comparable to the human disease, hereditary elliptocytosis, caused by dysfunctional or absent protein 4.1R. The *ret* encodes a mutated erythroid specific anion exchanger 1 (AE1 or band 3 protein) which produces a severe anemia similar to congenital dyserythropoietic anemia type II in humans.[72]

Heme synthesis is critical for hemoglobin synthesis. Mutations in this pathway cause severe defects in erythropoiesis. One mutant *sauternes* (*sau*), which encodes a mutated version of δ-aminolevulinate syntase (ALAS2)[73] represents the first animal model for congenital sideroblastic anemia characterized by excessive accumulation of iron in the mitochondria and defective heme synthesis. Two other mutants serve as models for human porphyria-like syndrome are *yquem* (*yqe*) and *dracula* (*dra*) that encode mutated enzymes, uroporphyrinogen decarboxylase and ferrochelatase, respectively.[74,75] One characteristic of the phenotypes is the presence of photosensitive red cells.

Several mutants characterized by anemia due to iron deficiency serve as models for human diseases, such as hemochromatosis and microcytic anemia. The *weissherbst* (*weh*) mutant encodes a mutated iron transporter, ferroportin-1 which functions to transport iron from the yolk sac to the circulation.[76] In human patients with an autosomal dominant form of hemochromatosis, mutations in ferroportin-1 are occasionally found.[77,78] Another mutant, *chadonnay* (*cdy*), is characterized by hypochromic, microcytic anemia and encodes a mutated divalent metal transporter-1 that mediates iron uptake at the gastrointestinal brush borders and release of the transported iron in the cytosol of red blood cells.[79] Recently, *chianti* (*cia*) was found to encode an erythroid-specific isoform of *transferrin receptor-1* (*trf1a*) required for iron uptake exclusively by developing erythrocytes.[80] *cia* displays hypochromic, microcytic anemia due to inability to obtain iron in erythrocytes. Since zebrafish has two isoforms of transferring receptor-1 (*trf-1a* and *trf-1b* in non-hematopoietic cells), *cia* provides a model for studying Trf-1 function in erythropoiesis without interfering with developmental processes in other organs.

Table 2. Zebrafish mutants for human hematopoietic disease models.

Genes	Mutants	Defective steps	Models for human diseases	References
GATA-1	*Vlad tepes*	Erythropoiesis	X-linked Anemia and Thrombocytopenia	66
Tif-1γ	*Moonshine*	Erythropoiesis	Diamond Blackfan Anemia	69
ALAS-E	*Sauternes*	Heme Synthesis	X-linked Sideroblastic Anemia	73
β-Spectrin	*Riesling*	Membrane Structure	Hereditary Spherocytosis (HS)	70
Protein 4.1R	*Merlot/Chablis*	Membrane Structure	Hereditary Elliptocytosis	71
Anion Exchanger-1	*Retsina*	Membrane Structure and Cytokinesis	Congenital Dyserythropoietic Anemia Type II/HS	72
Ferrochelatase	*Dracula*	Heme Biosynthesis	Erythropoietic Protoporphyria	75
URO-D	*Yquem*	Heme Biosynthesis	Hepatoerythropoietic Porphyria	74
Ferroportin-1	*Weissherbst*	Iron Metabolism	Hemochromatosis	76
DMT-1	*Chardonnay*	Iron Transport	Microcytic Anemia	79
Transferrin receptor	*Chianti*	Iron Transport	Microcytic Anemia	80
Globin	*Zinfandel*	Globin Synthesis	Thalassemia	81

Given that molecular events during hematopoiesis are highly conserved among vertebrates, these mutants are invaluable resources to dissect pathology of the diseases in humans, which will eventually contribute to drug design for prevention and treatment of such diseases.

Zebrafish Globins

The zebrafish genome contains multiple globin genes that are regulated in a development-stage specific manner in common with other vertebrates. The embryonic globin genes in zebrafish are strictly expressed in erythroid cell lineages during early development (Fig. 4).[81] Likewise, adult globin gene expression is only found in circulating adult erythroid cells, but not in embryonic erythroid cells.[82] An extensive analysis of the globin genes in another teleost species, medaka (*Oryzias latipes*), has recently been reviewed.[83]

There are two globin loci in the zebrafish, which differ in organization from the mammalian counterparts (Fig. 4).[81] First, one locus (linkage group 12; LG12) contains only the embryonic gene cluster and the other locus (LG3) contains a cluster of embryonic and adult globin genes. Second, these loci contain both α- and β-genes. Third, some globin genes show head to head orientation rather than sequential orientation arranged with early expressed gene in the $5'$ end as in the human globin loci. Therefore, the organization of the zebrafish globin loci features a unique character and provides an evolutionary insight leading to alternative mechanism for globin gene regulation.

Since the combination of 2α and 2β chains along with heme group produces a functional hemoglobin molecule, expression of globin genes during development is tightly regulated to ensure hemoglobin switching. Defects in such regulation have been shown to cause human diseases, such as thalassemias. One zebrafish mutant, *zinfandel* (*zin*), was mapped to LG3 near the major globin gene locus.[81] *zin* is characterized by decreased embryonic red cell production, recovery of anemia during adult life and dominant inheritance. The phenotype observed in the *zin* mutants may be due to

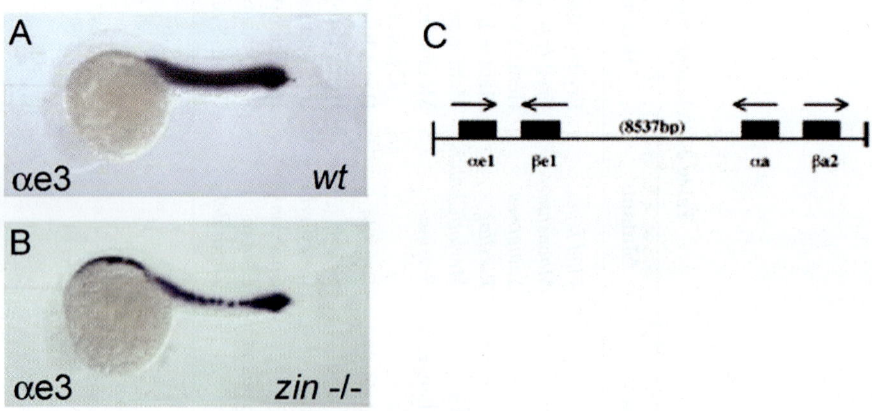

Fig. 4. Embryonic globin expression in zebrafish embryos and zebrafish globin locus. (A, B) Zebrafish embryos at 24 hpf processed by whole mount *in situ* hybridization show expression of the αe3 globin in the yolk sac equivalent [the intermediate cell mass (ICM)]. Expression of the αe3 globin gene is normal in ICM of wild type (A), but reduced in *zin* mutant (B). (C) An embryonic globin locus in zebrafish shows a unique character. This locus contains both α- and β-like counterparts in mammals and shows mixed gene orientation. *e* and *a* represent embryonic and adult globin genes, respectively. α and β represent α- and β-like clusters in mammals, respectively. Arrows above the globin locus indicate the direction of gene expression. Used with permission from Brownlie A *et al.*[81]

abnormalities in the regulatory sequence of the embryonic globin genes.[81] This suggestion has highlighted *zin* as a model for thalassemia in human (Fig. 4). Further identification and characterization of genes for hematopoietic mutants may improve our understanding of the mechanisms of balanced expression of globin genes in a time-dependent manner and advance the search for novel therapies in thalassemia and SCD.

Zebrafish Transgenesis to Study Hematopoiesis

Generation of transgenic animals that express a visible marker in a specific tissue has greatly facilitated studies for cell lineage development as well as genetic studies. Recently, several transgenic zebrafish have been established to investigate hematopoietic development. These transgenic animals provide a direct visualization of specific blood lineages. For erythroid lineage development, the *gata-1* promoter has been used to drive a marker gene expression, such as green fluorescent protein (GFP). These transgenic fish were generated to visualize erythrocytes *in vivo* (Fig. 5). The expression of the marker gene has been shown to recapitulate endogenous expression of *gata-1*, which is required for erythroid cell development.[84] For myeloid and lymphoid lineages, specific promoters such as *pu.1* and *rag-1*, respectively, have been utilized to drive GFP expression in lineage-restricted manner.[85–89] To trace thrombocytes *in vivo*, the *CD41* promoter has been used to induce GFP expression specifically in mature thrombocytes (Fig. 5).[90] These hematopoietic lineage-specific transgenic animals can also provide means to generate cell-type specific cDNA libraries. A lineage-specific zebrafish cDNA

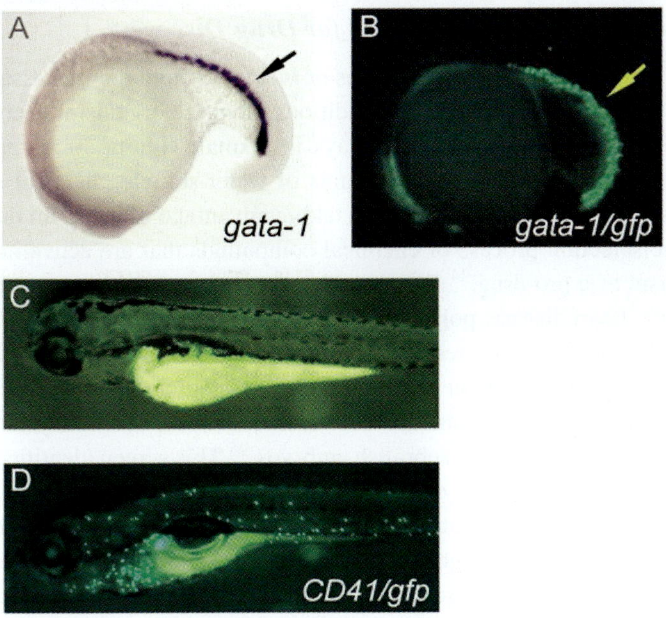

Fig. 5. Stable transgenic zebrafish expressing GFP in specific blood lineages. (A) Whole mount *in situ* hybridization using the *gata-1* RNA probe detects developing erythroid cells in the yolk sac equivalent (ICM) at 20 somite stage of the embryo. (B) A *gata-1/gfp* transgenic animal shows GFP expression in the same region shown in A.[84] Arrows indicate the ICM. (C) A wild type animal at 6 days post-fertilization (dpf) is shown. (D) A *CD41/gfp* transgenic animal at 6 dpf shows GFP-expressing thrombocytes in the circulation.[90] Figures have been generously provided by Wingert (A, B) and Shafizadeh (C, D).

library constructed from highly purified cells can be utilized to look for genes uniquely expressed in the lineage. In efforts to isolate hematopoietic-specific genes, one study has successfully utilized the *gata-1:gfp* transgenic zebrafish to construct cDNA library and found a novel zebrafish hematopoietic death receptor.[91] This study has validated the utility of transgenic zebrafish to identify candidate genes required for development of specific lineages.

Since one can introduce expression of a gene in a specific tissue by using a tissue-specific promoter, this has allowed the creation of disease models. Forced expression of the mouse *myc* oncogene under the regulation of *rag-2* promoter produces leukemic transformation in the T-cell lineage.[87] Further investigation showed that the leukemic cells infiltrate muscle, kidney and guts. This transgenic zebrafish model made it possible to systemically investigate the pathogenesis of T cell lymphoblastic leukemia and may serve as a screening tool to develop novel drugs to treat leukemia.

Transgenic animals can also be used for hematopoietic cell transplantation that may provide insights into hematopoiesis, stem cell biology and leukemogenesis. Recent studies have established zebrafish hematopoietic cell transplantation in which sub-lethally irradiated animals can be rescued by transplanting zebrafish hematopoietic cells.[88,92] In these studies, transgenic animals expressing fluorescent markers to identify different hematopoietic lineages have been utilized to identify repopulated cells derived from transplanted donor hematopoietic cells. Accordingly, zebrafish hematopoietic cell transplantation, along with ability to isolate major zebrafish hematopoietic lineages based on size and granularity of the cells,[93] allows one to test cell autonomy of hematopoietic mutant gene function and further provides tools to study hematopoietic stem cell biology.

High Throughput Small Molecule Screens for Drug Discovery

Conventional drug screens have utilized cell lines or *in vitro* protein binding assays. However, these assays may not be relevant to physiological conditions in whole organisms. Zebrafish embryos are optically transparent and can survive several days in a small volume of water and are therefore amenable to large-scale studies. Indeed, thousands of water-soluble chemical compounds can be applied to embryos arrayed in multi-well plates, taking advantage of a system that would be particularly suited for the selection process of chemical compounds that are active in a whole organism or require metabolism of a pro-drug.[94] Furthermore, many zebrafish mutants can be represented as models for congenital heart disease, polycystic kidney disease, cancer and anemia, and thus offer an excellent whole-system approach for testing candidate drugs for diverse human diseases.

One screen performed by Peterson and colleagues[95] tested compounds from a small molecule library for the ability to induce developmental defects in the central nervous system, the cardiovascular system, pigmentation and the ear in zebrafish embryos.[95] This screen identified many compounds with specific and lethal effect on the developing zebrafish embryos. This result has validated zebrafish as a screening tool to identify potent and specific compounds. Accordingly, such screens in search of chemical compounds that mimic genetic mutations or suppress mutant phenotypes have successfully been performed.[96–98] These screens isolated chemicals that induce specific developmental defects and one of the identified molecules is a potentially therapeutic drug that inhibits tubulin polymerization *in vivo*.[98] More recently similar screens have been performed for hematopoiesis using *gata-1:gfp* transgenic zebrafish to screen a library of 5000 synthetic compounds.[99] Taken together, these data illustrate that zebrafish provides a powerful system to screen chemical libraries in search of promising therapeutic molecules that may prevent or treat human diseases including the hemoglobin disorders.

Summary

Mouse models of globin gene switching and SCD and zebrafish models for hematopoiesis were described with an emphasis on the development of disease models for human hemoglobinopathies. SCD and β-thalassemia models in mice are quite advanced and are now employed in the testing of pharmacologic compounds and development of safe gene therapy approaches. As a complementary tool to the studies using mice, zebrafish provides many hematopoietic disease models to study molecular mechanisms underlying pathogenesis. Although most genes responsible for the mutant phenotypes described earlier have been identified as already known players resulting in human diseases, a few novel genes have also been discovered.[69,76,100] For example, the recent identification of glutaredoxin 5 (*grx5*), the gene disrupted in the hypochromic *shiraz* mutation, has afforded a mechanistic explanation to integrate mitochondrial Fe-S cluster biogenesis with heme metabolism.[100] These results have validated the utility of zebrafish as a model system to study organogenesis. In addition, ongoing cloning and characterization of mutated genes from the forward genetic screens will further identify additional novel genes. This gene identification process should improve with current advances in zebrafish genome sequence projects, as the density of genetic maps and genomics infrastructure continue. Therefore, with advances in genome informatics, available technologies and plenty of disease models accounting for hematopoietic processes, studies using mice and zebrafish should provide a vast amount of information that will lead to directions in the treatment as well as prevention of SCD and other human hematopoietic diseases.

References

1. Grosveld F, van Assendelft GB, Greaves DR, Kollias G (1987). Position-independent, high-level expression of the human β-globin gene in transgenic mice. *Cell* **51**:975–985.
2. Kaufman RM, Pham CT, Ley TJ (1999). Transgenic analysis of a 100-kb human β-globin cluster-containing DNA fragment propagated as a bacterial artificial chromosome. *Blood* **94**:3178–3184.
3. Peterson KR (2003). Transgenic mice carrying yeast artificial chromosomes. *Expert Rev Mol Med* **2003**: 1–25.
4. Katsantoni EZ, de Krom M, Kong-a-San J, Imam AM, Grosveld F, Anagnou NP, Strouboulis J (2004). An embryonic-specific repressor element located 3′ to the $^{A}\gamma$-globin gene influences transcription of the human β-globin locus in transgenic mice. *Exp Hematol* **32**:224–233.
5. Trimborn T, Gribnau J, Grosveld F, Fraser P (1999). Mechanisms of developmental control of transcription in the murine α- and β-globin loci. *Genes Dev* **13**(1):112–124.
6. Chada K, Magram J, Raphael K, Radice G, Lacy E, Costantini F (1985). Specific expression of a foreign β-globin gene in erythroid cells of transgenic mice. *Nature* **314**:377–380.
7. Kollias G, Wrighton N, Hurst J, Grosveld F (1986). Regulated expression of human $^{A}\gamma$, β-, and hybrid $\gamma\beta$-globin genes in transgenic mice: manipulation of the developmental expression patterns. *Cell* **46**:89–94.
8. Enver T, Raich N, Ebens AJ, Papayannopoulou T, Costantini F, Stamatoyannopoulos G (1990). Developmental regulation of human fetal-to-adult globin gene switching in transgenic mice. *Nature* **344**:309–313.
9. Hanscombe O, Whyatt D, Fraser P, Yannoutsos N, Greaves D, Dillon N, Grosveld F (1991). Importance of globin gene order for correct developmental expression. *Genes Dev* **5**:1387–1394.
10. Peterson KR, Stamatoyannopoulos G (1993). Role of gene order in developmental control of human γ- and β-globin gene expression. *Mol Cell Biol* **13**:4836–4843.
11. Dillon N, Grosveld F (1991). Human γ-globin genes silenced independently of other genes in the β-globin locus. *Nature* **350**:252–254.
12. Raich N, Papayannopoulou T, Stamatoyannopoulos G, Enver T (1992). Demonstration of a human ε-globin gene silencer with studies in transgenic mice. *Blood* **79**:861–864.

13. Behringer RR, Ryan TM, Palmiter RD, Brinster RL, Townes TM (1990). Human γ- to β-globin gene switching in transgenic mice. *Genes Dev* **4**:380–389.
14. Harju S, McQueen KJ, Peterson KR (2002). Chromatin structure and control of β-like globin gene switching. *Exp Biol Med* **227**:683–700.
15. de Laat W, Grosveld F (2003). Spatial organization of gene expression: the active chromatin hub. *Chromosome Res* **11**:447–459.
16. Higgs DR, Wood WG, Jarman AP, Sharpe J, Lida J, Pretorius IM, Ayyub H (1990). A major positive regulatory region located far upstream of the human α-globin gene locus. *Genes Dev* **4**:1588–1601.
17. Sharpe JA, Wells DJ, Whitelaw E, Vyas P, Higgs DR, Wood WG (1993). Analysis of the human α-globin gene cluster in transgenic mice. *Proc Natl Acad Sci USA* **90**:11262–11266.
18. Waggoner SA, Liebhaber SA (2003). Regulation of α-globin mRNA stability. *Exp Biol Med* **228**:387–395.
19. Gaensler KM, Kitamura M, Kan YW (1993). Germ-line transmission and developmental regulation of a 150-kb yeast artificial chromosome containing the human β-globin locus in transgenic mice. *Proc Natl Acad Sci USA* **90**:11381–11385.
20. Peterson KR, Clegg CH, Huxley C, Josephson BM, Haugen HS, Furukawa T, Stamatoyannopoulos G (1993). Transgenic mice containing a 248-kb yeast artificial chromosome carrying the human β-globin locus display proper developmental control of human globin genes. *Proc Natl Acad Sci USA* **90**:7593–7597.
21. Narayanan K, Williamson R, Zhang Y, Stewart AF, Ioannou PA (1999). Efficient and precise engineering of a 200 kb β-globin human/bacterial artificial chromosome in *E. coli* DH10B using an inducible homologous recombination system. *Gene Ther* **6**:442–447.
22. Nagel RL, Fabry ME (2001). The panoply of animal models for sickle cell anaemia. *Br J Haematol* **112**:19–25.
23. Rubin EM, Witkowska HE, Spangler E, Curtin P, Lubin BH, Mohandas N, Clift SM (1991). Hypoxia-induced in vivo sickling of transgenic mouse red cells. *J Clin Invest* **87**:639–647.
24. Popp RA, Popp DM, Shinpock SG, Yang MY, Mural JG, Aguinaga MP, Kopsombut P, Roa PD, Turner EA, Rubin EM (1997). A transgenic mouse model of hemoglobin S Antilles disease. *Blood* **89**:4204–4212.
25. Fabry ME, Costantini F, Pachnis A, Suzuka SM, Bank N, Aynedjian HS, Factor SM, Nagel RL (1992). High expression of human β^S- and α-globins in transgenic mice: erythrocyte abnormalities, organ damage, and the effect of hypoxia. *Proc Natl Acad Sci USA* **89**:12155–12159.
26. Fabry ME, Nagel RL, Pachnis A, Suzuka SM, Costantini F (1992). High expression of human β^S- and α-globins in transgenic mice: hemoglobin composition and hematological consequences. *Proc Natl Acad Sci USA* **89**:12150–12154.
27. Fabry ME, Sengupta A, Suzuka SM, Costantini F, Rubin EM, Hofrichter J, Christoph G, Manci E, Culberson D, Factor SM, Nagel RL (1995). A second generation transgenic mouse model expressing both hemoglobin S (HbS) and HbS-Antilles results in increased phenotypic severity. *Blood* **86**:2419–2428.
28. Trudel M, Saadane N, Garel MC, Bardakdjian-Michau J, Blouquit Y, Guerquin-Kern JL, Rouyer-Fessard P, Vidaud D, Pachnis A, Romeo PH *et al.* (1991). Towards a transgenic mouse model of sickle cell disease: hemoglobin SAD. *EMBO J* **10**:3157–3165.
29. Trudel M, De Paepe ME, Chretien N, Saadane N, Jacmain J, Sorette M, Hoang T, Beuzard Y (1994). Sickle cell disease of transgenic SAD mice. *Blood* **84**:3189–3197.
30. Ciavatta DJ, Ryan TM, Farmer SC, Townes TM (1995). Mouse model of human beta zero thalassemia: targeted deletion of the mouse β^{maj}- and β^{min}-globin genes in embryonic stem cells. *Proc Natl Acad Sci USA* **92**:9259–9263.
31. Yang B, Kirby S, Lewis J, Detloff PJ, Maeda N, Smithies O (1995). A mouse model for β^0-thalassemia. *Proc Natl Acad Sci USA* **92**:11608–11612.
32. Paszty C, Mohandas N, Stevens ME, Loring JF, Liebhaber SA, Brion CM, Rubin EM (1995). Lethal α-thalassaemia created by gene targeting in mice and its genetic rescue. *Nat Genet* **11**:33–39.
33. Paszty C, Brion CM, Manci E, Witkowska HE, Stevens ME, Mohandas N, Rubin EM (1997). Transgenic knockout mice with exclusively human sickle hemoglobin and sickle cell disease. *Science* **278**:876–878.
34. Ryan TM, Townes TM, Reilly MP, Asakura T, Palmiter RD, Brinster RL, Behringer RR (1990). Human sickle hemoglobin in transgenic mice. *Science* **247**:566–568.
35. Ryan TM, Ciavatta DJ, Townes TM (1997). Knockout-transgenic mouse model of sickle cell disease. *Science* **278**:873–876.

36. Chang JC, Lu R, Lin C, Xu SM, Kan YW, Porcu S, Carlson E, Kitamura M, Yang S, Flebbe-Rehwaldt L, Gaensler KM (1998). Transgenic knockout mice exclusively expressing human hemoglobin S after transfer of a 240-kb β^S-globin yeast artificial chromosome: a mouse model of sickle cell anemia. *Proc Natl Acad Sci USA* **95**:14886–14890.
37. Blouin MJ, Beauchemin H, Wright A, De Paepe M, Sorette M, Bleau AM, Nakamoto B, Ou CN, Stamatoyannopoulos G, Trudel M (2000). Genetic correction of sickle cell disease: insights using transgenic mouse models. *Nat Med* **6**:177–182.
38. Fabry ME, Suzuka SM, Weinberg RS, Lawrence C, Factor SM, Gilman JG, Costantini F, Nagel RL (2001). Second generation knockout sickle mice: the effect of HbF. *Blood* **97**:410–418.
39. Skow LC, Burkhart BA, Johnson FM, Popp RA, Popp DM, Goldberg SZ, Anderson WF, Barnett LB, Lewis SE (1983). A mouse model for β-thalassemia. *Cell* **34**:1043–1052.
40. Imren S, Payen E, Westerman KA *et al.* (2002). Permanent and panerythroid correction of murine β thalassemia by multiple lentiviral integration in hematopoietic stem cells. *Proc Natl Acad Sci USA* **99**: 14380–14385.
41. May C, Rivella S, Chadburn A, Sadelain M (2002). Successful treatment of murine β-thalassemia intermedia by transfer of the human β-globin gene. *Blood* **99**:1902–1908.
42. Rivella S, May C, Chadburn A, Riviere I, Sadelain M (2003). A novel murine model of Cooley anemia and its rescue by lentiviral-mediated human β-globin gene transfer. *Blood* **101**:2932–2939.
43. Sauvage C, Rouyer-Fessard P, Beuzard Y (1993). Improvement of mouse β thalassaemia by hydroxyurea. *Br J Haematol* **84**:492–496.
44. De Franceschi L, Brugnara C, Rouyer-Fessard P, Jouault H, Beuzard Y (1999). Formation of dense erythrocytes in SAD mice exposed to chronic hypoxia: evaluation of different therapeutic regimens and of a combination of oral clotrimazole and magnesium therapies. *Blood* **94**:4307–4313.
45. Fibach E, Prasanna P, Rodgers GP, Samid D (1993). Enhanced fetal hemoglobin production by phenylacetate and 4-phenylbutyrate in erythroid precursors derived from normal donors and patients with sickle cell anemia and β-thalassemia. *Blood* **82**:2203–2209.
46. Liakopoulou E, Blau CA, Li Q *et al.* (1995). Stimulation of fetal hemoglobin production by short chain fatty acids. *Blood* **86**:3227–3235.
47. Yang J, Kawai Y, Hanson RW, Arinze IJ (2001). Sodium butyrate induces transcription from the Gα(i2) gene promoter through multiple Sp1 sites in the promoter and by activating the MEK-ERK signal transduction pathway. *J Biol Chem* **276**:25742–25752.
48. Witt O, Monkemeyer S, Kanbach K, Pekrun A (2002). Induction of fetal hemoglobin synthesis by valproate: modulation of MAP kinase pathways. *Am J Hematol* **71**:45–46.
49. Pace B, Li Q, Peterson K, Stamatoyannopoulos G (1994). α-amino butyric acid cannot reactivate the silenced gamma gene of the β locus YAC transgenic mouse. *Blood* **84**:4344–4353.
50. Patton EE, Zon LI (2001). The art and design of genetic screens: zebrafish. *Nat Rev Genet* **2**:956–966.
51. Ackermann GE, Paw BH (2003). Zebrafish: a genetic model for vertebrate organogenesis and human disorders. *Front Biosci* **8**:1227–1253.
52. Ransom DG, Haffter P, Odenthal J, Brownlie A, Vogelsang E, Kelsh RN, Brand M, van Eeden FJ, Furutani-Seiki M, Granato M, Hammerschmidt M, Heisenberg CP, Jiang YJ, Kane DA, Mullins MC, Nusslein-Volhard C (1996). Characterization of zebrafish mutants with defects in embryonic hematopoiesis. *Development* **123**:311–319.
53. Weinstein BM, Schier AF, Abdelilah S, Malicki J, Solnica-Krezel L, Stemple DL, Stainier DY, Zwartkruis F, Driever W, Fishman MC (1996). Hematopoietic mutations in the zebrafish. *Development* **123**:303–309.
54. Bennett CM, Kanki JP, Rhodes J, Liu TX, Paw BH, Kieran MW, Langenau DM, Delahaye-Brown A, Zon LI, Fleming MD, Look AT (2001). Myelopoiesis in the zebrafish, *Danio rerio*. *Blood* **98**:643–651.
55. Lieschke GJ, Oates AC, Paw BH, Thompson MA, Hall NE, Ward AC, Ho RK, Zon LI, Layton JE (2002). Zebrafish SPI-1 (PU.1) marks a site of myeloid development independent of primitive erythropoiesis: implications for axial patterning. *Dev Biol* **246**:274–295.
56. Trede NS, Langenau DM, Traver D, Look AT, Zon LI (2004). The use of zebrafish to understand immunity. *Immunity* **20**:367–379.
57. Galloway JL, Zon LI (2003). Ontogeny of hematopoiesis: examining the emergence of hematopoietic cells in the vertebrate embryo. *Curr Top Dev Biol* **53**:139–158.

58. Yoder MC, Palis J (2001). In: Zon LI, editor. *Hematopoiesis: A Developmental Approach*, New York: Oxford University Press.
59. Lieschke GJ, Oates AC, Crowhurst MO, Ward AC, Layton JE (2001). Morphologic and functional characterization of granulocytes and macrophages in embryonic and adult zebrafish. *Blood* **98**:3087–3096.
60. Willett CE, Cortes A, Zuasti A, Zapata AG (1999). Early hematopoiesis and developing lymphoid organs in the zebrafish. *Dev Dyn* **214**:323–336.
61. Jagadeeswaran P, Sheehan JP, Craig FE, Troyer D (1999). Identification and characterization of zebrafish thrombocytes. *Br J Haematol* **107**:731–738.
62. Shafizadeh E, Paw BH (2004). Zebrafish as a model of human hematologic disorders. *Curr Opin Hematol* **11**:255–261.
63. Kishimoto Y, Lee KH, Zon L, Hammerschmidt M, Schulte-Merker S (1997). The molecular nature of zebrafish *swirl*: BMP2 function is essential during early dorsoventral patterning. *Development* **124**: 4457–4466.
64. Liao W, Bisgrove BW, Sawyer H, Hug B, Bell B, Peters K, Grunwald DJ, Stainier DY (1997). The zebrafish gene *cloche* acts upstream of a flk-1 homologue to regulate endothelial cell differentiation. *Development* **124**(2):381–389.
65. Thompson MA, Ransom DG, Pratt SJ, MacLennan H, Kieran MW, Detrich HW 3rd, Vail B, Huber TL, Paw B, Brownlie AJ, Oates AC, Fritz A, Gates MA, Amores A, Bahary N, Talbot WS, Her H, Beier DR, Postlethwait JH, Zon LI (1998). The *cloche* and *spadetail* genes differentially affect hematopoiesis and vasculogenesis. *Dev Biol* **197**:248–269.
66. Lyons SE, Lawson ND, Lei L, Bennett PE, Weinstein BM, Liu PP (2002). A nonsense mutation in zebrafish *gata1* causes the bloodless phenotype in *vlad tepes*. *Proc Natl Acad Sci USA* **99**:5454–5459.
67. Pevny L, Simon MC, Robertson E, Klein WH, Tsai SF, D'Agati V, Orkin SH, Costantini F (1991). Erythroid differentiation in chimaeric mice blocked by a targeted mutation in the gene for transcription factor GATA-1. *Nature* **349**:257–260.
68. Fujiwara Y, Chang AN, Williams AM, Orkin SH (2004). Functional overlap of GATA-1 and GATA-2 in primitive hematopoietic development. *Blood* **103**:583–585.
69. Ransom DG, Bahary N, Niss K *et al.* (2004). The zebrafish *moonshine* gene encodes transcriptional intermediary factor 1γ, an essential regulator of hematopoiesis. *PLoS Biol* **2**:E237.
70. Liao EC, Paw BH, Peters LL *et al.* (2000). Hereditary spherocytosis in zebrafish *riesling* illustrates evolution of erythroid β-spectrin structure, and function in red cell morphogenesis and membrane stability. *Development* **127**:5123–5132.
71. Shafizadeh E, Paw BH, Foott H *et al.* (2002). Characterization of zebrafish *merlot/chablis* as non-mammalian vertebrate models for severe congenital anemia due to protein 4.1 deficiency. *Development* **129**:4359–4370.
72. Paw BH, Davidson AJ, Zhou Y *et al.* (2003). Cell-specific mitotic defect and dyserythropoiesis associated with erythroid band 3 deficiency. *Nat Genet* **34**:59–64.
73. Brownlie A, Donovan A, Pratt SJ *et al.* (1998). Positional cloning of the zebrafish *sauternes* gene: a model for congenital sideroblastic anaemia. *Nat Genet* **20**:244–250.
74. Wang H, Long Q, Marty SD, Sassa S, Lin S (1998). A zebrafish model for hepatoerythropoietic porphyria. *Nat Genet* **20**:239–243.
75. Childs S, Weinstein BM, Mohideen MA, Donohue S, Bonkovsky H, Fishman MC (2000). Zebrafish *dracula* encodes ferrochelatase and its mutation provides a model for erythropoietic protoporphyria. *Curr Biol* **10**:1001–1004.
76. Donovan A, Brownlie A, Zhou Y, Shepard J, Pratt SJ, Moynihan J, Paw BH, Drejer A, Barut B, Zapata A, Law TC, Brugnara C, Lux SE, Pinkus GS, Pinkus JL, Kingsley PD, Palis J, Fleming MD, Andrews NC, Zon LI (2000). Positional cloning of zebrafish ferroportin1 identifies a conserved vertebrate iron exporter. *Nature* **403**:776–781.
77. Gordeuk VR, Caleffi A, Corradini E *et al.* (2003). Iron overload in Africans and African-Americans and a common mutation in the SCL40A1 (ferroportin 1) gene. *Blood Cells Mol Dis* **31**:299–304.
78. Montosi G, Donovan A, Totaro A, Garuti C, Pignatti E, Cassanelli S, Trenor CC, Gasparini P, Andrews NC, Pietrangelo A (2001). Autosomal-dominant hemochromatosis is associated with a mutation in the ferroportin (SLC11A3) gene. *J Clin Invest* **108**:619–623.

79. Donovan A, Brownlie A, Dorschner MO, Zhou Y, Pratt SJ, Paw BH, Phillips RB, Thisse C, Thisse B, Zon LI (2002). The zebrafish mutant gene *chardonnay* (*cdy*) encodes divalent metal transporter 1 (DMT1). *Blood* **100**:4655–4659.
80. Wingert RA, Brownlie A, Galloway JL, Dooley K, Fraenkel P, Axe JL, Davidson AJ, Barut B, Noriega L, Sheng X, Zhou Y, Zon LI (2004). The *chianti* zebrafish mutant provides a model for erythroid-specific disruption of transferrin receptor 1. *Development* **131**:6225–6235.
81. Brownlie A, Hersey C, Oates AC, Paw BH, Falick AM, Witkowska HE, Flint J, Higgs D, Jessen J, Bahary N, Zhu H, Lin S, Zon LI (2003). Characterization of embryonic globin genes of the zebrafish. *Dev Biol* **255**:48–61.
82. Chan FY, Robinson J, Brownlie A, Shivdasani RA, Donovan A, Brugnara C, Kim J, Lau BC, Witkowska HE, Zon LI (1997). Characterization of adult α- and β-globin genes in the zebrafish. *Blood* **89**:688–700.
83. Maruyama K, Yasumasu S, Iuchi I (2004). Evolution of globin genes of the medaka *Oryzias latipes* (Euteleostei; Beloniformes; Oryziinae). *Mech Dev* **121**:753–769.
84. Long Q, Meng A, Wang H, Jessen JR, Farrell MJ, Lin S (1997). GATA-1 expression pattern can be recapitulated in living transgenic zebrafish using GFP reporter gene. *Development* **124**:4105–4111.
85. Hsu K, Traver D, Kutok JL, Hagen A, Liu TX, Paw BH, Rhodes J, Berman JN, Zon LI, Kanki JP, Look AT (2004). The *pu.1* promoter drives myeloid gene expression in zebrafish. *Blood* **104**:1291–1297.
86. Ward AC, McPhee DO, Condron MM, Varma S, Cody SH, Onnebo SM, Paw BH, Zon LI, Lieschke GJ (2003). The zebrafish *spi1* promoter drives myeloid-specific expression in stable transgenic fish. *Blood* **102**:3238–3240.
87. Langenau DM, Traver D, Ferrando AA, Kutok JL, Aster JC, Kanki JP, Lin S, Prochownik E, Trede NS, Zon LI, Look AT (2003). Myc-induced T cell leukemia in transgenic zebrafish. *Science* **299**:887–890.
88. Langenau DM, Ferrando AA, Traver D, Kutok JL, Hezel JP, Kanki JP, Zon LI, Look AT, Trede NS (2004). In vivo tracking of T cell development, ablation, and engraftment in transgenic zebrafish. *Proc Natl Acad Sci USA* **101**:7369–7374.
89. Jessen JR, Willett CE, Lin S (1999). Artificial chromosome transgenesis reveals long-distance negative regulation of *rag1* in zebrafish. *Nat Genet* **23**:15–16.
90. Lin H-F, Traver D, Zhu H, Dooley K, Paw BH, Zon LI, Handin RI (2005). Analysis of thrombocyte development in CD41-GFP transgenic zebrafish. *Blood* **106**:3803–3810.
91. Long Q, Huang H, Shafizadeh E, Liu N, Lin S (2000). Stimulation of erythropoiesis by inhibiting a new hematopoietic death receptor in transgenic zebrafish. *Nat Cell Biol* **2**:549–552.
92. Traver D, Winzeler A, Stern HM, Mayhall EA, Langenau DM, Kutok JL, Look AT, Zon LI (2004). Effects of lethal irradiation in zebrafish and rescue by hematopoietic cell transplantation. *Blood* **104**:1298–1305.
93. Traver D, Paw BH, Poss KD, Penberthy WT, Lin S, Zon LI (2003). Transplantation and in vivo imaging of multilineage engraftment in zebrafish *bloodless* mutants. *Nat Immunol* **4**:1238–1246.
94. Stern HM, Zon LI (2003). Cancer genetics and drug discovery in the zebrafish. *Nat Rev Cancer* **3**:533–539.
95. Peterson RT, Link BA, Dowling JE, Schreiber SL (2000). Small molecule developmental screens reveal the logic and timing of vertebrate development. *Proc Natl Acad Sci USA* **97**:12965–12969.
96. Peterson RT, Fishman MC (2004). Discovery and use of small molecules for probing biological processes in zebrafish. *Methods Cell Biol* **76**:569–591.
97. Zon LI, Peterson RT (2005). In vivo drug discovery in the zebrafish. *Nat Rev Drug Discov* **4**:35–44.
98. Moon HS, Jacobson EM, Khersonsky SM, Luzung MR, Walsh DP, Xiong W, Lee JW, Parikh PB, Lam JC, Kang TW, Rosania GR, Schier AF, Chang YT (2002). A novel microtubule destabilizing entity from orthogonal synthesis of triazine library and zebrafish embryo screening. *J Am Chem Soc* **124**:11608–11609.
99. Shafizadeh E, Peterson RT, Lin S (2004). Induction of reversible hemolytic anemia in living zebrafish using a novel small molecule. *Comp Biochem Physiol C Toxicol Pharmacol* **138**:245–249.
100. Wingert RA, Galloway JL, Barut B, Foott H, Fraenkel P, Axe JL, Weber GJ, Dooley K, Davidson AJ, Schmid B, Paw BH, Shaw GC, Kingsley P, Palis J, Schubert H, Chen O, Kaplan J, Zon LI, Tubingen 2000 Screen Consortium (2005). Glutaredoxin 5 deficiency reveals Fe/S clusters are required for vertebrate heme synthesis. *Nature* **436**:1035–1039.

16

Stem Cell Biology

by Wei Li and Alan W. Flake

Introduction

Recent developments in stem cell biology require that specific nomenclature be used for the term "stem cell" to define hierarchies. The term Totipotent Stem Cell applies only to the zygote, and refers to a cell capable of giving rise to all of the tissues of the embryo, the extraembryonic membranes, and placenta; a cell capable of developing into an adult organism. The term Pluripotent Stem Cell refers to cells capable of providing stem cells of all embryonic tissues, but not of the extraembryonic tissues. Embryonic stem cells derived from the inner cell mass of the blastocyst are pluripotent. Multipotent stem cells are cells capable of self-renewal and of giving rise to all cell types of a specific tissue. There are several well characterized multipotent stem cell populations that have been identified from fetal and adult sources, such as the hematopoietic stem cells (HSCs).

A general working definition of a stem cell is a cell capable of self-renewal and differentiation into at least one type of mature progeny cell. The primary property distinguishing a stem cell from its progeny cell is the capacity for self-renewal. There is also the concept that a stem cell has the capacity for a vast number of symmetric (giving rise to two new stem cells) or asymmetric (one stem cell and one progeny cell) divisions, to maintain the stem cell compartment throughout the lifetime of the organism.

A HSC can therefore be defined as a single cell capable of self-renewal, and differentiation into all types of blood forming cells. Hematopoiesis can be defined as the process by which an adequate supply of blood cells are generated to maintain normal homeostasis, and/or to respond to physiologic challenges. The hematopoietic system under normal circumstances is dynamic, and profoundly robust, providing a lifetime of oxygen delivery, immune protection, and clotting capacity, among other functions, to the individual. The source of adult hematopoiesis is the HSC, and the observation that engraftment and clonal proliferation of a relatively small number of normal HSCs can sustain normal hematopoiesis provides the basis for hematopoietic cell transplantation (HCT). This chapter will discuss the advances made in our understanding of HSC biology over the past few decades, and the role they will play in the successful application of HCT.

Hematopoiesis

Primitive Hematopoiesis

Primitive hematopoiesis can be defined as the process required for producing progenitors that give rise to red cells with embryonic hemoglobin synthesis. While the yolk sac (YS) is the source of primitive hematopoiesis, and the first hematopoietic organ in the mammal, the earliest evidence of mesoderm commitment to endothelial and hematopoietic lineages (hemangioblasts) occurs prior to extraembryonic mesoderm migration to the proximal YS.[1] Recently, in the mouse, Flk-1^{+}Brachyury^{+} mesoderm-derived hemangioblast were detected in the posterior primitive streak, and colony assays detect primitive erythroid colony forming cells (EryP-CFC) by the mid-primitive streak stage (post-coital day 7.25 or E7.5). The integrin αIIb, cluster differentiation (CD) marker CD41, is perhaps the earliest marker for hematopoietic progenitors[2,3] and can be used to track the extraembryonic migration of primitive progenitors to the proximal yolk sac, where they form a circumferential band by the mid-late primitive streak stages (E7.5–E7.75). This band of blood cells comprises the classically-described blood islands of the YS and expresses several early hematopoietic genes, such as the transcription factors Scl,[4] Gata-1,[5] and embryonic globin genes (ε and ζ globin).[6] The YS remains active as a hematopoietic organ until approximately E11.5 in the murine fetus, well after initiation of circulation which occurs after the initiation of cardiac activity at E8.5. Corresponding stages in human hematopoietic development are less well delineated, but primordial blood islands are observed in the mesoderm of the YS at 16 days gestation, and hematopoiesis disappears from the human YS by 60 days gestation.

Definitive Hematopoiesis

Definitive hematopoiesis can be defined as hematopoiesis derived from HSC capable of providing all of the blood cell lineages required for reconstitution of the adult bone marrow. Hemoglobin production during definitive hematopoiesis is developmentally programmed in humans to switch from fetal to adult hemoglobin synthesis after birth. While, in the mouse definitive HSC can be identified in the YS after the fetal circulation is established, it is unclear whether they arise from the YS or circulate there from the Aorto-Gonadal-Mesonephric (AGM) region.[7] In the murine system it has become apparent that the placenta acts as a primary hematopoietic organ early in gestation, and may be a primary reservoir for maturation of HSC prior to migration to the FL.[8,9] After E16 in the murine fetus and 20 weeks gestation in humans, the bone marrow (BM) becomes the primary site for definitive hematopoiesis, and remains so throughout life.

Characterization of Hematopoietic Stem Cells

Cell Surface Markers

Cell surface markers represent antigenic proteins on the cell surface that can be identified by monoclonal antibodies. The restricted expression of cell surface marker genes to subsets of cells has been extensively utilized to define populations. HSCs are the prototype for stem cell isolation using cell surface markers. A single molecular marker that is expressed exclusively by HSCs has not been discovered. However, there are markers whose expression is gained (or lost) at different stages of hematopoietic cell differentiation. By targeting various combinations of markers, it has become possible to subdivide this functionally heterogeneous mixture of cells into more homogeneous subpopulations. Finally, the specific clusters of markers identifying an HSC are dependent upon the developmental stage and the source of the HSC, the species of origin, and the assay system utilized.

It is beyond the scope of this chapter to exhaustively review all cell surface phenotypes of HSC. Rather, an attempt will be made to discuss the more commonly used surface markers and to highlight the phenotypic differences between murine and human HSC populations.

Fetal Hematopoietic Stem Cells

In the mouse, the earliest definitive marker of committed HSCs is CD41.[2,3] Definitive HSCs in the AGM and YS are enriched by CD41 selection which continues to be expressed in a subpopulation of FL HSCs, however BM HSCs are $CD41^{lo/-}$. In addition to CD41, Sca-1 can be utilized to enrich for HSCs in the AGM region, placenta, and BM, but CD41 is also expressed in non-hematopoietic cells.[10,11] As the transition from primitive to definitive hematopoiesis occurs, HSCs capable of providing definitive hematopoiesis express the more classical $CD34^+$c-Kit^+ phenotype.[8,11,12] It is interesting that CD34 is expressed on the majority of murine fetal HSC, whereas quiescent HSC in the adult murine BM are $CD34^-$,[13] and only 10% of all adult murine HSCs express CD34.[14] Additional differences in surface marker expression between murine fetal and adult BM HSCs have been noted with alternative enrichment protocols. Weissman and associates were the first to characterize a highly purified population of murine HSCs from FL of C57BL6-Thy1.1 mice.

Human fetal HSC markers are less well defined at early gestation due to the limitations of available assay systems (described below). It appears that human fetal HSCs express Sca-1, CD34, and c-kit. In addition, $CD34^+lin^-$ HLA^-DR^+ populations from fetal BM and cord blood can be further enriched for HSCs by CD38 depletion. Thus the phenotype for characterized human fetal HSCs is $CD34^+$Sca-1^+c-kit^+lin^- HLA^-DR^+.

Adult Hematopoietic Stem Cells

Adult HSC are limited to the BM, and therefore do not demonstrate the evolution of phenotype observed during hematopoietic ontogeny. The first highly characterized murine HSC population was isolated from C57BL6-Thy1.1 mice, and expressed the surface phenotype $Thy1.1^{lo}$Sca-1^+lin^- with the subsequent addition of c-kit as a further enrichment marker. In Thy1.1 negative mice, cells carrying the Sca-1^+c-kit^+lin^- marker combination represent a highly enriched population that approaches single-cell repopulating capacity.[15] Similarly, the Sca-1^+c-$kit^+CD34^-lin^-$ population in adult murine BM is capable of single cell repopulation.[13] CD34 expression in murine cells has been shown to be not only developmentally regulated, but in adult cells, reversible and dependent upon activation status.[16] Whereas, in the murine fetus CD34 enriches for HSC, in the 10-week-old adult mouse CD34 is a marker of late adult progenitors, and its exclusion enriches for long-term repopulating cells. Other useful markers that are present on subsets of highly enriched murine HSCs include the fibroblast growth factor receptor (FGFR),[17] and CD105/endoglin.[18] Very recently, it was shown that Signaling Lymphocyte Activation Molecule (SLAM) family markers CD150 and CD48 could be used for enrichment of Sca-1^+c-kit^+lin^- cells from old or reconstituted mice, which are less efficient at repopulation than the same cells from younger mice,[19] and that SLAM family expression can discriminate HSC from primitive progenitors.[20]

The majority of human cells capable of producing multilineage hematopoietic engraftment in myeloablated recipients express CD34. This has been determined experimentally in non-human primate models, where CD34 expression is similarly regulated, and by transplantation of purified human $CD34^+$ cells into sublethally irradiated immunodeficient mice or preimmune fetal sheep.[21] The engraftment potential of enriched populations of human $CD34^+$ cells has also been demonstrated in numerous autologous and allogeneic human transplantation trials. Nevertheless, even in a highly purified population of $CD34^+$ cells, the frequency of HSCs is $<0.1\%$.[22] A number of markers are expressed on progenitors or more mature cells that can be depleted in combination with CD34 selection to enrich human HSCs. These include CD38, HLA-DR, CD33, CD71, and CD45RA. Other surface markers have been identified in human adult marrow that identify populations overlapping with, but not identical with, CD34. These include CD133 (also called AC133),[23] CDCP1 (CUB-domain containing protein),[24] c-kit (CD117),[25] KDR or Vascular Endothelial Growth Factor Receptor 2 (VEGFR-2),[26] and VEGFR-1.[27]

Although the $CD34^+$ fraction of adult BM clearly contains repopulating capacity, there are also $CD34^-$ HSCs. Thus far, the phenotype of $CD34^-$ HSCs has been characterized by the additional absence of CD38 and lineage-specific markers, as well as the expression of CD133. Only 0.2% of lin^-CD34^- cells in human cord blood were found to express CD133 which were capable of repopulating NOD/SCID mice.[28] There may be differential homing and engraftment capability for $CD34^-$ and $CD34^+$ HSCs, however. When the populations are injected directly into the BM space, greater repopulating capacity is observed in the CD34-population than after systemic injection.[29] This is an important example of how the assay system utilized may alter interpretation of the biology.

The different combinations of cell surface markers used to isolate HSCs underscore the lack of a definitive phenotype for their purification. Nevertheless, validated enrichment protocols can now provide highly purified HSC populations for experimental and clinical applications based on surface marker expression alone.

The Side Population

Apart from cell-surface antigen expression, primitive HSCs have been isolated based on their ability to efflux certain fluorescent dyes, such as Hoechst 33342 (Ho) or Rhodamine-123 (Rho). The most relevant cell population for this discussion is the so called "side population" or SP cell. Adult BM contains a rare population of $Ho^{-/lo}$ cells that have been designated the SP cell because they form a characteristic cluster of events on dual wavelength fluorescent activated cell sorting (FACS) analysis of Ho-stained cells.[30] SP cells have been identified in human cord blood and adult BM as well as hematopoietic populations from numerous other species.[31] Cells capable of engrafting NOD/SCID mice are found in the $CD34^+CD38^-$ HSC population that display a SP phenotype.[32] This suggests that transplantable cells might be highly purified by sorting for $CD34^+CD38^-SP^+$cells.

Methods for Isolating HSCs

The low frequency of HSCs and lack of unique markers to identify them poses a challenge to the development of techniques to isolate HSCs reproducibly. The most effective protocols require several separation steps that differ in capacity, selectivity, and choice of separation parameters.

Non-Antibody-Based Isolation

Centrifugation of HSC suspensions through density gradients such as percoll or mixtures of ficoll and hypaque is very commonly used to separate blood cells. The accompanying enrichment of HSCs is typically from 2- to 5-fold. Most antibody-mediated protocols are designed for low density cell suspensions, thus requiring a density separation step. Aldehyde dehydrogenase (ALDH) is selectively expressed in primitive HSCs. Fluorescent ALDH-substrates have been used to identify and isolate human HSCs by FACS.[33] ALDH expression overlaps with CD34 in adult BM cells; thus, sorting for $ALDH^+$ cells yields very primitive as well as lineage-committed progenitors. The highest expression of ALDH has been found on the $CD34^+CD38^-$ subset of human BM cells, and isolation of the $ALDH^{bright}$ subset of $CD34^+$ cells gives a 2-fold enrichment of long-term culture initial cells.[33]

Antibody-Mediated Isolation

Antibodies coupled to toxins to achieve selective cell killing, fluorochromes for cell sorting, solid matrices to allow selective cell adherence, and cell surface antigens with subsequent activation of the complement cascade to lyse specific cell populations, have been developed. The great advantage of FACS-based separations is that cell suspensions of high purity can be generated. However, FACS is limited by the time required to process large numbers of cells. Thus, this technology is applied to cell suspensions in which the HSC content has been increased by the removal of the more mature cells, and the sample volume reduced.

Positive selection involves recovery of cells with one or more antibodies against antigens expressed on the surface of the desired cells. By contrast, negative selection is the process of removal of unwanted cells, using antibodies to cell surface antigens not expressed on the desired cells. Primitive stem cells, including HSCs, do not express the lineage (lin) markers found on committed or mature progenitors. This fact can be used to distinguish immature from the more abundant differentiated cells. Selection of lin^- cells typically produces a 20- to 500-fold enrichment of HSCs, depending on the combination of markers used,[34] including glycophorin A, and CD2, CD3, CD4, CD8, CD14, CD15, CD16, CD19, CD20, CD56, and CD66b.[35] Isolation of $CD34^+$ or $CD133^+$ cells is the most common example of positive selection for primitive HSCs. Antibody/complement treatment and the use of immuno-toxins are by definition negative selection techniques, whereas FACS and the majority of batch-wise immuno-adsorption techniques can be adapted for either positive or negative selection.

Cell separation protocols designed to give HSC suspensions of high purity involve several steps that include positive and negative selection. High purity is the main advantage of positive selection methods. In addition, a single antibody is utilized whereas with negative selection numerous cell types are targeted for removal, requiring the use of several antibodies. The disadvantages of positive selection are the elimination of primitive cell subsets that do not express the selection marker[36] and the desired cells are labeled with antibody that must be removed later from the separation matrix (e.g., StemSep™, StemCell Technologies, and MACS, Miltenyi Biotec). However, recent advances in magnetic cell separation techniques have overcome the requirement for removing the desired cells from the separation matrix (EasySep™, StemCell Technologies).[37]

Stem Cell Assay Systems

HSCs are historically the most characterized cell, and many assay systems have been developed to determine the number of HSCs in various hematopoietic populations. These include *in vivo* and *in vitro* culture-based assay systems which will be discussed below.

In Vitro Assay Systems

In vitro assay systems are useful and allow relatively rapid comparison of HSC-containing populations; however, there is no *in vitro* assay that specifically detects HSCs. *In vitro* assays consist of colony-forming culture systems which measure the clonogenic capacity of cells in various conditions.

In semisolid media, HSCs divide poorly and have a different phenotype than progenitor cells able to form colonies of mature myeloid, erythroid, or megakaryocytic progeny. Colony-forming cells (CFCs) comprise a large, intermediate progenitor compartment that spans the entire stepwise process of lineage restriction. Long-term BM cultures demonstrated that HSCs could be maintained when placed on pre-established stromal layers, allowing close association with supportive stroma.

A similar assay with a different readout is the long-term culture initiating cell (LTC-IC) assay. This assay was originally developed for identifying human long-term repopulating cells, and measures the persistence of progenitor cells with colony-forming capacity after long-term culture on stromal layers. With critical analysis, it is clear that LTC-IC assays can detect some, but not all, HSCs,[21] and that there is considerable variability in these assays between laboratories.[38] Although clearly useful in some applications, the validity of *in vitro* assay systems, as truly representative of HSC content, has been repeatedly challenged, and remains controversial.

In Vivo Assay Systems

Implicit in the functional definition of HSCs is the requirement for repopulation of the hematopoietic system, which can only be validated by an *in vivo* assay system. The observations that BM could radioprotect lethally irradiated mice, that long-term multilineage engraftment was achieved, and that the donor cells could be transplanted into second or third generation recipients and provide reconstitution, were the seminal observations defining the presence of HSCs within the BM. Although these assays remain valuable qualitative assays for HSC presence in cell populations, more informative quantitative assay systems have since been developed for murine and human HSCs.

Competitive Repopulation

Competitive repopulation is the gold standard for quantitative assessment of HSC content. The assay is based on competition between a test cell population and a standard number of freshly isolated, unseparated congenic BM cells. The competitor population provides both short-term support as well as a defined number of normal HSCs against which the test population can be compared, and the number of "repopulating units" (RU) estimated. It is important to appreciate that a RU estimates the repopulating capacity of a test cell population relative to 1×10^5 competitor BM cells, rather than the function of an individual HSC. For quantitative assessment of the actual number of HSCs in a test population, limiting dilution and the application of Poisson statistical analysis makes possible estimation of the competitive repopulating units.

The limitation of competitive repopulating assays is the assumption that the donor cells can effectively compete with the normal competitor cells. In circumstances where the donor cells may be weakly competitive, they may be overwhelmed by the competitors, giving a negative readout even though cells capable of reconstitution are present. This problem has resulted in the development of several modifications of the competitive repopulation assay that involve weakly competitive or non-competitive host cells. The disadvantage of non-competitive assays, however, is high sensitivity, which can preclude limiting dilution quantification of HSC frequency.

In vivo assays for human HSCs required the development of surrogate animal models. The non-obese diabetic/SCID or NOD/SCID mouse model,[39,40] with or without additional engineered immunodeficiency[41] has been the most useful. The NOD/SCID background allows engraftment of human HSCs with multilineage expression for many weeks after engraftment. Limiting dilution has allowed the definition of a SCID repopulating unit as a basis for comparison of HSC content in different human cell populations.[22] An additional modification is to inject cells directly into the femoral intraosseous space.[29,42] This may increase the sensitivity of the assay, and removes the necessity for retention of homing capacity of the assayed cells. These models have now become the gold standard for human HSCs; *in vitro* assay results require validation *in vivo*.

A second approach for the development of a surrogate model of human hematopoiesis has been to take advantage of fetal immune tolerance in the sheep to engraft human cells. This model has now been used extensively to assay human hematopoietic populations[43] and to examine plasticity of stem cell populations.[44] The advantages are the size of the sheep, allowing expansion of transplanted cells into much larger total cell numbers than can be achieved in the mouse. In addition, the chimeric animals can be followed for years, an obvious advantage when examining human hematopoiesis. The sheep model is limited by expense, long gestation, limited pregnancies, and limited ability to quantify HSC frequency. Nevertheless, it is a valuable model for comparison to murine systems.

Regulation of Stem Cell Self-Renewal

Stem cell self-renewal is a tightly regulated process that maintains a pool of HSCs and a homeostatic supply of progenitors. The regulation of symmetric and asymmetric cell divisions of HSCs to maintain hematopoiesis, and allow appropriate responses to hematopoietic stress is poorly understood. Rather than the traditional reductionist approach, a true understanding of stem cell regulation will require a complex system analysis. Unfortunately, the complexity of most biological systems defies the current analytical tools. This is a limiting factor for the application of systems theory to exact modeling of stem cell behavior, but not for the development of a conceptual view of systems or network theory applied to stem cell biology.[45]

Recently, genetic regulatory programs involved in the self-renewal process have been identified. The homeobox family member HoxB4 was the first transcription factor shown to promote the expansion of HSCs *in vitro* and *in vivo*, while retaining the capacity to differentiate into lymphoid and myeloid cells. Engineered enforced expression of HoxB4 results in a competitive repopulation advantage over normal cells and restoration but not expansion of HSC pools in transplanted mice, indicating that HoxB4 expression does not override the normal regulatory mechanisms governing HSC self-renewal.[46] A similar function has been recently noted for HoxC4,[47] implicating that Hox4 paralogs may be redundant mediators of HSC renewal.

A second regulatory program involved in HSC renewal is the Wnt signaling pathway.[48] Wnt signaling results in stabilization and accumulation of β-catenin in the cytosol. Reya *et al.*[49] transduced HSCs with constitutively active β-catenin in a retrovirus construct. Enforced expression of β-catenin expanded the pool of HSCs in long-term cultures. In addition, Wnt signaling induces increased expression of HoxB4 and Notch1, genes that are implicated in the self-renewal of HSCs. Notch1 expression has been identified in murine and human progenitors where it participates in cell fate decisions and inhibits progenitor differentiation.[50]

Bmi1 is another protein implicated in HSCs regulation through cell cycle regulation. Bmi1 is highly expressed in murine and human HSCs,[51] and its absence results in progressive loss of all hematopoietic lineages. The interactions of Bmi1 with cell cycle associated genes could be central to the loss of HSCs in Bmi-$1^{-/-}$ mice; furthermore, cell cycle kinetics regulation could allow for accumulation of specific intracellular proteins and transcription factors that may be essential for self-renewal. Other genes involved in HSC regulation and progenitor cycling include the cyclin-dependent kinase inhibitors P21 and P27. Entry into the cell cycle for HSC is governed by the action of P21.[52]

In the absence of this inhibitor, the stem cell pool is larger, more actively cycling, and more sensitive to exhaustion.

Finally, another interesting genetic pathway involved in regulating multiple stem cell types involves the bone morphogenetic proteins (BMPs) which belong to the transforming growth factor β superfamily. BMPs play important roles in hematopoietic tissue induction during early embryonic development;[53] BMP4 maintains HSC reconstitution *in vitro*.[54] Interestingly, BMP has been implicated in the control of HSC pool size, via interactions with a subset of osteoblastic cells lining the bone surface. Analysis of a conditional Bmpr1a mutant mouse suggests the pool size of HSCs is controlled by the volume of trabecular bone,[55] the larger the volume, the greater the HSC number. This study suggests that BMP signaling mediated by Bmpr1a plays a role in controlling the HSC number through regulation of the BM niche size. This incomplete summary of known regulatory factors in HSC self-renewal demonstrates the complexity of interactions required to achieve normal hematopoiesis through its quarterback, the HSC.

Erythroid Lineage Commitment

HSCs become committed to specific cell lineages through a series of complex steps. Mechanisms of multipotent stem cell commitment to the erythroid lineage will provide a basis for novel therapeutics to modulate globin gene expression. Stem cell leukemia (SCL), a basic helix-loop-helix transcription factor, binds the E-box DNA elements (CAGGTG) as a heterodimer in a complex containing E12/E47. SCL interacts with GATA-1, through LIM-only protein 2 (LMO2) which acts as a bridge in a complex containing E2A and Lbd1.[56,57] During erythropoiesis SCL mediates proliferation and differentiation while repressing myeloid and lymphoid progenitors. SCL null embryonic stem cells fail to give rise to hematopoietic cells, suggesting that SCL is crucial for primitive hematopoiesis as well.[58] Another protein, c-Myb, is required for early definitive cellular expansion, and is down-regulated during terminal differentiation.[59] c-Myb null mice exhibit normal primitive but severely impaired erythrocyte production, while megakaryocytes, granulocytes, and monocytes develop normally.

All members of the GATA family of transcription factors contain two homologous zinc-finger domains, and bind to the GATA-consensus sequence (T/AGATAA/G) present in the regulatory elements of most erythroid genes.[60] GATA-2 promotes proliferation and blocks erythroid differentiation.[59] Expression of GATA-2 precedes that of GATA-1 and must decrease to enable erythroid differentiation. GATA-2 null mice are embryonic lethal, due to severe anemia during the early phase of YS hematopoiesis.[61] Multipotential progenitors arising from GATA-2 null ES cells proliferate poorly and undergo excessive apoptosis,[59,61] suggesting that GATA-2 is essential for appropriate expansion and survival of early HSCs.

GATA-1 null mice show complete ablation of embryonic erythropoiesis due to arrested maturation and apoptosis of erythroid precursors at the proerythroblast stage,[62] supporting its key role in erythroid/megakaryocytic commitment. A block in megakaryocyte development at mid-maturation also occurs by E11.5. Enforced GATA-1 expression in early myeloid progenitors promotes megakaryocytic differentiation, suggesting that GATA-1 may affect both lineages and late erythroid maturation.[59,63] The N-terminal zinc-fingers in GATA-1 is required for physical interaction with EKLF, LMO2, and CREB binding protein.[60,64]

FOG (Friend Of GATA) is a complex zinc-finger protein that associates with GATA-1 to promote erythroid and megakaryocytic differentiation.[65] Failure of GATA-1 to interact with FOG produces a failure of terminal erythroid maturation due to deregulated expression of α-globin, β-globin and band 3.[60] FOG null mice display failure of megakaryocytic lineage commitment versus a block in megakaryocyte maturation produced by GATA-1 knockouts. This supports a GATA-1 independent role for FOG during megakaryocyte commitment.[66] Familial X-linked dyserythropoietic anemia is caused by disruption of GATA-1:FOG1 interactions,[67] resulting in severe anemia and thrombocytopenia at birth.

Globin Gene Expression

Several additional transcription factors are required for normal erythropoiesis including erythroid Kruppel-like factor (EKLF), basic Kruppel-like factor (BKLF), Fli-1, PU.1, signal transducer and activator of transcription (STAT) 5, and hematopoietic RING finger 1. A brief discussion of the role of these factors in erythropoiesis will follow.

EKLF is a zinc finger protein synthesized in erythroid, megakaryocytic, and mast cells,[68] and plays an essential role in β-globin gene expression. Three C2H2 type zinc fingers are contained in the C-terminus of EKLF, which binds a CACC consensus-sequence located close to a GATA site in the promoter of erythroid-specific genes. GATA proteins interact with EKLF to regulate gene expression. EKLF favors binding to the human β-globin CACC element over the γ-globin CACC. Naturally-occurring mutations in the β-CACC result in reduced β-gene expression and severe β-thalassemia.[69] Enforced EKLF expression induces an earlier switch from fetal to adult type globin expression,[70] and EKLF-deficient mice carrying the human β-locus display elevated γ-globin mRNA. Moreover, elevated γ-globin levels have been reported in adults with point mutations in the β-globin CACC box,[69] suggesting a role for EKLF in γ-gene silencing. A protein complex that activates transcription from a chromatin-assembled β-globin gene, in an EKLF-dependent fashion, was purified and named EKLF co-activator remodeling complex-1 (E-RC1).[71] This suggests that EKLF functions as an activator of transcription and/or stabilizes the E-RC1 complex in the β-globin promoter.

Basic Kruppel-like factor (BKLF) is found in erythroid cells, fibroblasts, and brain.[65] It binds CACC motifs through three highly conserved C-terminal zinc fingers, and interacts with the co-repressor CtBP to compete for binding to the β-CACC box to inhibit gene activation *in vitro*.[72] EKLF-deficient mice express significantly reduced levels of BKLF[72] and elevated γ-globin mRNA, suggesting that BKLF represses the expression of embryonic and fetal-stage-specific globin genes.[68] Fetal KLF (FKLF) activates ε-globin expression, and to a lesser extent γ-globin through its interaction with CACC boxes, but fails to activate other CACC box-containing erythroid genes.[73] By contrast, murine FKLF-2 activates γ-globin expression.[74]

The Fli-1 oncogene is a member of the Ets family of transcription factors that regulate erythroid progenitor self-renewal. Fli-1 enforced expression inhibits erythroid differentiation and reduces GATA-1 levels.[75] The oncogene PU.1 is a hematopoietic-specific Ets protein that promotes differentiation of lymphoid and myeloid lineages. PU.1 enforced expression induces erythroleukemia in mice secondary to blockade of erythroid differentiation.[76] The PU.1 DNA binding and trans-activation domains are required for GATA-1 repression and inhibition of terminal differentiation in mouse erythroleukemia cells.[76] Hematopoietic RING finger 1 protein expression coincides with definitive erythropoiesis and its inhibition blocks terminal erythroid differentiation.[77]

Cell Signaling Pathways

The regulation of erythropoiesis is essential for both embryonic development and adult red cell production. In the past decade, through genetic and biochemical approaches, growth factors and cytokines have been extensively characterized as key players in regulating erythropoiesis.

Erythropoietin

Erythropoietin (Epo) is critical for normal red blood cell production achieved through its ability to regulate cell signaling pathways necessary for growth, differentiation and survival of erythroid progenitors.[78,79] During erythropoiesis, early progenitors committed to the erythroid lineage required Epo for differentiation as expression of the Epo receptor (EpoR) occurs.[80] The EpoR belongs to the class I cytokine receptor family (Fig. 1) and it initiates Epo-mediated signaling, by activating Janus kinase (JAK) 2, which binds to the Src homology 2 (SH2) conserved motif.[81] The SH2 domain in the EpoR also interacts with other signaling molecules such as signal transducers and activators of transcription 5A (STAT5A), STAT5B, SHP1, SHP2, phosphatidylinositol 3-kinase subunit (PI-3K), Grb2, Lyn tyrosine kinase, and suppressor of cytokine signaling 3.[82] Tyrosine 343 is required for STAT5 activation and induction of Bcl-xL expression.[83] Mice deficient in Bcl-xL expression have abnormal survival and terminal differentiation of erythroid progenitors.[84] Mutation of tyrosine 343 to phenylalanine eliminates the ability of the EpoR to recruit STAT5.[85] Quelle *et al.* reported no consequences of this mutation on EpoR function, while others have demonstrated reduced proliferation[85] and hemoglobin synthesis.[86] The precise role of tyrosine 343 in EpoR signaling remains elusive.

To clarify intracellular mechanisms, Kubota *et al.*[87] demonstrated that Epo-induced differentiation is dependent upon crosstalk between the Src[88] and the PI-3 kinase-signaling pathways. Cooperation between Src, Lyn and PI-3 kinase may play a prominent role in regulating murine and human erythropoiesis.[89] The signaling pathways activated via the Epo-R are summarized in Fig. 1.

The suppressor of cytokine signaling (SOCS) family of negative regulators have also been described.[90] CIS, the first SOCS family member, is an Epo-inducible molecule that is activated

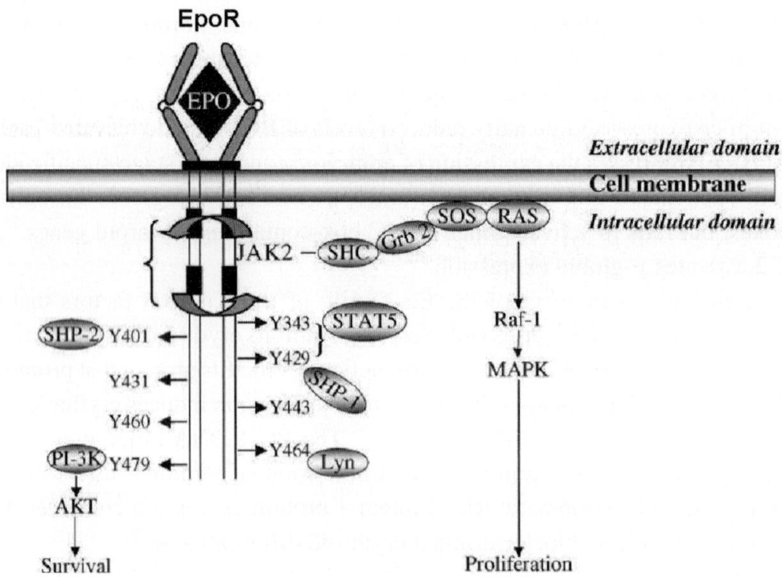

Fig. 1. Cell signaling in response to erythropoietin. After erythropoietin receptor (Epo) is bound, to the Epo receptor (EpoR) dimerizes and serves as a docking site for Janus kinase (JAK) 2 activation and the recruitment of multiple SH-2 domain-containing proteins to complete the process of targeted gene regulation. Abbreviation: PI-3K, phosphatidylinositol 3-kinase; MAPK, mitogen-activated protein kinase; STAT5, signal transducers and activators of transcription 5. Used with permission from Munugalavadla and Kapur.[88]

by STAT5.[91] while SOCS-1 inhibits JAK2.[90,92] Activated SOCS-3 can bind the EpoR and JAK2[93] to inhibit cell signaling initiated by Epo. This negative feedback loop provides a system to regulate cell growth.

Stem Cell Factor

Erythropoiesis is regulated by a number of growth factors including stem cell factor (SCF). Studies in White spotting (W) and Steel (Sl) mutant mice demonstrated erythroid defects due to inherited mutations within the c-Kit and SCF genes. When SCF is bound, its receptor c-Kit undergoes dimerization and autophosphorylation on several distinct cytoplasmic tyrosine residues which become binding sites for a variety of SH2 domain-containing enzymes and adaptor proteins such as phospholipase Cγ (PLCγ),[94] P85-PI3K,[95] and Grb2[96] (Fig. 2). This list is not exhaustive, and illustrates the level of complexity required to accomplish erythropoiesis. An in-depth review of the transcriptional networks that control this process is beyond the scope of this chapter however the signaling molecules activated via c-Kit in response to SCF are summarized in Fig. 2.

Hematopoietic Stem Cell Plasticity

Previously, adult HSCs were viewed as being committed to a hematopoietic cell fate, and incapable of contributing to repair or regeneration of other tissues. However, numerous studies have demonstrated that HSCs can give rise to other tissue types. The concept of plasticity challenges the principle of

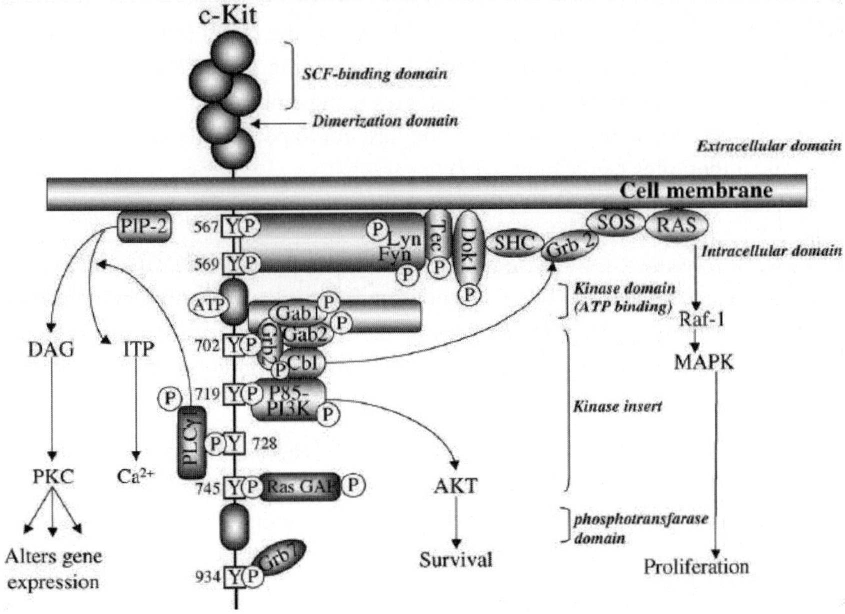

Fig. 2. Cell signaling in response to stem cell factor (SCF). The binding of SCF to c-Kit results in dimerization and autophosphorylation of the receptor. Various tyrosine residues in the cytoplasmic domain of c-Kit become binding sites for several SH-2 domain-containing proteins to activate or repress gene expression. Abbreviation: PLC-γ, phospholipase Cγ; P85-PI-3K, phosphatidylinositol 3-kinase p85 subunit; IP-2, 1 phosphatidylinositol 4,5 bisphosphate; DAG, diacylglycerol; PKC, protein kinase C. Used with permission from Munugalavadla and Kapur (Fig. 1).[88]

lineage restriction during morphogenesis; therefore, we will define plasticity as: 1) the differentiation of a single stem cell into progeny of different germ layer origin, 2) the ability of daughter cells to differentiate and function *in vitro* and *in vivo*, and 3) the ability of stem cells to undergo engraftment.

Other stem cells exist in BM, including mesenchymal stem cells which give rise to osteoblasts, adipocytes, and chondroblasts, and muscle cells[97]; endothelial stem cells contribute to vasculogenesis at the site of blood vessel injury and possibly the hemangioblast.[98] Recent evidence suggested that a subset of BM-derived stem cells have the potential to differentiate into multiple germ lines.[99]

Mechanisms of Stem Cell Plasticity

Many investigators have used BM to study HSC plasticity; BM contains several stem cell types that may be responsible for tissue engraftment. The controversial events of trans-differentiation and de-differentiation would require the nuclei of stem cells to undergo reprogramming of expressed genes. While there is evidence that such phenomena may occur, there is no rigorous *in vivo* evidence that plasticity of HSCs can be ascribed to such events. Studies documenting that such cells exist *in vivo* and are capable of physiologic plasticity are needed to determine if plasticity is an artifact of culture systems.[97]

Clinical Use of Hematopoietic Stem Cells

HSCs have been utilized clinically from a variety of sources to treat many disorders. These include BM transplantation for hematologic, immunodeficiency and lysosomal storage diseases. In addition, HSCs have been used extensively for BM rescue after cytotoxic chemotherapeutic regimens are given for solid and hematologic malignancies. The use of HSCs as the target for gene transfer technology will be a major therapeutic strategy in the future. Likewise, the use of HSCs in tissue engineering and regenerative medicine is being explored. Strategies using embryonic stem cells, or the transfer of post-mitotic somatic cell nuclei into enucleated oocytes, could potentially create an unlimited source of HSCs for medical therapies (see Chapter 17).

Summary

Understanding stem cell biology will be the foundation for clinical advances in the use of stem cells for replacement and regenerative therapy. Progress will be required before the full potential of cellular therapy is realized. There are fundamental questions in stem cell biology which when answered will have staggering clinical implications.

(1) Stem cell renewal and cycling — The ability to regulate stem cell cycling is one of the primary goals of stem cell biology. The ability to drive stem cell renewal without differentiation would allow expansion of therapeutically useful rare stem cell populations for transplantation or tissue engineering. The positive or negative regulation of stem cell cycling would have therapeutic implications for competitive engraftment.

(2) Stem Cell Differentiation — The ability to control lineage specification is the corollary to regulation of stem cell renewal. It would allow the "harvest" of differentiated cell populations from expanded stem cell cultures or the programming of transplanted cells towards appropriate lineages.

(3) Stem Cell Organization — Much of what we know and understand regarding stem cells is derived from reductionism, i.e. isolation of stem cells from their normal microenvironment.

However, stem cell function should be viewed in a network and its function determined by relationships (organization) within that network. The recognition that stem cell biology cannot be fully understood outside of its microenvironment is essential for progress in this field.

(4) Stem Cell Genetics — The genetic manipulation of stem cells provides numerous possibilities, from regulation of stem cell function to treatment of genetic disorders or gene-specific correction in germ line cells. Advances in targeted gene delivery and improved understanding of stem cell biology will ultimately make this possible.

Progress in these areas will expand the therapeutic application of stem cell therapy as well as allow current therapies to be safer and more efficacious.

References

1. Huber TL, Kouskoff V, Fehling HJ, Palis J, Keller G (2004). Haemangioblast commitment is initiated in the primitive streak of the mouse embryo. *Nature* **432**:625–630.
2. Ferkowicz MJ, Starr M, Xie X, Li W, Johnson SA, Shelley WC, Morrison PR, Yoder MC (2003). CD41 expression defines the onset of primitive and definitive hematopoiesis in the murine embryo. *Development* **130**:4393–4403.
3. Mikkola HK, Fujiwara Y, Schlaeger TM, Traver D, Orkin SH (2003). Expression of CD41 marks the initiation of definitive hematopoiesis in the mouse embryo. *Blood* **101**:508–516.
4. Silver L, Palis J (1997). Initiation of murine embryonic erythropoiesis: a spatial analysis. *Blood* **89**: 1154–1164.
5. Onodera K, Takahashi S, Nishimura S, Ohta J, Motohashi H, Yomogida K, Hayashi N, Engel JD, Yamamoto M (1997). GATA-1 transcription is controlled by distinct regulatory mechanisms during primitive and definitive erythropoiesis. *Proc Natl Acad Sci USA* **94**:4487–4492.
6. Belaoussoff M, Farrington SM, Baron MH (1998). Hematopoietic induction and respecification of A-P identity by visceral endoderm signaling in the mouse embryo. *Development* **125**:5009–5018.
7. Cumano A, Ferraz JC, Klaine M, Di Santo JP, Godin I (2001). Intraembryonic, but not yolk sac hematopoietic precursors, isolated before circulation, provide long-term multilineage reconstitution. *Immunity* **15**: 477–485.
8. Gekas C, Dieterlen-Lievre F, Orkin SH, Mikkola HK (2005). The placenta is a niche for hematopoietic stem cells. *Dev Cell* **8**:365–375.
9. Mikkola HK, Gekas C, Orkin SH, Dieterlen-Lievre F (2005). Placenta as a site for hematopoietic stem cell development. *Exp Hematol* **33**:1048–1054.
10. Ottersbach K, Dzierzak E (2005). The murine placenta contains hematopoietic stem cells within the vascular labyrinth region. *Dev Cell* **8**:377–387.
11. de Bruijn MF, Ma X, Robin C, Ottersbach K, Sanchez MJ, Dzierzak E (2002). Hematopoietic stem cells localize to the endothelial cell layer in the midgestation mouse aorta. *Immunity* **16**:673–683.
12. Yoder MC, Hiatt K, Dutt P, Mukherjee P, Bodine DM, Orlic D (1997). Characterization of definitive lymphohematopoietic stem cells in the day 9 murine yolk sac. *Immunity* **7**:335–344.
13. Osawa M, Hanada K, Hamada H, Nakauchi H (1996). Long-term lymphohematopoietic reconstitution by a single CD34-low/negative hematopoietic stem cell. *Science* **273**:242–245.
14. Ito T, Tajima F, Ogawa M (2000). Developmental changes of CD34 expression by murine hematopoietic stem cells. *Exp Hematol* **28**:1269–1273.
15. Ogawa M (2002). C hanging phenotypes of hematopoietic stem cells. *Exp Hematol* **30**:3–6.
16. Sato T, Laver JH, Ogawa M (1999). Reversible expression of CD34 by murine hematopoietic stem cells. *Blood* **94**:2548–2554.
17. de Haan G, Weersing E, Dontje B, van Os R, Bystrykh LV, Vellenga E, Miller G (2003). In vitro generation of long-term repopulating hematopoietic stem cells by fibroblast growth factor-1. *Dev Cell* **4**:241–251.

18. Chen CZ, Li M, de Graaf D, Monti S, Gottgens B, Sanchez MJ, Lander ES, Golub TR, Green AR, Lodish HF (2002). Identification of endoglin as a functional marker that defines long-term repopulating hematopoietic stem cells. *Proc Natl Acad Sci USA* **99**:15468–15473.
19. Yilmaz OH, Kiel MJ, Morrison SJ (2005). SLAM family markers are conserved among hematopoietic stem cells from old and reconstituted mice and markedly increase their purity. *Blood* [Epub ahead of print].
20. Kiel MJ, Yilmaz OH, Iwashita T, Yilmaz OH [sic], Terhorst C, Morrison SJ (2005). SLAM family receptors distinguish hematopoietic stem and progenitor cells and reveal endothelial niches for stem cells. *Cell* **121**:1109–1121.
21. Larochelle A, Vormoor J, Hanenberg H, Wang JC, Bhatia M, Lapidot T, Moritz T, Murdoch B, Xiao XL, Kato I, Williams DA, Dick JE (1996). Identification of primitive human hematopoietic cells capable of repopulating NOD/SCID mouse bone marrow: implications for gene therapy. *Nat Med* **2**:1329–1337.
22. Wang JC, Doedens M, Dick JE (1997). Primitive human hematopoietic cells are enriched in cord blood compared with adult bone marrow or mobilized peripheral blood as measured by the quantitative in vivo SCID-repopulating cell assay. *Blood* **89**:3919–3924.
23. Yin AH, Miraglia S, Zanjani ED, Almeida-Porada G, Ogawa M, Leary AG, Olweus J, Kearney J, Buck DW (1997). AC133, a novel marker for human hematopoietic stem and progenitor cells. *Blood* **90**:5002–5012.
24. Buhring HJ, Kuci S, Conze T, Rathke G, Bartolovic K, Grunebach F, Scherl-Mostageer M, Brummendorf TH, Schweifer N, Lammers R (2004). CDCP1 identifies a broad spectrum of normal and malignant stem/progenitor cell subsets of hematopoietic and nonhematopoietic origin. *Stem Cells* **22**:334–343.
25. Kawashima I, Zanjani ED, Almaida-Porada G, Flake AW, Zeng H, Ogawa M (1996). CD34+ human marrow cells that express low levels of Kit protein are enriched for long-term marrow-engrafting cells. *Blood* **87**:4136–4142.
26. Ziegler BL, Valtieri M, Porada GA, De Maria R, Muller R, Masella B, Gabbianelli M, Casella I, Pelosi E, Bock T, Zanjani ED, Peschle C (1999). KDR receptor: a key marker defining hematopoietic stem cells. *Science* **285**:1553–1558.
27. Hattori K, Heissig B, Wu Y, Dias S, Tejada R, Ferris B, Hicklin DJ, Zhu Z, Bohlen P, Witte L, Hendrikx J, Hackett NR, Crystal RG, Moore MA, Werb Z, Lyden D, Rafii S (2002). Placental growth factor reconstitutes hematopoiesis by recruiting VEGFR1(+) stem cells from bone-marrow microenvironment. *Nat Med* **8**: 841–849.
28. Gallacher L, Murdoch B, Wu DM, Karanu FN, Keeney M, Bhatia M (2000). Isolation and characterization of human CD34(−)Lin(−) and CD34(+)Lin(−) hematopoietic stem cells using cell surface markers AC133 and CD7. *Blood* **95**:2813–2820.
29. Wang J, Kimura T, Asada R, Harada S, Yokota S, Kawamoto Y, Fujimura Y, Tsuji T, Ikehara S, Sonoda Y (2003). SCID-repopulating cell activity of human cord blood-derived CD34− cells assured by intra-bone marrow injection. *Blood* **101**:2924–2931.
30. Goodell MA, Brose K, Paradis G, Conner AS, Mulligan RC (1996). Isolation and functional properties of murine hematopoietic stem cells that are replicating in vivo. *J Exp Med* **183**:1797–1806.
31. Goodell MA, Rosenzweig M, Kim H, Marks DF, DeMaria M, Paradis G, Grupp SA, Sieff CA, Mulligan RC, Johnson RP (1997). Dye efflux studies suggest that hematopoietic stem cells expressing low or undetectable levels of CD34 antigen exist in multiple species. *Nat Med* **3**:1337–1345.
32. Uchida N, Fujisaki T, Eaves AC, Eaves CJ (2001). Transplantable hematopoietic stem cells in human fetal liver have a CD34(+) side population (SP)phenotype. *J Clin Invest* **108**:1071–1077.
33. Storms RW, Trujillo AP, Springer JB, Shah L, Colvin OM, Ludeman SM, Smith C (1999). Isolation of primitive human hematopoietic progenitors on the basis of aldehyde dehydrogenase activity. *Proc Natl Acad Sci USA* **96**:9118–9123.
34. Thomas TE, Miller CL, Eaves CJ (1999). Purification of hematopoietic stem cells for further biological study. *Methods* **17**:202–218.
35. Wognum AW, Eaves AC, Thomas TE (2003). Identification and isolation of hematopoietic stem cells. *Arch Med Res* **34**:461–475.
36. Bonnet D, Bhatia M, Wang JC, Kapp U, Dick JE (1999). Cytokine treatment or accessory cells are required to initiate engraftment of purified primitive human hematopoietic cells transplanted at limiting doses into NOD/SCID mice. *Bone Marrow Transplant* **23**:203–209.

37. Woodside SM *et al.* (2002). Advances in magnet and magnetic particle technologies enable development of a more robust and simple method of CD34 positive cell selection. *Blood* **100**:830a.
38. Olesen G, Tonder H, Holm MS, Hokland P (2001). Long-term culture of hematopoietic stem cells — validating the stromal component of the CAFC assay. *Cytotherapy* **3**:107–116.
39. Bhatia M, Wang JC, Kapp U, Bonnet D, Dick JE (1997). Purification of primitive human hematopoietic cells capable of repopulating immune-deficient mice. *Proc Natl Acad Sci USA* **94**:5320–5325.
40. Dick JE, Bhatia M, Gan O, Kapp U, Wang JC (1997). Assay of human stem cells by repopulation of NOD/SCID mice. *Stem Cells* **15**:199–203 [discussion 204-197].
41. Kerre TC, De Smet G, De Smedt M, Zippelius A, Pittet MJ, Langerak AW, De Bosscher J, Offner F, Vandekerckhove B, Plum J (2002). Adapted NOD/SCID model supports development of phenotypically and functionally mature T cells from human umbilical cord blood CD34(+) cells. *Blood* **99**:1620–1626.
42. Mazurier F, Doedens M, Gan OI, Dick JE (2003). Rapid myeloerythroid repopulation after intrafemoral transplantation of NOD-SCID mice reveals a new class of human stem cells. *Nat Med* **9**:959–963.
43. Zanjani ED, Almeida-Porada G, Livingston AG, Flake AW, Ogawa M (1998). Human bone marrow CD34– cells engraft in vivo and undergo multilineage expression that includes giving rise to CD34+ cells. *Exp Hematol* **26**:353–360.
44. Almeida-Porada G, Porada CD, Chamberlain J, Torabi A, Zanjani ED (2004). Formation of human hepatocytes by human hematopoietic stem cells in sheep. *Blood* **104**:2582–2590.
45. Flake AW (2004). The conceptual application of systems theory to stem cell biology: a matter of context. *Blood Cells Mol Dis* **32**:58–64.
46. Thorsteinsdottir U, Sauvageau G, Humphries RK (1999). Enhanced in vivo regenerative potential of HOXB4-transduced hematopoietic stem cells with regulation of their pool size. *Blood* **94**:2605–2612.
47. Daga A, Podesta M, Capra MC, Piaggio G, Frassoni F, Corte G (2000). The retroviral transduction of HOXC4 into human CD34(+) cells induces an in vitro expansion of clonogenic and early progenitors. *Exp Hematol* **28**:569–574.
48. Reya T (2003). Regulation of hematopoietic stem cell self-renewal. *Recent Prog Horm Res* **58**:283–295.
49. Reya T, Duncan AW, Ailles L, Domen J, Scherer DC, Willert K, Hintz L, Nusse R, Weissman IL (2003). A role for Wnt signalling in self-renewal of haematopoietic stem cells. *Nature* **423**:409–414.
50. Stier S, Cheng T, Dombkowski D, Carlesso N, Scadden DT (2002). Notch1 activation increases hematopoietic stem cell self-renewal in vivo and favors lymphoid over myeloid lineage outcome. *Blood* **99**:2369–2378.
51. Park IK, Qian D, Kiel M, Becker MW, Pihalja M, Weissman IL, Morrison SJ, Clarke MF (2003). Bmi-1 is required for maintenance of adult self-renewing haematopoietic stem cells. *Nature* **423**:302–305.
52. Cheng T, Rodrigues N, Dombkowski D, Stier S, Scadden DT (2000). Stem cell repopulation efficiency but not pool size is governed by p27(kip1). *Nat Med* **6**:1235–1240.
53. Dzierzak E (2003). Ontogenic emergence of definitive hematopoietic stem cells. *Curr Opin Hematol* **10**: 229–234.
54. Bhatia M, Bonnet D, Wu D, Murdoch B, Wrana J, Gallacher L, Dick JE (1999). Bone morphogenetic proteins regulate the developmental program of human hematopoietic stem cells. *J Exp Med* **189**:1139–1148.
55. Zhang J, Niu C, Ye L, Huang H, He X, Tong WG, Ross J, Haug J, Johnson T, Feng JQ, Harris S, Wiedemann LM, Mishina Y, Li L (2003). Identification of the haematopoietic stem cell niche and control of the niche size. *Nature* **425**:836–841.
56. Wadman IA, Osada H, Grutz GG, Agulnick AD, Westphal H, Forster A, Rabbitts TH (1997). The LIM-only protein Lmo2 is a bridging molecule assembling an erythroid, DNA-binding complex which includes the TAL1, E47, GATA-1 and Ldb1/NLI proteins. *EMBO J* **16**:3145–3157.
57. Vyas P, Ault K, Jackson CW, Orkin SH, Shivdasani RA (1999). Different sequence requirements for expression in erythroid and megakaryocytic cells within a regulatory element upstream of the GATA-1 gene. *Development* **126**:2799–2811.
58. Porcher C, Swat W, Rockwell K, Fujiwara Y, Alt FW, Orkin SH (1996). The T cell leukemia oncoprotein SCL/tal-1 is essential for development of all hematopoietic lineages. *Cell* **86**:47–57.
59. Shivdasani RA, Orkin SH (1996). The transcriptional control of hematopoiesis. *Blood* **87**:4025–4039.
60. Crispino JD, Lodish MB, MacKay JP, Orkin SH (1999). Use of altered specificity mutants to probe a specific protein–protein interaction in differentiation: the GATA-1:FOG complex. *Mol Cell* **3**:219–228.
61. Shivdasani RA (1997). Stem cell transcription factors. *Hematol Oncol Clin North Am* **11**:1199–1206.

62. Fujiwara Y, Browne CP, Cunniff K, Goff SC, Orkin SH (1996). Arrested development of embryonic red cell precursors in mouse embryos lacking transcription factor GATA-1. *Proc Natl Acad Sci USA* **93**: 12355–12358.
63. Shivdasani RA, Fujiwara Y, McDevitt MA, Orkin SH (1997). A lineage-selective knockout establishes the critical role of transcription factor GATA-1 in megakaryocyte growth and platelet development. *EMBO J* **16**:3965–3973.
64. Trainor CD, Omichinski JG, Vandergon TL, Gronenborn AM, Clore GM, Felsenfeld G (1996). A palindromic regulatory site within vertebrate GATA-1 promoters requires both zinc fingers of the GATA-1 DNA-binding domain for high-affinity interaction. *Mol Cell Biol* **16**:2238–2247.
65. Perry C, Soreq H (2002). Transcriptional regulation of erythropoiesis. Fine tuning of combinatorial multidomain elements. *Eur J Biochem* **269**:3607–3618.
66. Tsang AP, Fujiwara Y, Hom DB, Orkin SH (1998). Failure of megakaryopoiesis and arrested erythropoiesis in mice lacking the GATA-1 transcriptional cofactor FOG. *Genes Dev* **12**:1176–1188.
67. Nichols KE, Crispino JD, Poncz M, White JG, Orkin SH, Maris JM, Weiss MJ (2000). Familial dyserythropoietic anaemia and thrombocytopenia due to an inherited mutation in GATA1. *Nat Genet* **24**:266–270.
68. Turner J, Crossley M (1999). Basic Kruppel-like factor functions within a network of interacting haematopoietic transcription factors. *Int J Biochem Cell Biol* **31**:1169–1174.
69. Perkins A (1999). Erythroid Kruppel like factor: from fishing expedition to gourmet meal. *Int J Biochem Cell Biol* **31**:1175–1192.
70. Tewari R, Gillemans N, Wijgerde M, Nuez B, von Lindern M, Grosveld F, Philipsen S (1998). Erythroid Kruppel-like factor (EKLF) is active in primitive and definitive erythroid cells and is required for the function of 5′HS3 of the beta-globin locus control region. *EMBO J* **17**:2334–2341.
71. Armstrong JA, Bieker JJ, Emerson BM (1998). A SWI/SNF-related chromatin remodeling complex, E-RC1, is required for tissue-specific transcriptional regulation by EKLF in vitro. *Cell* **95**:93–104.
72. Crossley M, Whitelaw E, Perkins A, Williams G, Fujiwara Y, Orkin SH (1996). Isolation and characterization of the cDNA encoding BKLF/TEF-2, a major CACCC-box-binding protein in erythroid cells and selected other cells. *Mol Cell Biol* **16**:1695–1705.
73. Asano H, Li XS, Stamatoyannopoulos G (1999). FKLF, a novel Kruppel-like factor that activates human embryonic and fetal beta-like globin genes. *Mol Cell Biol* **19**:3571–3579.
74. Asano H, Li XS, Stamatoyannopoulos G (2000). FKLF-2: a novel Kruppel-like transcriptional factor that activates globin and other erythroid lineage genes. *Blood* **95**:3578–3584.
75. Athanasiou M, Mavrothalassitis G, Sun-Hoffman L, Blair DG (2000). FLI-1 is a suppressor of erythroid differentiation in human hematopoietic cells. *Leukemia* **14**:439–445.
76. Rekhtman N, Radparvar F, Evans T, Skoultchi AI (1999). Direct interaction of hematopoietic transcription factors PU.1 and GATA-1: functional antagonism in erythroid cells. *Genes Dev* **13**:1398–1411.
77. Harada H, Harada Y, O'Brien DP, Rice DS, Naeve CW, Downing JR (1999). HERF1, a novel hematopoiesis-specific RING finger protein, is required for terminal differentiation of erythroid cells. *Mol Cell Biol* **19**:3808–3815.
78. Wojchowski DM, Gregory RC, Miller CP, Pandit AK, Pircher TJ (1999). Signal transduction in the erythropoietin receptor system. *Exp Cell Res* **253**:143–156.
79. Cheung JY, Miller BA (2001). Molecular mechanisms of erythropoietin signaling. *Nephron* **87**:215–222.
80. Richmond TD, Chohan M, Barber DL (2005). Turning cells red: signal transduction mediated by erythropoietin. *Trends Cell Biol* **15**:146–155.
81. Jiang N, He TC, Miyajima A, Wojchowski DM (1996). The box1 domain of the erythropoietin receptor specifies Janus kinase 2 activation and functions mitogenically within an interleukin 2 beta-receptor chimera. *J Biol Chem* **271**:16472–16476.
82. Wojchowski DM, Gregory RC, Miller CP, Pandit AK, Pircher TJ (1999). Signal transduction in the erythropoietin receptor system. *Exp Cell Res* **253**:143–156.
83. Pircher TJ, Geiger JN, Zhang D, Miller CP, Gaines P, Wojchowski DM (2001). Integrative signaling by minimal erythropoietin receptor forms and c-Kit. *J Biol Chem* **276**:8995–9002.
84. Motoyama N, Kimura T, Takahashi T, Watanabe T, Nakano T (1999). bcl-x prevents apoptotic cell death of both primitive and definitive erythrocytes at the end of maturation. *J Exp Med* **189**:1691–1698.

85. Gobert S, Chretien S, Gouilleux F, Muller O, Pallard C, Dusanter-Fourt I, Groner B, Lacombe C, Gisselbrecht S, Mayeux P (1996). Identification of tyrosine residues within the intracellular domain of the erythropoietin receptor crucial for STAT5 activation. *EMBO J* **15**:2434–2441.
86. Gregory RC, Jiang N, Todokoro K, Crouse J, Pacifici RE, Wojchowski DM (1998). Erythropoietin receptor and STAT5-specific pathways promote SKT6 cell hemoglobinization. *Blood* **92**:1104–1118.
87. Kubota Y, Tanaka T, Kitanaka A, Ohnishi H, Okutani Y, Waki M, Ishida T, Kamano H (2001). Src transduces erythropoietin-induced differentiation signals through phosphatidylinositol 3-kinase. *EMBO J* **20**: 5666–5677.
88. Munugalavadla V, Kapur R (2005). Role of c-Kit and erythropoietin receptor in erythropoiesis. *Crit Rev Oncol Hematol* **54**:63–75.
89. Huddleston H, Tan B, Yang FC, White H, Wenning MJ, Orazi A, Yoder MC, Kapur R, Ingram DA (2003). Functional p85alpha gene is required for normal murine fetal erythropoiesis. *Blood* **102**:142–145.
90. Starr R, Willson TA, Viney EM, Murray LJ, Rayner JR, Jenkins BJ, Gonda TJ, Alexander WS, Metcalf D, Nicola NA, Hilton DJ (1997). A family of cytokine-inducible inhibitors of signaling. *Nature* **387**:917–921.
91. Verdier F, Rabionet R, Gouilleux F, Beisenherz-Huss C, Varlet P, Muller O, Mayeux P, Lacombe C, Gisselbrecht S, Chretien S (1998). A sequence of the CIS gene promoter interacts preferentially with two associated STAT5A dimers: a distinct biochemical difference between STAT5A and STAT5B. *Mol Cell Biol* **18**:5852–5860.
92. Naka T, Narazaki M, Hirata M, Matsumoto T, Minamoto S, Aono A, Nishimoto N, Kajita T, Taga T, Yoshizaki K, Akira S, Kishimoto T (1997). Structure and function of a new STAT-induced STAT inhibitor. *Nature* **387**:924–929.
93. Cacalano NA, Sanden D, Johnston JA (2001). Tyrosine-phosphorylated SOCS-3 inhibits STAT activation but binds to p120 RasGAP and activates Ras. *Nat Cell Biol* **3**:460–465.
94. Gommerman JL, Sittaro D, Klebasz NZ, Williams DA, Berger SA (2000). Differential stimulation of c-Kit mutants by membrane-bound and soluble steel factor correlates with leukemic potential. *Blood* **96**: 3734–3742.
95. Ueda S, Mizuki M, Ikeda H, Tsujimura T, Matsumura I, Nakano K, Daino H, Honda Zi Z, Sonoyama J, Shibayama H, Sugahara H, Machii T, Kanakura Y (2002). Critical roles of c-Kit tyrosine residues 567 and 719 in stem cell factor-induced chemotaxis: contribution of src family kinase and PI3-kinase on calcium mobilization and cell migration. *Blood* **99**:3342–3349.
96. Thommes K, Lennartsson J, Carlberg M, Ronnstrand L (1999). Identification of Tyr-703 and Tyr-936 as the primary association sites for Grb2 and Grb7 in the c-Kit/stem cell factor receptor. *Biochem J* **341**:211–216.
97. Javazon EH, Beggs KJ, Flake AW (2004). Mesenchymal stem cells: paradoxes of passaging. *Exp Hematol* **32**:414–425.
98. Bailey AS, Jiang S, Afentoulis M, Baumann CI, Schroeder DA, Olson SB, Wong MH, Fleming WH (2004). Transplanted adult hematopoietic stems cells differentiate into functional endothelial cells. *Blood* **103**: 13–19.
99. Lakshmipathy U, Verfaillie C (2005). Stem cell plasticity. *Blood Rev* **19**:29–38.

17

Bone Marrow Transplantation

by Robert I. Raphael and Mark C. Walters

Introduction

Since the enactment of a national effort to establish and sustain comprehensive centers for sickle cell anemia 30 years ago, the vast majority of infants diagnosed with sickle cell anemia by state newborn screening programs and enrolled in comprehensive care programs will survive to adulthood, at a rate that appears indistinguishable from African-Americans who do not inherit this disorder.[1] Against this backdrop, attitudes about the role of hematopoietic cell transplantation (HCT) are also shifting.[2] Initially performed in those who had co-existent hematological malignancies, the first cases served primarily to demonstrate that allogeneic transplantation had the potential to cure sickle cell anemia.[3] The next phase of investigations targeted symptomatic patients for enrollment and focused on characterizing the consequences of transplantation and the quality of cure among those who survived free of sickle cell disease (SCD).[4–6] Follow-up studies after transplantation confirm the sustained benefit of donor erythropoiesis which improves significantly the quality of life among those with stable engraftment of donor cells.[5–7] Today there is little doubt about the potential for cure, since greater than 90% of recipients survive, and approximately 85% of children survive free of the underlying disease after human leukocyte antigen (HLA)-identical sibling donor HCT. Rather, the debate today revolves around the question of who is eligible for transplantation, and when to intervene.

As occurs for malignant disorders, there are ongoing efforts to better define the role of transplantation for SCD as novel supportive and targeted gene therapies become available. These analyses are difficult in the absence of randomized, prospective trials that compare therapeutic alternatives. Thus, current clinical research in transplantation is focused on three important objectives. The first is to identify those patients who have the greatest risk of developing sickle-related complications, and who are most likely to benefit from HCT. The second is to reduce transplant-related complications by minimizing the short- and long-term toxicities of HCT, but in such a way that does not diminish the likelihood of a successful transplantation outcome. The third is to increase the availability of HCT to potential recipients by expanding the pool of suitable donors, either by using alternate sources of hematopoietic stem cells or by overcoming HLA disparity and its formidable barrier to donor-host immunological tolerance.

Current Indications for Hematopoietic Cell Transplantation (HCT)

In standard practice, HCT for SCD currently is reserved almost exclusively for patients with clinical features that portend a poor outcome or significant sickle-related morbidity, in part due to the toxicity of this intensive therapy.[8] These clinical indications, which were adapted from the multicenter investigation of bone marrow transplantation for SCD, are listed in Table 1. In addition, these criteria have been applied almost exclusively to children, where the risk-benefit ratio is most advantageous in terms of years-of-life gained among those who survive with sustained engraftment of donor cells. Less certain is how to apply inclusion criteria to adults with sickle cell anemia, where the experience of transplantation is limited, but for whom the risk of significant transplantation-related toxicity remains substantial. For all patients, clinicians must carefully weigh therapeutic alternatives to HCT, with particular attention to safety, efficacy, availability, and the cost of intervention.[9–12] Unfortunately, prospective clinical trials that compare HCT to other therapeutic interventions for SCD have not been conducted, thus therapeutic decision-making is hampered by this limitation.

Current Results of HCT for Sickle Cell Disease (SCD)

The worldwide experience of transplantation for SCD is summarized in Table 2.[17–19] In the collective experiences of these studies, the transition of HCT from an experimental intervention reserved for severely affected patients, to one in which younger children with early signs of sickle-related morbidity are targeted, has been observed. Several series in Europe and North America have reported very similar results after HLA-identical sibling transplantation.[5,6,20] The principal aim of these multicenter clinical studies was to define more completely the risks and benefits of this therapy, and to characterize the natural history of those surviving free of SCD. The results of transplantation were best when performed in children with SCD who had HLA-identical sibling donors. Even though many children who received allografts had significant sickle-related complications such as stroke and recurrent episodes of acute chest syndrome, the disease-free survival was 80–85% in several series. However, 5–10% of patients died of complications related to transplantation, with GVHD and its treatment the leading cause of death.

In the multicenter investigation of HCT for SCD, 59 children who ranged in age between 3.3 and 15.9 (median 9.9) years received HLA-identical sibling allografts between September 1991 and April 2000. Patients received a myeloablative preparative regimen of BU, CY, and horse ATG, and most received a combination of MTX and CSA after HCT to prevent graft versus host disease.

Table 1. Indications of Hematopoietic Cell Transplantation (HCT) for Sickle Cell Disease (SCD).

Patients with SCD (SS or HbSβ^0-thalassemia) less than 16 years of age
One or more of the following complications:
Stroke or CNS event lasting longer than 24 hours
Recurrent acute chest syndrome
Recurrent vaso-occlusive painful episodes or recurrent priapism
Impaired neuropsychologic function with abnormal cerebral MRI and angiography
Stage I or II sickle lung disease
Sickle nephropathy (GFR 30–50% of predicted normal)

Abbreviations: HCT, hematopoietic cell transplantation; CNS, central nervous system; GFR, glomerular filtration rate; MRI, magnetic resonance imaging.

Table 2. HCT for SCD.[1]

	Minimal-toxicity conditioning regimen	Reduced-intensity conditioning regimen	Conventional myeloablative conditioning regimen
No. of patients	11†	12‡	201
Patient age (median, yrs)	10 (range, 3–28)	22 (range, 5–56)	— (range, 0.9–22)
Conditioning regimen (dose) [no. of patients]	Flu(90–150)/TBI (200) [5]; Flu(125–150)/ATG/ TBI (200) [6]	Flu(175)/BU(8)/ ATG/TLI (500) [5]; Flu(120)/Mel(140)/ ATG [2]; Flu(120)/CY(120) [1]; Flu(120)/Mel(140)/ Campath [2]; Flu(120)/BU(3.2) [2]	BU/CY/ATG [133] BU/CY [59] BU/CY/TLI [6] CY/TBI [3]
Source of stem cells (no. of patients)	Marrow (9); PBHC (2)	Marrow (6)*; CBSC (1); PBHC (6)	Marrow
Induction of mixed chimerism	Yes (transient in 10)	Yes	Yes, in 11%
No. with graft rejection/disease recurrence	10	3	16 (8%)
No. with GVHD	Acute 1 (Gr I), chronic, none	Acute 4 (Gr II-IV), chronic, 3 (2 fatal)	25% aGVHD/12% cGVHD
No. of deaths	1 (after a 2nd HCT)	2 (17%)	20 (10%)
No. with event-free survival	1 (9%)	7 (58%)	166 (83%)

[1]Published reports (3–6, 64–70, 79–85).
†Includes two patients with thalassemia major.
‡Includes one patient with thalassemia major.
*One patient received a combination of marrow and umbilical cord blood from the same donor.
Abbreviations: ATG, antithymocyte globulin; CY, cyclophosphamide; Flu, fludarabine; Gr, grade; GVHD, graft-versus-host disease; Mel, melphalan; No., number; PBHC, peripheral blood stem cells; TBI, total body irradiation; TLI, total lymphoid irradiation; CBSC, umbilical cord blood.

Fifty of 55 children experienced disease-free survival. The Kaplan–Meier probabilities of survival and disease-free survival are 93% and 84%, respectively (Fig. 1). Results among 101 patients who received myeloablative HLA-identical sibling HCT in Europe were remarkably similar, with an overall survival probability of 88% and disease-free survival of 80%. A recent report of 60 French patients with severe SCD who received myeloablative HCT showed similar results, with overall and event-free survival probabilities of 90% and 82%, respectively. There was no graft rejection observed in patients who received rabbit ATG before HCT, but donor-host chimerism heralded graft rejection in 4 of 12 patients who did not receive antithymocyte globulin (ATG).[21] A retrospective review of 24 Belgian patients with SCD analyzed the impact of pre-transplant ATG and hydroxyurea on outcome. The latter before

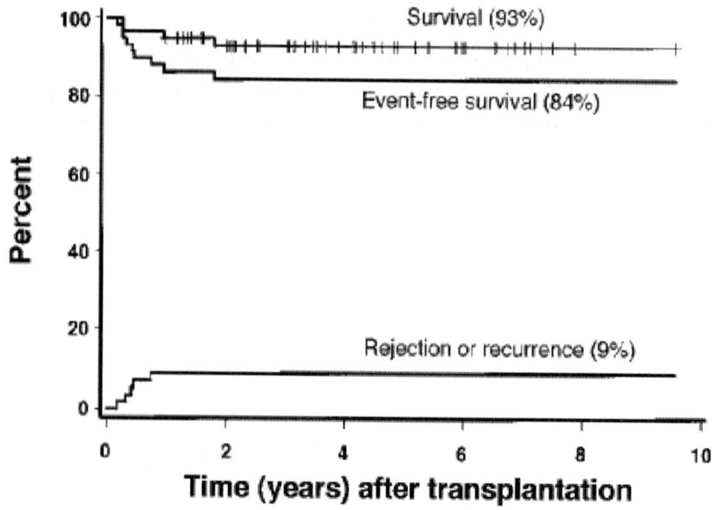

Fig. 1. Kaplan–Meier estimates of survival and event-free survival after bone marrow transplantation for SCD — multi-center investigation outcome after transplantation for 59 children with advanced symptomatic SCD.[5] Kaplan–Meier estimates for survival and event-free survival following bone marrow transplantation are shown. An event was defined as death, graft rejection, or recurrence of sickle cell disease. A cumulative incidence curve for graft rejection and return of SCD is also depicted. Used with permission from Walters.[5]

HCT was associated with a significantly lower rate of graft rejection, compared to patients who did not receive hydroxyurea, but the administration of ATG before transplantation did not affect the rate of engraftment. Overall 20 of 24 patients had stable engraftment of donor cells after HCT, and 23 patients experienced 79% disease-free survival.[22]

Complications after HCT

Disease recurrence after HCT remains a key obstacle to success. The cumulative incidence of graft rejection accompanied by autologous recovery was approximately 10% in several published series.[6,7,20] Autologous reconstitution is usually associated with a renewed risk of vaso-occlusive complications, including stroke.[26] However, selected individuals benefited from elevated fetal hemoglobin levels after autologous recovery, and remained free of clinical vaso-occlusive symptoms for several years after HCT.[32] Unfortunately, in most cases the fetal hemoglobin elevation was not sustained in the long-term. In the multicenter investigation of transplantation for SCD, in a univariate analysis no significant association was found between risk of graft rejection and factors including age, sex, donor genotype, number of transfusions, or erythrocyte alloimmunization. There was a significant association between graft rejection and pre-transplant administration of iron chelation therapy, suggesting that pre-transplant exposures to blood products promoted an immunological response that interfered with donor engraftment.[7] This contrasts with graft rejection rates that are very low among recipients with hematological malignancies who undergo HLA-identical sibling transplantation.[33] Thus, the high frequency of graft rejection observed in sickle cell patients implies that immunological barriers that might include suppression of host natural killer cells and T-lymphocytes are not reliably overcome by myeloablative pre-transplantation therapy. In a murine model of allograft rejection after HCT, sickle mice demonstrated heterologous immunity with memory T-cells directed against major histocompatibiltity-mismatched donor alloantigens, even in the absence of prior exposure to these antigens. The minority of sickle mice that had stable donor engraftment demonstrated loss of alloreactivity to these donor antigens. These experiments suggest that heterologous immunity might reflect in part, the influence of inflammation that accompanies SCD, and might generate a cytokine milieu that promotes graft rejection.[34]

The problem of rejection might be approached by several different strategies. If, in fact, sensitization to minor histocompatibility antigens mediated by host memory T-cells is a key determinant of rejection, then targeting younger patients with few or even no transfusion exposures might improve outcomes. When transfusions are necessary in transplant candidates, it is important to consider the use of gamma-irradiated, leukocyte-reduced blood products to reduce the risk of sensitization to minor histocompatibility antigens.[35] Alternatively, it is possible that targeted immunosuppressive therapy that inhibits the host-versus-graft reaction before and after transplantation, either with or in lieu of myeloablative pre-transplantation conditioning, might be employed to promote engraftment. Ex vivo immune co-stimulation, blockade of alloreactive T-cells by anti-CD40 ligand and cytotoxic T-lymphocyte antigen-4 (CTLA-4) immunoglobulin has been applied successfully in murine models of major histocompatibility incompatibility and in murine models of SCD.[36,37] Immunosuppressive medications such as sirolimus and mycephenophelate mofetil may promote apoptosis of alloreactive host T-cells, compared with more traditional agents which non-specifically block T-cell activation by Interleukin-2 inhibition.[38,39] Post-transplantation use of alkylators such as cyclophosphamide might also be employed to selectively destroy activated alloreactive host T-cells; this approach has been reported to reduce the radiation requirement for engraftment in mice, and has been utilized in two series of adults receiving haploidentical and partially HLA-mismatched HCT for hematologic malignancy after non-myeloablative conditioning.[40–42] The identification and enrichment of donor cellular populations that facilitate engraftment without causing GVHD is another strategy that could be pursued. Very high doses of $CD34^+$ stem cells may theoretically facilitate engraftment by competition with surviving host hematopoietic cells, as well as by other mechanisms. "Megadose" T-cell depleted HCT promoted engraftment in histo-incompatible mice after sublethal radiation, and in one clinical trial using 10-fold higher doses of $CD34^+$ cells for haploidentical HCT in patients with hematologic malignancies, 16 of 17 recipients engrafted compared with none of five historical controls given conventional doses of $CD34^+$ cells.[43,44] Another approach employs ex vivo photochemical treatment (PCT) of donor T-cells with psoralen and ultraviolet light activation, to impair the proliferative capacity of alloreactive donor cells and minimize GVHD, while retaining the capacity of the donor lymphocytes to facilitate engraftment. In a murine model of major histocompatibility — mismatched HCT, the addition of PCT-treated T-cells to T-depleted bone marrow facilitated donor engraftment and complete chimerism without causing GVHD.[45] In a murine β-thalassemia model, lethally-irradiated thalassemic animals received bone marrow with added PCT-treated or control T-cells from non-thalassemic major histocompatibility — mismatched donors. Recipients of PTC-treated T-cells showed stable chimerism with resolution of thalassemic features including anemia and splenomegaly, while recipients of control T-cells died of GVHD by 14 days after transplantation.[46] A similar approach is being investigated in a murine model of SCD.

One complication distinctive of SCD and HCT is an increased risk of adverse neurologic events after HCT (Table 2). In particular, patients with a history of stroke had an increased risk of intracranial hemorrhage.[47] Four of seven initial patients in the multicenter investigation had adverse events including two episodes of intracranial hemorrhage, which led to incorporation of measures to prevent central nervous system complications after transplantation. These measures included anti-convulsant prophylaxis with phenytoin initiated with BU dosing and continued for 6 months following transplantation (or until CSP was discontinued), strict control of hypertension, prompt repletion of magnesium deficiency, and maintenance of hemoglobin concentrations between 9 and 11 gm/dl and platelet counts $>$50,000/mm^3. By implementing these measures, the problem of intracranial hemorrhage was eliminated in the multicenter trial, however patients continued to experience self-limited seizures, particularly in the setting of hypertension, relative hyperviscosity caused by transfusion, or if receiving CSP. There were no long-term sequelae associated with these events.[48,49] Anti-convulsant prophylaxis in a European series similarly did not appear to reduce the seizure incidence after HCT (38% versus 35%) among those who received, or did not receive prophylaxis, respectively.[6]

Chronic GVHD was the major cause of transplant-related morbidity and mortality late after transplantation.[8] Chronic GVHD was reported in 20 of 201 patients after HCT for SCD and resulted in or contributed to the death of eight patients. In the multicenter trial, 11 of 59 patients (19%) developed acute or chronic GVHD, which was the cause of death in three patients.[7] Of note, fatal bronchiolitis obliterans organizing pneumonia (BOOP) as a consequence of chronic GVHD has been reported in two patients after HCT for recurrent ACS, and it is speculated that pre-existing lung injury may predispose to this type of inflammation after engraftment.[50] GVHD by definition occurs when an immunocompetent graft is transplanted into an immunodeficient host, stimulating an inflammatory cytokine cascade resulting in apoptosis of epithelial and endothelial cells, manifest primarily with involvement of the skin, liver, and gastrointestinal tract. Risk factors include older age and HLA disparity between donor and recipient. This condition may have some benefit in the setting of HCT for malignancy by producing an immunologic graft-versus-leukemia effect, but is of no therapeutic value in HCT for non-malignant conditions such as SCD. GVHD prophylaxis regimens have varied among institutions, but most commonly consist of a combination of MTX and CSA, while treatment for GVHD usually involves prednisone, sometimes accompanied by other immunosuppressive agents. Attempts to reduce the toxicity of conditioning regimens and create stable mixed chimerism may work to abrogate GVHD. Notably, in the multicenter investigation, none of 13 patients with stable mixed chimerism developed acute or chronic GVHD.[7]

The risk of malignancy after conditioning regimens using chemotherapy rather than total body irradiation was reported to be less than 1% in the cohort transplanted for β-thalassemia major.[51,52] However, the development of myelodysplastic syndrome and myeloid leukemia in one SCD patient after HCT was associated with the administration of intensive immunosuppression for chronic GVHD.[6]

Growth and Development after HCT for SCD

A recent analysis by the multicenter trial of transplantation for SCD compared linear growth after HCT to three cohorts of children with SCD. Data from the Cooperative Study of Sickle Cell Disease (CSSCD) and from a Phase I-II pediatric hydroxyurea trial (HUG-KIDS) before and during treatment at the maximum tolerated dose were used for comparison. A Hierarchical Linear Model was used to model height and weight over the period of follow-up. At the baseline, there were no significant height or weight differences between the HCT and comparison groups among females; however, males in the HUG-KIDS (Pre) group were approximately 4 cm taller than males in the HCT group. In all comparisons except that of males in the HCT and CSSCD groups, there were no significant differences in the linear growth velocity. However, males treated by HCT grew on average 0.67 cm/year more than males in the CSSCD group, and gained on average 1.0 kg/year more weight than males in the CSSCD, and these differences reached statistical significance. These data support the conclusion that growth after HCT for SCD is not impaired compared to children who receive standard therapy or hydroxyurea. Moreover, in males, growth was more rapid after HCT than among males enrolled in the CSSCD. This analysis supports the notion that HCT for SCD offers a significant therapeutic benefit without a negative impact on growth.[53]

Gonadotropin and sex hormone levels of surviving patients were evaluated after HCT, confirming the toxic effect of busulfan on gonadal function. Among seven surviving females in the multicenter study who were older than 13 years of age after HCT, an interim analysis showed that 5 had primary amenorrhea and 5 had corresponding elevated lutenizing and follicle stimulating hormone levels that were associated with decreased serum estradiol levels in 4. One individual receiving hormonal replacement therapy had elevated lutenizing and follicle stimulating hormone levels and a normal serum estradiol. One post-pubertal female had normal serum follicle stimulating hormone and estradiol levels. Of the 7 males who were more than 13 years of age, none of the 4 tested had elevated serum lutenizing and follicle stimulating hormone levels. However, two males who were 14 and 16 years of age had low testosterone levels that were correlated with gonadotropin levels in the pre-pubertal range. Among 6 pre-pubertal girls in the Belgian cohort, 5 had primary amenorrhea with elevated serum lutenizing and follicle stimulating hormone.[6] Two post-pubertal females developed secondary amenorrhea. Testicular function was also adversely affected in 4 of the 6 boys who demonstrated decreased testosterone and elevated follicle stimulating hormone levels. It is anticipated that many if not most of the females will require hormonal replacement therapy after HCT.

Stable Mixed Chimerism after HCT

Stable mixed hematopoietic chimerism has been observed after conventional myeloablative transplantation for hemoglobinopathies. This condition has the potential for considerable ameliorative effect, well documented for β-thalassemia major and other hereditary disorders.[54–58] Approximately 10% of children with SCD and thalassemia major developed stable mixed chimerism after conventional HLA-identical sibling HCT.[7,54] In the multicenter investigation of BMT for sickle cell anemia, 13 of 50 patients with successful allografts developed stable mixed chimerism. The levels of donor chimerism, measured $\geq$6 months after transplantation in peripheral blood, varied between 90% and 99% in eight patients. Five additional patients had a lower proportion of donor cells (range, 11–74%). Among these 5, hemoglobin levels varied between 11.2 and 14.2 g/dL (median, 11.3; mean, 12.0). In patients who had donors with a normal hemoglobin genotype, the sickle hemoglobin (HbS) fractions were $<$7%, which corresponded to donor chimerism levels of 67%, 74%, and 11% (Fig. 2). Among patients who had donors with sickle trait, the HbS fractions were 36% and 37%, which corresponded to donor chimerism levels of 25% and 60%, respectively. Thus, allograft recipients with stable mixed chimerism had HbS levels similar to donor levels, and only 1 patient required a RBC transfusion beyond 90 days after transplantation. None of the patients experienced painful events or other clinical complications related to SCD after transplantation. These observations are consistent with the idea that chimerism even with a minority of donor cells might have a curative effect, and that full engraftment of donor cells is not a requirement for successful HCT.

Murine models of sickle cell anemia have been utilized in non-myeloablative HCT investigations that appear to support the notion that this approach affords a significant benefit even when stable mixed chimerism develops after HCT.[59,60] These models were used to investigate the fraction of

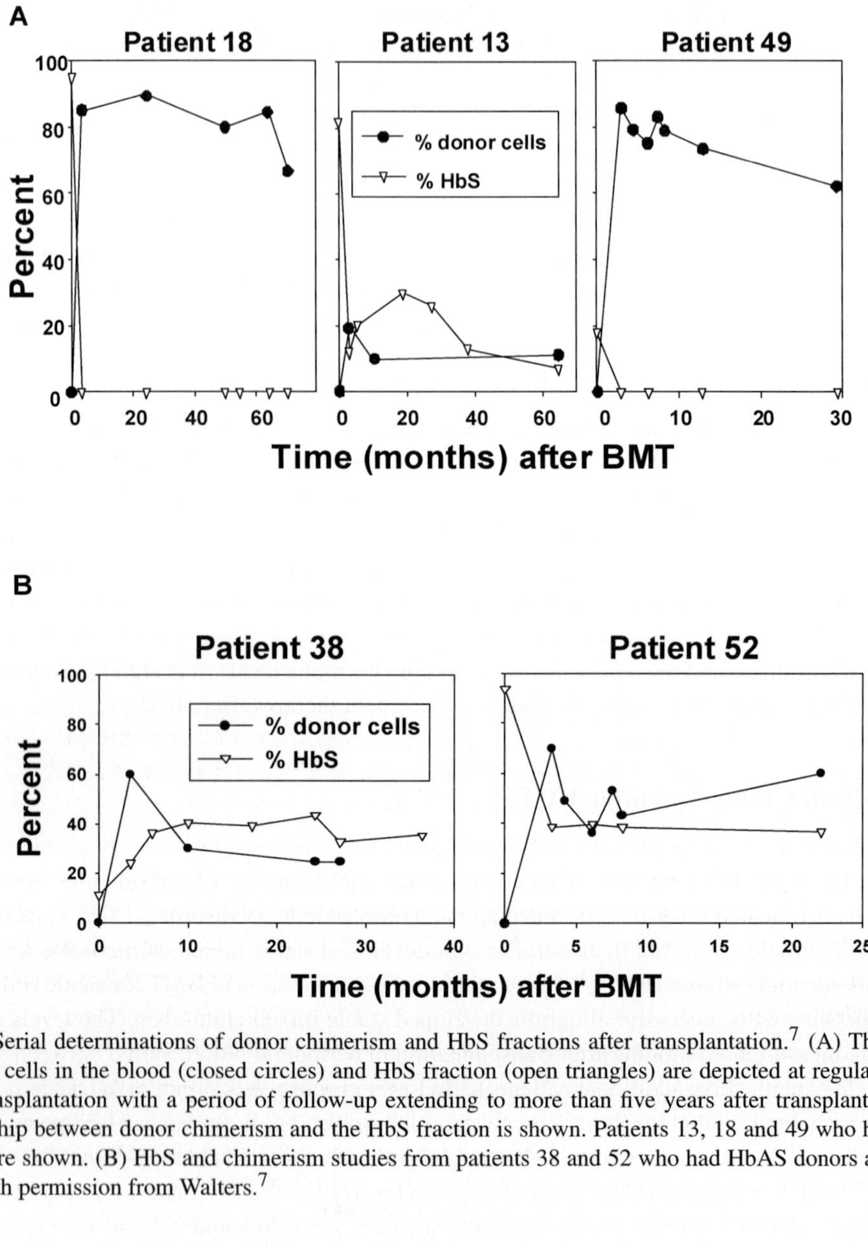

Fig. 2. Serial determinations of donor chimerism and HbS fractions after transplantation.[7] (A) The fraction of donor cells in the blood (closed circles) and HbS fraction (open triangles) are depicted at regular intervals after transplantation with a period of follow-up extending to more than five years after transplantation. The relationship between donor chimerism and the HbS fraction is shown. Patients 13, 18 and 49 who had HbAA donors are shown. (B) HbS and chimerism studies from patients 38 and 52 who had HbAS donors are shown. Used with permission from Walters.[7]

donor chimerism sufficient to effect a correction of sickle erythropoiesis. Using non-myeloablative conditioning by T-cell co-stimulation blockade, Kean *et al.*[60] created a panel of SCD mice that had a spectrum of white blood cell and red blood cell chimerism after transplantation that correlated with the size of the donor marrow inoculums. At all levels of chimerism evaluated, the donor red cell fraction in the blood was greater than donor leukocyte and progenitor cell fractions in the bone marrow. This apparent enrichment of donor red bells compared to donor white cells in sickle mice after transplantation reflected an improved survival of donor compared to native sickle erythrocytes. There was no apparent selective advantage of donor red cell precursors in the bone marrow.[60] Most evidence of sickle pathophysiology was eliminated after a threshold of 70% donor hemoglobin was achieved, although histological changes in the spleen and marrow were not eliminated until 100% donor chimerism was established.[36] In another murine model, sickle mice were treated *in utero* by donor hematopoietic cell injection to achieve donor hemoglobin chimera that ranged less than 30%. In a subsequent experiment, mice grafted *in utero* later received a postnatal, non-myeloablative hematopoietic stem cell "boost", which increased donor chimerism from 30% to 80–90%. The extent of red cell sickling observed *in vitro* during a challenge of hypoxia was proportional to the level of donor chimerism, and spleen size varied inversely with the level of donor chimerism. In addition, mice in the first group died of pulmonary sequestration within 15 minutes after initiating a hypoxic challenge, suggesting that a threshold of 30% donor chimerism was necessary to avoid this phenomenon.[61]

Non-Myeloablative HCT for Sickle Cell Disease

Building upon the concept that stable mixed chimerism might generate a significant ameliorative if not curative effect, several groups have attempted to extend the early encouraging results after non-myeloablative HCT for hematological malignancies. However, the clinical application of non-myeloablative HCT for SCD remains limited and incompletely defined, particularly in the application of transplantation for adults with SCD. There have been two approaches, one utilizing a minimally toxic regimen first developed in a large animal model and translated successfully into human trials for older adult patients with hematological malignancies, and a second, reduced-intensity regimen that retains a moderate degree of the myelosuppressive effect of transplantation.[62,63] Representative examples of the regimens are presented in Table 2. The former causes minimal myelosuppression and thus can be administered in the outpatient setting. This minimally toxic regimen relies on post-grafting immunosuppression to prevent GVH and host-versus-graft (HVG) reactions and thereby promotes engraftment of donor cells. In contrast, the latter approach relies on reduced-intensity preparation to suppress the HVG reaction and promote engraftment. This reduced-intensity regimen is associated with hospitalization and accompanied by a risk of regimen-related toxicity, albeit at a reduced level.

A survey of the experience to date utilizing alternative non-myeloablative regimens is presented in Table 2.[64–69] All patients received HLA-identical related donor allografts, utilizing either granulocyte colony-stimulating factor-mobilized peripheral blood hematopoietic cells or bone marrow, in one case combined with cord blood cells. Ten of the 11 recipients of a minimally toxic regimen developed mixed chimerism initially, but tolerance was not sustained after post-grafting immunosuppression was discontinued in 10 of the 11 patients. However, most patients were treated in the outpatient setting, none developed life-threatening transplant-related complications, and GVHD occurred infrequently and was mild in this predominantly pediatric cohort. Patients reverted to a sickle cell anemia phenotype after non-fatal graft rejection. Alternatively, those who received reduced-intensity conditioning regimens benefited from augmented pre-grafting immunosuppression that facilitated engraftment of donor cells, and 3 of 12 recipients experienced graft rejection. Acute and chronic GVHD occurred more frequently in this older cohort, affecting 4 of 12 patients that were fatal in 2 cases. Thus, the problem of transplant-related mortality was not eliminated by the reduced-intensity conditioning regimen, especially among older recipients. In contrast to adults, the administration of a reduced-intensity regimen in children consisting busulfan, FLU, ATG, and TLI was successful in establishing donor chimerism. Five patients who were between 6 and 18 years of age received HLA-identical sibling bone marrow transplantation with CSA and mycophenolate mofetil for post-grafting immunosuppression. Treatment-related toxicity was minimal, and 1 patient developed mild GVHD. All the patients had evidence of donor engraftment, and there was no graft rejection or recurrent SCD symptoms.[70] These encouraging results might reflect the younger age of this cohort, and suggest that future application of non-myeloablative conditioning regimens should focus on the enrollment of younger patients.

The observation of mixed chimerism after non-myeloablative preparation also confirmed the clinical benefit of this condition, even when it was transitory. This benefit is illustrated by 2 patients who had initial engraftment of donor cells that resulted in a mixed hematopoietic chimera, although the contribution from donor cells declined after post-grafting immunosuppression was withdrawn.[71] Despite the decline, the HbS fraction remained <30% during the period when there was a low level of donor chimerism. Serial blast forming unit-erythroid and colony forming unit-granulocyte macrophage colonies were obtained from bone marrow samples and donor contribution was determined. In both patients, there was overrepresentation of donor erythroid progenitors compared with myeloid counterparts. These findings suggested that the clinical benefits of mixed chimerism after non-myeloablative transplantation for sickle cell anemia resulted from an extended life span of mature donor erythrocytes and a selective advantage for donor erythroid progenitors in the bone marrow. This was confirmed in a series reported by Wu *et al.*[73] in which there was a 2-fold higher expression of donor β-globin RNA compared with total genomic DNA in the blood. Direct bone marrow analysis revealed ineffective erythropoiesis of native homozygous SS erythroblasts, with progressive enrichment of donor chimerism during erythrocyte maturation.[72,73] These findings were associated with clinical improvement after transplantation, and with improvements in hemolysis, endothelial function, and nitric oxide bioavailability. In 1 patient who experienced graft rejection, these parameters subsequently returned to pre-transplantation levels.[72,73]

Alternative Stem Cell Sources

The utilization of alternative stem cell sources is an obvious but unproven method to expand HCT for SCD. Following eligibility criteria described by the multicenter collaborative study, 38% of patients meet eligibility criteria for HCT, however only 14% of patients will have an HLA-matched sibling donor.[4] Almost all HCT cases to date have utilized bone marrow from HLA-identical siblings, and thus the experience of unrelated and HLA-mismatched related donor HCT for sickle cell anemia is very limited. The potential for unacceptable regimen-related toxicity and GVHD remain significant obstacles, but the development of new strategies to reduce toxicity and promote immunological tolerance has stimulated interest in unrelated and HLA-mismatched donor HCT for hemoglobinopathies. To date, there is a single report of 3 patients with SCD who received unrelated donor HCT,[74] although the experience in thalassemia major is growing more rapidly. A recent analysis of donor searches in the National Marrow Donor Program (NMDP) involving 77 putative transplantation recipients reported that 59.7% had at least one potential unrelated donor or umbilical cord blood unit that was matched for 6 of 6 HLA-antigens, and all 77 had a donor mismatched for a single HLA antigen.[75] Thus, there is potential for expanding HCT donors beyond the current limit of HLA-identical siblings.

Cord blood is a source of hematopoietic stem cells undergoing investigation to support HCT for SCD. Cord blood stem cells have several unique properties that make them useful in this setting. A lower incidence of severe GVHD is associated with HLA-mismatched cord blood transplantation, however, abrogation of the allogeneic effect combined with a limiting number of hematopoietic stem cells might act to increase the risk of engraftment failure.

Early reports of successful related donor cord blood transplantation for hemoglobinopathies have been extended and confirmed by recent experience from the Eurocord registry.[76] Forty-four patients ranging from 1 to 20 years of age, 33 with thalassemia and 11 with SCD received related donor cord blood HCTs. All but two donors (who were mismatched at a single HLA-antigen) were HLA-identical sibling donors. Twenty-six patients received BU and CY alone or in combination with ATG or antilymphocyte globulin. In 17 patients, the pre-conditioning regimen was augmented by thiotepa and/or fludarabine There were no deaths and 36 of 44 children experienced disease-free survival at a median follow-up of 24 months post-transplantation. Four patients experienced grade II acute GVHD and 2 of 39 patients at risk developed limited chronic GVHD. The 2-year probabilities of event-free survival were 79% and 90% among patients with thalassemia and SCD, respectively (Fig. 3).

One patient with SCD and 7 of 33 patients with thalassemia experienced disease recurrence after cord blood transplantation. Among the 8 patients with graft failure, 1 received an autologous marrow infusion and 3 out of 5 patients had sustained donor engraftment after a second conventional HCT from the same sibling donor. The impact of the pre-conditioning regimen and the use of MTX to prevent GVHD on disease-free survival were evaluated in a univariate analysis which showed that patients given MTX had a significantly lower probability of event-free survival compared to those who did not receive MTX (55% versus 90%, respectively). In addition, pre-conditioning with BU and CY alone was associated with a lower probability of engraftment. Among thalassemia patients, the combinations of BU and TT, with either CY or FLU as a third agent were associated with a higher probability of event-free survival. These results suggest that outcomes after cor blood transplantation from sibling donors are similar to those observed after BMT, with potentially a lower rate of GVHD. The results also suggest that the problem of graft rejection is affected by the composition of pre- and post-engraftment immunosuppressive regimens. To test these hypotheses, a prospective multicenter clinical trial of cord blood hematopoietic cell transplantation from HLA-identical donors has been initiated.

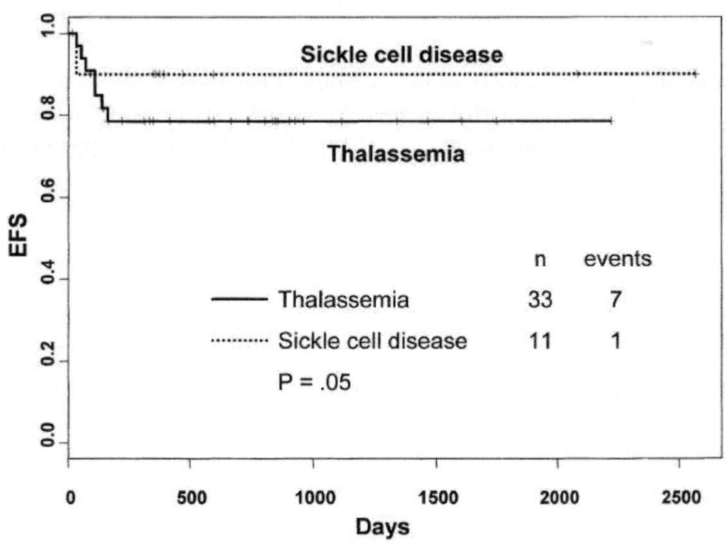

Fig. 3. Kaplan–Meier estimate of the probability of event-free-survival (EFS) after sibling donor cord blood hematopoietic cell (CBHC) transplantation for sickle cell anemia and thalassemia.[76] Eleven patients with sickle cell anemia and 33 with thalassemia major received sibling donor CBSC grafts after myeloablative conditioning therapy. The event-free survival probabilities are depicted.

A Sibling Donor Cord Blood Program was initiated in 1998 as a resource to collect, characterize, and store cord blood units from families affected by malignant and non-malignant disorders treatable by transplantation. To date, 1617 collections have been processed in families with malignancies, hemoglobinopathies, and other rare hematologic conditions. Three hundred eighty-nine units (28%) were collected from families with SCD. Among these 23% were HLA-identical donor-recipient pairs. Twenty-two SCD or thalassemia patients have received umbilical cord blood transplantation, supporting the continued use of sibling donor cord blood units as a suitable resource for transplantation.[77]

Experience with unrelated cord blood transplantation has been limited, but might be considered for high-risk patients without a suitable sibling-donor. A recent report of unrelated cord blood transplantation in three children with stroke and chronic transfusions was published.[78] They received unrelated cord blood units mismatched at two HLA antigens after myeloablative using BU, CY, and ATG. All children developed acute GVHD and one had extensive chronic GVHD however two patients have complete donor chimerism with no detectable HbS, and one experienced graft rejection with autologous recovery.[78]

Summary

Since the first report of successful hematopoietic cell transplantation for sickle cell anemia 20 years ago, this curative therapy has emerged as an important option. While its curative potential distinguishes it from therapeutic alternatives such as red blood cell transfusions and hydroxyurea, the acute and late toxicities of transplantation raise concerns about its wider use. These concerns are being addressed by ongoing efforts to develop non-myeloablative transplantation regimens for sickle cell anemia. However, an optimal regimen has not yet been identified. Ultimately, the future of transplantation for SCD will revolve around efforts to better understanding and overcoming the barriers to engraftment. The possibility of transplantation from alternate stem cell sources such as cord blood will become routine, thereby making transplantation available to a larger number of families.

References

1. The Centers for Disease Control and Prevention (1998). Mortality among children with sickle cell disease identified by newborn screening during 1990–1994 — California, Illinois, and New York. *JAMA* **279**: 1059–1060.
2. van Besien K, Koshy M, Anderson-Shaw L, Talishy N, Dorn L, Devine S, Yassine M, Kodish E (2001). Allogeneic stem cell transplantation for sickle cell disease. A study of patients' decisions. *Bone Marrow Transplant* **28**:545–549.
3. Johnson FL, Look AT, Gockerman J, Ruggiero MR, Dalla-Pozza L, Billings FT III (1984). Bone-marrow transplantation in a patient with sickle-cell anemia. *N Engl J Med* **311**:780–783.
4. Walters MC, Patience M, Leisenring W, Eckman JR, Scott JP, Mentzer WC, Davies SC, Ohene-Frempong K, Bernaudin F, Matthews DC, Storb R, Sullivan KM (1996). Bone marrow transplantation for sickle cell disease [see comments]. *N Engl J Med* **335**:369–376.
5. Walters MC, Storb R, Patience M, Leisenring W, Taylor T, Sanders JE, Buchanan GE, Rogers ZR, Dinndorf P, Davies SC, Roberts IA, Dickerhoff R, Yeager AM, Hsu L, Kurtzberg J, Ohene-Frempong K, Bunin N, Bernaudin F, Wong WY, Scott JP, Margolis D, Vichinsky E, Wall DA, Wayne AS, Pegelow C, Redding-Lallinger R, Wiley J, Klemperer M, Mentzer WC, Smith FO, Sullivan KM (2000). Impact of bone marrow transplantation for symptomatic sickle cell disease: an interim report. *Blood* **95**:1918–1924.
6. Vermylen C, Cornu G, Ferster A, Brichard B, Ninane J, Ferrant A, Zenebergh A, Maes P, Dhooge C, Benoit Y, Beguin Y, Dresse MF, Sariban E (1998). Haematopoietic stem cell transplantation for sickle cell anaemia: the first 50 patients transplanted in Belgium. *Bone Marrow Transplant* **22**:1–6.

7. Walters MC, Patience M, Leisenring W, Rogers ZR, Aquino VM, Buchanan GR, Roberts IA, Yeager AM, Hsu L, Adamkiewicz T, Kurtzberg J, Vichinsky E, Storer B, Storb R, Sullivan KM (2001). Multicenter Investigation of Bone Marrow Transplantation for Sickle Cell Disease. Stable mixed hematopoietic chimerism after bone marrow transplantation for sickle cell anemia. *Biol Blood Marrow Transplant* **7**:665–673.
8. Hoppe CC, Walters MC (2001). Bone marrow transplantation in sickle cell anemia. *Curr Opin Oncol* **13**:85–90.
9. Wayne AS, Schoenike SE, Pegelow CH (2000). Financial analysis of chronic transfusion for stroke prevention in sickle cell disease. *Blood* **96**:2369–2372.
10. Nietert PJ, Abboud MR, Silverstein MD, Jackson SM (2000). Bone marrow transplantation versus periodic prophylactic blood transfusion in sickle cell patients at high risk of ischemic stroke: a decision analysis. *Blood* **95**:3057–3064.
11. Nietert PJ, Silverstein MD, Abboud MR (2002). Sickle cell anaemia: epidemiology and cost of illness. *Pharmacoeconomics* **20**:357–366.
12. Woods K, Karrison T, Koshy M, Patel A, Friedmann P, Cassel C (1997). Hospital utilization patterns and costs for adult sickle cell patients in Illinois. *Public Health Rep* **112**:44–51.
13. Adams RJ, McKie VC, Brambilla D, Carl E, Gallagher D, Nichols FT, Roach S, Abboud M, Berman B, Driscoll C, Files B, Hsu L, Hurlet A, Miller S, Olivieri N, Pegelow C, Scher C, Vichinsky E, Wang W, Woods G, Kutlar A, Wright E, Hagner S, Tighe F, Waclawiw MA *et al.* (1998). Stroke prevention trial in sickle cell anemia. *Control Clin Trials* **19**:110–129.
14. Bernaudin F (1999). Resultats indications actuelles de l'allogreffe de moelle dans la drepanocytose. *Pathologie Biologie* **47**:59–64.
15. Peterson DA (2004). Umbilical cord blood cells and brain stroke injury: bringing in fresh blood to address an old problem. *J Clin Invest* **114**:312–314.
16. Taguchi A, Soma T, Tanaka H, Kanda T, Nishimura H, Yoshikawa H, Tsukamoto Y, Iso H, Fujimori Y, Stern DM, Naritomi H, Matsuyama T (2004). Administration of CD34+ cells after stroke enhances neurogenesis via angiogenesis in a mouse model. *J Clin Invest* **114**:330–338.
17. Atkins RC, Walters MC (2003). Haematopoietic cell transplantation in the treatment of sickle cell disease. *Expert Opin Biol Ther* **3**:1215–1224.
18. Cornu G, Vermylen C, Ferster A, Brichard B, Ninane J, Ferrant A, Zenebergh A, Maes P, Dhooge C, Benoit Y, Beguin Y, Dresse MF, Sariban E (1999). Hematopoietic stem cell transplantation in sickle cell anemia. *Arch Pediatr* **6**:345s–347s.
19. Mentzer WC (2000). Bone marrow transplantation for hemoglobinopathies. *Curr Opin Hematol* **7**:95–100.
20. Bernaudin F SG, Vannier JP *et al.* (1997). Report of the French experience concerning 26 children transplanted for severe sickle cell disease. *Bone Marrow Transplant* **2**:112–115.
21. Vernant JP, Vilmer E, Souillet G *et al.* (2002). Results of myeloablative allogeneic stem cell transplantation (SCT) for severe sickle cell disease (SCD) in France. *Blood* **100**:5a.
22. Brachet C, Azzi N, Demulder A, Devalck C, Gourdin A, Gulbis B, Klein A, Le PQ, Loop M, Sariban E, Ferster A (2004). Hydroxyurea treatment for sickle cell disease: impact on haematopoietic stem cell transplantation's outcome. *Bone Marrow Transplant* **33**:799–803.
23. Hernigou P, Bernaudin F, Reinert P, Kuentz M, Vernant JP (1997). Bone-marrow transplantation in sickle-cell disease. Effect on osteonecrosis: a case report with a four-year follow-up. *J Bone Joint Surg Am* **79**:1726–1730.
24. Ferster A, Bujan W, Corazza F, Devalck C, Fondu P, Toppet M, Verhas M, Sariban E (1993). Bone marrow transplantation corrects the splenic reticuloendothelial dysfunction in sickle cell anemia. *Blood* **81**: 1102–1105.
25. Walters MC, Patience M, Edwards S *et al.* (2004). Hematopoietic cell transplantation for sickle cell disease: updated results of the multicenter trial. *Blood* **104**:33a.
26. Walters MC, Patience M, Leisenring W *et al.* (2002). Updated results of Bone Marrow Transplantation (BMT) for Sickle Cell Disease (SCD): impact on CNS disease. *Blood* **100**:45a.
27. Steen RG, Helton KJ, Horwitz EM, Benaim E, Thompson S, Bowman LC, Krance R, Wang WC, Cunningham JM (2001). Improved cerebrovascular patency following therapy in patients with sickle cell disease: initial results in 4 patients who received HLA-identical hematopoietic stem cell allografts. *Ann Neurol* **49**:222–229.

28. Softley T, Groney P, Bradley W *et al.* (2002). Follow-up neuropsychological assessment after bone marrow transplantation in children with sickle cell anemia disease. *Proceedings of the 30th Meeting of the National Sickle Cell Disease Program and the Sickle Cell Disease Association of America*, p. 32.
29. van Besien K, Bartholomew A, Stock W, Peace D, Devine S, Sher D, Sosman J, Chen YH, Koshy M, Hoffman R (2000). Fludarabine-based conditioning for allogeneic transplantation in adults with sickle cell disease. *Bone Marrow Transplant* **26**:445–449.
30. van Besien K, Smith S, Stock W *et al.* (2002). Allogeneic stem cell transplantation with fludarabine, melphalan, and Campath in adults with advanced sickle cell disease. Preliminary experience with a well tolerated regimen. *Proceedings of the 30th Meeting of the National Sickle Cell Disease Program and the Sickle Cell Disease Association of America*, 40.
31. Wu CJ, Rogers SA, Kutok JL *et al.* (2003). Erythroid chimerism in peripheral blood exceeds bone marrow following non-myeloablative stem cell transplantation in adults with sickle cell anemia. *Blood* **102**:763a.
32. Bernaudin F, Vernant JP, Vilmer E *et al.* (2002). Results of allogeneic stem cell transplant for severe sickle cell disease (SCD) in France and consideration about conditioning regimen and indications in 2002. *Proceedings of the 30th Meeting of the National Sickle Cell Disease Program and the Sickle Cell Disease Association of America*, p. 33.
33. Anasetti C, Amos D, Beatty PG, Appelbaum FR, Bensinger W, Buckner CD, Clift R, Doney K, Martin PJ, Mickelson E *et al.* (1989). Effect of HLA compatibility on engraftment of bone marrow transplants in patients with leukemia or lymphoma. *N Engl J Med* **320**:197–204.
34. Kean L, Hamby K, Perry J *et al.* (2003). Evidence for heterologous immunity in sickle-mediated allograft rejection. *Blood* **102**:112a.
35. Bean MA, Graham T, Appelbaum FR, Deeg HJ, Schuening F, Sale GE, Storb R (1994). Gamma-irradiation of pretransplant blood transfusions from unrelated donors prevents sensitization to minor histocompatibility antigens on dog leukocyte antigen-identical canine marrow grafts. *Transplantation* **57**:423–426.
36. Kean LS, Manci EA, Perry J, Balkan C, Coley S, Holtzclaw D, Adams AB, Larsen CP, Hsu LL, Archer DR (2003). Chimerism and cure: hematologic and pathologic correction of murine sickle cell disease. *Blood* **102**:4582–4593.
37. Wekerle T, Sayegh MH, Hill J, Zhao Y, Chandraker A, Swenson KG, Zhao G, Sykes M (1998). Extrathymic T cell deletion and allogeneic stem cell engraftment induced with costimulatory blockade is followed by central T cell tolerance. *J Exp Med* **187**:2037–2044.
38. Iannone R, Ohene-Frempong K, Fuchs EJ, Casella JF, Chen AR (2005). Bone marrow transplantation for sickle cell anemia: progress and prospects. *Pediatr Blood Cancer* **44**:436–440.
39. Antin JH, Kim HT, Cutler C, Ho VT, Lee SJ, Miklos DB, Hochberg EP, Wu CJ, Alyea EP, Soiffer RJ (2003). Sirolimus, tacrolimus, and low-dose methotrexate for graft-versus-host disease prophylaxis in mismatched related donor or unrelated donor transplantation. *Blood* **102**:1601–1605.
40. Luznik L, Jalla S, Engstrom LW, Iannone R, Fuchs EJ (2001). Durable engraftment of major histocompatibility complex-incompatible cells after non-myeloablative conditioning with fludarabine, low-dose total body irradiation, and post-transplantation cyclophosphamide. *Blood* **98**:3456–3464.
41. Luznik L, Engstrom LW, Iannone R, Fuchs EJ (2002). Post-transplantation cyclophosphamide facilitates engraftment of major histocompatibility complex-identical allogeneic marrow in mice conditioned with low-dose total body irradiation. *Biol Blood Marrow Transplant* **8**:131–138.
42. O'Donnell PV, Luznik L, Jones RJ, Vogelsang GB, Leffell MS, Phelps M, Rhubart P, Cowan K, Piantados S, Fuchs EJ (2002). Non-myeloablative bone marrow transplantation from partially HLA-mismatched related donors using post-transplantation cyclophosphamide. *Biol Blood Marrow Transplant* **8**:377–386.
43. Bachar-Lustig E, Rachamim N, Li HW, Lan F, Reisner Y (1995). Megadose of T cell-depleted bone marrow overcomes MHC barriers in sublethally irradiated mice. *Nat Med* **1**:1268–1273.
44. Aversa AF, Tabilio A, Velardi A, Cunningham I, Terenzi A, Falzetti F, Ruggeri L, Barbabietola G, Aristei C, Latini P, Reisner Y, Martelli MF (1998). Treatment of high-risk acute leukemia with T-cell-depleted stem cells from related donors with one fully mismatched HLA haplotype. *N Engl J Med* **339**:1186–1193.
45. Truitt RL, Johnson BD, Hanke C, Talib S, Hearst JE (1999). Photochemical treatment with S-59 psoralen and ultraviolet A light to control the fate of naive or primed T lymphocytes in vivo after allogeneic bone marrow transplantation. *J Immunol* **163**:5145–5156.

46. Kuypers F, Sage E, Talib S, Watson G, Hearst J (2000). Stem cell transplantation with S-59 photochemically treated T-cell add-backs to establish allochimerism in hemoglobinopathies. *Proceedings of the 30th Meeting of the National Sickle Cell Disease Program and the Sickle Cell Disease Association of America*, p. 37.
47. Walters MC, Sullivan KM, Bernaudin F, Souillet G, Vannier JP, Johnson FL, Lenarsky C, Powars D, Bunin N, Ohene-Frempong K *et al.* (1995). Neurologic complications after allogeneic marrow transplantation for sickle cell anemia [see comments]. *Blood* **85**:879–884.
48. Abboud MR, Jackson SM, Barredo J, Holden KR, Cure J, Laver J (1996). Neurologic complications following bone marrow transplantation for sickle cell disease. *Bone Marrow Transplant* **17**:405–407.
49. Ferster A, Christophe C, Dan B, Devalck C, Sariban E (1995). Neurologic complications after bone marrow transplantation for sickle cell anemia [letter; comment]. *Blood* **86**:408–409.
50. Woodard JP, Cunningham JM, Sulliva KM, Kamani NR, Walters MC (2003). Bronchiolitis obliterans organizing pneumonia (BOOP) after bone marrow transplantation (BMT) for sickle cell disease (SCD): association with acute chest syndrome (ACS). *Blood* **102**:460b.
51. Baronciani D, Angelucci E, Polchi P, Martinelli F, Mariotti E, Marzocchi A, Longo M, Poggi S, Pileri S, Lucarelli G (1998). An unusual marrow transplant complication: cardiac myxoma. *Bone Marrow Transplant* **21**:825–827.
52. Gaziev D, Lucarelli G, Galimberti M *et al.* (1997). Malignancies after bone marrow transplantation for thalassemia. *Bone Marrow Transplant* **19**(Suppl 2):142.
53. Walters MC, Patience M, Edwards S *et al.* (2004). Growth after hematopoietic cell transplantation for sickle cell disease. *Blood* **104**:33a.
54. Andreani M, Nesci S, Lucarelli G, Tonucci P, Rapa S, Angelucci E, Persini B, Agostinelli F, Donati M, Manna M (2000). Long-term survival of ex-thalassemic patients with persistent mixed chimerism after bone marrow transplantation. *Bone Marrow Transplant* **25**:401–404.
55. Amrolia PJ, Vulliamy T, Vassiliou G, Lawson S, Bryon J, Kaeda J, Dokal I, Johnston R, Veys P, Darbyshire P, Roberts IA (2001). Analysis of chimaerism in thalassaemic children undergoing stem cell transplantation. *Br J Haematol* **114**:219–225.
56. Le Deist F, Blanche S, Keable H, Gaud C, Pham H, Descamp-Latscha B, Wahn V, Griscelli C, Fischer A (1989). Successful HLA non-identical bone marrow transplantation in three patients with the leukocyte adhesion deficiency. *Blood* **74**:512–516.
57. Fischer A, Landais P, Friedrich W, Gerritsen B, Fasth A, Porta F, Vellodi A, Benkerrou M, Jais JP, Cavazzana-Calvo M *et al.* (1994). Bone marrow transplantation (BMT) in Europe for primary immunodeficiencies other than severe combined immunodeficiency: a report from the European Group for BMT and the European Group for Immunodeficiency. *Blood* **83**:1149–1154.
58. Horwitz ME, Barrett AJ, Brown MR, Carter CS, Childs R, Gallin JI, Holland SM, Linton GF, Miller JA, Leitman SF, Read EJ, Malech HL (2001). Treatment of chronic granulomatous disease with non-myeloablative conditioning and a T-cell-depleted hematopoietic allograft. *N Engl J Med* **344**:881–888.
59. Iannone R, Luznik L, Engstrom LW, Tennessee SL, Askin FB, Casella JF, Kickler TS, Goodman SN, Hawkins AL, Griffin CA, Noffsinger L, Fuchs EJ (2001). Effects of mixed hematopoietic chimerism in a mouse model of bone marrow transplantation for sickle cell anemia. *Blood* **97**:3960–3965.
60. Kean LS, Durham MM, Adams AB, Hsu LL, Perry JR, Dillehay D, Pearson TC, Waller EK, Larsen CP, Archer DR (2002). A cure for murine sickle cell disease through stable mixed chimerism and tolerance induction after non-myeloablative conditioning and major histocompatibility complex-mismatched bone marrow transplantation. *Blood* **99**:1840–1849.
61. Abdulmalik O, Hayashi S, Chen Q, Asakura T, Flake AW (2002). Evaluation of the efficacy of mixed chimerism in transgenic sickle mice that were treated with in utero hematopoietic stem cell transplantation. *Proceedings of the 30th Meeting of the National Sickle Cell Disease Program and the Sickle Cell Disease Association of America*, p. 32.
62. Maris M, Storb R (2003). The transplantation of hematopoietic stem cells after non-myeloablative conditioning: a cellular therapeutic approach to hematologic and genetic diseases. *Immunol Res* **28**:13–24.
63. Storb R (2004). Non-myeloablative allogeneic transplantation — state-of-the-art. *Pediatr Transplant* **8**(5 Suppl):12–18.

64. Wu CJ, Rogers SA, Kutok JL *et al.* (2004). Mixed chimerism following non-myeloablative stem cell transplantation in patients with severe sickle cell anemia reveals ineffective erythropoiesis in recipient precursor cells. *Proceedings of the 27th Meeting of the National Sickle Cell Disease Program*, p. 23.
65. Schleuning M, Stoetzer O, Waterhouse C, Schlemmer M, Ledderose G, Kolb HJ (2002). Hematopoietic stem cell transplantation after reduced-intensity conditioning as treatment of sickle cell disease. *Exp Hematol* **30**:7–10.
66. Krishnamurti L, Blazar BR, Wagner JE (2001). Bone marrow transplantation without myeloablation for sickle cell disease. *N Engl J Med* **344**:68.
67. Iannone R, Casella JF, Fuchs EJ, Chen AR, Jones RJ, Woolfrey A, Amylon M, Sullivan KM, Storb RF, Walters MC (2003). Results of minimally toxic non-myeloablative transplantation in patients with sickle cell anemia and beta-thalassemia. *Biol Blood Marrow Transplant* **9**:519–528.
68. Horan JT, Liesveld JL, Fenton P, Blumberg N, Walters MC (2005). Hematopoietic stem cell transplantation for multiply transfused patients with sickle cell disease and thalassemia after low-dose total body irradiation, fludarabine, and rabbit anti-thymocyte globulin. *Bone Marrow Transplant* **35**:171–177.
69. Krishnamurti L, Venkataramanan R, Wu C (2004). A pilot study of hematopoietic transplantation for sickle cell disease following a non-myeloablative conditioning regimen. *Proceedings of the 27th Meeting of the National Sickle Cell Disease Program*, p. 172.
70. Krishnamurti L, Wu C, Baker S, Goyal R, Yeager AM, Wagner JE (2005). Hematopoietic stem cell transplantation from a matched sibling donor for high risk patients with sickle cell disease using a reduced intensity conditioning regimen is well tolerated and can lead to long term stable engraftment. *Proceedings of the 28th Meeting of the National Sickle Cell Disease Program*, p. 21.
71. Walters M, Woolfrey A, Torok-Storb B, Zaucha JM, Little MT, Trachtenberg E, Sullivan KM, Storb R (2001). Enrichment of donor erythroid cells after non-myeloablative Bone Marrow Transplantation (BMT) for Sickle Cell Anemia (SCA). *Blood* **98**:490a.
72. Wu CJ, Gladwin MT, Krishnamurti L *et al.* (2004). Mixed chimerism following nonmyeloablative stem cell transplantation for sickle cell disease prevents intravascular hemolysis and restores endothelial function. *Blood* **104**:467a.
73. Wu CJ, Krishnamurti L, Kutok JL, Biernacki M, Rogers S, Zhang W, Antin JH, Ritz J (2005). Evidence for ineffective erythropoiesis in patients with severe sickle cell anemia. *Blood* **106**:3639–3645.
74. Maiers M, Flesch S (2002). *Unrelated Donor Hematopoietic Stem Cell Transplantation for Hemoglobinopathies Facilitated by NMDP*. Minneapolis: National Marrow Donor Program.
75. Krishnamurti L, Abel S, Maiers M, Flesch S (2003). Availability of unrelated donors for hematopoietic stem cell transplantation for hemoglobinopathies. *Bone Marrow Transplant* **31**:547–550.
76. Locatelli F, Rocha V, Reed W, Bernaudin F, Ertem M, Grafakos S, Brichard B, Li X, Nagler A, Giorgiani G, Haut PR, Brochstein JA, Nugent DJ, Blatt J, Woodard P, Kurtzberg J, Rubin CM, Miniero R, Lutz P, Raja T, Roberts I, Will AM, Yaniv I, Vermylen C, Tannoia N, Garnier F, Ionescu I, Walters MC, Lubin BH, Gluckman E, Eurocord Transplant Group (2003). Related umbilical cord blood transplant in patients with thalassemia and sickle cell disease. *Blood* **101**:2137–2143.
77. Walters MC, Quirolo L, Trachtenberg ET, Edwards S, Hale L, Lee J, Morton-Wiley J, Quirolo K, Robertson S, Saba J, Lubin B (2005). Sibling donor cord blood transplantation for thalassemia major: experience of the sibling donor cord blood program. *Annals NY Acad Sci* **1054**:206–213.
78. Adamkiewicz TV, Mehta PS, Boyer MW, Kedar A, Olson TA, Olson E, Chiang KY, Maurer D, Mogul MJ, Wingard JR, Yeager AM (2004). Transplantation of unrelated placental blood cells in children with high-risk sickle cell disease. *Bone Marrow Transplant* **34**:405–411.
79. Abboud MR, Jackson SM, Barredo J, Beatty J, Laver J (1994). Bone marrow transplantation for sickle cell anemia. *Am J Pediatr Hematol Oncol* **16**:86–89.
80. Kalinyak KA, Morris C, Ball WS, Ris MD, Harris R, Rucknagel D (1995). Bone marrow transplantation in a young child with sickle cell anemia. *Am J Hematol* **48**:256–261.
81. Johnson FL, Mentzer WC, Kalinyak KA, Sullivan KM, Abboud MR (1994). Bone marrow transplantation for sickle cell disease. The United States experience. *Am J Pediatr Hematol Oncol* **16**:22–26.
82. Giardini CGM, Lucarelli G, Polchi P, Angelucci E, Baronciani D, Gaziev D, Erer B, Agostinelli F, Giorgi C, Muretto P (1997). Bone marrow transplantation in sickle cell disorders in Pesaro. *Bone Marrow Transplant* **19**(2 Suppl):106–109.

83. Bernaudin F , Vilmer E, Souillet G, Vannier JP, Bordigoni P, Thuret I, Fischer A, Bertrand Y, Lutz P, Duval M, Kuentz M, Loic L, Plouvier E, Stephan JL, Margueritte G, de Montalembert M, Benkerrou M, Bories D, Gluckman E (2002). Results of myeloablative allogenic stem cell transplant (SCT) for severe sickle cell disease (SCD) in France. *Blood* **100**:5a.
84. Mentzer WC, Packman S, Wara W *et al.* (1990). Successful bone marrow transplant in a child with sickle cell anemia and Morquio's disease. *Blood* **76**:69a.
85. Milpied N HJ, Garand R, David A (1988). Bone-marrow transplantation in a patient with sickle-cell anaemia. *Lancet* **2**:328–329.

18

Genetically Engineered Cures: Gene Therapy for Sickle Cell Disease

by Punam Malik and Philippe Leboulch

Introduction

The only curative therapy currently available for patients with sickle cell disease (SCD) is bone marrow or hematopoietic stem cell (HSC) transplantation from an HLA-matched sibling donor. Unfortunately, few patients have such donors and those transplanted face the risk of increased morbidity and mortality from graft-versus-host disease (GVHD), and potential sterility and secondary malignancy from chemotherapeutic preparative regimens. Transplantation with cord blood cells, "adult" CD34+ cells from matched unrelated donors or highly purified HSCs is still experimental. Gene therapy provides a potential viable alternative for permanent correction of the β^S mutation in autologous HSCs. This approach can circumvent the problem of donor shortage and avoid complications related to GVHD.

Retroviral vectors, which integrate permanently into the host genome, have been widely explored for gene transfer into HSCs, thereby resulting in permanent phenotypic correction of progeny including red blood cells (RBCs). Although many complex technical hurdles have made this road arduous for the last 20 years, tremendous strides have been made recently. We will review these advances in this chapter with emphasis on the complex issues remaining towards the path to clinical trials.

Complexity of the β-Globin Gene

The human β-globin gene is expressed in erythroid cells postnatally. Therefore, all β-chain defects including SCD clinically manifest shortly after birth when the γ- to β-globin switch occurs. Expression of the β-globin gene is highly lineage-specific, and globins are expressed at very high levels in the erythroid lineage, and is actively translated long after the nucleus is extruded, representing approximately 95% of all proteins in reticulocytes.

An essential regulatory element of the β-globin gene cluster is the locus control region (LCR) composed of four DNase I hypersensitive (HS) sites. The LCR is located 6 to 20 kb upstream of ε-globin gene on chromosome 11[1] and is required for high-level position-independent transcription of all β-like globin genes in experimental models, although true position-independent expression may be limited even with large LCR constructs. The detailed discussion of how the LCR enhances the rate of transcription over large distances in a developmentally regulated is discussed in detail in Chapter 13. Additional reviews are referenced for further reading.[2–18] Even when placed under the control of the LCR, the β-globin gene remains a highly intron-dependent gene[19–20] with a unique role attributed to the second intron for mRNA stability, cleavage/polyadenylation and nucleo-cytoplasmic export.[21–24]

Prerequisites for Effective Gene Therapy of SCD

In addition to the general hurdles of the field of gene therapy for inherited hematopoietic disorders, its application to SCD poses greater challenges: (1) β-globin is an intron-dependent gene that requires an antisense configuration in integrating RNA virus vectors, which imparts instability to vectors; (2) Large distal control LCR elements are required to achieve high level gene expression, a cargo that is often too large for most viral vectors; (3) There is no selective advantage for the corrected HSCs, which may require an selection strategy to enriched modified HSCs, although one anticipates beneficial extension of the life-span of corrected RBCs and amelioration of ineffective erythropoiesis; and (4) SCD originates from an abnormal β-globin gene rather than an absent β-globin protein, which either requires its replacement by a normal globin protein or its inhibition by other globin protein with anti-sickling properties.

Difficulties Related to Human β-Globin Gene Transfer Using Oncoretroviral Vectors

In an order to prevent splicing of the β-globin gene from RNA retroviral vectors, a common approach is to insert the gene in reverse orientation, so that the natural splicing signals of the β-globin introns are not recognized by the splicing machinery on the plus strand viral transcript. Dzierzak *et al.*[25] were the first to design such recombinant mouse oncoretroviral vector capable of transferring an intact human β-globin gene. This vector did not include LCR elements, as it was constructed before they were discovered. Although sustained, erythroid-specific expression of the human β-globin gene was achieved in mice transplanted with *in vitro* transduced syngenic marrow, expression of transferred gene was well below therapeutic levels ($<1\%$).

After the discovery of the LCR, numerous investigators[26–29] attempted to add segments encompassing the LCR-HS sites to similar vectors in various combinations in an effort to increase globin expression to therapeutic levels. Unfortunately, these vectors were produced in low titers and were very unstable with multiple rearrangements of the transferred proviral structure, even when the minimal "core" elements of the HS sites were utilized.

It appeared that viral RNA splicing and ineffective nucleo-cytoplasmic export were key determinants of the shortfall of β-globin/LCR oncoretroviral vectors. In early studies,[27] we examined vector sequences in search of deleterious elements. We concluded that the β-globin gene in a reverse or antisense orientation in viruses (to avoid introns splicing during packaging of the viral transcript) may have inadvertently created RNA splice sites in the sense open reading frame of the vector genomic transcript. By mutating the most "dangerous" potential splice sites on the basis of consensus algorithms, while being cautious to avoid alterations in coding regions or critical *cis*-acting elements, we were able to obtain stable β-globin/LCR vectors at titers sufficiently high to achieve long-term engraftment with genetically modified HSCs in mouse transplant experiments. Deletion of a 372 bp fragment of the β-globin IVS-2, which contains most of the satellite DNA-like AT rich segment and three of five polypurine tracts,[27] was also performed, although this deletion alone was not sufficient for vector stability. Although long-term human globin gene expression was obtained in transplanted mice and human cells, expression remained sub-therapeutic and variegated.

The Sadelain group also achieved stable proviral transmission of β-globin/LCR oncoretroviral vectors by rearranging the orientation of HS sites. Although stably transmitted, this vector showed variable expression in mouse erythroleukemia cell clones.[29] It is conceivable that these modifications were effective because they prevented cryptic splicing.

Since viral RNA splicing appeared to be a key determinant of instability, and the titers of β-globin/LCR oncoretroviral vectors remained low, use of RNA nucleo-cytoplasmic elements that included the Rev/Rev responsive element (RRE) system of HIV, to obtain a greater amount of full-length vector RNA available for packaging led to the advances recently achieved using lentiviral vectors.

Renaissance of Gene Therapy for the β-Hemoglobinopathy Disorders with HIV-Based Lentiviral Vectors

Major progress came from the use of HIV-derived vectors. The success of these vectors for β-globin gene therapy is likely due their capacity to actively export unspliced RNA from the nucleus of packaging cells in a Rev/RRE-dependent manner. Additionally, lentiviral vectors are capable of transferring a larger "cargo" ($\sim$9–10 kb), allowing insertion of larger LCR elements into vectors. This was not possible with oncoretroviral or AAV based vectors, with a capacity to carry 4–5 kb fragments. Another limitation of oncoretroviral vectors is that their pre-integration complex cannot enter an intact nucleus because of their lack of active nuclear import machinery through the nuclear membrane, which only allows for vector integration during mitosis.[30] By contrast, HIV-based lentiviral vectors overcome this hurdle and are able to enter the intact nucleus of non-dividing cells. This is a critical capability since a substantial fraction of HSCs are maintained in a quiescent state at any given time.

The Sadelain and Leboulch laboratory groups undertook to design β-globin/LCR HIV-1-derived lentiviral vectors for gene therapy for β-thalassemia and SCD, respectively. These studies resulted in the first publications of the correction of β-thalassemia intermedia[31] and SCD,[32] in mouse models, ushering in a new era for gene therapy strategies as described below.

Production of HIV-1 Based Lentiviral Vectors

HIV-1 viruses carry two structural genes (gag, pol, env), two regulatory genes (tat and rev) and four accessory genes (vpr, vpu, nef and vif) between its 5′ and 3′ long terminal repeats (LTRs). The untoward possibility of producing a replication competent lentivirus (RCL) has been reduced or eliminated by deleting the HIV env gene, using a multi-plasmid expression system and removing the non-essential accessory genes. One such plasmid encodes an alternative envelope protein for pseudotyping, the G glycoprotein of vesicular stomatitis virus (VSV-G). This modification broadens host cell infectivity and allows viral particle concentration by physical methods while eliminating the possibility of reforming HIV. A second plasmid carries the transfer vector that comprises the therapeutic transgene together with viral *cis*-acting sequences required for encapsidation, reverse transcription and integration. The viral elements include the LTRs, the packaging signal and the RRE. The remaining plasmid(s) express the gag and pol genes with or without the regulatory genes tat and rev.

Various packaging systems have been designed, including a two-plasmid packaging system that was used for the initiation of anti-HIV gene therapy in the first lentiviral human clinical trial for HIV infected patients.[33] In this latter design, no RCL contamination was detected during large-scale vector manufacturing in spite of the low degree of separation of packaging functions.[33] The generation of lentiviral packaging system has since improved on biosafety by further deleting the regulatory genes (so-called "second" and "third" generation vector systems) and/or splitting packaging functions among multiple plasmids, with a variable impact on gene transfer efficiency.

Transfer vector design was also improved by replacing the viral enhancer/promoter elements from the left LTR by a heterologous promoter or by a self-inactivating (SIN) transfer vector design.[34,35] In a SIN vector, the promoter/enhancer is deleted from the distal 3′ LTR. This deletion is copied over to the proximal 5′ LTR when the vector integrates into the host genome, thereby removing the viral transcriptional elements from both ends of the provirus (the integrated form of the transfer vector). In more recent vector designs, the central polypurine tract (cPPT)/DNA flap of the pol gene has been added to the transfer vector.[36,37] The latter allowed efficient second strand synthesis, slightly improved viral titers and transduction of quiescent cells.[34] Further modifications of the SIN design have been performed, by inserting chromatin insulator elements in place of the deleted LTR, reviewed in detail below.

Gene Therapy of Murine β-Thalassemia

Sadelain and colleagues[31] were the first to show stable transmission and high-level β-globin expression using a first generation lentiviral vector containing the β-globin gene and a large human LCR cassette in a mouse model of β-thalassemia intermedia. They tested two lentiviral vectors termed RNS1 (carrying minimal core LCR elements) and TNS9 (with large LCR fragments encompassing HS2, HS3 and HS4 of approximately 3.2 kb). Cells transduced with the TNS9 vector produced higher human β-globin transcript levels than RNS1; the same vector achieved marked improvement in hematocrit, RBC, reticulocyte counts, and hemoglobin levels in β-thalassemia mice ($Hbb^{th3/+}$ mice).[38,39] This correction was sustained in secondary transplanted mice as well.

In a separate study, they also transplanted β-globin null ($Hbb^{th3/th3}$) fetal liver cells into lethally irradiated C57/BL6 mice and established an elegant model of adult murine β^0-thalassemia major with baseline hemoglobin levels of 2–3 g/dL.[40] In this model, the TNS9 vector increased the hemoglobin up to 6.5 ± 2.9 g/dL in six long-term chimeras with an average copy number of 1.6. One chimera achieved a hemoglobin level of 12 g/dl with a vector copy number of 2.2, but most mice were converted to a severe thalassemia intermedia phenotype. These studies underscore the effectiveness of β-globin/LCR lentiviral vectors in curing thalassemia intermedia. However, they also highlighted the need for vectors with reduced variability in expression (chromosomal position effect variegation, (PEV)) and higher β-globin expression for the treatment of thalassemia major.

Imren *et al.*[41] used a similar lentiviral vector, termed Lenti-β^A vector.[32] Features that were distinct in this vector were as follows: it contained a shorter 2.7 kb HS2, HS3 and HS4 LCR segment, a shorter 254 bp β-globin promoter and a cPPT element. As described in Section 6c below, this vector was designed by Leboulch and colleagues[32] and was shown in a preceding study to correct mouse models of SCD using a β-globin gene modified with an anti-sickling mutation. The Lenti-β^A vector was packaged using a five-plasmid system. Transduction was sustained for more than seven months in both primary and secondary transplants, at which time approximately 95% of RBCs in all mice contained human β-globin contributing to 32% of all β-like globin chains. Hematological parameters approached complete phenotypic correction, assessed by hemoglobin levels, and reticulocyte and RBC counts. Free α-globin chains were completely cleared from RBC membranes, splenomegaly abated, and iron deposits were significantly eliminated in liver sections. Of note, no correction was observed if there was a single proviral integrant per cell. This was attributed to chromatin position effects. This may have reflected an initial statistical bias, since a larger series of subsequently transplanted mice now show phenotypic correction even with a mean of 0.1 vector copy per cell with the same vector (Philippe Leboulch, personal communication).

Persons and colleagues[42] have used a different approach with a third generation self-inactivating (SIN)-lentiviral transfer vector carrying the human γ-globin gene under the control of the β-globin promoter and a 1.7 kb LCR fragment in $Hbb^{th3/+}$ mice. The rationale for using γ-globin was to achieve correction of SCD and β-thalassemia using the same vector. They showed an increase of hemoglobin by 2.5 gm/dL, and normalization of RBC morphology with approximately two copies of the provirus per cell. While this was the highest level of γ-globin expression reported in adult (post-natal) RBCs, the vectors produced much less hemoglobin/vector copy than reported in the previous studies.[31,41] As had occurred in other groups, γ-globin vectors also showed significant PEV. They subsequently used a γ-globin SIN lentiviral vector carrying a larger 3.2 kb LCR with site specific mutations that prevented cryptic splicing, and showed higher γ-globin expression (a 2.2 gm/dl increase in hemoglobin F/vector copy) with reduced PEV effects, the latter attributed to the larger LCR fragment.[43]

Gene Therapy of SCA in Transgenic Mouse Models

Sickle hemoglobin polymerization occurs when a hemoglobin S ($\alpha_2\beta_2^S$) molecule interacts with a hydrophobic pocket, formed by phenylalanine[85] and leucine[88] in the β_2^s chain of another hemoglobin molecule. Fetal hemoglobin (HbF; $\alpha_2\gamma_2$) is a natural anti-sickling molecule, because glutamine[87] of γ-globin aligns with threonine[87] of β^S globin, resulting in mixed tetramers ($\alpha_2\gamma\beta^S$) that do not participate in polymer formation.[44] Several synthetic anti-sickling β-globin molecules have been designed based on this principle, which carry a mutation at the 87th position. Additional mutations that are additive to the anti-sickling effect, generated by the Townes group,[45–47] are also attractive targets for gene therapy of SCD. However, in contrast to β-thalassemia, gene therapy of SCD requires expression of the therapeutic gene in the majority of RBCs to prevent them from sickling and causing vaso-occlusion.

Pawliuk *et al.*[32] were the first to express an anti-sickling β-globin from a lentivirus. They used a lentiviral vector optimized to express the anti-sickling protein β^{T87Q}-globin at therapeutic levels in circulating RBCs in two different transgenic mouse models of SCD with long-term phenotypic correction of the disease.[32] The first mouse model, the SAD[48] mice, expresses human α-globin together with a "super-sickling" β^{SAD} globin (composed of β-globin chains carrying the β^S, $\beta^{Antilles}$ and $\beta^{D\text{-}Punjab}$ mutations). The second mouse model, the BERK mice,[48] express human α and human β^S globins in the absence of murine globin expression because of complete disruption of both mouse α- and β-globin gene loci. The phenotype of BERK mice is more severe than SAD mice, although the BERK mice have some degree of associated β-thalassemia from sub-optimal expression of the transgenic human β^S gene.[48] This was the first report of correction of the SCD phenotype with approximately three proviral copies/cell for both transgenic mouse models.

Ryan *et al.*[49] had also generated transgenic sickle mice, simultaneous to the BERK mice, which exclusively produce human α and β^S globins. In an elegant study published subsequently, the same group used a SIN lentiviral anti-sickling construct lenti/β^{AS3}, where the β globin had three amino acid mutations that imparted a highly efficacious antisickling effect in transgenic sickle mice.[50] A rise in hemoglobin level of 4.6 g/dl was achieved with a proviral copy number of approximately two copies per cell.

Globin Gene Transfer into Human Hematopoietic Cells

Connie Eaves, Keith Humphries and colleagues[51] subsequently reported high level expression of human β^{87} globin from a SIN β-globin/LCR version of the anti-sickling lentiviral vector made by the Leboulch group, in erythroid progeny produced from human $CD34^+$ cord blood cells and transplanted into sublethally irradiated NOD/LtSz-scid/scid mice. On the basis of data from 17 mice analyzed 6 months post-transplantation in five experiments, the overall average transduction efficiency of long-term NOD/LtSz-scid/scid mouse–repopulating cells was 45% with approximately two integrated proviral copies per cell. High pressure liquid chromatography analysis of RBC lysates of human cells harvested from transplanted NOD-SCID mice showed β^{87}-globin protein to represent 35–59% of total β-globin chains. Similar results were achieved with bone marrow-derived $CD34^+$ cells from human SCD patients (Phillipe Leboulch, personal communication).

Malik and colleagues[52] simultaneously reported on a SIN β-globin/LCR vector, termed BG-I, carrying a 3.2 kb LCR cassette. To improve safety and reduce PEV for effective gene therapy of human thalassemia major, they inserted the chicken hypersensitive site-4 insulator element flanking the integrated provirus. BGI was used to transduce $CD34^+$ cells isolate from human bone marrow of four thalassemia major patients (TM). We were able to achieve high gene transfer efficiency of 80–90% in human colonies derived from sublethally irradiated $\beta 2M^{null}$ NOD-SCID mice four months following transplantation of human $CD34^+$ cells. Erythroid differentiation and expansion of BGI-TM $CD34^+$ cells were indistinguishable from normal $CD34^+$ cells, with complete restoration of effective erythropoiesis. The amount of β-globin chains produced by the BGI vector were 75% in TM-BGI (with about two vector copies/cell) compared to 65–75% of all β-like globins in normal, and $7.3 \pm 7\%$ in the erythroid progeny of untransduced TM $CD34^+$ cells. These results were confirmed three to four months after transplant ($n = 28$ mice); there was multilineage cell engraftment with human hemoglobin A production only in normal and TM-BGI xenografts. Furthermore, Malik and colleagues were able to show circulating human hemoglobin A expressing cells in the blood of normal and TM-BGI xenografts. Taken together, these data comprised the first report of complete phenotypic and functional correction of human thalassemia major *in vitro* and in a xenograft model *in vivo*, with normal human β-globin production and effective erythropoiesis with a SIN-lentiviral vector flanked by the chicken hypersensitive site-4 insulator.

Safety from Randomly Integrating Lentiviral Vectors

Expression from onco-retroviruses is transcriptionally silenced in embryonic and hematopoietic stem cells[53] and is subject to chromatin position effects.[54] Although this phenomenon is less pronounced with lentiviral vectors, PEV is still observed with [globin/LCR] lentiviral vectors in spite of the inclusion of β- or α-globin regulatory elements.[41,42,55] In addition, a worrisome feature of all retroviruses (onco-retrovirus and lentivirus vectors) is random integration into the host genome, which can result in gene disruption or untoward *cis*-activation of surrounding genes including tumor suppressor genes and proto-oncogenes. This has occurred recently, with the report of insertional oncogenesis in three children, initially cured of X-linked severe combined immunodeficiency disease by gene therapy, where the vector provirus integrated within or in the vicinity of the LMO2 proto-oncogene, resulting in T-cell leukemia.[56] In this study the common immunoglobulin cDNA was driven by the viral LTR, and thus expressed ubiquitously. This protein chain is shared by several interleukin receptors (IL-2, IL-7, IL-9, IL-15) and signaling via these interleukin receptors is necessary for normal lymphopoieisis. Recent evidence points to the oncogenic role of dysregulated expression of the common γ-cDNA in T cell precursors.

The advent of powerful methods to explore the human genome has made mapping of vector provirus integration sites possible. These studies have shown a slight propensity of both oncoretroviral and lentiviral vectors to integrate preferentially within or around actively transcribed genes.[57,58] While the Murine Moloney Leukemia oncoretroviruses/vectors tend to integrate near the 5′ end near transcriptional start sites or promoter regions,[57] HIV vectors tend to integrate into genes with no preference for the 5′ or 3′ end of genes.[58] Recently, a lentiviral vector carrying the anti-sickling β^{87}-globin made by the Leboulch group was studied in human HSC from cord blood.[51] The above findings were confirmed, with 86% of insertions occurring within genes with one insertion in the vicinity of MLL, a known proto-oncogene.[51] Taken together, safer designs of randomly integrating retroviruses are necessary to bring these vectors safety into human trials.

Lentiviral Vectors Modifications to Improve Safety and Efficacy: SIN Deletions and Chromatin Insulators

Unlike MLV vectors that have been used in clinical trials, where the transgene is driven by the viral LTR and expressed constitutively in HSC and all its progeny, the safety feature inherent in β/γ-globin/LCR vectors is highly restricted expression in the erythroid lineage. Furthermore, in an effort to decrease untoward *cis*-activation of proto-oncogenes at the site of chromosomal integration of the provirus, lentiviral vectors have been modified by several groups to generate viral enhancer vectors. Efforts to decrease PEV and further improve safety included the insertion of chromatin insulators in the 3′ LTR, to flank the expression cassette. Although prevention of gene disruption by provirus integration cannot be predicted, the oncogenic risk produced by haplo-insufficiency in somatic cells is likely to be low. Nevertheless, the risk of activation of a cellular oncogene in the vicinity is considerably reduced by the SIN design, lineage restricted expression and insertion of chromatin insulator elements.

Chromatin Insulators

Felsenfeld and colleagues[62] have demonstrated that the chicken β-globin locus hypersensitive site 4 (cHS4) behaves as a chromatin insulator and found the cHS4 to have features of a boundary element (thereby resisting silencing effects of surrounding heterochromatin) and enhancer blocking activity. Emery and colleagues[54] have studied this element in erythroid-specific oncoretroviral vectors carrying reporter genes and the human γ-globin gene. Higher and more consistent transgene expression was observed with the insulated vectors. In one study with a lentiviral vector carrying a green fluorescent protein reporter gene, insertion of the cHS4 insulator element in the 3′ LTR did not alter the mean fluorescence intensity in pools or single copy clones in cell lines or primary $CD34^+$ cells.[63] However, recent data suggest that cHS4 may display chromatin insulator activity in lentiviral vectors, as seen with oncoretroviral vectors.[64] Puthenveetil *et al.*[52] have designed a SIN lentiviral vector containing a 1.2 kb fragment of the cHS4 insulator in place of the SIN deletion. They demonstrated that the

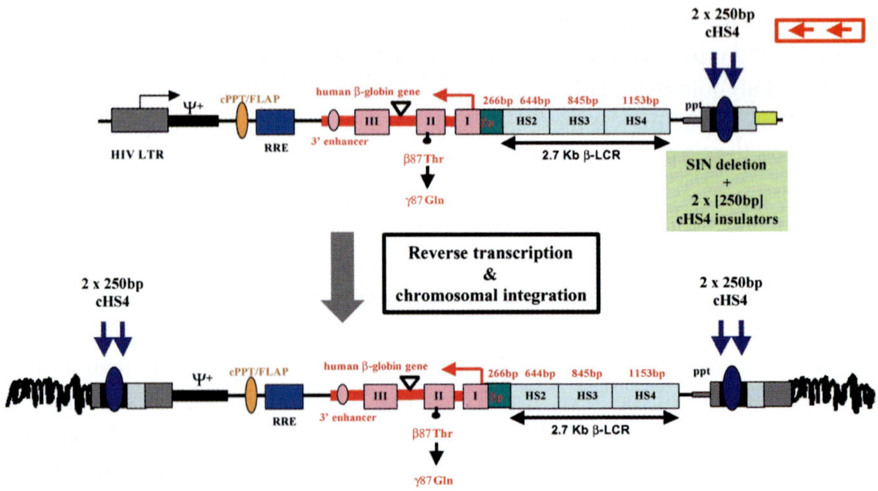

Fig. 1. Diagram of the lenti/globin vector, HPV569, before and after chromosomal integration of the provirus. The 3′β-globin enhancer, the 372 bp IVS2 deletion, the βA-T87Q mutation (ACA Thr to CAG Gln) and DNase I Hypersensitive Sites (HS) 2, HS3 and HS4 of the human β-globin Locus Control Region (LCR) are indicated. Safety modifications including 2 stop codons in the $\psi+$ signal, the 400 bp deletion in the U3 of the right HIV LTR, the rabbit β-globin polyA signal and the 2×250 bp cHS4 chromatin insulator are indicated. HIV LTR, human immunodeficiency type-1 virus long terminal repeat; $\Psi+$, packaging signal; cPPT/flap, central polypurine tract/DNA flap; RRE, Rev-responsive element; βp, human β-globin promoter; ppt, polypurine tract.

vector expressed normal amounts of human β-globin in erythroid cells from four β-thalassemia major patients *in vitro* with correction of the disease phenotype.[52]

Similar results have been obtained by the Leboulch group with their lentiviral vector, referred to as HPV569 or LentiGlobin^{TM} (Genetix Pharmaceuticals, Cambridge, MA), which expresses the anti-sickling β^{T87Q}-globin mutant, in the context of SIN LTRs and a doublet of the 250 bp cHS4 chromatin insulator core on each side of the SIN provirus (Fig. 1). Mouse toxicology studies were performed with this vector in long-term primary and secondary transplants of β-thalassemia mice and mice transgenic for over expression of Spi-1, a commonly activated oncogene in Friend-mediated mouse erythroleukemia; the purpose for the Spi-1 transgenic is to sensitize these mice to the emergence of leukemias. No vector-bearing leukemia have been detected in any of the transplanted mice (Philippe Leboulch, personal communication).

HSC Selection Strategies

It is no surprise that the first genetic disorders of the hematopoietic system for which long-term correction was achieved by *ex vivo* gene transfer are those for which a spontaneous selection advantage exists *in vivo* for the corrected HSCs. This was the case for the severe combined immunodeficiency syndromes caused by a defect of the common γ-chain of the cytokine receptors. In the vast majority of genetic diseases of the hematopoietic system, including the β-hemoglobinopathy disorders, no spontaneous advantage exists at the HSC level, although beneficial extension of RBC life-span and amelioration of ineffective erythropoiesis are anticipated. It remains possible, that the proportion of transduced HSCs and corrected circulating RBCs may be insufficient in the absence of a complementary selection strategy. In addition, *in vivo* selection approaches may prove less toxic than current myeloablative regimens.

Ex Vivo Expansion and Selection of HSC

There have been numerous attempts at inducing HSC expansion using various cytokine cocktails, but it remains unclear whether the most primitive HSCs capable of long-term hematopoietic reconstitution have been substantially expanded. A homeobox transcriptional factor, HoxB4, was recently shown to have the potential for increasing HSC proliferation and engraftment in both mouse transplants and assays for human HSCs.[65–68] A membrane permeable recombinant TAT-HoxB4 fusion protein, in particular, may hold the potential for "safe" HSC expansion either alone or in combination with other co-factors.

Several approaches have been developed to purify transduced HSCs *ex vivo* on the basis of expressed cell surface molecules, fluorescent markers and other dominant selectable genes linked with the therapeutic gene within the vector. However, the duration and amount of *ex vivo* manipulation is quite impractical and has raised concerns for maintenance of HSC quality.

In Vivo Expansion and Selection of HSC

An effective strategy for selectively amplifying genetically modified HSCs *in vivo* is to produce a homogeneous distribution of expression of anti-sickling therapeutic gene may remove the need for myeloablation, thereby increasing the safety of clinical gene therapy. Substantial amplification has been documented in mice using a variant dihydrofolate reductase gene and a combination of trimetexate and a nucleoside transporter inhibitor for selection, but this selection system has less effective in primate models.[69] Another system based on the multidrug resistance gene (MDR) has been evaluated in human trials,[70] however evidence of MDR-induced myeloproliferative syndrome in mice has raised concerns with this approach.[70]

Gerson and colleagues[71] have developed the variant O^6-benzylguanine (BG) resistant, methyl guanine methyl transferase (MGMT) selection system. MGMT is an alkyl transferase, which repairs DNA damage at the O^6 position of guanine produced by methylating agents such as 1,3-bis(2-chloroethyl)-1-nitrosourea or Temozolomide. Persons and colleagues have transferred an MGMT gene into normal HSCs and transplanted these into syngenic β-thalassemia mice. They showed complete correction of thalassemia by selecting genetically MGMT-expressing HSCs.[72] They have recently extended these findings with a γ-globin lentiviral vector that co-expresses MGMT although the efficiency and safety of such selection systems requires evaluation in a non-human primate model. Chemical inducers of dimerization have been established to selectively expand transduced HSCs.[73,74] Human cells showed a poorer response which was biased towards the erythroid lineage.

Target Number of HSCs for Genetic Correction of SCD

There are few cases of patients with SCD who received a myeloablative allogeneic bone marrow transplant but went on to develop a stable mixed chimerism. Post transplant these patients are free of vaso-occlusive symptoms with improvement in blood vessel damage and lung functions.[75] However, myeloablative regimens often have significant morbidity and mortality that often outweighs the disease morbidity, except in patients with severe disease. Studies in sickle mice have shown that bone marrow chimerism of 15–25% normal cells results in 80% normal RBC and hemoglobin concentrations, due to the longer life-span of normal RBC.[76–78] However, there was no selective advantage for the early RBC precursors, which corresponded to the donor's stem cells and white blood cell chimerism. Organ damage was not improved unless greater than 80–100% donor chimerism was achieved,[77,78] although these mouse studies do not reconcile with observations collected with SCD patients. With current regimens, gene therapy may be able to achieve at least 15–20% correction of HSCs, and one can be optimistic that this level of gene transfer will suffice to correct acute symptoms and anemia. Whether it will also be able to prevent or reverse organ damage remains to be established.

First Clinical Trial of a Lentiviral Gene Therapy Vector for Hemoglobinopathy Disorders

A Phase I/II clinical trial in β-thalassemia major and SCD has been recently approved by the regulatory authorities in Europe based on the *ex vivo* transduction of autologous bone marow-derived $CD34^+$ cells with the SIN, cHS4 insulated version of the anti-sickling lentiviral vector from the Leboulch group (Fig. 1) following conditioning with the single agent Busulfex. We are expecting this trial to be initiated at the end of 2005.

Alternative Vectors and Gene Therapy Strategies

Adeno-associated virus (AAV), a non pathogenic DNA virus, has also been explored. Stable and regulated globin gene expression from AAV-2 vectors has been shown by several groups.[79,80] However the main disadvantage of AAV vectors is the small cargo they can carry (approximately 5 kb), restricting them to incorporation of only core LCR elements. There are also contradictory reports as to the ability of AAV vectors to achieve stable integration at high efficiency into HSCs.[80–82] Another retroviral vector, based on the non-pathogenic human foamy virus, a member of the spumavirus family[83] is also been evaluated in mouse models for this application.

Other groups have attempted to reduce β^S-globin expression using either hammer-head ribozymes against the mutant β^S-globin,[84] or anti-sense oligonucleotides.[85–87] Vacek *et al.*[88] have used a novel approach targeted to specific splice mutations in the β-globin gene.

A very attractive option for gene therapy of SCD is homologous recombination/DNA repair, reverting the mutant T $\rightarrow$ A. This has been attempted by several groups using ingenious techniques such as DNA-RNA hybrids,[89] which remain controversial, trans-splicing ribozymes,[90] short fragment homologous recombination,[91] and more recently, homologous recombination mediated by chimeric zinc finger nucleases.[92] These attempts have been met with different levels of success, and may eventually be the most physiological correction of SCD by gene therapy. However, currently, efficient delivery systems and gene conversion efficiency are the limiting features that have prevented these approaches from coming to the forefront.

Impact of the Human Genome Project (HGP)

The Human Genome Project (HGP) has greatly facilitated studies on the integration pattern of retroviruses, with recent discovery of the propensity of oncoretroviruses and lentiviruses to integrate in and around genes. The HGP is particularly vital for SCD research since diverse phenotypic variation occurs despite the existence of the same single base single nucleotide polymorphism. This heterogeneity is attributed to the effects of several "modifier" genes. Microarray technology is being utilized to determine gene profiles of individuals with mild versus severe SCD disease, so that a genetic profile of patients with a high probability of developing severe disease may become available. This will allow gene therapy to be targeted to patients who most need it before severe organ damage occur.

Summary

Devising an effective gene therapy approach for the long-term correction of the β-hemoglobinopathy disorders which involves a "gene addition" strategy has proven especially challenging. Major advances have been made recently to achieve by *ex vivo* transfer of therapeutic normal globin gene variants into HSCs using lentiviral vectors. Safety features were subsequently introduced into vector systems to avoid contamination with a replication-competent lentivirus during vector manufacturing and prevent oncogenesis by vector insertional mutagenesis. These added features include self-inactivating deletions of the viral enhancer/promoter and inclusion of chromatin insulators. Preclinical studies in mouse models resulted in long-term correction of disease phenotype without untoward side effects. Complementary studies in human cells showed effective transduction of primitive human HSCs and high expression levels of therapeutic proteins. The first human clinical trial for gene therapy of SCD and β-thalassemia by means of a therapeutic lentiviral vector is likely to be initiated in Europe before the end of 2005. Other complimentary or alternative approaches, which include selection/expansion of transduced cells, other viral vectors, use of diseased RNA degradation and gene correction, are also being investigated.

References

1. Stamatoyannopoulos G, Nienhuis AW (1994). Hemoglobin switching. In: Stamatoyannopoulos G, Neinhuis AW, Majerus PW, Varmus H, editors. *The Molecular Basis of Blood Diseases*. Philadelphia: W.B. Saunders Company, pp. 107–155.
2. Kim A, Dean A (2004). Developmental stage differences in chromatin subdomains of the beta-globin locus. *Proc Natl Acad Sci USA* **101**:7028–7033.
3. Kim A, Dean A (2003). A human globin enhancer causes both discrete and widespread alterations in chromatin structure. *Mol Cell Biol* **23**:8099–8109.
4. Cook PR (2003). Nongenic transcription, gene regulation and action at a distance. *J Cell Sci* **116**:4483–4491.
5. Palstra RJ, Tolhuis B, Splinter E, Nijmeijer R, Grosveld F, de Laat W (2003). The beta-globin nuclear compartment in development and erythroid differentiation. *Nat Genet* **35**:190–194.
6. Sawado T, Halow J, Bender MA, Groudine M (2003). The beta-globin locus control region (LCR) functions primarily by enhancing the transition from transcription initiation to elongation. *Genes Dev* **17**:1009–1018.
7. Tolhuis B, Palstra RJ, Splinter E, Grosveld F, de Laat W (2002). Looping and interaction between hypersensitive sites in the active beta-globin locus. *Mol Cell* **10**:1453–1465.
8. Bulger M, Groudine M (2002). TRAPping enhancer function. *Nat Genet* **32**:555–556.
9. Li Q, Peterson KR, Fang X, Stamatoyannopoulos G (2002). Locus control regions. *Blood* **100**:3077–3086.
10. Harju S, McQueen KJ, Peterson KR (2002). Chromatin structure and control of beta-like globin gene switching. *Exp Biol Med* (Maywood) **227**:683–700.
11. Tang Y, Liu DP, Liang CC (2002). Further understanding of the beta-globin locus regulation at the molecular level: looping or linking models? *Genes Cells* **7**:889–900.
12. Levings PP, Bungert J (2002). The human beta-globin locus control region. *Eur J Biochem* **269**:1589–1599.
13. Bulger M, Sawado T, Schubeler D, Groudine M (2002). ChIPs of the beta-globin locus: unraveling gene regulation within an active domain. *Curr Opin Genet Dev* **12**:170–177.
14. Simon I, Tenzen T, Mostoslavsky R, Fibach E, Lande L, Milot E, Gribnau J, Grosveld F, Fraser P, Cedar H (2001). Developmental regulation of DNA replication timing at the human beta globin locus. *EMBO J* **20**:6150–6157.
15. Gribnau J, Diderich K, Pruzina S, Calzolari R, Fraser P (2000). Intergenic transcription and developmental remodeling of chromatin subdomains in the human beta-globin locus. *Mol Cell* **5**:377–386.
16. Gribnau J, de Boer E, Trimborn T, Wijgerde M, Milot E, Grosveld F, Fraser P (1998). Chromatin interaction mechanism of transcriptional control in vivo. *EMBO J* **17**:6020–6027.

17. Vieira KF, Levings PP, Hill MA, Crusselle VJ, Kang SH, Engel JD, Bungert J (2004). Recruitment of transcription complexes to the beta-globin gene locus in vivo and in vitro. *J Biol Chem* **279**:50350–50357.
18. Tanimoto K, Liu Q, Bungert J, Engel JD (1999). Effects of altered gene order or orientation of the locus control region on human beta-globin gene expression in mice. *Nature* **398**:344–348.
19. Buchman AR, Berg P (1988). Comparison of intron-dependent and intron-independent gene expression. *Mol Cell Biol* **8**:4395–4405.
20. Collis P, Antoniou M, Grosveld F (1990). Definition of the minimal requirements within the human beta-globin gene and the dominant control region for high level expression. *EMBO J* **9**:233–240.
21. Dominski Z, Kole R (1993). Restoration of correct splicing in thalassemic pre-mRNA by antisense oligonucleotides. *Proc Natl Acad Sci USA* **90**:8673–8677.
22. Nesic D, Zhang J, Maquat LE (1995). Lack of an effect of the efficiency of RNA 3′-end formation on the efficiency of removal of either the final or the penultimate intron in intact cells. *Mol Cell Biol* **15**:488–496.
23. Lewis JD, Izaurralde E (1997). The role of the cap structure in RNA processing and nuclear export. *Eur J Biochem* **247**:461–469.
24. Huang S, Spector DL (1996). Dynamic organization of pre-mRNA splicing factors. *J Cell Biochem* **62**: 191–197.
25. Dzierzak EA, Papayannopoulou T, Mulligan RC (1988). Lineage-specific expression of a human beta-globin gene in murine bone marrow transplant recipients reconstituted with retrovirus-transduced stem cells. *Nature* **331**:35–41.
26. Karlsson S, Bodine DM, Perry L, Papayannopoulou T, Nienhuis AW (1988). Expression of the human beta-globin gene following retroviral-mediated transfer into multipotential hematopoietic progenitors of mice. *Proc Natl Acad Sci USA* **85**:6062–6066.
27. Leboulch P, Huang GM, Humphries RK, Oh YH, Eaves CJ, Tuan DY, London IM (1994). Mutagenesis of retroviral vectors transducing human beta-globin gene and beta-globin locus control region derivatives results in stable transmission of an active transcriptional structure. *EMBO J* **13**:3065–3076.
28. Raftopoulos H, Ward M, Leboulch P, Bank A (1997). Long-term transfer and expression of the human beta-globin gene in a mouse transplant model. *Blood* **90**:3414–3422.
29. Sadelain M, Wang CH, Antoniou M, Grosveld F, Mulligan RC (1995). Generation of a high-titer retroviral vector capable of expressing high levels of the human beta-globin gene. *Proc Natl Acad Sci USA* **92**: 6728–6732.
30. Lewis PF, Emerman M (1994). Passage through mitosis is required for oncoretroviruses but not for the human immunodeficiency virus. *J Virol* **68**:510–516.
31. May C, Rivella S, Callegari J, Heller G, Gaensler KM, Luzzatto L, Sadelain M (2000). Therapeutic haemoglobin synthesis in beta-thalassaemic mice expressing lentivirus-encoded human beta-globin. *Nature* **406**:82–86.
32. Pawliuk R, Westerman KA, Fabry ME, Payen E, Tighe R, Bouhassira EE, Acharya SA, Ellis J, London IM, Eaves CJ, Humphries RK, Beuzard Y, Nagel RL, Leboulch P (2001). Correction of sickle cell disease in transgenic mouse models by gene therapy. *Science* **294**:2368–2371.
33. Humeau LM, Binder GK, Lu X, Slepushkin V, Merling R, Echeagaray P, Pereira M, Slepushkina T, Barnett S, Dropulic LK, Carroll R, Levine BL, June CH, Dropulic B (2004). Efficient lentiviral vector-mediated control of HIV-1 replication in CD4 lymphocytes from diverse HIV+ infected patients grouped according to CD4 count and viral load. *Mol Ther* **9**:902–913.
34. Zufferey R, Dull T, Mandel RJ *et al.* (1998). Self-inactivating lentivirus vector for safe and efficient in vivo gene delivery. *J Virol* **72**:9873–9880.
35. Miyoshi H, Blomer U, Takahashi M, Gage FH, Verma IM (1998). Development of a self-inactivating lentivirus vector. *J Virol* **72**:8150–8157.
36. Zennou V, Petit C, Guetard D, Nerhbass U, Montagnier L, Charneau P (2000). HIV-1 genome nuclear import is mediated by a central DNA flap. *Cell* **101**:173–185.
37. Follenzi A, Ailles LE, Bakovic S, Geuna M, Naldini L (2000). Gene transfer by lentiviral vectors is limited by nuclear translocation and rescued by HIV-1 pol sequences. *Nat Genet* **25**:217–222.
38. Ciavatta DJ, Ryan TM, Farmer SC, Townes TM (1995). Mouse model of human beta zero thalassemia: targeted deletion of the mouse beta maj- and beta min-globin genes in embryonic stem cells. *Proc Natl Acad Sci USA* **92**:9259–9263.

39. Yang B, Kirby S, Lewis J, Detloff PJ, Maeda N, Smithies O (1995). A mouse model for beta 0-thalassemia. *Proc Natl Acad Sci USA* **92**:11608–11612.
40. Rivella S, May C, Chadburn A, Riviere I, Sadelain M (2003). A novel murine model of Cooley anemia and its rescue by lentiviral-mediated human beta-globin gene transfer. *Blood* **101**:2932–2939.
41. Imren S, Payen E, Westerman KA *et al.* (2002). Permanent and panerythroid correction of murine beta thalassemia by multiple lentiviral integration in hematopoietic stem cells. *Proc Natl Acad Sci USA* **99**: 14380–14385.
42. Persons DA, Hargrove PW, Allay ER, Hanawa H, Nienhuis AW (2003). The degree of phenotypic correction of murine beta-thalassemia intermedia following lentiviral-mediated transfer of a human gamma-globin gene is influenced by chromosomal position effects and vector copy number. *Blood* **101**:2175–2183.
43. Hanawa H, Hargrove PW, Kepes S, Srivastava DK, Nienhuis AW, Persons DA (2004). Extended beta-globin locus control region elements promote consistent therapeutic expression of a gamma-globin lentiviral vector in murine beta-thalassemia. *Blood* **104**:2281–2290.
44. Nagel RL, Bookchin RM, Johnson J, Labie D, Wajcman H, Isaac-Sodeye WA, Honig GR, Schiliro G, Crookston JH, Matsutomo K (1979). Structural bases of the inhibitory effects of hemoglobin F and hemoglobin A2 on the polymerization of hemoglobin S. *Proc Natl Acad Sci USA* **76**:670–672.
45. McCune SL, Reilly MP, Chomo MJ, Asakura T, Townes TM (1994). Recombinant human hemoglobins designed for gene therapy of sickle cell disease. *Proc Natl Acad Sci USA* **91**:9852–9856.
46. Nagel RL (1995). Disorders of hemoglobin function and stability. In: Handin RI, Lux SE, Stossel TP, editors. *Blood: Principles and Practice of Hematology*. Philadelphia: J.B. Lippincott Company, pp. 1591–1646.
47. Levasseur DN, Ryan TM, Reilly MP, McCune SL, Asakura T, Townes TM (2004). A recombinant human hemoglobin with anti-sickling properties greater than fetal hemoglobin. *J Biol Chem* **279**:27518–27524.
48. Nagel RL, Fabry ME (2001). The panoply of animal models for sickle cell anaemia. *Br J Haematol* **112**: 19–25.
49. Ryan TM, Ciavatta DJ, Townes TM (1997). Knockout-transgenic mouse model of sickle cell disease [see comments]. *Science* **278**:873–876.
50. Levasseur DN, Ryan TM, Pawlik KM, Townes TM (2003). Correction of a mouse model of sickle cell disease: lentiviral/antisickling beta-globin gene transduction of unmobilized, purified hematopoietic stem cells. *Blood* **102**:4312–4319.
51. Imren S, Fabry ME, Westerman KA, Pawliuk R, Tang P, Rosten PM, Nagel RL, Leboulch P, Eaves CJ, Humphries RK (2004). High-level beta-globin expression and preferred intragenic integration after lentiviral transduction of human cord blood stem cells. *J Clin Invest* **114**:953–962.
52. Puthenveetil G, Scholes J, Carbonell D, Qureshi N, Xia P, Zeng L, Li S, Yu Y, Hiti AL, Yee JK, Malik P (2004). Successful correction of the human beta-thalassemia major phenotype using a lentiviral vector. *Blood* **104**:3445–3453.
53. Challita P, Kohn D (1994). Lack of expression from a retroviral vector after transduction of murine hematopoietic stem cells is associated with methylation in vivo. *Proc Natl Acad Sci USA* **91**:2567–2571.
54. Emery DW, Yannaki E, Tubb J, Nishino T, Li Q, Stamatoyannopoulos G (2002). Development of virus vectors for gene therapy of beta chain hemoglobinopathies: flanking with a chromatin insulator reduces gamma-globin gene silencing in vivo. *Blood* **100**:2012–2019.
55. Mohamedali A, Moreau-Gaudry F, Richard E *et al.* (2004). Self-inactivating lentiviral vectors resist proviral methylation but do not confer position-independent expression in hematopoietic stem cells. *Mol Ther* **10**:249–259.
56. Hacein-Bey-Abina S, von Kalle C, Schmidt M *et al.* (2003). LMO2-associated clonal T cell proliferation in two patients after gene therapy for SCID-X1. *Science* **302**:415–419.
57. Wu X, Li Y, Crise B, Burgess SM (2003). Transcription start regions in the human genome are favored targets for MLV integration. *Science* **300**:1749–1751.
58. Schroder AR, Shinn P, Chen H, Berry C, Ecker JR, Bushman F (2002). HIV-1 integration in the human genome favors active genes and local hotspots. *Cell* **110**:521–529.
59. Ginn SL, Fleming J, Rowe PB, Alexander IE (2003). Promoter interference mediated by the U3 region in early-generation HIV-1-derived lentivirus vectors can influence detection of transgene expression in a cell-type and species-specific manner. *Human Gene Ther* **14**:1127–1137.

60. Moreau-Gaudry F, Xia P, Jiang G, Perelman NP, Bauer G, Ellis J, Surinya KH, Mavilio F, Shen CK, Malik P (2001). High-level erythroid-specific gene expression in primary human and murine hematopoietic cells with self-inactivating lentiviral vectors. *Blood* **98**:2664–2672.
61. Richard E, Mendez M, Mazurier F, Morel C, Costet P, Xia P, Fontanellas A, Geronimi F, Cario-Andre M, Taine L, Ged C, Malik P, de Verneuil H, Moreau-Gaudry F (2001). Gene therapy of a mouse model of protoporphyria with a self-inactivating erythroid-specific lentiviral vector without preselection. *Mol Ther* **4**:331–338.
62. Burgess-Beusse B, Farrell C, Gaszner M, Litt M, Mutskov V, Recillas-Targa F, Simpson M, West A, Felsenfeld G (2002). The insulation of genes from external enhancers and silencing chromatin. *Proc Natl Acad Sci USA* **99**:16433–16437.
63. Ramezani A, Hawley TS, Hawley RG (2003). Performance- and safety-enhanced lentiviral vectors containing the human interferon-beta scaffold attachment region and the chicken beta-globin insulator. *Blood* **101**:4717–4724.
64. Xia P, Mohamedali A, Puthenveetil P *et al.* (2003). Erythroid-specific self-inactivating (SIN) lentiviral vectors express long-term but do not confer position-independent expression: role of an insulator element. *Blood* **102**:992a [abstract].
65. Krosl J, Austin P, Beslu N, Kroon E, Humphries RK, Sauvageau G (2003). In vitro expansion of hematopoietic stem cells by recombinant TAT-HOXB4 protein. *Nat Med* **9**:1428–1432.
66. Brun AC, Fan X, Bjornsson JM, Humphries RK, Karlsson S (2003). Enforced adenoviral vector-mediated expression of HOXB4 in human umbilical cord blood CD34+ cells promotes myeloid differentiation but not proliferation. *Mol Ther* **8**:618–628.
67. Amsellem S, Pflumio F, Bardinet D, Izac B, Charneau P, Romeo PH, Dubart-Kupperschmitt A, Fichelson S (2003). Ex vivo expansion of human hematopoietic stem cells by direct delivery of the HOXB4 homeoprotein. *Nat Med* **9**:1423–1427.
68. Sawai N, Persons DA, Zhou S, Lu T, Sorrentino BP (2003). Reduction in hematopoietic stem cell numbers with in vivo drug selection can be partially abrogated by HOXB4 gene expression. *Mol Ther* **8**:376–384.
69. Persons DA, Allay JA, Bonifacino A, Lu T, Agricola B, Metzger ME, Donahue RE, Dunbar CE, Sorrentino BP (2004). Transient in vivo selection of transduced peripheral blood cells using antifolate drug selection in rhesus macaques that received transplants with hematopoietic stem cells expressing dihydrofolate reductase vectors. *Blood* **103**:796–803.
70. Bunting KD, Galipeau J, Topham D, Benaim E, Sorrentino BP (1998). Transduction of murine bone marrow cells with an MDR1 vector enables ex vivo stem cell expansion, but these expanded grafts cause a myeloproliferative syndrome in transplanted mice. *Blood* **92**:2269–2279.
71. Allay JA, Dumenco LL, Koc ON, Liu L, Gerson SL (1995). Retroviral transduction and expression of the human alkyltransferase cDNA provides nitrosourea resistance to hematopoietic cells. *Blood* **85**:3342–3351.
72. Persons DA, Allay ER, Sawai N, Hargrove PW, Brent TP, Hanawa H, Nienhuis AW, Sorrentino BP (2003). Successful treatment of murine beta-thalassemia using in vivo selection of genetically modified, drug-resistant hematopoietic stem cells. *Blood* **102**:506–513.
73. Richard RE, De Claro RA, Yan J, Chien S, Von Recum H, Morris J, Kiem HP, Dagarno DC, Heimfeld S, Clackson T, Andrews R, Blau CA (2004). Differences in F36VMpl-based in vivo selection among large animal models. *Mol Ther* **10**(4):730–740.
74. Neff T, Horn PA, Valli VE, Gown AM, Wardwell S, Wood BL, von Kalle C, Schmidt M, Peterson LJ, Morris JC, Richard RE, Clackson T, Kiem HP, Blau CA (2002). Pharmacologically regulated in vivo selection in a large animal. *Blood* **100**(6):2026–2031.
75. Walters MC, Patience M, Leisenring W, Rogers ZR, Aquino VM, Buchanan GR, Roberts IA, Yeager AM, Hsu L, Adamkiewicz T, Kurtzberg J, Vichinsky E, Storer B, Storb R, Sullivan KM (2001). Stable mixed hematopoietic chimerism after bone marrow transplantation for sickle cell anemia. *Biol Blood Marrow Transplant* **7**:665–673.
76. Iannone R, Luznik L, Engstrom LW, Tennessee SL, Askin FB, Casella JF, Kickler TS, Goodman SN, Hawkins AL, Griffin CA, Noffsinger L, Fuchs EJ (2001). Effects of mixed hematopoietic chimerism in a mouse model of bone marrow transplantation for sickle cell anemia. *Blood* **97**:3960–3965.

77. Kean LS, Manci EA, Perry J, Balkan C, Coley S, Holtzclaw D, Adams AB, Larsen CP, Hsu LL, Archer D (2003). Chimerism and cure: hematologic and pathologic correction of murine sickle cell disease. *Blood* **102**:4582–4593.
78. Kean LS, Durham MM, Adams AB, Hsu LL, Perry JR, Dillehay D, Pearson TC, Waller EK, Larsen CP, Archer DR (2002). A cure for murine sickle cell disease through stable mixed chimerism and tolerance induction after non-myeloablative conditioning and major histocompatibility complex-mismatched bone marrow transplantation. *Blood* **99**:1840–1849.
79. Zhou SZ, Li Q, Stamatoyannopoulos G, Srivastava A (1996). Adeno-associated virus 2-mediated transduction and erythroid cell-specific expression of a human beta-globin gene. *Gene Ther* **3**:223–229.
80. Goodman S, Xiao X, Donahue RE, Moulton A, Miller J, Walsh C, Young NS, Samulski RJ, Nienhuis AW (1994). Recombinant adeno-associated virus-mediated gene transfer into hematopoietic progenitor cells. *Blood* **84**(5):1492–1500.
81. Malik P, McQuiston SA, Yu X-J, Pepper KA, Krall WJ, Podsakoff GM, Kurtzman GJ, Kohn DB (1997). Recombinant adeno-associated virus mediates a high level of gene transfer but less efficient integration in the K562 humna hematopoietic cell line. *J Virol* **71**:1776–1783.
82. Bertran J, Miller JL, Yang Y, Fenimore-Justman A, Rueda F, Vanin EF, Nienhuis AW (1996). Recombinant adeno-associated virus-mediated high-efficiency, transient expression of the murine cationic amino acid transporter (ecotropic retroviral receptor) permits stable transduction of human HeLa cells by ecotropic retroviral vectors. *J Virol* **70**:6759–6766.
83. Josephson NC, Trobridge G, Russell DW (2004). Transduction of long-term and mobilized peripheral blood-derived NOD/SCID repopulating cells by foamy virus vectors. *Human Gene Ther* **15**:87–92.
84. Alami R, Gilman JG, Feng YQ, Marmorato A, Rochlin I, Suzuka SM, Fabry ME, Nagel RL, Bouhassira EE (1999). Anti-beta s-ribozyme reduces beta s mRNA levels in transgenic mice: potential application to the gene therapy of sickle cell anemia. *Blood Cells Mol Dis* **25**:110–119.
85. Pace BS, Qian X, Ofori-Acquah SF (2004). Selective inhibition of beta-globin RNA transcripts by antisense RNA molecules. *Cell Mol Biol* (Noisy-le-grand) **50**:43–51.
86. Xu L, Ferry AE, Monteiro C, Pace BS (2000). Beta globin gene inhibition by antisense RNA transcripts. *Gene Ther* **7**:438–444.
87. Suwanmanee T, Sierakowska H, Lacerra G, Svasti S, Kirby S, Walsh CE, Fucharoen S, Kole R (2002). Restoration of human beta-globin gene expression in murine and human IVS2-654 thalassemic erythroid cells by free uptake of antisense oligonucleotides. *Mol Pharmacol* **62**:545–553.
88. Vacek MM, Ma H, Gemignani F, Lacerra G, Kafri T, Kole R (2003). High-level expression of hemoglobin A in human thalassemic erythroid progenitor cells following lentiviral vector delivery of an antisense snRNA. *Blood* **101**:104–111.
89. Yoon K, Cole Strauss A, Kmiec EB (1996). Targeted gene correction of episomal DNA in mammalian cells mediated by a chimeric RNA.DNA oligonucleotide. *Proc Natl Acad Sci USA* **93**:2071–2076.
90. Byun J, Lan N, Long M, Sullenger BA (2003). Efficient and specific repair of sickle beta-globin RNA by trans-splicing ribozymes. *RNA* **9**:1254–1263.
91. Goncz KK, Prokopishyn NL, Chow BL, Davis BR, Gruenert DC (2002). Application of SFHR to gene therapy of monogenic disorders. *Gene Ther* **9**:691–694.
92. Porteus MH, Baltimore D (2003). Chimeric nucleases stimulate gene targeting in human cells. *Science* **300**:763.

19

Sickle Cell Disease: The Past, Present and Future Social and Ethical Dilemmas

by Vence L. Bonham, Jr., Carlton Haywood, Jr. and Vanessa Northington Gamble

Introduction

A historical analysis of sickle cell disease (SCD) reveals that sickle cell research has not occurred in a vacuum, but within a broader social and ethical context. This remains true even with the renaissance of SCD research in the genome era. As social scientists Karl Atkin and Waqar Ahmad remind us, "The new genetics is not enacted in a neutral, value-free space, but represents an historical construction, given meaning and realised [sic] through social practices".[1] This chapter examines the history of SCD, focusing on ethical, social, and political implications. It uses SCD as a prism for understanding the historical connections between genetic diseases and racial diseases, and the consequences of these connections.

SCD provides a foundation for understanding race and genetics in the genome era, and for comprehending the attitudes of African-Americans toward genetic research. This chapter also includes an examination of the complex relationship of trust, racial and ethnic minorities, and health care. Social science research has demonstrated that racial and ethnic minorities have lower levels of trust towards both medical institutions and medical research. Thus, in order to respond to the needs of patients with SCD, it is critical that issues of distrust be addressed, and a trustworthy health care system for racial and ethnic minorities to be realized.

The Creation of a "Black Genetic Disease"

The identification of SCD as a racial disease began in the early 20th century.[2–5] In 1904, 20-year-old Walter Clement Noel, a black dental student from Grenada, sought medical care at Presbyterian Hospital in Chicago, Illinois, for symptoms of weakness, dizziness, and joint pain. Four years later, the attending physician, James Herrick, published his clinical observations.[6] He noted that Noel's symptoms included leg ulcers, headache, fever, cough, and joint pain, and that his blood smear contained "peculiar elongated and sickle shaped red blood corpuscles and a case of severe anemia". Although at the time Herrick did not comprehend the link between the sickle cells and Noel's symptoms, his article is recognized as the first clinical description of SCD.

Within 10 years of the publication of Herrick's article, SCD had become identified as a racial disease. In 1922, Verne Mason named the blood disorder "sickle cell anemia". He argued that the disease was a racial disease because "up to the present, the malady has been seen only in the negro, and so far as could be ascertained, it is the only disease peculiar to that race".[7] A 1947 editorial in the *Journal of the American Medical Association* called sickle cell anemia "a race specific disease" and proclaimed, "The most significant feature of sickle cell anemia is the fact that it is apparently the only known disease that is completely confined to a single race".[8] As medical historian Keith Wailoo has argued, the disease became synonymous with "Negro blood" and a "genetic marker of segregation".[9]

SCD became recognized as a "black disease" despite scientific evidence that demonstrated it was not restricted to black people. Reports of sickle cell anemia in families that were identified as white or Caucasian first appeared in the 1920s. In one of the earliest reports about the disease, Verne Mason described five cases — two of which occurred in Caucasians. Physicians tried to explain such contradictions to their race theories by suggesting that sickle cell anemia occurred in Caucasians as a result of miscegenation or racial passing.[5] In a 1943 article in the *Archives of Internal Medicine*, pathologist M.A. Ogden presented a case report of an 8-year-old white boy with sickle cell anemia. Initially the child's father identified himself as of German ancestry and his mother identified herself as of Indian and Scotch ancestry, but after Ogden found evidence of sickling in her blood, the mother admitted that she was passing and indeed had African ancestry. Thus he concluded that the "presence of the sickling trait in a white person is a definite proof of admixture of Negro blood in the immediate or remote ancestry".

At the time, many physicians saw "Negro blood" as a scientific term that carried with it clear biological, social, and public health implications.[9] Blood carried race, racial traits, and racial diseases. Ogden argued that the racial specificity of SCD provided scientific evidence for legal segregation and anti-miscegenation measures. He contended that interracial marriages between blacks and whites posed a serious threat to the health of whites, because they could lead to the transmission of the sicklemic trait. Therefore, he called for federal law to prohibit miscegenation.[10] As early as 1929, some researchers began to question the race-specific model of SCD.[11] However, the definition of SCD as a black disease — even to this day — has proven difficult to eradicate.

An examination of the history of SCD in the first decades of the 20th century clearly illustrates how social, political, and cultural factors — not just medical factors — influenced the conceptualization of the disease. Contemporary medical journals contained several articles that analyzed the "unique" health problems of African-Americans. Many of these articles focused on high rates of syphilis and tuberculosis. Physicians frequently attributed the high rates of these diseases to purported racial characteristics such as large penises and small brains, and to alleged immoral behaviors such as excessive sexuality and debauchery. Such theories demonstrate the then-widely held belief — even in medical circles — that African-Americans were inferior to white Americans.[8–10] Clinical descriptions of sickle cell anemia did not escape the influence of social prejudices. A 1930 *Southern Medical Journal* article labeled those with the disease as "subnormal"; the authors stated, "Our own impression has been that the sickler who presents even mild anemia is a subnormal individual and even though he may not be regarded as an active case of sickle cell anemia, he is still ill equipped to withstand the vicissitudes of life".[12]

Eliminating Racial and Ethnic Disparities in Health and Health Care: The Role of SCD

During the 1970s, SCD became deeply entrenched in the politics of race in the United States. At the beginning of the decade, Robert B. Scott's article, "Health Care Priority and Sickle Cell Anemia", appeared in *JAMA*. In this article, he harshly criticized the lack of federal and private funding that sickle cell anemia had received in comparison to childhood genetic diseases affecting primarily white children[13]; the article implied that the disease had been neglected because it was a "black disease". This inattention to the disease, especially by the federal government, prompted a broad segment of the black community, including media, actors, physicians, politicians, and organizations ranging from the National Medical Association to the Black Panthers, to take action. They launched a campaign to promote disease awareness, screening, and research funding.[2,3,5]

The activism of the black community saw almost immediate results. By February 1971, President Richard M. Nixon had pledged $6 million for SCD research. Announcing this five-fold increase in federal funding, the President gave credence to the accusations of the black community when he stated, "It is a sad and shameful fact that the causes of this disease have been largely neglected throughout our history... We cannot rewrite this history of neglect but we can reverse it".[14] The Senate Subcommittee on Health held hearings on the disease in November 1971; many black health and advocacy organizations testified at these hearings and pushed for increased financial support to combat the disease, seeing this funding as a demonstration of the federal government's commitment to addressing the needs of the black community.[3] In May 1972, the President signed the Sickle Cell Anemia Control Act (PL 92–294), which appropriated $115 million over a three-year period to fund screening, research, and treatment programs for sickle cell anemia. This legislation legitimized the role of the federal government in providing funds to address a disease that is prevalent in a specific racial group, and to remedy a health disparity.

States also responded to these federal initiatives and to grass roots efforts. In 1971, Connecticut became the first state to pass a sickle cell anemia screening law. By 1974, 10 states had mandatory screening procedures, and four more had laws for voluntary screening programs. However, many states that enacted these screening laws failed to appropriate funds to support education and counseling programs.[15]

The intense lobbying by the black community had indeed played the major role in increasing the visibility and funding for SCD, yet this strong support began to erode even before President Nixon had signed the Sickle Cell Anemia Control Act. Documented abuses from the sickle cell programs, including mandatory screening, inappropriate counseling, job discrimination, and misinformation, fueled this erosion. The programs also raised the specter of eugenics: for example, the legislation to provide research support for the disease was originally entitled "National Sickle Cell Anemia *Prevention* Act". Black constituency groups decried the title of this legislation because, as they accurately pointed out, the only way that the disease could be prevented was for people not to reproduce. Following this outcry, the legislation was renamed "National Sickle Cell Anemia *Control* Act".

Clearly the original title of the Senate Bill communicated a message that harked back to the eugenics movement of the early 20th century. At a time when governmental birth control programs became controversial because of coercive sterilizations, and the United States Public Health Service (USPHS) Syphilis Study at Tuskegee (commonly, though inappropriately, referred to as the Tuskegee Study) became public, the sickle cell anemia programs spread misinformation about the disease and raised community concerns of genocide against black people.[16]

Race and Disease in the Genome Era: Lessons from the History of Sickle Cell Disease

With the completion of the Human Genome Project in 2003, the National Institutes of Health (NIH), National Human Genome Research Institute (NHGRI), hundreds of scientists, and members of the public across the globe participated in more than a dozen workshops and numerous individual consultations to establish a vision for the future of genomics research. This vision is formulated into three major themes: genomics to biology, genomics to health, and genomics to society.[17] Each theme includes a series of grand challenges — "bold, ambitious, research targets for the scientific community". One grand challenge is to understand the relationships between genomics, race, and ethnicity, and the consequences of uncovering these relationships. As the history of SCD clearly demonstrates, it has evolved into the touchstone by which both the scientific and lay communities discuss issues of race, genetics, and disease. Furthermore, sickle cell is a genetic disease that has been defined as a racial disease. Thus an historical examination of the disease and its ethical, legal, and social implications can serve as a guide for addressing this grand challenge.

In April 1993, a panel assembled by the federal Agency for Health Care Policy and Research (now the Agency for Healthcare Research and Quality) issued guidelines for the screening, diagnosis, management, and counseling of newborns and infants with SCD. They recommended that all newborns, regardless of race or ethnicity, be screened for SCD. Difficulties in defining race in the increasingly heterogeneous American population had been the major factor prompting the panel's action, and its recommendation underscores historian Evelyn Brooks Higginbotham's assessment of racial definition. She contends, "When we talk about the concept of race, most people believe that they know it when they see it but arrive at nothing short of confusion when pressed to define it".[18]

Scientists have long sought to define race, and to categorize humans into racial groups. In the 18th century, scientists hoping to categorize humans taxonomically in the same way that they categorized other species began to create racial categories. However, the categories were not constant. For example, Carolus Linneaus asserted in 1758 that all humans belonged to four groups; by 1795, Johann Blumenbach declared that there were five such groups. These scientists attached hierarchical designations to their categorizations, claiming that differences in skin color, physiognomy, and geography were associated with scientifically measurable differences in character, aptitude, and temperament.[19] Today, genomic research refutes the use of racist theories to separate humans; current research concludes that there are no gene variants that are present in all individuals of one population group and in no individuals of another. No sharp genetic boundaries can be drawn between human population groups, and frequencies of genetic variants and haplotypes differ across the world.[19]

But the mapping of the human genome has not ended debates and controversies about racial categories and the biological meaning of race, and indeed has added new complexity. Cooper *et al.*[20] contend, "Race is a thoroughly contentious topic, as one might expect of an idea that intrudes on the everyday life of so many people . . . Into the storm of controversy rides genomics. With the acknowledgement that race is the product of a marriage of social and biologic influences, it has been proposed that genomics now at least offers the opportunity to put its biologic claims to an objective test". They conclude that race, at the continental level, has not been shown to provide a useful categorization of genetic information about the response to drugs, diagnosis, or cause of disease.[20] In a provocative 2001 *New England Journal of Medicine* editorial, Robert Schwartz declared "instruction in medical genetics should emphasize the fallacy of race as a scientific concept and the dangers inherent in practicing race-based medicine".[21] Sociologist Troy Duster, in a recent essay in *Science*, raised questions about whether the growing sophistication of genomic technology to provide single nucleotide polymorphism (SNP) profiles of populations will solve the dilemma. He argued, "When the phenotype distinguishing these populations is race, the likelihood of committing the fallacy of misplaced concreteness in science is nearly overwhelming".[22]

However, such views are not universally accepted.[23] Other scientists believe that race is biologically important, and that there is great validity in racial/ethnic self-categorizations in conducting biomedical and genetic research. Neil Risch and colleagues[24] argue, "If biological is defined as genetic, then . . . a decade or more of population genetics research has documented genetic and therefore biological, differentiation among the races". There is mounting evidence that approximately 10–15% of total genetic variation in humans is explained by geographical continental populations such as sub-Saharan Africans, Northern Europeans, and East Asians: "This means that individuals from different geographic populations are, on average, slightly more different from one another than individuals from the same population".[25]

The advent of genomic medicine has made clear the need to find common ground on the thorny issue of race and human genetic variation; a new vocabulary to discuss human genetic variation is also required. One suggestion has been to replace discussions of race and ethnicity with discussions of geographic origins and ancestry. As Francis Collins,[26] Director of the NHGRI, has stated, "Increasing scientific evidence ... indicates that genetic variation can be used to make a reasonably accurate prediction of geographic origin of an individual, at least if the individual's grandparents all came from the same part of the world. As those ancestral origins in many cases have a correlation albeit often imprecise with self-identified race or ethnicity, it is not strictly true that race or ethnicity has no biological connection.

SCD provides an important example of our understanding of the relationships of race, ancestry and genetics. Feldman *et al.*[27] also support the use of ancestry to discuss human variation. They note that SCD is not an African trait, but rather is characteristic of geographic areas where malaria was endemic. They argue, "To use genotype effectively in making diagnostic and therapeutic decisions, it is not race that is relevant, but both intra- and trans-continental contributions to a person's ancestry". Rotimi contends that the mislabeling of sickle cell disease as a "black disease" has "rendered the distribution of sickle cell anemia invisible in other populations, leading to erroneous understanding of the geographical distribution of the underlying genetic variants". He points out that there is a town in central Greece where the rate of sickle cell anemia is twice that of African-Americans. Furthermore, he notes that black South Africans do not carry the sickle cell trait.[28] The disease is better described as one of individuals with ancestry from select geographic regions where malaria is endemic. Ancestral information gathered through the collection of a family health history, not merely racial information, provides useful and comprehensive information on disease susceptibility.

Ancestry and family history matter in determining risk of disease.[27] Collecting family history is not without its difficulties, as many people in the United States do not know the health history of their parents or grandparents, or the geographic origins of their ancestors.[29] "Although advances arising from the Human Genome Project and related efforts are already adding important new genomic tools, the family history will remain highly relevant for years to come".[30] The widespread collection and use of a three-generation family health history in primary care settings should be widely adopted.[31]

History makes plain that a focus on biological difference between groups — no matter the terms used — has frequently had negative consequences, such as discrimination and stereotyping, for members of racial and ethnic minorities. In the age of genomic medicine, the potential for abuse has not diminished. For example, controversial fringe researchers and scholars have interpreted the research and commentaries of mainstream scientists and clinicians, such as Risch and Burchard, to support their views of biologic and genetic differences between the races, and to advance particular agendas.[32] These contemporary uses of genetic theories to promote social and political ideologies clearly illustrate what lawyer and bioethicist Patricia King calls "the danger of difference". King warns, "In a racist society that incorporates beliefs about the inherent inferiority of African-Americans in contrast to the superior status of whites, any attention to the questions of differences that may exist is likely to be pursued in a manner that burdens rather than benefits African-Americans".[33]

Ethical, Legal and Social Implications Research of SCD in the Genome Era

The complex ethical, legal, and social issues raised by the history of SCD are representative of the types of issues that must be addressed in this era of genomic research. The planners of the International Human Genome Project recognized this, and in 1989 the Program Advisory Committee established a working group to plan an ethical, legal, and social implications (ELSI) research program as part of the Human Genome Project. The original objectives of the ELSI program were: (1) to anticipate and address the implications for individuals and society of mapping and sequencing the human genome; (2) to examine the ethical, legal, and social consequences of mapping and sequencing the human genome; (3) to stimulate public discussion of the issues; and (4) to develop policy options which would assure that the information be used to benefit individuals and society.[34] Today, the NHGRI is the largest funder of genetic and genomic ELSI research in the world, while a growing number of institutes fund ELSI-related research.

Since the beginning of the ELSI program there have been studies funded to investigate the ethical, legal, and social issues involving SCD.[35] In February 2005, the NHGRI issued a notice to the scientific community that the NHGRI ELSI research program was soliciting studies on the implications of research, technologies, and information related to SCD and other hemoglobinopathies.[36] This program notice was informed by a landmark conference at the NIH of members of the SCD, genomics, and ELSI research communities held November 19–21, 2003.

This invitation-only conference, "New Directions for Sickle Cell Therapy in the Genome Era", brought together more than 120 individuals from the United States and abroad. The goal of the conference was to consider how the new tools and techniques of genomics might be applied, both to understand more fully the biology of SCD, and to develop more effective therapeutic and preventative strategies for the disease. The conference planners recognized the importance of including ELSI issues on the agenda, and conference participants recommended that the NIH should fund cultural and ELSI research on SCD.[37]

SCD Research: Trust, Trustworthiness, & Distrust

The genome era promises to bring a renaissance of SCD research, with a rapid increase in the scale and scope of new research. The future of research to enhance the quality of life for persons with SCD is bright, but many barriers to fulfilling the promise must be overcome. Social science research has demonstrated that trust is the cornerstone of the provider–patient relationship, the foundation for quality health care delivery and outcomes, and an element of constructive relationships between investigators and participants in biomedical research. Research has also documented that racial and ethnic minorities have lower levels of trust in medical providers and in medical institutions than do white Americans.[38,39] Distrust can adversely affect health outcomes and contribute to racial and ethnic disparities in health and health care; it leads to decreased patient satisfaction, low enrollment in clinical trials, greater reluctance to seek medical care, and poorer patient adherence to treatment recommendations. An often-cited factor for the distrust that many African-Americans hold toward genetic initiatives is the failure of the sickle cell screening programs of the 1970s.[40] Thus, in order to provide optimal care and achieve positive health outcomes for patients with SCD, it is critical that researchers and providers understand the complex relationship of trust and distrust in racial and ethnic minorities.

Trust has been defined as the specific expectation that another's actions will be beneficial and not detrimental.[41] It is an essential component of patient–physician and research participant–investigator relationships because of the vulnerability of the patient–research participant, and the imbalance of knowledge and power between patient–physician and research participant–investigator.[41,42] Illness inherently places patients (and oftentimes research participants) at risk. Seeking care from a physician, and participating in a therapeutic trial, entail that patients and research participants submit their bodies and personal information about themselves to a physician–investigator. At risk of harm to their bodies and their privacy, patients and research participants rely upon the knowledge and skill of the physician or investigator to work to the patient's benefit (or at least to not contribute to the patient's detriment). In this way, a trusting relationship has intrinsic value.[41] The greater the risk from illness, the greater the vulnerability, and thus the greater the potential for trust (or distrust) within the clinical or biomedical research relationship.[41,43]

Trustworthiness is a morally valuable trait of character.[44] Through the display of certain other qualities, such as compassion, altruism, honesty, and reliability, and through open communication with patients or research participants, physicians and investigators may demonstrate that they are deserving of the trust placed in them.[42] Because of the intrinsic value of trust in the patient–physician and research participant–investigator relationship, it is the obligation of physicians and investigators to demonstrate their trustworthiness and earn the trust of the patient–research participant.[41]

Distrust can be understood as either a passive or active notion. Passively, distrust may be thought of as the absence of trust.[43] Actively speaking to distrust is to hold an anxious or pessimistic view of the motivations of or the expected results from relations with others.[43] Passive distrust may be merely the result of unfamiliarity with another; active distrust can be shaped by prior personal experience, the experiences of friends or loved ones, and — in the case of minorities in the United States — knowledge of the historical interactions between the biomedical system and minority groups.[41,42,45]

Race, Trust, and Medical Research

Several studies suggest that minorities in the United States, particularly African-Americans, possess higher levels of distrust in medical research than other groups. In a national survey of attitudes towards physicians and participation in clinical research, Corbie-Smith *et al.*[46] found that African-American respondents were more likely than white respondents not to trust that their physicians would fully explain research participation, and also more likely to believe that someone like them would be used as a guinea pig in research without consent. Additionally, they were more likely to believe that physicians sometimes exposed them to unnecessary risks, and less likely to believe that they could freely ask questions of their physicians. In focus groups conducted with African-Americans concerning attitudes towards research, Corbie-Smith *et al.*[47] found that participants had limited understanding of the informed consent process, and many believed that this process itself was designed to protect doctors and medical institutions from liability.

Freimuth *et al.*[48] also conducted focus groups with African-Americans on their views toward research, and found that their participants generally believed that African-Americans should avoid involvement with medical research, given knowledge of previous abuses. Focus group participants in this study expressed the view that providing informed consent was equivalent to "signing away your rights", and that the consent process was designed to protect hospitals and doctors from legal responsibility. Several participants expressed doubt that participation in medical research served any benefit to African-Americans.[48]

Shavers *et al.*[49] conducted a regional survey of attitudes toward research participation and knowledge of the United States Public Health Service Syphilis Study at Tuskegee. They found that among respondents who stated that knowledge of the USPHS study would affect their willingness to participate in medical studies, more African-Americans than whites said they would be unwilling to participate. Among respondents who believed that minorities bore the greatest burden of risk from medical research, fewer African-Americans than whites stated that they would be willing to participate in research.[49]

Race, Trust, and Genetics

The literature also suggests that African-Americans and other minority groups possess concerns about genetics and genetic research. Reporting on the results of a regional survey, L. Allen Furr found that African-Americans expressed a slightly higher, but statistically significant, belief that genetic screening posed more harm than good to society.[50] Schulz *et al.*[51] reported on focus group data collected as part of the Communities of Color Genetics Policy Project (CCGPP), which was designed to engage African-American and Latino communities in dialogues pertaining to genetics research, and to develop policy recommendations regarding genetics research from these dialogues. These authors found that participants in their focus groups recognized the potential benefits to be gained from genetics research, but struggled with ethical and social concerns regarding it; they expressed concerns over a perceived lack of moral or spiritual guidance governing our use of genetics research, fear over the inequitable distribution of the risks and benefits of research participation, fear that genetics research would serve to re-inscribe racial inequality and group stigmatization, and concern about lack of community control over genetics information gained through research.[51]

Ilana Mittman and Marian Secundy[52] reported on deliberations on genetics issues by a group of experts, genetics consumers, scientists, providers, and community representatives participating in a two-day conference. Conference participants were charged with the task of identifying problems pertaining to minority issues and genetics, the mechanisms contributing to those problems, and recommended solutions. Concerns about genetics expressed at the conference included a lack of trust in biomedical research, scientists' responses to the perceived distrust, the distribution of the risks and benefits of genetics research, and the lack of minority participation as genetics researchers and in setting the genetics research and public policy agenda.[52]

Attitudes toward Research Participation among Persons with SCD

The studies reviewed above give insight into the views of racial and ethnic minorities toward medical and genetics research in general. Unfortunately, studies of the attitudes of persons with SCD towards participation in research are scant, and anecdotal evidence of this group's willingness to participate in medical studies is equivocal. At present, there are no studies which analyze the attitudes toward genetic research of persons with SCD, but there are a few that are suggestive.

Investigators from the Multicenter Study of Hydroxyurea in Sickle Cell Anemia (MSH), the largest clinical trial concerning adults with SCD in the United States to date, reported that the number of participating sites had to be doubled from the original design of the study in order to meet recruitment goals. This increase in sites, though, did not cause a significant delay in the progress of the study.[53] In contrast, Gaston *et al.*[54] reported that recruitment for the federally-funded Cooperative Study of Sickle Cell Disease (CSSCD) was accomplished successfully. While enrollment for the CSSCD was completed over 27 months instead of the originally planned 24 months, this extension of the enrollment period did not significantly impede the progress or increase costs of the study. Factors believed to contribute to the successful enrollment figures for the CSSCD included the relationships between potential subjects and investigators that had existed prior to the study, the majority African-American makeup of recruitment staff, the efforts of an active recruitment committee, and a long-standing working relationship between the clinics involved in the study and community sickle cell groups.[54]

Other factors have been identified as leading to effective recruitment of people with SCD as research participants. Francine Jones and Marion Broome[55] examined the views of adolescents with SCD about successful recruitment and retention strategies for interventional research; they emphasized the use of honest and straightforward communication, the potential to learn disease management techniques, and the presence of investigators who understood what adolescents with the disease experienced. Jones and Broome's research underscores the importance of seeking input from affected communities when policies and programs are created. Nield-Anderson *et al.*[56] provide an anecdotal account of their experiences with a pilot study examining the efficacy of a relaxation videotape as a pain management strategy for sickle cell, noting that two adults with the disease expressed concerns over their assignment to the non-intervention group. This small and anecdotal study suggests that sickle cell patient attitudes towards participation in research involving random assignment to a placebo or other non-intervention arm should be explored on a larger scale.

British hematologists attempting to conduct their own studies of hydroxyurea and SCD in the United Kingdom have found great difficulty in recruiting enough patients to take part in research.[57] Olujohungbe *et al.*[57] report that studies in four institutions in the United Kingdom had to be discontinued because they failed to recruit enough patients, and that patient support groups in the United Kingdom had openly campaigned against the use of hydroxyurea; the authors, however, did not report the reasons behind this opposition. While patients' fears about developing secondary malignancies may also have played a role, Olujohungbe *et al.*[57] speculate that patient reluctance to participate was more complex, perhaps also having stemmed from wariness in patient–doctor relationships, peer group criticism, and anxiety about change. The experience of those United Kingdom researchers makes clear the need to understand internationally the attitudes of patients with SCD.

Given the findings of the literature reviewed above specific to SCD, it is obvious that more systematic research is needed to understand the sickle cell patient's decision of whether or not to participate in research. Social sciences researchers, clinical researchers, and patient groups must collaborate in the design and conduct of studies which seek to gain a better understanding of the structural barriers and attitudinal factors that may promote or hinder the sickle cell patient's involvement in research. The involvement of patient groups in the design and conduct of these studies is vital, as the issues that sickle cell patients consider are likely to be multifaceted and could be unique to this patient population. In addition to leading to better designed studies, patient group involvement at the very beginning of the research process may also have the more significant and ethically optimal outcome of engendering trust in the community between sickle cell patients and medical researchers. This engagement of patient groups may empower the patient community to be more directive in the future course of clinical treatment and research for this disease.

Lessons for the Conduct of SCD Research in the Genome Era

As researchers and providers work to advance sickle cell research and care in the genome era, there are several lessons that can be learned from an analysis of the historical, ethical, and social contexts of SCD. First, it is clear that researchers need to clarify and define how they use the term "race" in their research. Is it a proxy for social class? Is it a proxy for human genetic variation? How was the race of the patients and research participants ascertained? Researchers also need to understand the negative ramifications of SCD being classified as a "black disease", and continue efforts to dethrone this categorization.

Another lesson is that initiatives that help staff and institutions to provide care in a culturally-competent fashion need to be advanced and supported. Cultural competence education can help providers better understand the histories, culture, experiences, preferences, and health behaviors of their patients. Increasing physicians' knowledge of the African-American experience, including the USPHS Study at Tuskegee and the history of coercive sterilization in this country, would help them better understand the perceptions of some of their African-American patients. The growing importance of taking extensive family histories is also a critical component of culturally-appropriate education.

As our examination of the literature on trust demonstrates, research in the role of trust in improving the quality of care for patients with SCD is in its infancy, especially with regard to minority populations and patients with SCD. More studies, for example, need to be done to develop validated measurement instruments, and to better elucidate the mechanisms of how trust influences health care. The research on trust underscores the importance of social science research in advancing care for patients with SCD. Thus it is critical that research which examines the ethical, legal, and social aspects of genetic diseases be actively supported. It is also important that social scientists not be marginalized, but be considered as crucial members of interdisciplinary research teams.

Health care providers and institutions can also improve the levels of care for patients with SCD by supporting efforts to diversify the health care profession, and to eliminate racial and ethnic disparities in health care. Research has demonstrated that racial and ethnic concordance between patient and provider is associated with more positive satisfaction with the care received.[58] Patients believed that these visits were more participatory and supportive, and less discriminatory — in other words, more trustworthy.

It is critical that programs to increase the number of clinicians and researchers who are from underrepresented minorities be supported. Likewise, it is important that health care providers and health care institutions be visible supporters of efforts to eliminate racial and ethnic disparities in health care. Such advocacy would demonstrate to racial and ethnic minority patients that health care providers and health care institutions acknowledge the disparate treatment that they may receive in the health care system, and are prepared to address the problem.

The final lesson learned from our overview is that researchers and providers need to have mechanisms for community consultation, evaluation, and collaboration. They must develop strong relationships with members of minority communities, so that community members can assist them in determining both sources of distrust and activities to address and overcome them. For example, researchers may think that the source of distrust is the USPHS Syphilis Study, when it may in fact be a local or institutional breach of trust.

In addition, a community-based approach to research also needs to be adopted whenever possible. This may include supporting efforts to involve communities in the research setting, designing the research agenda, and controlling research results. In the context of SCD, this would entail acknowledging patients with SCD as experts. As an example: if a researcher wanted to create a study to find out why patients with SCD were not participating in clinical trials, the patients could be involved in developing the focus group questions. By consulting patients with SCD and their families, medical researchers and health care providers would clearly demonstrate that they respect the needs, thoughts, and experiences of these patients and their families.

The Sankofa bird is a mythic bird of the Akan people of West Africa; it moves forward while looking backward, and has come to symbolize the wisdom of learning from the past in order to build for the future. This chapter illustrates the importance of the concept of Sankofa — we have looked backward in order to provide knowledge that will help build a better future for patients with SCD.

Summary

This chapter examines the history of SCD, focusing on ethical, social, and political implications and its importance for the genome era. It also examines the complex relationship of trust, trustworthiness and distrust of medical institutions and the biomedical research enterprise and its role in the participation of individuals with SCD in genomic research. It uses SCD as a prism to understand the historical connections between genetic diseases and racial diseases and the consequences of these connections. SCD provides a foundation for understanding race and genetics in the genome era and for comprehending the attitudes of African-Americans toward participation in genetic research.

Acknowledgments

"Research for this chapter was supported [in part] by the Intramural Research Program of the National Human Genome Research Institute, National Institutes of Health. Carlton Haywood Jr. was supported by the National Research Service Award Minority Predoctoral Fellowship # 1 F31 HL082037-01 from the National Heart, Lung, and Blood Institute, National Institutes of Health. We gratefully acknowledge Adebola Odunlami, MPH, for her research assistance".

References

1. Atkin K, Ahmad WI (1998). Genetic screening and haemoglobinopathies: ethics, politics and practice. *Soc Sci Med* **46**(3):445–458.
2. Wailoo K (2001). *Dying in the City of the Blues: Sickle Cell Anemia and the Politics of Race And Health.* Chapel Hill and London: The University of North Carolina Press.
3. Tapper M (1999). *In the Blood: Sickle Cell Anemia and the Politics of Race.* Philadelphia: University of Pennsylvania Press.
4. Savitt TL, Goldberg MF (1989). Herrick's 1910 case report of sickle cell anemia. The rest of the story. *JAMA* **261**(2):266–271.
5. Gamble VN (1999). Race and the new genetics: a historical perspective. In: Hager M, editor. *The Implications of Genetics for Health Professionals.* Josiah Macy Foundation.

6. Herrick JB (1910). Peculiar elongated and sickle-shaped red corpuscles in a case of severe anemia. *Arch Intern Med* **6**:517–521.
7. Mason VR (1985). Landmark article Oct. 14, 1922: sickle cell anemia. By V. R. Mason. *JAMA* **254**(14): 1955–1957.
8. Editorial (1997). Sickle cell anemia: a race specific disease. *JAMA* **133**:33–34.
9. Wailoo K (1996). Genetic marker of segregation: sickle cell anemia, thalassemia, and racial ideology in American medical writing 1920–1950. *Hist Philos Life Sci* **18**(3):305–320.
10. Ogden MA (1943). Sickle cell anemia in the white race. *Arch Intern Med* **71**:169–182.
11. Cooley TB, Lee P (1929). Sickle cell anemia in a Greek family. *Am J Dis Child* **38**:105–106.
12. Graham GS, McCarty SH (1930). Sickle cell (Meniscocytic) anemia. *South Med J* **23**(7):600.
13. Scott RB (1970). Health care priority and sickle cell anemia. *JAMA* **214**(4):731–734.
14. Nixon RM (1971). *National Health Strategy. The President's Message to the Congress Proposing a Comprehensive Health Policy for the Seventies.* Weekly Compilation of Presidential Documents 7, 254.
15. Rutkow IM, Lipton JM (1974). Some negative aspects of sickle cell anemia screening programs (negative aspects of sickle cell screening). *Conn Med* **38**(5):257–262.
16. Bowman JE (1960). Genetic screening programs and public policy. *Phylon* **38**(2):117–142.
17. Collins FS, Green ED, Guttmacher AE, Guyer MS (2003). A vision for the future of genomics research. *Nature* **422**(6934):835–847.
18. Higginbotham EB (1992). African-American womens history and the metalanguage of race. *Signs* **17**(2):251–274.
19. Bonham VL, Warshauer-Baker E, Collins FS (2005). Race and ethnicity in the genome era: the complexity of the constructs. *Am Psychol* **60**(1):9–15.
20. Cooper RS, Kaufman JS, Ward R (2003). Race and genomics. *N Engl J Med* **348**(12):1166–1170.
21. Schwartz RS (2001). Racial profiling in medical research. *N Engl J Med* **344**(18):1392–1393.
22. Duster T (2005). Medicine. Race and reification in science. *Science* **307**(5712):1050–1051.
23. Burchard EG, Ziv E, Coyle N, Gomez SL, Tang H, Karter AJ, Mountain JL, Perez-Stable EJ, Sheppard D, Risch N (2003). The importance of race and ethnic background in biomedical research and clinical practice. *N Engl J Med* **348**(12):1170–1175.
24. Risch N, Burchard E, Ziv E, Tang H (2002). Categorization of humans in biomedical research: genes, race and disease. *Genome Biol* **3**(7):1–12.
25. Bamshad M (2005). Genetic influences on health: does race matter? *JAMA* **294**(8):937–946.
26. Collins FS (2004). What we do and don't know about 'race', 'ethnicity', genetics and health at the dawn of the genome era. *Nat Genet* **36**(11 Suppl):13–15.
27. Feldman MW, Lewontin RC, King MC (2003). Race: a genetic melting-pot. *Nature* **424**(6947):374.
28. Rotimi CN (2004). Are medical and non-medical uses of large-scale genomic markers conflating genetics and 'race'? *Nat Genet* **36**(11 Suppl):S43–S47.
29. Condit CM, Parrott R, Harris TM (2002). Lay understandings of the relationship between race and genetics: development of a collectivized knowledge through shared discourse. *Public Underst Sci* **11**:373–387.
30. Guttmacher AE, Collins FS, Carmona RH (2004). The family history — more important than ever. *N Engl J Med* **351**(22):2333–2336.
31. Rich EC, Burke W, Heaton CJ, Haga S, Pinsky L, Short MP, Acheson L (2004). Reconsidering the family history in primary care. *J Gen Intern Med* **19**(3):273–280.
32. Rushton JP, Jensen AR (2005). Thirty years of research on race differences in cognitive ability. *Psychol Public Policy Law* **11**(2):235–294.
33. King PA (1992). The dangers of difference. *Hastings Cent Rep* **22**:35–38.
34. National Human Genome Research Institute Website. Available: http://www.genome.gov/10001754. Accessed September 21, 2005.
35. National Human Genome Research Institute Website. Available: http://www.genome.gov/10001796. Accessed September 21, 2005.
36. National Institutes of Health. Notice HG-05-002 Release, February 4, 2005, Issued by the National Human Genome Research Institute.
37. National Human Genome Research Institute Website. New Directions for Sickle Cell Therapy in the Genome Era, Conference Summary. Available: http://www.genome.gov/11509561.

38. Boulware LE, Cooper LA, Ratner LE, LaVeist TA, Powe NR (2003). Race and trust in the health care system. *Public Health Rep* **118**(4):358–365.
39. Voils CI, Oddone EZ, Weinfurt KP, Friedman JY, Schulman KA, Bosworth HB (2005). Who trusts healthcare institutions? Results from a community-based sample. *Ethn Dis* **15**(1):97–103.
40. Dula A (1994). African-American suspicion of the healthcare system is justified: what do we do about it? *Camb Q Healthc Ethics* **3**(3):347–357.
41. Goold SD (2001). Trust and the ethics of health care institutions. *Hastings Cent Rep* **31**(6):26–33.
42. Crawley LM (2001). African-American participation in clinical trials: situating trust and trustworthiness. *J Natl Med Assoc* **93**(12 Suppl):14S–17S.
43. Hall MA, Dugan E, Zheng B, Mishra AK (2001). Trust in physicians and medical institutions: what is it, can it be measured, and does it matter? *Milbank Q* **79**(4):613–639.
44. Beauchamp TL, Childress JF (2001). *Principles of Biomedical Ethics*, 5th edition. New York: Oxford University Press.
45. Gamble VN (1997). Under the shadow of Tuskegee: African-Americans and health care. *Am J Public Health* **87**(11):1773–1778.
46. Corbie-Smith G, Thomas SB, St George DM (2002). Distrust, race, and research. *Arch Intern Med* **162**(21):2458–2463.
47. Corbie-Smith G, Thomas SB, Williams MV, Moody-Ayers S (1999). Attitudes and beliefs of African-Americans toward participation in medical research. *J Gen Intern Med* **14**(9):537–546.
48. Freimuth VS, Quinn SC, Thomas SB, Cole G, Zook E, Duncan T (2001). African-Americans' views on research and the Tuskegee Syphilis Study. *Soc Sci Med* **52**(5):797–808.
49. Shavers VL, Lynch CF, Burmeister LF (2002). Racial differences in factors that influence the willingness to participate in medical research studies. *Ann Epidemiol* **12**(4):248–256.
50. Furr LA (2002). Perceptions of genetics research as harmful to society: differences among samples of African-Americans and European-Americans. *Genet Test* **6**(1):25–30.
51. Schulz A, Caldwell C, Foster S (2003). "What are they going to do with the information?" Latino/Latina and African-American perspectives on the Human Genome Project. *Health Educ Behav* **30**(2):151–169.
52. Mittman IS, Secundy MG (1998). A national dialogue on genetics and minority issues. *Community Genet* **1**(3):190–200.
53. Charache S, Terrin ML, Moore RD, Dover GJ, McMahon RP, Barton FB, Waclawiw M, Eckert SV (1995). Design of the multicenter study of hydroxyurea in sickle cell anemia. Investigators of the Multicenter Study of Hydroxyurea. *Control Clin Trials* **16**(6):432–446.
54. Gaston M, Smith J, Gallagher D, Flournoy-Gill Z, West S, Bellevue R, Farber M, Grover R, Koshy M, Ritchey AK *et al.* (1987). Recruitment in the Cooperative Study of Sickle Cell Disease (CSSCD). *Control Clin Trials* **8**(4 Suppl):131S–140S.
55. Jones FC, Broome ME (2001). Focus groups with African-American adolescents: enhancing recruitment and retention in intervention studies. *J Pediatr Nurs* **16**(2):88–96.
56. Nield-Anderson L, Dixon JK, Lee K (1999). Random assignment and patient choice in a study of alternative pain relief for sickle cell disease. *West J Nurs Res* **21**(2):266–274.
57. Olujohungbe A, Cinkotai KI, Yardumian A (1998). Hydroxyurea therapy for sickle cell disease in Britain. Disappointing recruitment despite promising results. *Br Med J* **316**(7146):1689.
58. Cooper LA, Roter DL, Johnson RL, Ford DE, Steinwachs DM, Powe NR (2003). Patient-centered communication, ratings of care, and concordance of patient and physician race. *Ann Intern Med* **139**(11):907–915.

20

It Takes a Village to Cure Sickle Cell Disease

by Rosie Peterson and Denise Davis-Maye

Introduction

This chapter will get to the heart of the matter and ask the paramount question: How excited is the lay community about the prospect of someone "messing with their genes" and making people "guinea pigs"? Here we will explore specific issues peculiar to the African-American experience, related to previous research practices, strategies for regaining trust in the community, and the role of the community in making informed decisions about genetic testing. Issues related to the adolescent and young adult African-American populations, often excluded from the decision-making process, will be addressed; suggestions will be offered for improving the disbursement of information at the community level and keeping the general public educated about advances in healthcare delivery and gene therapy efforts, stressing that education is the key. Finally, we will explore ways to achieve a successful and well-accepted gene therapy program through the establishment of a sense of ownership, responsibility, and empowerment of the lay community to control its own destiny as relationships are forged with the research community.

Social and Cultural Context of Sickle Cell Disease (SCD)

It is not a new presumption that people of color, when they are challenged with a chronic disease, have little to offer in terms of understanding and managing the components of their care. However, African-Americans faced with managing chronic disease, and given the opportunity to do so, have been quite resourceful in establishing supportive structures and creating strategies to alleviate the stress created by the disease.

Sickle cell anemia is a disease which primarily affects persons of African descent. The condition has special further implications for African-Americans, in that although they are afflicted with sickle cell anemia at a rate of 1 out of 500, 1 out of 12 African-Americans is a carrier of the sickle cell trait.

How disease and health issues have been managed historically in African-American communities deserves consideration. Much of the management of healthcare for African-Americans in a segregated society has involved experiences of maltreatment, unethical studies, and segregated and less technologically advanced medical settings.[1] As a result of this tenuous relationship with the healthcare system, some African-Americans would intentionally seek care from African-American or Jewish physicians. Those who made this special effort believed they could trust physicians who were perceived as sharing similar challenges with discrimination, racism, and oppression. If they were unsuccessful in securing the services of an ethnic physician, some African-Americans would rely on self-diagnosis and self-treatment. Today, this inequitable system of healthcare is somewhat improved, but the image still persists — that is, many perceive a dual system where European-Americans and upper and middle income people receive adequate to superior healthcare, while the poor and people of color receive mediocre healthcare at best.

Traditionally, African-Americans have: (1) provided care for sick and dying members of their families in the home; (2) provided the primary care-giving for ailing or transitioning loved ones not in the immediate family; and (3) provided support, assistance, and caretaking tasks for community members who were not blood relatives.

There is an accepted adage that the hope of African-Americans "goes forward on the feet of little children." The concept that the youth of the population would realize the hopes and achievements of a people is one which is passed along with each generation. Sickle cell anemia is usually diagnosed in childhood. Before the more advanced screening processes and vast health promotion campaigns, there were families who were unfamiliar with the disease, beyond the fact that a family member who had experienced similar symptoms would die as a result of cerebral hemorrhage or shock. Families would thus begin to enter stages of mourning soon after the first presentation of symptoms; community members would provide long-term mourning assistance, including preparation of meals, sitting (respite) services, and end of life support (hospice).

Conceptualization of the magnitude of health concern was and is filtered through the lens of religiosity and spirituality. The integral role which faith plays for many African-Americans is an important factor in understanding the acceptance and management of disease. Though there are increasing numbers of African-American adherents of Eastern Orthodoxy, Buddhism, Islam, Judaism and other faiths, the overwhelming majority of African-Americans identify themselves as Christians; African-American Baptists actually make up the fourth largest religious denomination in the United States and the largest religious group among African-Americans, with Methodist groups the second largest. The ethos of this faith heavily influences both individual and family views of physical health challenges.

Baptist tradition promotes a deep spiritual conviction, which may yield one of two perceptions: First, if one actively serves God and participates in the faith's identified religious practices, the faithful believe that pain and suffering will one day be transcended; they are simply a "test of faith." This perception is rationalized in statements like "God won't put any more on you than you can bear"; "God is powerful medicine for sick souls"; "Jesus is my medicine"; and "God won't move your mountains, but he will help you get over them."

The second perception is one related to the Biblical scripture, which suggests that the sins of the father will be visited upon subsequent generations. So a family's level and expression of spirituality guides its perception of the dying process and death. Some may approach the expectation of death as a joyful transition to another stage of existence or to a "heavenly home," while others may approach death with fear and emotional difficulty.[2]

Inconsistencies in Information Dissemination

Disparities in the quality of healthcare, provider and patient education, access, and services received by adults raise problems in caring for individuals with a chronic illness. Low literacy level patients are faced with health information, materials, and procedures that are difficult or impossible for them to understand and apply. But there are other groups of patients with whom it is also difficult for healthcare professionals to communicate effectively. These hard-to-reach patients include those who face language and cultural barriers, and those who have difficulties processing health information because of physical or cognitive disabilities.[3]

Published studies have suggested a strong link between literacy and healthcare that many providers have not yet recognized. Examples of this linkage include patients having the ability to follow medication-label instructions, compliance with written instructions on care, and unwillingness to participate in clinical trials.[4] Many patients with chronic diseases are poorly educated and socio-economically disadvantaged; for these reasons, it is important to include healthcare literacy as a part of chronic disease management. "Improving healthcare communication is one way in which we can get information to people that will help them change their behavior, but it cannot be the only intervention."[5] Self-management and compliance play a vital role in any treatment plan.

Healthcare Communication Needs Assessment Surveys

Demographics: The following is a report of 54 Healthcare Literacy Needs Assessment Surveys collected in the South. Respondents were 38.6% male, 61.4% female. Ages ranged from 14 to 56; 52.3% of respondents were between ages 14–18, and 72.7% were age 23 and under. Data Summary: 20.5% (± 11.93%) indicated they would rather practice skills alone; 77.3% (± 12.38%) indicated they would rather practice skills with a healthcare worker. 81.8% (± 11.40%) indicated they use a computer; 88.6% (± 9.39%) indicated they liked using a computer. 95.5% (± 6.13%) indicated that using a computer helps them to learn. 88.6% (± 9.39%) indicated support for a library or other place of information, while 13.6% (± 10.13%) preferred healthcare workers take care of this. 40.9% (± 14.53%) indicated that they believed more training was needed for doctors, nurses, and other healthcare workers to communicate better with sickle cell patients.[6]

Inclusion of Adolescents and Young Adults in the Healthcare Process

African-American adolescents faced with sickle cell anemia have to manage a period in their lives which is already filled with massive cognitive and emotional growth, while fielding community and peer-based expectations at the same time. Having to cope simultaneously with a chronic disease is exigent at the very least.[7] Delayed development of primary and secondary sexual characteristics, decreased coping strategies, lower self-esteem, and strained parent–child relationships (as primary psychosocial aspects) are frequently overlooked in managing the care of adolescents with sickle cell disease.

Delay in attaining pubescent height and weight development, and often cessation altogether, places adolescents with SCD in an especially precarious position. During a life phase fraught with intensified peer relationships, goals of independence from caregivers, and desires to pursue self-control, adolescents with SCD must also manage pain, alter coping strategies, and face the possibility of morbidity. Rosie B. Pinckney and Gail W. Stuart[7] cite a study by Burlew *et al.*, which concluded that psychosocial factors could negatively influence how adolescents manage SCD. They further suggested that psychosocial factors have a more significant influence on adolescents' well-being than do the biomedical factors of SCD and other chronic illnesses.

Britto *et al.*[8] argues that there are sharp contrasts between how adults and adolescents rate healthcare quality. They suggest that many of adolescents' psychosocial challenges can be addressed through their physicians' practice methods and philosophy. They also report that healthcare for adolescents with chronic illness must be significantly more patient-centered. In studying adolescents with chronic illness and their healthcare preferences, Britto *et al.*[8] found that adolescents rated interpersonal interaction and care as important in their assessment of quality of healthcare. Further the adolescent respondents ranked physician honesty and attention to pain as two of the most important issues. Britto *et al.*[8] cited numerous studies, which demonstrated that adolescents thought physician communication using "straight talk," avoiding false promises or giving them "high hopes," was pertinent to their perception of honesty.

Because adolescents with SCD experience considerable pain, they are especially concerned about additional pain that may be associated with medical procedures. Physicians' acknowledgement of patients' pain, and their willingness to proactively discuss and educate adolescents on traditional and nouveau pain management methods, are also considered important.

The Impact of Previous Healthcare and Research Practices

The overall quality of healthcare delivery has improved for individuals with SCD over time, as manifested by the recently reported increase in life expectancy. This is largely related to efforts stemming from Comprehensive Sickle Cell Centers, which have changed the standards of care. These efforts include mandatory newborn screening, penicillin prophylaxis, improved immunization standards, hydroxyurea therapy, and parent and physician education.[9] But while their life expectancy has increased, these individuals are also experiencing more chronic complications associated with sickle cell disease, such as kidney failure, chronic lung disease, avascular necrosis of the hip and shoulder joints, and chronic leg ulcers.[9] Despite the efforts of many to improve its delivery, disparities in healthcare remain a problem in the sickle cell community for a number of reasons. In many rural and urban areas in the United States that are predominantly African-American and socio-economically poor, there is still a shortage of healthcare manpower.[10] Also, a basic distrust continues to exist between healthcare providers and patients, while adequate health insurance coverage is not available for disabled adults with a chronic illness such as SCD. Identification of the problems outlined above is not enough. Without rectification, they are likely to increase healthcare cost, thereby adversely affecting both quality of care and length of life for individuals with SCD.[11]

The Tuskegee Syphilis Study

During the years 1932 to 1972, the United States Public health Service (PHS) conducted an experiment on 399 African-American men in the late stages of syphilis. These men were never told what disease they were suffering from or how serious it was; they were informed only that they had "bad blood." As they were sharecroppers from one of the poorest counties in Alabama, for the most part these men were considered illiterate. Data for the experiment was to be collected from their autopsies, a fact never revealed to the participants; they were not informed that their doctors had no intentions of achieving a cure.[12] The tendency of African-Americans to avoid participating in medical research can be directly associated with the history of the Tuskegee Syphilis Study.

Dr. Vanessa Northington Gamble, a physician and medical historian, chaired the Presidential Committee on the legacy of Tuskegee that secured an apology from the government. In the commentary she wrote to mark the 30th anniversary of the news report that unmasked the study, Dr. Gamble said that many African-Americans pointed to the syphilis study as a reason why they won't participate in clinical trials today,[13] or donate organs and, as in the more recent case of workers at the Brentwood Post Office in Washington, D.C., are wary of being vaccinated against anthrax.

Mistrust of the Healthcare Community

Although others point to more complex reasons for the trend of mistrust in the African-American population, ranging from socioeconomic factors to a lack of community-based outreach educational programs, Peter Clarke disagrees.[14] This medical ethics expert and theologian contends that culture plays a major role in what he refers to as a "legacy of mistrust" within the African-American community toward both the medical profession and public health authorities. Such mistrust encompasses two areas: healthcare delivery and medical research. Doctors, medical researchers and patients' advocates agree that a pervasive mistrust of the medical establishment exists in the African-American community, which has caused a significant number of African-Americans with HIV/AIDS to forego valuable protease inhibitor treatment.[15]

The Tuskegee experiment was the most salient symbolic source of black suspicion in the 20th century, but the debacle of sickle cell screening in the 1970s is another factor that has bred mistrust.[16] In 1910, Dr. James Herrick described sickle cell anemia in a West Indian student[17]; however, argues that SCD was not made a high priority until President Richard Nixon announced this in his special message to Congress on healthcare on March 2, 1972. Nixon included SCD in his strategy to improve healthcare and have it delivered at a reasonable cost to every citizen, regardless of income or area of residence. In 1972, he allocated $10 million to attack the problem of SCD.

An important goal of the national sickle cell screening initiative was to get African-Americans to change their mating behavior, and focus was on the sickle cell trait carrier to accomplish this goal.[16] However, due to lack of clarity about disease versus sickle cell trait, the initiative promoted confusion. Some publications did not know the difference between sickle cell anemia and sickle cell trait. Such erroneous information circulated throughout the community; this prompted instances of discrimination leading to misunderstanding and resentment, further encouraging the African-American community not to trust the medical profession. Medical experts lament the levels of mistrust that are keeping potentially life-saving drugs out of the hands of a population that is disproportionately affected by AIDS and HIV.[15]

The Role of National Institutes of Health (NIH) in Promoting Participation in Clinical Trials

Human clinical trials are critical to improving medical treatment and finding cures for chronic, debilitating diseases.[18] The American Association of Medical Colleges (AAMC) believes that clinical research is a component of medical health research intended to produce knowledge valuable to understanding human disease, preventing and treating illness, and health promotion. In a National Clinical Research Summit (AAMC) stakeholders described "clinical research as the neck of the scientific bottle, through which all basic biomedical research discoveries must pass if they are to be transformed into cures."[19] The biggest obstacle in encouraging individuals to participate in clinical research studies is lack of awareness about the role these studies play in the development of new drugs.[20]

The National Institutes of Health (NIH) has an established record of funding, designing, reviewing, and monitoring clinical trials; it has the scientific expertise and the staffing to assume responsibility for the appropriateness of clinical trial designs.[21] From a community perspective, NIH's most important role is to monitor the safety of individuals participating in various trials. Every clinical trial in the United States must be approved and monitored by an Institutional Review Board (IRB) to make sure that risks are as low as possible and worth any potential benefits.

An IRB is an independent committee of physicians, statisticians, community advocates, and others which ensures that a clinical trial is ethical, and that the rights of study participants are protected. All institutions that conduct or support biomedical research involving humans must, by federal regulation, have an IRB to initially approve and periodically review the research.[22] In the past, however, minorities were exploited and abused in several clinical trials, which has created a great deal of resentment and suspicion of researchers among such individuals. This in turn results in fewer minorities volunteering for clinical trials. Obtaining the trust of minorities (and women are included here) so that they will participate in clinical trials, is a task proving problematic and challenging to many investigators.[22]

Even though it is a major challenge to include these populations in clinical trials, many sponsors of clinical trials have realized the value of their inclusion. More specifically, the NIH developed a policy stating that "women and members of minority groups and their subpopulations must be included in all NIH-supported biomedical and behavioral research projects involving human subjects, unless a clear and compelling rationale and justification establishes to the satisfaction of the relevant Institute/Center Director that inclusion is inappropriate with respect to the health of the subjects or the purpose of the research."[22]

The protection of individual confidential health information is another way by which NIH promotes participants' confidence in clinical trials. NIH issues certificates of confidentiality (CoC) to protect the privacy of research subjects. These prevent investigators and institutions from being compelled to release information that might be used. Such research allows the investigator, and others who have access to research records, to refuse to disclose identifying information in any civil, criminal, administrative, legislative, or other proceeding, whether at the federal, state, or local level. The availability of such a document frequently helps an investigator to recruit participants for protocols that addresses issues of a particularly sensitive nature.[23]

Summary

Much can be done to reduce the burden of chronic diseases, such as SCD, and achieve *Healthy People 2010* objectives, through a multifaceted approach that includes clinical research, community-based initiatives, comprehensive care, self-care, and other professional care. In recent years, national organizations, funding agencies, and researchers have all called for renewed focus on an approach to public health research recognizing the importance of social, political, and economic systems to health behaviors and outcomes. The need for this renewed focus has risen from many converging factors, including our increased understanding of the many and complex issues that affect health; the importance of both qualitative and quantitative research methods; and the need to translate the findings of basic, interventional, and applied research into practice and policy changes.[24]

Breakthroughs in basic biomedical sciences, including human genomics, stem cell biology, biomedical engineering, molecular biology and immunology, over the past five decades have provided an unprecedented supply of information for improving human health.[25] Translating the information gained through these basic discoveries into knowledge that will affect clinical practice and, ultimately, human health requires research involving human subjects and human populations as well as development of improved health services based on a particular research interest.[25] To achieve this goal researcher will need human subjects to meet the next challenge. It is also important that African-American be included as a part of the various studies since they are represented in many major chronic diseases. Although the infamous Tuskegee Study of untreated syphilis has plagued the African-American population for many years, leading to the belief that African-Americans are used as guinea pigs for research, this belief has continued throughout the decades and has served as an obstacle to providing care for many African-Americans across America. In a report in the National Medical Association Journal, Shavers *et al.*,[11] referred to the Tuskegee study as a prominent source of mistrust of the health researchers among African-Americans. However, Shavers supports the need for healthcare providers to be aware of this problem and confront the issue of distrust of healthcare providers. The belief is the best tactic to reverse these feelings of mistrust is to create an environment with the patient, providers and community that will foster positive life experiences through education on the benefits of positive clinical outcomes.

Medical research is the backbone of medical innovations enjoyed today as standard medical practice. Most medications and devices physicians use today have been developed through former research. Without medical research, we would be unable to advance medicine and improve the care we provide to patients. To accomplish what lies ahead... it will take a village.

References

1. Sheham M (1973). *Blacks and American Medical Care*. Minneapolis: University of Minnesota Press.
2. Mitchell H, Lewter N (1986). *Soul Theology: The Heart of American Black Culture*. San Francisco: Harper & Row.
3. Shohet L, Malo D, Langlais O (2001). *Report on the Needs Assessment, Informed Consent Form*. The Centre for Literacy of Quebec.
4. Andrus M, Roth M (2002). A review, pharmacotherapy. *Health Lit* **3**:291.
5. Weiss BD (2001). Health literacy: an important issue for communicating health information to patients. *Chin Med J [Zhonghua Yi Xue Za Zhi] (Taipei)* **64**(11):603–608.
6. Peterson RM, Owens W, Kitt D (2003–2005). *Education is the Key to Successful Newborn Screening Follow-up, Adult Transition to Adult Care and Understanding Healthcare Literacy Project* (MCHB Newborn Screening Grant).
7. Pinckney RB, Stuart GW (2004). Adjustment difficulties of adolescents with sickle cell disease. *J Child Adolesc Psychiatr Nurs* **17**(1):5–12.
8. Britto MT, DeVellis RF, Hornun RW, DeFriese GH, Atherton HD, Slap G (2004). Health care preferences and priorities of adolescents with chronic illnesses. *Pediatrics* **114**(5):1272–1280.
9. National Institutes of Health, National Heart, Lung and Blood Institute (Revised December 1995). Management and therapy for sickle cell disease. *NIH Publication* **95**(2117):7–12.
10. Tremback T, Dreger V (2002). Optimize patient health by treating literacy and language barriers. *ACORN J* **75**:282–287.
11. Shavers VL, Lynch CF, Burmeister LF (2000). Knowledge of the Tuskegee Study and its impact on the willingness to participate in medical research studies. *J Nat Med Assoc* **92**(12):563–572.
12. Jones H (1993). *Bad Blood: The Tuskegee Syphillis Experiment — A Tragedy of Race and Medicine*, Expanded Edition. New York: Free Press.
13. Gamble V (1993). A legacy of distrust: African-Americans and medical research. *Am J Prevent Med* **9**(35).
14. Clark PA (1998). A legacy of mistrust: African-Americans, the Medical Profession and Aids, *Linacre Quart* **65**(1):66–88.
15. Richardson L (1994, November 12). Black Images Stir Up Strong Emotions. *New York Times*, p. C14.
16. Dula AP (1994). African-American suspicion of the healthcare system is justified: what do we do about it? *Cambridge Quart Healthcare Ethics* **3**:347–357.

17. Savitt TL, Goldberg MF (1989). Herrick's 1910 case report of sickle cell anemia. The rest of the story. *JAMA* **261**(2):266–271.
18. Weijer C (1999). Selecting subjects for participation on clinical research: one sphere of justice. *J Med Ethics* **25**:31–36.
19. Cohen JJ (2000). National clinical research summit unites stakeholders, sparks legislative agenda. *AAMC Reporter* **25**:4.
20. *Faster Cures: Patients Helping Doctors* (2006). Retrieved January 1, 2006, from http://www.patientshelpingdoctors.org/
21. Drazen JM (2003). Controlling research trials. *N Engl J Med* **348**(14).
22. US Department of Health and Human Services (1994). National Institute of Health guidelines on the inclusion of women and minorities as subjects in clinical research. *Fed Regist* **59**(14508).
23. US Department of Health and Human Services (2002). National Institute of Health Statement of Certificate of Confidentiality, Public Health Service Act 42. [*USC* 241(d) section 301(d)].
24. US Department of Health and Human Services (2002). *Healthy People 2010 Objectives*. Hyattsville MD: Centers for Disease Control and Prevention, National Center for Health Statistics.
25. Sung NS, Crowley WF Jr, Genel M, Salber P, Sandy L, Sherwood LM, Johnson SB, Catanese V, Tilson H, Getz K, Larson EL, Scheinberg D, Reece EA, Slavkin H, Dobs A, Grebb J, Martinez RA, Korn A, Rimoin D (2003). Central challenges facing the national clinical research enterprise. *JAMA* **10**:1278–1287.

21

Beyond National Borders: A Global Perspective on Advances in Sickle Cell Disease Research and Management, and New Challenges in the Genome Era

by Solomon F. Ofori-Acquah and Kwaku Ohene-Frempong

Introduction

The genetic disorders of hemoglobin sickle cell disease (SCD) and thalasaemia are the most common single gene diseases in man.[1,2] They are inherited in an autosomal recessive mode, and pose a major global public health problem. Worldwide estimates of the frequency of these disorders have been published since the early 1980s by the World Health Organization (WHO), and are updated by the working group on hemoglobin disorders under the auspices of the Hereditary Disease Program.[3] These reports indicate a growing burden of SCD on health services worldwide, one that is poised to continue given the current trend in world population growth.

The clinical expression of SCD is highly variable. Therefore, effective preventative and control programs will depend on establishing protocols with markedly lower risk-benefit ratios than the disorders. While the mutation is most prevalent in Africa, population migration has spread it to other geographic regions, so it is now distributed throughout America and Europe.[4] Several asymptomatic strategies have been developed and successfully implemented in these immigrant populations that may find widespread use in sub-Saharan Africa. Genetic control programs in the Middle East emphasizing pre-marital counseling have reduced incidence of SCD, while information from the Human Genome Project offers prospects for predicting disease phenotype and individualizing clinical management. Advancement in personalized treatment and prediction of disease phenotype will depend on significant improvement of our understanding of the clinical heterogeneity of this condition, which can reasonably be achieved by studying large cohorts of patients of various genetic backgrounds. Such population projects will offer the opportunity for international collaborations, critically important for developing globally equitable health service planning for SCD.

Global Health Burden of Sickle Cell Disease (SCD)

It is estimated that 6–7% of the world's population are carriers of hemoglobin disorder, and that nearly 0.35 million babies are born each year with severe forms of this condition.[2,5] A large proportion of affected births occur in sub-Saharan Africa due to the high frequency of the sickle cell mutation in this region. Indeed, because of the endemic gene frequencies of this mutation, SCD accounts for three-quarters of global hemoglobin disorders (Fig. 1).

Africa

There is three epicenters of the sickle cell mutation in Africa, each with an outlet to the Atlantic Ocean and flanked by landlocked regions of low gene frequencies (Fig. 1). The most densely populated epicenter lies northward of the Gulf of Guinea, in present day Benin, Ghana, Nigeria and Togo, where about 91,000 babies are born each year with SCD. The population in this sub-region will increase to about 374 million in 2050, when Nigeria will become the fifth most populated nation on earth with 307 million people (Table 1).[6]

A national survey of 16,000 teenagers and adults from various ethnic groups confirmed historical estimates that 25% of the Nigerian people carry the sickle cell mutation,[4] and a Ghanaian newborn screening of 178,254 births revealed a heterozygote frequency of 13.3% (Ohene-Frempong, unpublished) (Table 2). The data from both studies underscore the importance of micro-mapping the sickle gene in individual countries to more accurately evaluate the health burden of the condition in each country.

Evidence from mitochondria DNA shows that individuals from Atlantic West Africa and Central West Africa have been apart from each other for over 40,000 years.[7] A second epicenter of the mutation is in central West Africa, bordered on the Atlantic coast by Angola and extending northward to Gabon and eastward to the Democratic Republic of Congo and Republic of the Congo. An estimated 26,000 affected babies are born annually in this sub-region. The Democratic Republic of Congo has the highest frequency of the sickle gene in this area, although wide variations exist among different

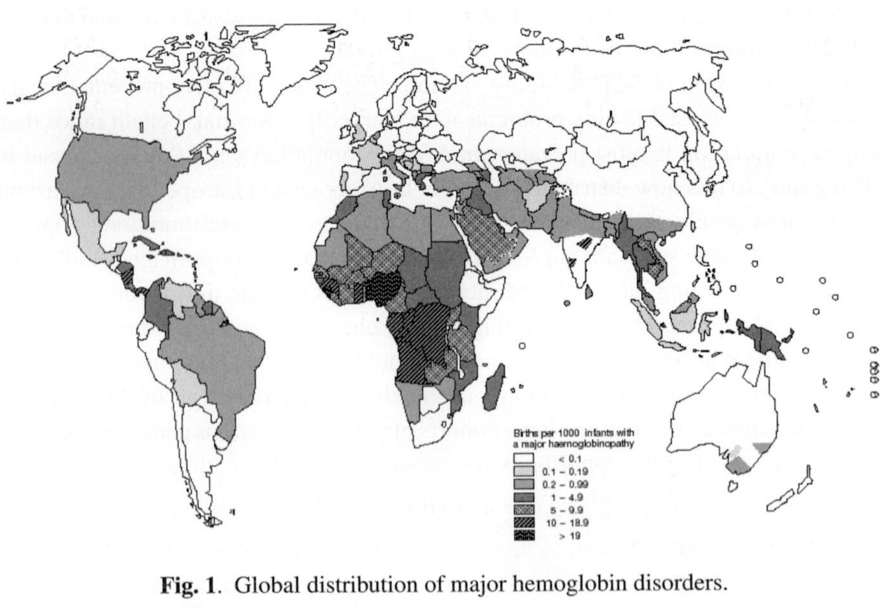

Fig. 1. Global distribution of major hemoglobin disorders.

Table 1. Burden of health indicators for the WHO African region.

Country	Population* (million)	Fertility rate	Ann. births (× 1000)	Het. freq. (%)	No. of births affected
Algeria	32.3	2.8	960	1.0	58
Angola	13.3	7.2	356	25	5554
Benin	7.3	5.7	177	22	2251
Botswana	1.7	3.7	44	1	0
Burkino Faso	13.6	6.7	304	5–25	682
Burundi	6.2	6.8	187	10	468
Camerron	16.1	4.7	375	5–25	2111
Centr. Afr. Rep	3.7	5	107	12	385
Chad	9.5	6.7	205	1–20	513
Comoros	0.7	4.9	18	4	7
Congo	3.8	6.3	72	25	1123
Côte d'Ivoire	16.9	4.8	407	2–15	1669
Dem. Rep. Congo	58.3	6.7	1219	5–40	19,016
Equ. Guinea	0.5	5.9	16	20	160
Ethiopia	72.4	6.2	1632	0	0
Gabon	1.4	4	18	25	280
Gambia	1.5	4.8	30	15	170
Ghana	21.4	4.2	592	12	2131
Guinea	9.2	5.9	243	10–30	2957
Guinea Bis.	1449	7.1	24	2–24	48
Kenya	32.4	4.1	961	3	1180
Liberia	3.5	6.8	105	15.0	851
Madagascar	17.5	5.7	415	2–20	1040
Malawi	11.9	6.1	304	8	490
Mali	13.4	7	317	10.0	770
Mauritania	3.0	5.8	87	5.0	55
Mauritius	1.2	2	22	2	5
Mozambique	19.2	5.7	395	1–40	1240
Namibia	1.9	4.6	32	0–9	7
Niger	12.4	8	290	5–24	1595
Nigeria	137.3	5.5	4103	10–30	62,728
Rwanda	8.4	5.8	269	7.0	330
Senegal	10.9	5	331	5–20	1589
Sierra Leone	5.2	6.5	165	17–30	2523
Tanzania	36.1	5.2	885	5–20	4980
Togo	5.6	5.4	131	6–26	459
Uganda	26.1	7.1	628	9–20	4020
Zambia	10.9	5.7	280	10–21	1580
Zimbabwe	12.7	4	357	2–6	140

*Estimated published 2004.

ethnic groups in that country. Population growth in this region is expected to double in 20 years to 136 million, and this trend will continue until 2050.

The third epicenter is on the western tip of the continent, lying northwest from the trans-Atlantic focal point. It is located in present day Gambia, Guinea, Guinea Bisau, Liberia and Sierra Leone, and

Table 2. Hemoglobin phenotypes in a newborn screening in Ghana.*

Phenotype+	F	FA	FAS	FAC	FC	FCA	FS	FSA	FSC	Total
No.	18	134,201	23,640	15,155	1903	6	1847	6	1478	178,254
%	0.0	75.29	13.27	8.50	1.07	0.0	1.04	0.0	0.83	100

*Cumulative data from 1995–2004.
+In cases with multiple hemoglobin types, the sequence of the types shown represents a decreasing proportion of the total hemoglobin found by electrophoresis.

the heterozygote frequency reaches a maximum 30% in this sub-region. An estimated 11,000 babies are born each year with SCD in this sub-region.

The sickle mutation is found at a lower frequency in North Africans compared to sub-Saharan Africans. However, co-inheritance with β-thalasemia mutations, which are slightly more prevalent in this region, results in over 3000 births annually that are affected with some form of SCD. Presence of the sickle mutation in North Africans is likely due to gene flow during the ancient north-south trans-Saharan trade and the Arab slave trade.[8,9] The same mechanism likely accounts for the presence of the sickle mutation among native populations of Israel,[10] Greece,[11] Turkey,[12] Sicily[13] and Portugal.[14] In addition to the three epicenters of the mutation, there are several other countries in sub-Saharan Africa with at least 5.5 births per 1000 affected with SCD (Fig. 1). Current estimates show that 90% of the increase in world population projected by 2050 will occur in developing countries, highlighting an inevitable increase in the global health burden of SCD.

Despite the magnitude of the burden of sickle disease on several countries in Africa (Fig. 2), its effect on healthcare services and national productivity in this region remains virtually unknown. A major reason for the lack of healthcare planning for SCD, and indeed other genetic disorders in sub-Saharan Africa, is due to the overwhelming influence of infectious diseases, which remain the major health concern in most health ministries. Moreover, the majority of fatalities among children with SCD are caused by complications of malaria. WHO estimates indicate infection will remain the major cause of mortality in Africa at least until 2025. As a result, until infectious diseases and in some cases malnutrition are brought under control, the burden of SCD in Africa will remain obscure. Currently many African countries do not have national programs for management of SCD, and therefore its incidence will continue to increase with population growth.

In the absence of nationally coordinated programs, financially endowed patients in sub-Saharan Africa seek healthcare services in the United States and Europe; therefore there is a financially-driven healthcare disparity in SCD in Africa.

America

Formal description of SCD was first reported in the United States; the high frequency of the mutation among Americans of African descent is due to gene flow during the 400 years of the Atlantic slave trade, when an estimated 10 million Africans forcibly migrated to the United States.[15] It is estimated that 70,000 to 100,000 individuals are affected with SCD in this country.

Several government-sponsored programs have been established, mostly in major United States cities, and a pilot study in southern Alabama has demonstrated the feasibility of extending routine clinical consultation to those unable to travel to comprehensive centers in the major cities. However, patients in rural areas do not have access to potentially life-saving sophisticated diagnostics, such as transcranial doppler screening, located in major city centers. Nonetheless, the telemedicine experience

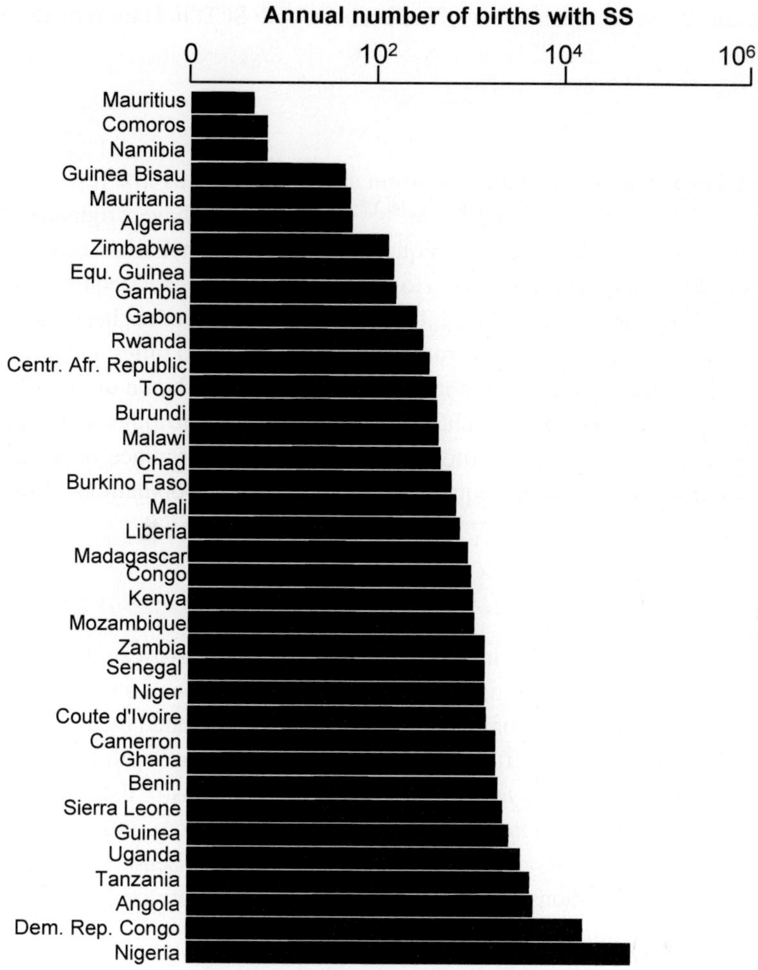

Fig. 2. Annual number of births with homozygous SS disease in the World Health Organization Africa region.

in rural Alabama has been successful and can be adopted by inter-government agencies to extend basic clinical expertise to other communities in American. For instance, Hispanic telemedicine services may be an attractive resource for clinical programs in Central American countries with high frequencies of the sickle cell mutation. (The heterozygote frequency in this region ranges from 8% in Costa Rica to 15% in Panama.)

A significant percentage (2.7% to 10%) of the populations in Brazil, Colombia, French Guyana, Guyana, Peru, Surinam, and Venezuela carry the sickle gene. The highest numbers of affected births in South America occur in Brazil and Colombia, where 1500 to 3000 children are born each year with SCD.[3] A program for standardizing diagnosis and managing the condition has been established in Brazil, and may provide a blueprint for other countries in the region.

Prevalence of SCD is high in the Caribbean islands, where heterozygote frequencies range from 3% to 7% in different regions of Cuba and up to 14% in the Bahamas.[3] Despite these high gene frequencies, relatively few children are born with SCD annually in most Caribbean islands because of their small population sizes. Affected births in Cuba have fallen since establishment of a sickle cell prevention program; however, the number of babies born with SCD in Haiti remains high, estimated at 1000 per year.

Asia

A fourth major focus for the sickle mutation originates in Asia. It was first reported in India among people in Nilgiris, in the state of Tamil-Nadu[16,17] and shown to have originated independent of the mutations in Africa.[18,19] The highest frequencies are found in India's central states, such as Madhya Pradesh (20%) and Orissa (25%). However, significant gene frequencies are also found in western and south-western states. Indeed, the sickle cell mutation has been identified in several widely dispersed tribal groups, separated from each other by ~2000 miles and with no present day contact. This finding substantiates the hypothesis that the ancestral home of the tribal communities may have been at the margins of the Indus river which is close to other populations who carry the sickle cell mutation, such as the Shiite Arabs of the eastern province of Saudi Arabia.[20] The sickle mutation is also found at high frequencies in central Asian populations of the former Soviet Union.

Europe

The risk factor for SCD is 2% to 5% in immigrant populations of African, Caribbean and Turkish ethnicity living in the United Kingdom, France, Netherlands, Belgium, and Germany. These populations are typically concentrated in major European cities, as illustrated by the demographics in the United Kingdom. The frequency of affected births is 0.2 per 1000 in South East London, compared to that in outlying counties which is less than 0.01% per 1000 births.[21] Estimates validated against community screening programs indicate that in the United Kingdom, more than 3000 babies are born annually with the sickle trait, and nearly 200 babies with the disease.[21] The major problem for these immigrant ethnic populations in Europe is access to health services, since virtually all levels of healthcare are available through national health programs.

Middle East

Given the relatively low gene frequency and historical records of trade with Africans, it is likely that gene flow accounts for the presence of the sickle gene in the Middle East. The number of children born annually with SCD in this region is estimated at 6000, at least 50% of these in Saudi Arabia.[3] A mandatory pre-marital screen for the sickle cell mutation for all aspiring couples has been established in Saudi Arabia, intended primarily to promote awareness of SCD among the youth, with the aim of reducing the number of at-risk pregnancies.[22,23] The highest heterozygote frequency (11.0%) in this region is in Bahrain, although the number of affected newborns in that country has fallen dramatically, from 2.1% of the total population in the 1980s to 0.9% in 2005. This decrease is due to an intensive preventative and control program involving antenatal screening, newborn testing, premarital counseling, and student screening projects.

Micromapping

From its native origins in Africa and India, the sickle gene is now widely distributed throughout the world, and has become a major public health concern in many countries. The gene frequencies used in this chapter are WHO estimates that may not accurately reflect the genetic burden in individual countries. Nonetheless, they provide a useful reference for planning health services. Some of these data were derived from immigrant populations living in developed countries; since these populations do not always reflect the diversity of ethnic groups in their native countries, such data need to be treated with caution. Indeed, we have found consistently lower numbers of affected births in the Kumasi metropolitan area in the Ashanti region of Ghana, compared to the estimates for Ghana as a whole derived from historical studies performed in other regions and abroad (Table 2).

These discrepancies underscore the importance of micromapping the sickle mutation in individual countries. There are, however, ethical, legal and social implications (ELSI) in genetic screening that need to be recognized by individual investigators and government-sponsored programs. Genetic counseling and capacity-building programs need to be incorporated into sickle cell research. Developing countries have well-established administrative infrastructures for controlling infectious diseases that can be expanded to include genetic testing at birth. But while recognizing that information gathered from such studies will serve an important community need in the planning of healthcare services, individual genotypes need to be accorded privacy protection.

Genetic Background of the Sickle Cell Mutation

First documented cases of sickle cell anemia in Africa were described in children from two ethnic populations in the western and eastern corners of the continent. Using recombinant DNA technology, it is now possible to identify ethnic-specific origins of the sickle cell mutation in the laboratory.

The β-globin gene is located on the short arm of chromosome 11, along with 134 million other bases. Variations in non-coding nucleotides flanking the globin genes have been used to track the origin of the sickle cell mutation and its flow to other geographical regions in the world. At the base level, these variations are known as single nucleotide polymorphism (SNP), and blocks of SNPs constitute a haplotype. A haplotype consisting of eight SNPs distinguishable by restriction length polymorphism, spanning $\sim$30,000 bases of the β-globin gene cluster has been studied extensively, and was used to determine the ancestral origin of the sickle cell mutation in individuals from different parts of Africa and Asia.[24,25] These studies led to identification of five common haplotypes named Senegal, Benin, Central African Republic/Bantu, Cameroon, and Asia/Indian/Saudi-Arabia.[24,25]

"Hybrid" and Functional Haplotypes

Individual SNPs in the sickle β-globin haplotype block lie in close proximity, and therefore have a very low probability of separating from each other during DNA replication or repair. However, several sickle haplotypes have been identified with SNPs typically associated with multiple ancestral chromosomes. The prevalence of these "hybrid" haplotype chromosomes is not known, but there is emerging consensus that they are useful for dissecting phenotypic traits associated with individual ancestral chromosomes.

We addressed this issue by studying cohorts of sickle cell patients living in Jamaica and South East London, the latter being recent immigrants from West Africa. Single strand conformational polymorphism was used in this study to identify variations in two SNPs at -1280 (G/A) and -1224 (T/C) in the γ globin promoter, and two microsatellites in the β-globin locus control region and β-globin promoter.[26,27] These variations are located in regulatory DNA elements that influence gene expression; therefore, the term "functional haplotype" was coined to describe their pattern of polymorphisms.[28] The patient majority in the South East London cohort was homozygous for functional alleles typically found on chromosomes with the Benin haplotype. Interestingly, the same frequency of the Benin functional haplotype was found in the Jamaican cohort, confirming that Black Jamaicans share the same ancestral chromosomes as compared to recent immigrants from West Africa. The Jamaicans had a significantly higher number of "hybrid" chromosomes with discordant polymorphic functional alleles at the LCR and the globin promoters, and while most of these hybrids harbored markers of African origin, a small minority contained functional alleles not found in African populations. This suggests other populations that migrated from Africa many years ago may have unique variations of the sickle chromosome. These mosaic sickle chromosomes are not expected to influence the natural history of SCD in different communities because of their relatively small numbers; however, they may uniquely ameliorate the severity of the disease in individual patients.

Progress in Health Management: From the Outside Looking in

Our understanding of the pathophysiology of SCD has improved; with this knowledge and training of specialist health workers, and the establishment of comprehensive centers, there have been significant advances in the management of individuals with SCD in the developed world. But despite these improvements, life expectancy of African-Americans with SCD is markedly lower than their counterparts with a normal hemoglobin genotype: It is 42 years for males, and 48 years for females, among sickle cell anemia patients in the United States. The comprehensive sickle cell centers are arguably the major sources of improvement for patient care in the United States, and there are several indications that validated United States programs can be exported to other nations once critically important cultural alterations have been made.

Pharmacologic Intervention

Children with SCD now receive prophylactic therapy with oral penicillin, which has reduced childhood mortality in the United States. However, the survival benefit of penicillin in countries where pneumococcal septicemia is not the major cause of bacteria in children with sickle cell remains to be determined.

Adult mortality has been reduced in severely affected patients by the administration of hydroxyurea, a chemotherapeutic agent that augments fetal hemoglobin expression and reduces white blood cell counts. Because of its myelotoxic property, patients on this drug are monitored carefully and receive routine hematology check-ups. This is an essential aspect of the management that needs to be widely publicized in other countries where government-backed, coordinated programs for hydroxyurea do not exist, yet where access to hydroxyurea may be readily available through on-line vendors. There is a need for national governments and international agencies to establish educational programs through the media to educate the public about the potential dangers of self-medicating hydroxyurea.

Blood Transfusions

An estimated 0.7% of children with SCD develop stroke in the first 20 years of life, with a peak incidence in those aged 5–10 years. Blood transfusion therapy has reduced the frequency of stroke in the United States, although there is controversy about the use of transfusion for maintenance therapy in other developed countries. The WHO estimates that 80% of the world population living mostly in developing countries has access to only 20% of the world supply of blood, whereas 17% of the population living in developed counties has access to 60% of the world blood supply. Moreover, the majority of donors in the developing world are family or replacement donors, who have a higher incidence of transmitting transfusion-borne infections than voluntary non-remunerated donors in developed countries.

Estimates from the WHO indicate that 5–10% of HIV infections worldwide are transmitted through transfusion of infected blood products. Therefore, the risks of acquiring serious infections such as HIV from transfusions detract from the potential benefit of transfusion therapy in developing countries. There is global understanding that blood must be safe from contamination in every country to prevent pandemic transmission of infectious diseases through the high level of human and commercial traffic across national borders. The WHO has established an international forum for Global Collaboration for Blood Safety that offers advice and consultation to all member states, and provides training assistance to developing countries. However, there will need to be significant shifts in attitudes toward donating blood in developing countries for blood transfusion to become a viable global therapeutic option for SCD.

Transplantation

Bone marrow transplantation has shown success in small numbers of patients with HLA-compatible donors, and offers potential cure for SCD, but it is very cost-intensive and has not found widespread application in rich countries; based on this experience, it is unlikely to be the choice for managing SCD in developing countries. The limitations imposed by HLA-compatibility can be overcome by using unrelated donor cord blood, although transplantation with this tissue is more successful in children than in adults. Cord blood can also be transplanted in utero prior to development of immunological competence. Indeed, in utero transplantation using this strategy has immense potential, and could be developed as a universal cure for SCD. However, it will require early prenatal diagnosis to take advantage of fetal tolerance.

There have been attempts to diagnose SCD in early gestation using fetal cells obtained by ultrasound-guided ceolocentesis, but this procedure currently carries a high rate of spontaneous miscarriages and will need to be substantially improved for routine application.[29] Enrichment of fetal genetic material from other sources, including extracellular DNA in maternal plasma, is being optimized for early prenatal diagnosis of other disorders; these may find useful application in prenatal transplantation in SCD.

Global Collaborative Research

A 21st century challenge in SCD, global and regional collaborative research efforts have played significant roles in advancing major landmarks in SCD; now there is a need for balanced international collaborations to address two major challenges in SCD in the 21st century. First, there is mounting pressure to predict the clinical course of SCD as part of prenatal diagnosis and prior to embarking on high-risk therapeutic interventions, and this demand will grow as clinical decision-making based on individual genomes become a reality. Second, there is increasing evidence implicating genetic heterogeneity in drug metabolism and efficacy, suggesting the outcome from conventional and experimental therapy may be predictive in the near future. To advance these emerging challenges will require significant improvement in our understanding of the phenotypic diversity of the disease, and the varied genetic backgrounds of the mutation. These efforts and future studies for mapping genes that influence the phenotype of SCD would benefit from studying large and genetically homogenous pedigrees that can reasonably be found only in sickle endemic countries.

Capacity-Building in Epicenters of the Sickle Cell Mutation

There are now a number of well-trained scientists, and the emergence of research-oriented sickle cell clinics in Africa, auguring well for more equitable collaborations with investigators in developed countries. Indeed, the prospects for such international collaborations have been boosted through joint sponsorships of workshops and conferences by investigators from developed countries and African National Health Ministries. These international meetings have significantly increased awareness of SCD among the general population and local politicians in several sub-Saharan African countries, and led to important new initiatives in neighboring countries. A formal arrangement between various organizations in these countries will help establish sustainable programs in Africa.

There is little doubt that the human genome project has created a genetically savvy population throughout the world. Many investigators in developed countries expect the post-genome era to help eradicate misconceptions about etiology of SCD and genetic disorders generally, create a more receptive mass audience for educational campaigns, and help attract both local and international funds to develop basic infrastructure for genetic services and research.

Perhaps the most pertinent expectation by geneticists in developed countries is use of the genome in clinical decision-making. Importantly, SCD is recognized as a prototype for these aspirations, and there is also a realization that a significant improvement in our understanding of the variable pathogenesis of this condition might be a prerequisite for genome-based clinical management. International collaborations involving networks of sickle cell clinics in sub-Saharan Africa with research programs in developed countries offer an opportunity to realize these regionally defined aspirations.

Each of the major national research agencies in the United States and Europe has played a significant role in training scientists from developing countries in sickle cell research. However, research scientists returning home face an impossible task to establish independent programs; consequently, the full benefit of their training and the investment made by these agencies has not yet been realized. This is a major cause of "brain drain" to the United States and Europe, which ironically means that several developing countries of low national income are currently subsidizing scientific infrastructure in richer developed countries. The challenge in this field is to convince national governments in developing countries and international grant agencies in the United States and Europe, that there is a mutual benefit in creating programs to attract more scientists from developing countries to return home and build highly competitive national sickle cell research institutions in Africa. Such programs can be used as cornerstones to support molecular genetic services for other causes, including infectious diseases.

A database of accredited investigators in developing countries with interest in sickle cell research will help advance international collaborations. The American Society of Hematology (ASH) has programs for promoting global collaborations through its International Outreach Initiative. This program currently provides educational material free of charge to institutions in developing countries with a GNI per capita of U.S. $6000 or less. ASH also operates a free web service for locating practicing hematologists worldwide; it can be expanded to include the names of potential investigators with interest in SCD in developing countries. Major research organizations in developing countries have expertise in peer-review and should be encouraged to invite applications from qualified investigators registered with such accredited databases. There are already several capacity-building and start-up programs such as the European Young Investigator Award (EURYI), a competitive peer-reviewed award created by 15 national research organizations to minimize brain-drain from Europe. These can be used as templates to tailor specific programs for developing countries.

The HapMap Project and International Collaborations

The Human Genome Project produced a complete catalogue of human DNA sequence, providing the first comprehensive template for comparative analysis of DNA sequence among individuals from different communities, tribes, and nationalities. Completion of this project has initiated several spin-off projects with potentially powerful socioeconomic, clinical and ethical implications in diverse daily life events. The HapMap project is a collaboration between scientists and national research agencies in Canada, China, Nigeria, United Kingdom, United States and Japan, and is by far the most global. It is funded by national and non-government research organizations in the collaborating countries, and aims to build a haplotype or a map of the pattern of common nucleotide variations in the human genome. This will provide information on specific nucleotide variations, where they occur in the genome, and their distribution among people of the same nationalities and among populations in different parts of the world.

To achieve this goal a total of 270 individuals comprising two sets of 90 related individuals (father, mother, child trios) of African and European descent, and two sets of 45 unrelated individuals of Asian origin, are being studied. The Africans are Yorubas living in Ibadan, Nigeria; Asians are from Tokyo, Japan and Beijing, China, while the European samples are from United States residents with northern and western European ancestry from the Centre d'Etude du Polymorphisme Humain (CEPH) database. Twenty-six researchers and research organizations are involved in various aspects of the HapMap project, with responsibilities for public consultation, sample collection, genotyping and analysis of DNA sequence. Information generated from this study is available freely on the World-Wide Web, and can be used for linking new genetic variants with clinical phenotypes and therapeutic outcome in SCD.

Several populations affected with SCD, including African-Americans, Arabs and Indians, are not represented in the HapMap population and will therefore need to be genotyped in separate studies. Moreover, the sickle cell mutation exists on at least five major genetic backgrounds, and each may interact uniquely with genetic variants identified by the HapMap project. There is also heterogeneity at the clinical level that will dictate the relative importance of individual genetic variants in different populations. Nonetheless, the HapMap project provides for the first time thousands of new polymorphic markers on every chromosome, thus expanding substantially the panel of SNPs that can be used to identify genetic determinants of disease pathophysiology. Indeed, several studies have already exploited the information from the HapMap project; in one such study, SNPs predictive of stroke in the general population has been validated in African-Americans with SCD.[30]

Summary

Population size, infant mortality, carrier gene frequency, pregnancies at risk, annual birth rates, affected births per year and life expectancy are health indicators for assessing the burden of SCD. Some of these indicators are not available in epicenters of this condition because of the overwhelming influence of infectious disease and malnutrition. The mutation has spread to most parts of the world by gene flow, including in developed countries where significant advances in healthcare services have impacted positively on asymptomatic management of SCD. This population, however, represents less than 1% of the global sickle cell population. With its expansive distribution throughout the world, its clinical manifestation that begins at birth and persists throughout adult life, and its myriad of psychosocial problems with varied cultural nuances, SCD exacts a heavy price on the health of the global community. The human genome project and its subsidiaries, including the HapMap project, offer a mutually beneficial opportunity in sickle cell research to bridge the global divide in biomedical research and genetic service.

References

1. Angastiniotis M, Modell B (1998). Global epidemiology of hemoglobin disorders. *Ann NY Acad Sci* **850**:251–269.
2. Weatherall DJ, Clegg JB (2001). Inherited haemoglobin disorders: an increasing global health problem. *Bull World Health Organ* **79**:704–712.
3. World Health Organization (1985). Update of the progress of hemoglobinopathies control. Report of the third and fourth annual meetings of the WHO working group for the community control of hereditary anemias. *WHO HMG/WG/858.*
4. Angastiniotis M, Modell B, Boulyjenko V (1995). Prevention and control of haemoglobinopathies. *Bull World Health Organ* **73**:375–386.
5. Weatherall DJ (2005). The global problem of genetic disease. *Ann Hum Biol* **32**:117–122.
6. Population Reference Bureau (2004). World population data sheet.
7. Stine OC, Dover GJ, Zhu D, Smith KD (1992). The evolution of two west African populations. *J Mol Evol* **34**:336–344.
8. Mears JG, Beldjord C, Benabadji M, Belghiti Y, Baddou MA, Labie D, Nagel RL (1981). The sickle gene polymorphism in North Africa. *Blood* **58**:599–601.
9. Mears JG, Lachman HM, Cabannes R, Amegnizin KP, Labie D, Nagel RL (1981). Sickle gene. Its origin and diffusion from West Africa. *J Clin Invest* **68**:606–610.
10. Rund D, Kornhendler N, Shalev O, Oppenheim A (1990). The origin of sickle cell alleles in Israel. *Hum Genet* **85**:521–524.
11. Boussiou M, Loukopoulos D, Christakis J, Fessas P (1991). The origin of the sickle mutation in Greece: evidence from beta S globin gene cluster polymorphisms. *Hemoglobin* **15**:459–467.
12. Aluoch JR, Kilinc Y, Aksoy M, Yuregir GT, Bakioglu I, Kutlar A, Kutlar F, Huisman TH (1986). Sickle cell anaemia among Eti-Turks: haematological, clinical and genetic observations. *Br J Haematol* **64**:45–55.
13. Ragusa A, Lombardo M, Sortino G, Lombardo T, Nagel RL, Labie D (1988). Beta S gene in Sicily is in linkage disequilibrium with the Benin haplotype: implications for gene flow. *Am J Hematol* **27**:139–141.
14. Lavinha J, Goncalves J, Faustino P, Romao L, Osorio-Almeida L, Peres MJ, Picanco I, Martins MC, Ducrocq R, Labie D *et al.* (1992). Importation route of the sickle cell trait into Portugal: contribution of molecular epidemiology. *Hum Biol* **64**:891–901.
15. Curtin P (1969). The Atlantic slave trade: a census. Madison, USA: University of Wisconsin Press.
16. Lehmann H, Cutbush M (1952). Sub-division of some Southern Indian communities according to the incidence of sickle-cell trait and blood groups. *Trans R Soc Trop Med Hyg* **46**:380–383.
17. Lehmann H, Cutbush M (1952). Sickle-cell trait in southern India. *Br Med J* **1**:404–405.
18. Kar BC, Satapathy RK, Kulozik AE, Kulozik M, Sirr S, Serjeant BE, Serjeant GR (1986). Sickle cell disease in Orissa State, India. *Lancet* **2**:1198–1201.
19. Kulozik AE, Wainscoat JS, Serjeant GR, Kar BC, Al-Awamy B, Essan GJ, Falusi AG, Haque SK, Hilali AM, Kate S *et al.* (1986). Geographical survey of beta S-globin gene haplotypes: evidence for an independent Asian origin of the sickle-cell mutation. *Am J Hum Genet* **39**:239–244.
20. Padmos MA, Roberts GT, Sackey K, Kulozik A, Bail S, Morris JS, Serjeant BE, Serjeant GR (1991). Two different forms of homozygous sickle cell disease occur in Saudi Arabia. *Br J Haematol* **79**:93–98.
21. Hickman M, Modell B, Greengross P, Chapman C, Layton M, Falconer S, Davies SC (1999). Mapping the prevalence of sickle cell and beta thalassaemia in England: estimating and validating ethnic-specific rates. *Br J Haematol* **104**:860–867.
22. El-Hazmi MA (2004). The natural history and the national pre-marital screening program in Saudi Arabia. *Saudi Med J* **25**:1549–1554.
23. El-Hazmi MA (2004). Ethics of genetic counseling — basic concepts and relevance to Islamic communities. *Ann Saudi Med* **24**:84–92.

24. Antonarakis SE, Boehm CD, Serjeant GR, Theisen CE, Dover GJ, Kazazian HH Jr (1984). Origin of the beta S-globin gene in blacks: the contribution of recurrent mutation or gene conversion or both. *Proc Natl Acad Sci USA* **81**:853–856.
25. Pagnier J, Mears JG, Dunda-Belkhodja O, Schaefer-Rego KE, Beldjord C, Nagel RL, Labie D (1984). Evidence for the multicentric origin of the sickle cell hemoglobin gene in Africa. *Proc Natl Acad Sci USA* **81**:1771–1773.
26. Ofori-Acquah SF, Lalloz MR, Layton DM (1999). Localisation of cis regulatory elements at the beta-globin locus: analysis of hybrid haplotype chromosomes. *Biochem Biophys Res Commun* **254**:181–187.
27. Ofori-Acquah SF, Lalloz MR, Serjeant G, Layton DM (2004). Dominant influence of gamma-globin promoter polymorphisms on fetal haemoglobin expression in sickle cell disease. *Cell Mol Biol (Noisy-le-grand)* **50**:35–42.
28. Ofori-Acquah SF, Lalloz MRA, Antoniou M, Serjeant GR, Layton DM (1998). Functional role of polymorphic enhancer elements on fetal haemoglobin expression in sickle cell anaemia. *Br J Haematol* **101**:26–26.
29. Jurkovic D, Jauniaux E, Campbell S, Mitchell M, Lees C, Layton M (1995). Detection of sickle gene by coelocentesis in early pregnancy: a new approach to prenatal diagnosis of single gene disorders. *Hum Reprod* **10**:1287–1289.
30. Sebastiani P, Ramoni MF, Nolan V, Baldwin CT, Steinberg MH (2005). Genetic dissection and prognostic modeling of overt stroke in sickle cell anemia. *Nat Genet* **37**:435–440.

INDEX